Young-Seuk Park and Soon-Jin Hwang (Eds.)

Ecological Monitoring, Assessment, and Management in Freshwater Systems

MDPI

This book is a reprint of the Special Issue that appeared in the online, open access journal, *Water* (ISSN 2073-4441) from 2015–2016, available at:

http://www.mdpi.com/journal/water/special_issues/eco-freshwater-sys

Guest Editors
Young-Seuk Park
Department of Life and Nanopharmaceutical Sciences, and Department of Biology
Kyung Hee University
Korea

Soon-Jin Hwang
Department of Environmental Health Science
Konkuk University
Korea

Editorial Office
MDPI AG
St. Alban-Anlage 66
Basel, Switzerland

Publisher
Shu-Kun Lin

Managing Editor
Cherry Gong

1. Edition 2016

MDPI • Basel • Beijing • Wuhan • Barcelona • Belgrade

ISBN 978-3-03842-266-2 (Hbk)
ISBN 978-3-03842-267-9 (PDF)

Table of Contents

List of Contributors

Kwang-Guk An Department of Bioscience and Biotechnology, Chungnam National University, Daejeon 305-764, Korea.

Kyoung-Jin An Department of Environmental Planning, Konkuk University, 120 Neungdong-ro, Kwangjin-gu, Seoul 05029, Korea.

Jens Arle Federal Environmental Agency (Umweltbundesamt), Section II 2.4 Inland Waters, Wörlitzer Platz 1, 06844 Dessau-Roßlau, Germany.

Mi-Jung Bae Freshwater Biodiversity Research Division, Nakdonggang National Institute of Biological Resources, Gyeongsangbuk-do 37242, Korea.

Seung-Ho Baek Department of Biology Education, Kongju National University, Gongju 32588, Korea.

Paweł Burandt Department of Water Resources, Climatology and Environmental Management, University of Warmia and Mazury in Olsztyn, Plac Łódzki 2, Olsztyn 10-719, Poland.

Yongcan Chen State Key Laboratory of Hydroscience and Engineering, Tsinghua University, Beijing 100084, China.

Aron Chmielewski Interfaculty Studies of Environmental Conservation, Warsaw University of Life Sciences—SGGW, ul. Nowoursynowska 159, Warsaw 02-776, Poland.

Byungwoong Choi School of Civil and Environmental Engineering, Yonsei University, Seoul 120-749, Korea.

Sung-Uk Choi School of Civil and Environmental Engineering, Yonsei University, Seoul 120-749, Korea.

Tae-Soo Chon Ecology and Future Research Association, Busan 609-735, Korea; Department of Integrated Biological Sciences, Pusan National University, Busan 609-735, Korea.

Jarosław Chormański Division of Hydrology and Water Resources, Warsaw University of Life Sciences, Nowoursynowska str. 159, Warszawa 02-776, Poland.

Jung Hwa Chun Forest Ecology Division, National Institute of Forest Science, Seoul 02455, Korea.

Kun-Woo Chun Department of Forest Resources, College of Forest & Environmental Sciences, Kangwon National University, Chuncheon 200-701, Korea.

Darren Drapper Drapper Environmental Consultants (DEC), 12 Treetops Avenue, Springfield Lakes 4300, QLD, Australia.

Julita Dunalska Department of Water Protection Engineering, University of Warmia and Mazury in Olsztyn, Prawochenskiego str. 1, Olsztyn 10-719, Poland.

Magdalena Frąk Department of Environmental Improvement, Warsaw University of Life Sciences—SGGW, ul. Nowoursynowska 159, Warsaw 02-776, Poland.

Katarzyna Glińska-Lewczuk Department of Water Resources, Climatology and Environmental Management, University of Warmia and Mazury in Olsztyn, Plac Łódzki 2, Olsztyn 10-719, Poland.

Magdalena Grabowska Department of Hydrobiology, Institute of Biology, University of Białystok, Ciołkowskiego 1J, Białystok 15-245, Poland.

Mateusz Grygoruk Department of Hydraulic Engineering, Warsaw University of Life Sciences—SGGW, ul. Nowoursynowska 159, Warsaw 02-776, Poland.

MuyoungHeo Department of Physics, Pusan National University, Busan 609-735, Korea.

Andy Hornbuckle SPEL Environmental (SPEL), 96 Cobalt St, Carole Park 4300, QLD, Australia.

Soon-Jin Hwang Department of Environmental Health Science, Konkuk University, 120 Neungdong-ro, Kwangjin-gu, Seoul 05029, Korea.

Sun-Ah Hwang Gyeonggi Research Institute, 1128 Gyeongsu-daero, Suwon, Gyeonggi-do 16207, Korea.

Min-Ho Jang Department of Biology Education, Kongju National University, Gongju 32588, Korea.

Karpjoo Jeong Institute for Ubiquitous Information Technology and Applications, and Department of Internet & Multimedia Engineering, Konkuk University, 120 Neungdong-ro, Gwangjin-gu, Seoul 05029, Korea.

Meilan Jiang Department of Advanced Technology Fusion, Konkuk University, 120 Neungdong-ro, Gwangjin-gu, Seoul 05029, Korea.

Yung-Chul Jun Department of Environmental Science, Konkuk University, Seoul 143-701, Korea.

Hojeong Kang School of Civil and Environmental Engineering, Yonsei University, Seoul 120-749, Korea.

Dong-Hwan Kim Department of Integrated Biological Sciences, Pusan National University, Busan 609-735, Korea; Department of Life and Nanopharmaceutical Sciences and Department of Biology, Kyung Hee University, Seoul 130-701, Korea.

Heui-Soo Kim Department of Integrated Biological Sciences, Pusan National University, Busan 609-735, Korea.

Jeong-Hui Kim Department of Biology Education, Kongju National University, Gongju 32588, Korea.

Ji Yoon Kim Department of Bioscience and Biotechnology, Chungnam National University, Daejeon 305-764, Korea.

Nan-Young Kim Department of Environmental Health Science, Konkuk University, 120 Neungdong-ro, Gwangjin-gu, Seoul 05029, Korea.

Sang-Hun Kim Watershed Ecology Research Team, Water Environment Research Department, National Institute of Environmental Research, Hwangyong-ro 42, Seogu, Incheon 22689, Korea.

Ingo Kirst Federal Environmental Agency (Umweltbundesamt), Section II 2.4 Inland Waters, Wörlitzer Platz 1, 06844 Dessau-Roßlau, Germany.

Szymon Kobus Department of Water Resources, Climatology and Environmental Management, University of Warmia and Mazury in Olsztyn, Plac Łódzki 2, Olsztyn 10-719, Poland.

Dong-Soo Kong Department of Biology, Kyonggi University, Suwon 443-760, Korea.

Roman Kujawa Department of Lake and River Fisheries, University of Warmia and Mazury in Olsztyn, Oczapowskiego str. 5, Olsztyn 10-718, Poland.

Gyu-Suk Kwak Ecology and Future Research Institute, Busan 609-802, Korea | Department of Integrated Biological Sciences, Pusan National University, Busan 609-735, Korea.

Hyun-Ju Lee Department of Forest Environment Protection, College of Forest & Environmental Sciences, Kangwon National University, Chuncheon 200-701, Korea.

Jae-An Lee Water Environment Research Department, Watershed Ecology Research Team, Incheon 22689, Korea.

Jin-Woong Lee Department of Biology Education, Kongju National University, Gongju 32588, Korea.

Sang-Bin Lee Ecology and Future Research Association, Busan 609-735, Korea | Department of Integrated Biological Sciences, Pusan National University, Busan 609-735, Korea.

Sang-Woo Lee Department of Environmental Planning, Konkuk University, 120 Neungdong-ro, Kwangjin-gu, Seoul 05029, Korea.

Sylwia Lew Department of Microbiology, University of Warmia and Mazury in Olsztyn, Oczapowskiego str. 5, Olsztyn 10-719, Poland.

Bin Li Department of Integrated Biological Sciences, Pusan National University, Busan 609-735, Korea.

Zhaowei Liu State Key Laboratory of Hydroscience and Engineering, Tsinghua University, Beijing 100084, China.

Volker Mohaupt Federal Environmental Agency (Umweltbundesamt), Section II 2.4 Inland Waters, Wörlitzer Platz 1, 06844 Dessau-Roßlau, Germany.

Krystian Obolewski Department of Hydrobiology, University of Kazimierz Wielki in Bydgoszcz, Jana Karola Chodkiewicza 30, Bydgoszcz 85-064, Poland.

Ji-Hyung Park Department of Environmental Science & Engineering, Ewha Womans University, Seoul 120-750, Korea.

Jung-Hwan Park Department of Environmental Health Science, Konkuk University, 120 Neungdong-ro, Gwangjin-gu, Seoul 05029, Korea.

Sang-Hyeon Park Bio Monitoring Center, Daejeon 34576, Korea.

Se-Rin Park Landscape Ecological Lab, Department of Environmental Planning, Konkuk University, 120 Neungdong-ro, Kwangjin-gu, Seoul 05029, Korea.

Young-Seuk Park Department of Life and Nanopharmaceutical Sciences, and Department of Biology, Kyung Hee University, Korea.

Christopher L. Shope Department of Hydrology, University of Bayreuth, Bayreuth D-95440, Germany.

Haoran Wang State Key Laboratory of Hydroscience and Engineering, Tsinghua University, Beijing 100084, China.

Kozo Watanabe Department of Civil and Environmental Engineering, Ehime University, Matsuyama 790-8577, Japan.

Ju-Duk Yoon Biological Resource Research Center, Kongju National University, Gongju 32588, Korea.

Young-Jin Yun Department of Bioscience and Biotechnology, Chungnam National University, Daejeon 305-764, Korea.

Dejun Zhu State Key Laboratory of Hydroscience and Engineering, Tsinghua University, Beijing 100084, China.

About the Guest Editors

Young-Seuk Park, Ph.D., is a professor at the Department of Biology, Kyung University, Seoul, Korea. After earning his Ph.D. at Pusan National University. His laboratory studies the effects of environmental changes on biological systems at different hierarchical levels from molecules, individuals, populations, and communities through ecological modelling and ecological informatics approaches. In particular, his research is focused on the effects of global changes and alien species on ecosystems, and ecological monitoring and assessment for sustainable ecosystem management. He is interested in computational approaches such as machine learning techniques and advanced statistical methods. He serves as the president of the Korean Society for Mathematical Biology. He is an associate editor of two scientific journals *Annales de Limnologie—International Journal of Limnology* and the *Journal of Ecology and Environment*, and he is also on the editorial boards of other journals including *Ecological Informatics*. He has been a guest editor for several international scientific journals including *Ecological Modelling*, *Ecological Informatics*, *Annales de Limnologie—International Journal of Limnology*, *Inland Waters*, and *Water*.

Soon-Jin Hwang, Ph.D., is a professor in the Department of Environmental Health Science, Konkuk University, Seoul, Korea. He is a lead scientist of an institutional program for national aquatic ecological monitoring supervised by the Korean Ministry of Environment. He has over 25 years of professional experience in freshwater research, education, and outreach. He has worked with aquatic ecosystems and used objective science for the management of aquatic ecosystems for the past 17 years. His research specialty is freshwater ecology, including freshwater ecosystem health assessment and restoration, harmful algal bloom dynamics, and periphyton ecology in lakes and streams. He is particularly interested in pelagic–benthic coupling of harmful cyanobacterial blooms, the effects of hydrology and watershed land use on aquatic biota and water quality, and genetic approaches on secondary metabolites produced by harmful cyanobacteria. He has served as the president of the Korean Society of Limnology and the Editor-in-Chief for the Korean *Journal of Limnology*. He has been an Associate Editor for several international peer-reviewed journals, including *Limnology* and *Paddy and Water Environment*. He has also been a Guest Editor for the *International Journal of Limnology*, *Paddy and Water Environment*, and *Water*.

Preface to "Ecological Monitoring, Assessment, and Management in Freshwater Systems"

The survival of humanity, flora and fauna depends on freshwater ecosystems. However, freshwater supply is limited and subjected to unprecedented anthropogenic threats all over the world. Climate change is a potent driving force that exacerbates the quality and condition of freshwater resources. Our planet is facing many kinds of emerging ecosystems caused by accumulated stresses with erratic system behaviors. Therefore, conserving and reviving freshwater health and biodiversity is becoming more and more important globally to ensure ecosystem integrity and freshwater sustainability. With the advent of the new millennium, the notion of ecosystem integrity has increasingly gained societal value, particularly in terms of a new paradigm of "ecosystem health (eco-health)". This paradigm is multidiscipline; it is elaborately used in a metaphorical sense with an ultimate goal to achieve well-being of the biosphere, for humans and nature. Scientific approaches coupled with societal values create the need to assess ecosystem health, determine the consequences of current behavior, and find options for changing courses. Achieving ecosystem health is the goal. To realize this goal, we need indicators, methods, and techniques.

This Special Issue disseminates recent works on current challenges faced by practicing scientists and engineers in ecological monitoring, assessment, and management in freshwater ecosystems. Seventeen papers were selected through a rigorous peer review process with the aim to broadly disseminate research results, developments, and application to both academia and practitioners. Original research papers addressed the following topics, including one invited review on German surface water monitoring: stream ecological health assessment, effect of watershed land use on aquatic biota, extensible data management and applications in monitoring programs, hydro-ecology, restoration, and field evaluation.

There is strong recognition of the importance of surface water health to meet future water sustainability. Firm institutional and governance frameworks for freshwater resources management also play a pivotal role in surface water health. The approaches discussed in this Special Issue of *Water* provides important information to the field of freshwater science and engineering. We understand that there are many valuable related works elsewhere with complementary and necessary analyses. We hope to see these works in the near future. We are grateful to the authors and reviewers for contributing to the improvement of our understanding of the monitoring, assessment and management of freshwater ecosystems with the ultimate goal to achieve healthy and sustainable freshwater ecosystems.

Young-Seuk Park and Soon-Jin Hwang
Guest Editors

Ecological Monitoring, Assessment, and Management in Freshwater Systems

Young-Seuk Park and Soon-Jin Hwang

Abstract: Ecological monitoring and assessment is fundamental for effective management of ecosystems. As an introduction to this Special Issue, this editorial provides an overview of "Ecological Monitoring, Assessment, and Management in Freshwater Systems". This issue contains a review article on monitoring surface waters, and research papers on data management, biological assessment of aquatic ecosystems, water quality assessment, effects of land use on aquatic ecosystems, etc. The papers in this issue contribute to the existing scientific knowledge of freshwater ecology. They also contribute to the development of more reliable biological monitoring and assessment methods for sustainable freshwater ecosystems and ecologically acceptable decision-making policies, and establishment of practices for effective ecosystem management and conservation.

Reprinted from *Water*. Cite as: Park, Y.-S.; Hwang, S.-J. Ecological Monitoring, Assessment, and Management in Freshwater Systemsr. *Water* **2016**, *8*, 324.

1. Introduction

The reliable monitoring and assessment of water resources is fundamental for effective management of water quality and aquatic ecosystems [1]. Traditionally, physicochemical parameters have been used to assess the quality of water resources. However, they have a limitation in grasping the wholeness of water systems, particularly with reference to ecosystem health and integrity [2]. Various approaches are applicable to ecosystem health assessment at different levels of the biological hierarchy, from genes to ecosystems.

Many countries conduct nationwide monitoring programmes on aquatic organisms for effective freshwater ecosystem management. For example, in Europe, such programmes are carried out under the Water Framework Directive (WFD) [3]. The WFD monitoring programme aims at collecting data for status assessment and controlling the efficiency of the applied water protection measures [4]. In the USA, two major national biomonitoring programmes exist which are funded through the US Environmental Protection Agency (National Aquatic Resources Survey; NARS (previously called EMAP)) and the US Geological Survey (National Water Quality Assessment; NAWQA) [5]. In China, there are three national monitoring programmes supported by the Ministry of Water Resource (National River and Lake Health Program), Ministry of Environment Protection (Watershed Health

Condition Assessment), and Chinese Major Science and Technology Program for Water Pollution Control and Management (Ten Important Rivers and Lakes Health Assessment) [6]. In Korea, the National Aquatic Ecological Monitoring Program (NAEMP) is conducted to assess the ecological health status of stream ecosystems based on biological indices, using benthic diatoms, macroinvertebrates, fish, and aquatic plants. The NAEMP funded by the Ministry of Environment was established in 2007, and since then, the number of sampling sites has increased from 540 to 960, covering the entire nation [7,8]. By 2018, the total number of monitoring sites will gradually increase to 3000.

We designed this special issue to improve the scientific understanding for monitoring, assessment, and management of freshwater aquatic ecosystems. The following section summarises the individual contributions.

2. Contributions

The WFD, established in 2000, provides the current basis for monitoring surface waters and ground water in the countries of European Union. Arle et al. [4] reviewed the monitoring of surface waters in Germany under the WFD. They considered monitoring methods, selection of monitoring sites, and monitoring frequencies. Furthermore, they examined the changes in water monitoring in Germany over the past 16 years and summarised the monitoring results from German surfaces waters under the WFD.

The datasets obtained in the monitoring programmes provide many opportunities for various advanced comparative and synthetic studies, policy-making, and ecological management [9]. In order to realise the potentials and opportunities, Jiang et al. [9] developed a RESTful API-based data management system called OSAEM (the Open, Sharable, and Extensible Data Management System for Aquatic Ecological Monitoring).

Choi et al. [10] presented the transferability of monitoring data from neighbouring streams in a physical habitat simulation. They examined similarities in the data related to channel geometry and in the observed distribution of the target species, and constructed habitat suitability curves using the gene expression programming model. They performed the physical habitat simulations with the proposed generalised habitat suitability curves. Their results indicated that the use of data from a neighbouring stream in the same watershed could result in large errors in the prediction of composite suitability index, and the proposed generalised habitat suitability curves increased the predictability of the composite suitability index in the physical habitat simulation.

Li et al. [11] implemented a self-organizing map (SOM) to detect outlier loci in the amplified fragment length polymorphism band presence/absence data, and demonstrated that genetic diversity adaptively responds to environmental constraints. Specifically, they characterised overall loci composition patterns according to the SOM, revealed environmental responsiveness according to altered input data based on SOM recognition, and addressed associations between outlier loci and environmental variables.

Benthic macroinvertebrates are commonly used for biological assessment of aquatic ecosystems owing to their taxonomic diversity, sedentariness in habitat range, and suitable lifespan [12,13]. Jun et al. [14] studied nationwide distribution patterns of benthic macroinvertebrates and important environmental factors affecting their spatial distribution using the data obtained from the NAEMP. They classified 720 sampling sites into five clusters according to the pollution levels from fast-flowing, less-polluted streams with low electrical conductivity to moderately or severely polluted streams with high electrical conductivity and low water velocity. Their analysis revealed that altitude, water velocity, and streambed composition are the most important determinants for explaining the variation in macroinvertebrate assemblage patterns.

Grygoruk et al. [15] studied the effects of dredging on the benthic macroinvertebrates in agricultural rivers. They demonstrated that the total abundance of riverbed macroinvertebrates in the dredged stretches of the rivers analysed was approximately 70% lower than that in non-dredged areas, and concluded that the dredging of small rivers in agricultural landscapes seriously affects their ecological status by negatively influencing the concentrations and species richness of benthic macroinvertebrates.

Mountainous and headwater streams are characterised by diverse microhabitats that help protect macroinvertebrates from competition, predation, and natural disturbances, and therefore support a rich regional biodiversity [16,17]. Lee et al. [18] examined the water chemistry data collected at headwater streams on different timescales to establish a monitoring programme optimised for identifying potential risks to stream water quality arising from rainfall variability and extremes. Their results suggested that routine monitoring, based on weekly to monthly sampling, is valid only in addressing general seasonal patterns or long-lasting phenomena such as drought effects.

Wang et al. [19] quantified the impacts of the run-of-river scheme on the instream habitat and macroinvertebrate community in a mountain river. They demonstrated that flow diversion at the 75% level and an in-channel barrier, due to the run-of-river scheme, are likely to lead to poor habitat conditions and decrease both the abundance and the diversity of macroinvertebrates in reaches influenced by water diversion.

Bae et al. [17] studied the structure and function of benthic macroinvertebrate communities in four headwater streams at two different spatial scales over three seasons of the year. They showed that the differences between samples were accounted for by seasonal variation more than spatial differences at the individual stream scale, and site differences became more important when performing an ordination within a single season.

Kim et al. [20] examined the effects of land use types on community structure patterns of benthic macroinvertebrates in streams of urban areas. They found that species composition patterns are mainly influenced by both the gradient of physicochemical variables such as altitude, slope, and conductivity, and the proportion of forest area. Community structure patterns were further correlated to the proportion of urbanisation and to biological indices such as diversity and number of species.

Hwang et al. [21] examined the relationships between urban land use and water quality in Korea. They analysed the data derived from NAEMP by using linear and generalised additive models. Their results showed that the generalised additive models had a better fit and suggested a non-linear relationship between urban land use and water quality.

Yun and An [22] assessed the influence of land use patterns on nutrient contents and N:P ratios in stream ecosystems, and determined the empirical relationships between N:P ratios and nutrients and sestonic algal biomass. Their results indicated that land use patterns in the study watersheds are a key factor regulating nutrient contents and N:P ratios in ambient water, and influenced empirical relationships between N:P ratios and sestonic chlorophyll.

An et al. [23] examined the non-stationary relationship between the ecological condition of streams and the proportions of forest and developed land in watersheds by using geographically weighted regression (GWR). They found that the GWR model had superior performance compared with the ordinary least squares method model.

Kim and An [24] evaluated the ecological health of Nakdong River in Korea by using an integrated health responses model based on chemical water quality, physical habitat, and biological parameters. They found that the key stressors were closely associated with nutrient enrichment (N and P) and organic matter pollutions from domestic wastewater disposal plants and urban sewage.

Glińska-Lewczuk et al. [25] evaluated the influence of habitat connectivity and local environmental factors on the distribution and abundance of functional fish groups in 10 floodplain lakes. Their results indicated that the composition and abundance of fish communities are determined by lake isolation gradient, physicochemical parameters, and water stage, suggesting that lateral connectivity between the main channel and floodplain lakes is of utmost importance.

Kim et al. [26] investigated the effectiveness of the nature-like fishway installed at a weir on the Nakdong River in Korea by using traps and passive integrated transponder telemetry. Moreover, they presented measures to improve the efficiency of the fishway by analysing the correlation between the upstream water level and fishway use data.

Drapper and Hornbuckle [27] presented a field evaluation of a stormwater treatment train with pit baskets and filter media cartridges. Their results were significantly different for the filters, but not the pit baskets. In addition, they identified the significant influence of analytical variability on performance results, specifically when influent concentrations are near the limits of detection.

3. Conclusions

We believe that the papers in this special issue contribute to scientific knowledge of freshwater ecology concerning the monitoring, assessment, and management of freshwater ecosystems. They also contribute to developing more reliable biological monitoring and assessment methods for sustainable freshwater ecosystems, and ecologically acceptable decision-making policies, and establishing practices for effective ecosystem management and conservation.

Acknowledgments: This study was performed under the project of "National Aquatic Ecosystem Health Survey and Assessment" in Korea, and was supported by the Ministry of Environment and the National Institute of Environmental Research, Korea. As Guest-Editors of this special issue, we wish to thank the journal editors for the generosity of their time and resources, the authors of the 17 papers published in this issue, and the many referees who contributed to the improved versions of these published papers.

Author Contributions: Young-Seuk Park and Soon-Jin Hwang reviewed and summarized papers published in the special issue. Young-Seuk Park wrote the draft of the manuscript and Soon-Jin Hwang revised the manuscript.

Conflicts of Interest: The authors declare no conflict of interest.

References

1. Park, Y.-S. Aquatic Ecosystem assessment and management. *Ann. Limnol. Int. J. Limnol.* **2016**, *52*, 61–63.
2. Park, Y.-S.; Chon, T.-S. Editorial: Ecosystem assessment and management. *Ecol. Inform.* **2015**, *29*, 93–95.
3. Water Framework Directive, European Union (WFD E.U.). *Establishing a Framework for Community Action in the Field of Water Policy*; Directive 2000/60/EC of the European Parliament and of the Council of 23 October 2000; European Union (EU): Brussels, Belgium, 2000.
4. Arle, J.; Mohaupt, V.; Kirst, I. Monitoring of Surface Waters in Germany under the Water Framework Directive—A Review of Approaches, Methods and Results. *Water* **2016**, *8*, 217.

5.   Buss, D.F.; Carlisle, D.M.; Chon, T.-S.; Culp, J.; Harding, J.S.; Keizer-Vlek, H.E.; Robinson, W.A.; Strachan, S.; Thirion, C.; Hughes, R.M. Stream biomonitoring using macroinvertebrates around the globe: A comparison of large-scale programs. *Environ. Monit. Assess.* **2015**, *187*, 4132.
6.   Qu, X. Personal Communication, Institute of Water Resource and Hydropower Research: Beijing, China, 2016.
7.   Hwang, S.J.; Lee, S.-W.; Park, Y.-S. Editorial: Ecological monitoring, assessment, and restoration of running waters in Korea. *Ann. Limnol. Int. J. Limnol.* **2011**, *47*, S1–S2.
8.   Lee, S.-W.; Hwang, S.-J.; Lee, J.-K.; Jung, D.-I.; Park, Y.-J.; Kim, J.-T. Overview and application of the National Aquatic Ecological Monitoring Program (NAEMP) in Korea. *Ann. Limnol. Int. J. Limnol.* **2011**, *47*, S3–S14.
9.   Jiang, M.; Jeong, K.; Park, J.-H.; Kim, N.-Y.; Hwang, S.-J.; Kim, S.-H. Open, Sharable, and Extensible Data Management for the Korea National Aquatic Ecological Monitoring and Assessment Program: A RESTful API-Based Approach. *Water* **2016**, *8*, 201.
10.   Choi, B.; Choi, S.-U.; Kang, H. Transferability of Monitoring Data from Neighboring Streams in a Physical Habitat Simulation. *Water* **2015**, *7*, 4537–4551.
11.   Li, B.; Watanabe, K.; Kim, D.-H.; Lee, S.-B.; Heo, M.; Kim, H.-S.; Chon, T.-S. Identification of Outlier Loci Responding to Anthropogenic and Natural Selection Pressure in Stream Insects Based on a Self-Organizing Map. *Water* **2016**, *8*, 188.
12.   Park, Y.-S.; Song, M.-Y.; Park, Y.-C.; Oh, K.-H.; Cho, E.; Chon, T.-S. Community patterns of benthic macroinvertebrates collected on the national scale in Korea. *Ecol. Model.* **2007**, *203*, 26–33.
13.   Bae, M.-J.; Li, F.; Kwon, Y.-S.; Chung, N.; Choi, H.; Hwang, S.-J.; Park, Y.-S. Concordance of diatom, macroinvertebrate and fish assemblages in streams at nested spatial scales: Implications for ecological integrity. *Ecol. Indic.* **2014**, *47*, 89–101.
14.   Jun, Y.-C.; Kim, N.-Y.; Kim, S.-H.; Park, Y.-S.; Kong, D.-S.; Hwang, S.-J. Spatial Distribution of Benthic Macroinvertebrate Assemblages in Relation to Environmental Variables in Korean Nationwide Streams. *Water* **2016**, *8*, 27.
15.   Grygoruk, M.; Frąk, M.; Chmielewski, A. Agricultural Rivers at Risk: Dredging Results in a Loss of Macroinvertebrates. Preliminary Observations from the Narew Catchment, Poland. *Water* **2015**, *7*, 4511–4522.
16.   Meyer, J.L.; Strayer, D.L.; Wallace, J.B.; Eggert, S.L.; Helfman, G.S.; Leonard, N.E. The contribution of headwater streams to biodiversity in river networks. *J. Am. Water Resour. Assoc.* **2007**, *43*, 86–103.
17.   Bae, M.-J.; Chun, J.H.; Chon, T.-S.; Park, Y.-S. Spatio-Temporal Variability in Benthic Macroinvertebrate Communities in Headwater Streams in South Korea. *Water* **2016**, *8*, 99.
18.   Lee, H.-J.; Chun, K.-W.; Shope, C.L.; Park, J.-H. Multiple Time-Scale Monitoring to Address Dynamic Seasonality and Storm Pulses of Stream Water Quality in Mountainous Watersheds. *Water* **2015**, *7*, 6117–6138.

19. Wang, H.; Chen, Y.; Liu, Z.; Zhu, D. Effects of the "Run-of-River" Hydro Scheme on Macroinvertebrate Communities and Habitat Conditions in a Mountain River of Northeastern China. *Water* **2016**, *8*, 31.

20. Kim, D.-H.; Chon, T.-S.; Kwak, G.-S.; Lee, S.-B.; Park, Y.-S. Effects of Land Use Types on Community Structure Patterns of Benthic Macroinvertebrates in Streams of Urban Areas in the South of the Korea Peninsula. *Water* **2016**, *8*, 187.

21. Hwang, S.-A.; Hwang, S.-J.; Park, S.-R.; Lee, S.-W. Examining the Relationships between Watershed Urban Land Use and Stream Water Quality Using Linear and Generalized Additive Models. *Water* **2016**, *8*, 155.

22. Yun, Y.-J.; An, K.-G. Roles of N:P Ratios on Trophic Structures and Ecological Stream Health in Lotic Ecosystems. *Water* **2016**, *8*, 22.

23. An, K.-J.; Lee, S.-W.; Hwang, S.-J.; Park, S.-R.; Hwang, S.-A. Exploring the Non-Stationary Effects of Forests and Developed Land within Watersheds on Biological Indicators of Streams Using Geographically-Weighted Regression. *Water* **2016**, *8*, 120.

24. Kim, J.Y.; An, K.-G. Integrated Ecological River Health Assessments, Based on Water Chemistry, Physical Habitat Quality and Biological Integrity. *Water* **2015**, *7*, 6378–6403.

25. Glińska-Lewczuk, K.; Burandt, P.; Kujawa, R.; Kobus, S.; Obolewski, K.; Dunalska, J.; Grabowska, M.; Lew, S.; Chormański, J. Environmental Factors Structuring Fish Communities in Floodplain Lakes of the Undisturbed System of the Biebrza River. *Water* **2016**, *8*, 146.

26. Kim, J.-H.; Yoon, J.-D.; Baek, S.-H.; Park, S.-H.; Lee, J.-W.; Lee, J.-A.; Jang, M.-H. An Efficiency Analysis of a Nature-Like Fishway for Freshwater Fish Ascending a Large Korean River. *Water* **2016**, *8*, 3.

27. Drapper, D.; Hornbuckle, A. Field Evaluation of a Stormwater Treatment Train with Pit Baskets and Filter Media Cartridges in Southeast Queensland. *Water* **2015**, *7*, 4496–4510.

Monitoring of Surface Waters in Germany under the Water Framework Directive— A Review of Approaches, Methods and Results

Jens Arle, Volker Mohaupt and Ingo Kirst

Abstract: The European Commission Water Framework Directive (WFD) was established 16 years ago and forms the current basis for monitoring surface waters and groundwater in Europe. This legislation resulted in a necessary adaptation of the monitoring networks and programs for rivers, lakes, and transitional and coastal waters to the requirements of the WFD at German and European levels. The present study reviews the most important objectives of both the monitoring of surface waters and the principles of the WFD monitoring plan. Furthermore, we look at the changes water monitoring in Germany has undergone over the past sixteen years and we summarize monitoring results from German surfaces waters under the WFD. Comparisons of European approaches for biological assessments, of standards set for physical and chemical factors and of environmental quality standards for pollutants reveal the necessity for further European-wide harmonization. The objective of this harmonization is to improve comparability of the assessment of the ecological status of waters in Europe, and thus also to more coherently activate action programs of measures.

Reprinted from *Water*. Cite as: Arle, J.; Mohaupt, V.; Kirst, I. Monitoring of Surface Waters in Germany under the Water Framework Directive—A Review of Approaches, Methods and Results. *Water* **2016**, *8*, 217.

1. Introduction

Water management poses a major challenge in many densely populated countries throughout the world. In Europe, and due to the WFD [1], stewardship of water resources is of paramount importance now and in the future. The major aim of the WFD is to reach good water quality in all European waters by managing water bodies, *i.e.*, lakes, rivers, groundwater bodies, transitional waters and coastal waters by 2027 at the latest. Official implementation of the WFD started on 22 December 2000 and marked the beginning of a new era in European water management. The WFD declares a unified and harmonized water protection framework for all European countries. Unified in this context means that European waters have been consolidated into large river basin districts managed collaboratively by the Member States (MS) concerned. The successful management of such river basin districts across national boundaries necessitates efficient collaboration in a spirit of partnership between all MS concerned. Hence the WFD aims at harmonized water protection regulations within the European Union (EU). The monitoring and management unit of the WFD is the "water body". It is defined as a discrete and significant element of surface water, which is uniform in type and status.

2. The Monitoring Program and Its Objectives under the WFD

The WFD monitoring program aims at collecting data for a status assessment and at controlling the efficiency of water protection measures applied. This is the reason why monitoring results have to facilitate resilient and reproducible statements. Annexes II and V of the WFD specify a comprehensive assessment and monitoring plan for waters. Annex V of the WFD specifies in depth the minimum requirements of the monitoring itself. Key aspects here include the monitoring types and objectives, the choice of monitoring sites, the quality elements (QEs) to be monitored, and the required monitoring frequencies ([1] compare Annex V 1.3). Pursuant to Article 7 WFD, the MS have to ensure that monitoring programs are set up to allow for a continuous and comprehensive view of the status of waters. By 22 December 2006, applicable monitoring programs had to be produced. The results of the pollution inventory of 2004 were the basis for drafting the first monitoring programs. For the purpose of ensuring consistent monitoring programs all over Germany, the German Working Group on Water Issues of the Federal States and the Federal Government (called Bund/Länder Arbeitsgemeinschaft Wasser, LAWA) compiled a conceptual framework for drafting monitoring programs and assessing the status of surface waters (called "Rahmenkonzeption Monitoring", RAKON). The essentials of this assessment and monitoring approach were implemented through the Ordinance for the Protection of Surface Waters ([2] Oberflächengewässerverordnung, OGewV) as of 20 July 2011 (Federal Gazette I, page 1429). This ordinance, which will be updated in 2016, sets up, among others, environmental quality standards (EQSs) for certain substances and outlines monitoring programs. It furthermore specifies sampling sites across the different water categories, determines how and how often samples have to be taken and sets the assessment rules for water status on the basis of the monitoring results. The monitoring results are presented in a management plan that is submitted to the EU-Commission. If the aims of the WFD, *i.e.*, the "good ecological status" or "good ecological potential" and the "good chemical status" are not reached, measures have to be planned and subsequently implemented (*cf.* [3,4]). German status assessments for surface water bodies are based on data from the monitoring programs. For surface water bodies, ecological status or ecological potential is to be assessed using different assessment methods in accordance with the biological quality elements (BQEs): Fish fauna, benthic invertebrates, macroalgae, phytobenthos and phytoplankton. Finally, the worst assessment result for the BQEs is used as the overall assessment result (the "one-out-all-out" principle, meaning that the worst assessment result for a BQE determines the overall assessment result). The classification scheme for the ecological status of water bodies includes five status classes: 1: very good, 2: good; 3: moderate; 4: poor and 5: bad. For Classes 3 to 5, measures need to be implemented to reach the WFD objectives. Heavily modified and artificial waters are distinguished from natural waterbodies by the WFD. These were either created artificially (e.g., a canal), or their structure has been modified so extensively that a "good ecological status" can no longer be achieved without significantly impairing an existing, economically significant water use that cannot be achieved by other means. For such waters, the environmental objective of a "good ecological potential" has been defined, which requires improvements to be made to the hydromorphological pressures without impairing non-substitutable water uses. In Germany, 50% of all surface water bodies were classified as heavily modified (35%) or artificial water bodies (15%). The WFD objective "good chemical status" applies to natural, artificial and heavily modified water bodies and the chemical status is determined by compliance with Environmental Quality Standards (EQSs) for several pollutants with European-wide significance Moreover, together with the BQEs so called supporting QEs are assessed: River basin-specific pollutants (RBSPs), physico-chemical QEs (e.g., temperature, oxygen, pH, nutrient conditions) and hydromorphological QEs. These supporting QEs are classified as "good" or as "less than good", according to good-moderate boundaries, which were defined for most of these QEs for different water body types in each water category. The requirements for achieving the overall "good ecological status" are (1) that all BQEs must reach the "good status" (Class 2 or better); (2) all EQSs (with defined threshold concentrations) of RBSPs should not be exceeded ("good") and (3) values for other physico-chemical supporting QEs and hydromorphology must fall within a range that allows for good ecosystem functionality ("good ecological status").

Chemical status of water bodies is classified as "good" or "not good". The chemical status is determined from the defined EU-wide EQSs for the 33 priority substances listed in the WFD and 8 other substances regulated on a European-wide basis under the older Directive on pollution caused by certain dangerous substances discharged into the aquatic environment (Directive 2006/11/EC, formerly: 76/464. The provisions of the Environmental Quality Standard Directive 2008/105/EC and the Nitrates Directive were adopted into Annex 7 of the Surface Waters Ordinance in 2011. The environmental quality standards Directive 2008/105/EG was updated in 2013 (2013/39/EU), and now regulates a total of 45 priority substances, which shall be included into the German Surface Waters Ordinance by 2016. The standards for the new priority substances will come into force in 2018. Additional standards for eleven "old" substances have been amended. Currently a prioritization process under the WFD is going on to select 12 to 15 new substances for the priority substance list. The selection criteria for substances are:

- an EU-wide relevance, this means an EQS exceedance in more than three MS in surface waters and
- a detailed substance dossier take into account state of science and the criteria for EQS derivation from the CIS Technical Guidance Document No. 27 for deriving EQSs [5].

The prioritization process based on three aspects:

- the monitoring based exercise [6],
- modeling based exercise [7] and
- substances which are high ranked in the last prioritization but not finial prioritized because of a lack of evidence or monitoring data.

One of the major challenges in the prioritization was the quality of monitoring data from the different sources (MS, scientific research and online databases). The minimum data requirements were substance name, CAS-No., unit, date, limit of quantification (LoQ), limit of detection (LoD), water category (river, lake, coastal), matrix of measurement (biota, sediment or surface water) and the identity of monitoring station. Furthermore, additional criteria like "LOQ is lower than EQS" which was only fulfilled by less than 20% of the monitoring data, were used. The main problem were missing LoQ and LoD values in the monitoring data set. The reasons for missing data in the data set was not evaluated until now.

With this framework the monitoring programs of the WFD essentially aim at:

- controlling the water status and level, up to which environmental objectives are being complied with;
- observing long-term natural and anthropogenic developments and identifying trends;
- ascertaining extent and effects of pollution and changes;
- creating the basis for planning measures, reporting and the efficiency review of measures implemented;
- gauging the effectiveness of water protection measures on the basis of quality data;
- preventing potential dangers to human health.

The above aims require different monitoring methods, which are uniformly regulated throughout the Federal Republic of Germany by the Surface Waters Ordinance [2]. According to their purpose, these methods imply differences in monitoring density, in the number of parameters to be investigated, in the spatial area

11

and in the measurement frequency. Article 7 of the WFD differentiates between the following monitoring methods: **Surveillance monitoring, operational monitoring, and investigative monitoring.**

Surveillance monitoring predominantly ensures the assessment of the overall surface water status within a river basin or sub-basin of a river basin district. Thus, it is especially suited for identifying large-scale and long-term trends in the development of water quality. The German Federal States specified more than 500 monitoring stations for surveillance monitoring of surface waters (Table 1). The surveillance monitoring network is wide-meshed (with a catchment area of up to 2500 km^2 per monitoring site), but must be representative of the assigned hydrological unit and must be permanent in time. The selected monitoring sites are designed to provide an integrative view of the overall status of the assigned hydrological unit and enable researchers to gauge target achievement in the region. Mostly, they are located in the main flows of the major rivers and at the inflow of major tributaries. In case they represent a pressure to the relevant water body, biological, hydromorphological and physico-chemical QEs, RBSPs and substances relevant for the classification of the chemical status of waters have to principally be measured at the monitoring sites for surveillance monitoring [8]. Surveillance monitoring sites usually measure all the QEs of ecological status, *i.e.*, biological, hydromorphological, chemical and physico-chemical QEs. BQEs are investigated at least twice during each six yearly river basin management plan period. RBSPs must be monitored if discharged in significant quantities. Furthermore, priority substances that are relevant for the classification of chemical waterbody status must also be measured if discharged into the respective waterbody.

Table 1. Number of monitoring stations for different monitoring types and categories of surface waters in Germany. Data source: Federal Environment Agency (Umweltbundesamt) data from the Working Group on Water Issues of the Federal States and the Federal Government (LAWA). Data origin: Reporting portal WasserBLIcK/BfG as of 22 March 2010 after Mohaupt *et al.*, 2012 [9] and reporting portal WasserBLIcK/BfG as of November 2015.

Type of Monitoring and Year		Streams & Rivers	Lakes	Transitional Waters	Coastal Waters
Surveillance	2010	290	67	5	32
monitoring	2014	318	127	42	76
Operational	2010	7252	449	20	100
monitoring	2014	12,342	711	13	76
Investigative	2010	375	0	0	0
monitoring	2014	1074	25	0	0

Operational monitoring is the tool for assessing the status of those water bodies that probably may not meet the environmental objectives. It is also used to control whether measures have been successfully implemented. Germany has delineated nearly 10,000 water bodies out of its rivers, lakes, transitional and coastal waters and more than 13,000 monitoring stations for the operational monitoring of surface waters have been specified (Table 1).

Hence, operational monitoring is the main focus of surface water monitoring. Along rivers and streams, an average of one monitoring site is to be found every 10–15 kilometers and the average size of the delineated stream and river water bodies is 15.2 km (median = 8.7 km; min < 1 km; max = 242 km). This means that there may be several monitoring sites along one water body. The operational monitoring generally deals with those biological, chemical and physico-chemical QEs that indicate the presence of pressures significant to the status of the water bodies being assessed [9] and that are indicative of the cause of pollution. In Germany, the operational monitoring sites are analyzed as set out in Table 2.

Investigative monitoring is necessary if the reasons as to why the water quality of a particular water body could not be assessed as "very good" or "good", or in order to ascertain the magnitude and spatial scale of impacts of accidental pollution. It applies, e.g., to unforeseen accidental discharges of pollutants or to sudden fish mortality in the water body. This is the reason why there are currently only 1074 sites of this kind installed in German surface waters.

Table 2. Minimum requirements for monitoring frequencies and intervals for the quality elements under the ecological status and the chemical status as defined by the EC Water Framework Directive. OM = operational monitoring; SM = surveillance monitoring; p.a. = per annum, p.m. = per month.

Status— Component	Quality Element	Water Category			
		Streams & Rivers	Lakes	Transitional Waters	Coastal Waters
Biological quality elements (BQE)					
Ecological status	Phytoplankton	6× p.a. (relevant vegetation period); every 1 to 3 years per site (SM) and every 3 years per site (OM), monitored in large rivers only	6× p.a. (relevant vegetation period); every 1 to 3 years per site (SM) and every 3 years (OM)	6× p.a. (relevant vegetation period); every 1 to 3 years per site (SM) and every 3 years (OM)	6× p.a. (relevant vegetation period); every 1 to 3 years per site (SM) and every 3 years (OM)
	Large algae/angiosperms	-	-	1 to 2 times p.a., every 1 to 3 years per site (SM) and every 3 years (OM)	1 to 2 times p.a., every 1 to 3 years per site (SM) and every 3 years (OM)
	Macrophytes/phytobenthos	1 to 2 times p.a., every 1 to 3 years per site (SM) and every 3 years (OM)	1 to 2 times p.a., every 1 to 3 years per site (SM) and every 3 years (OM)	-	-
	Macroinvertebrates	1 to 2 times p.a.; every 1 to 3 years per site (SM) and every 3 years (OM)	1× p.a., every 1 to 3 years per site (SM) and every 3 years (OM)	1× p.a., every 1 to 3 years per site (SM) and every 3 years (OM)	1× p.a., every 1 to 3 years per site (SM) and every 3 years (OM)
	Fish	1 to 2 times p.a.; every 1 to 3 years per site (SM) and every 3 years (OM)	1 to 2 times p.a.; every 1 to 3 years per site (SM) and every 3 years (OM)	1 to 2 times p.a.; every 1 to 3 years per site (SM) and every 3 years (OM)	BQE not used under the WFD
Hydromorphological quality elements					
	Continuity	1 time, needs-based monitoring, updated every 6 years	-	-	-
	Hydrology	continuously	1 time per month	-	-
	Hydromorphology	1 time, needs-based monitoring, updated every 6 years	1 time, needs-based monitoring, updated every 6 years	1 time, needs-based monitoring, updated every 6 years	1 time, needs-based monitoring, updated every 6 years

Table 2. *Cont.*

Status— Component	Quality Element	Water Category			
		Streams & Rivers	Lakes	Transitional Waters	Coastal Waters
Physical and chemical quality elements					
	General physical and chemical elements	Depending on the parameters; for most parameters 4 to 13 times p.a.; at least 1× in 6 years (SM), at least 1× in 3 years (OM)	Depending on the parameters; for most parameters 4 to 13 times p.a.; at least 1× in 6 years (SM), at least 1× in 3 years (OM)	Depending on the parameters; for most parameters 4 to 13 times p.a.; at least 1× in 6 years (SM), at least 1× in 3 years (OM)	Depending on the parameters; for most parameters 4 to 13 times p.a.; at least 1× in 6 years (SM), at least 1× in 3 years (OM)
	River basin-specific pollutants	4 to 13 times p.a.; at least 1× in 6 years (SM), at least 1× in 3 years (OM)	4 to 13 times p.a.; at least 1× in 6 years (SM), at least 1× in 3 years (OM)	4 to 13 times p.a.; at least 1× in 6 years (SM), at least 1× in 3 years (OM)	4 to 13 times p.a.; at least 1× in 6 years (SM), at least 1× in 3 years (OM)
Chemical Status	Chemical quality elements (45 priority pollutants with European-wide consistent ecological quality standards)	For priority pollutants in the water phase: 12 times p.a.; at least 1× in 6 years (SM), at least 1× in 3 years (OM); For priority pollutants in biota: 1–2 times p.a.; at least 1× in 6 years (SM), at least 1× in 3 years (OM). A reduced level of monitoring is required for the so-called ubiquitous, widespread substances and substances on a mandatory watch list of the EU Commission.			

3. Selection of Monitoring Sites

In general, the monitoring sites are selected by regional water managers. The selection of the monitoring sites is based on estimates with respect to the representativeness of a monitoring site for the specific water body. The term "representativeness" is not quantitatively defined. With respect to selecting monitoring sites, the following questions had and still have to be answered or continuously checked:

- How many monitoring sites are necessary to obtain reliable assessment results for each water body?
- Where to place monitoring sites to be sure that they really are representative of an entire water body?
- What assessment uncertainties can be expected and to what extent will they appear?
- In how far does the natural variability of biocenoses influence the assessment results?
- Going beyond the minimum requirements of the WFD, should the number of monitoring sites and the frequency of measurement be adjusted to the predominant pressures on a water body?

Many of these questions persist to date. The challenge now is to answer these questions as to the basis of experiences gained from monitoring and on the basis of additional analyses of monitoring data, both being part of an adaptive management process.

4. Monitoring Frequencies

Minimum requirements for monitoring frequencies and intervals for the quality elements of the ecological status and the chemical status as defined by the WFD are listed in Table 2. BQEs relevant for the assessment of the ecological status are generally reviewed at least every three years per monitoring site during operational monitoring. Hydromorphological and physico-chemical QEs are only used as supporting elements in the assessment of the ecological status and to provide supporting indications of major pressures in a specific water body (*i.e.*, hydromorphological degradation, organic pollution or eutrophication), which is important for the subsequent determination of measures. The basic idea of this assessment philosophy is that the "biological community", as represented by the BQE´s, is seen as the all-integrating element that reflects abiotic conditions and their interplay (natural factors and anthropogenic pressures). RBSPs are assessed within the context of a classification of ecological status. They are defined as pollutants that are discharged in significant quantities. The MS must derive EQSs to protect the aquatic community on the basis of longer-term ecotoxicological effect data

([1] compare Annex V, 1.2.6). In Germany, substances discharged into freshwaters leading to concentrations of more than half the EQS at representative monitoring sites were defined as "significant" and legally binding EQSs have been specified for a total of 162 RBSPs. Up to now, compliance with EQSs is verified using annual averages. EQSs for the ecological status of surface waters are defined on the basis of an EU chemical assessment as prescribed in Annex V, 1.2.6 of the WFD [1]. Valid long-term tests regarding the substance's effects on organisms at different levels of the aquatic food chain, *i.e.*, algae, invertebrates and fish, are compiled, and the most sensitive of these values is selected. However, as organisms in nature may be even more sensitive than those used to perform the laboratory tests, this value is divided by a safety factor in order to calculate the EQS. If valid long-term toxicity tests are available for all levels of the aquatic food chain, this factor is generally 10. If data are missing, it will be 100 or greater. The quality element RBSPs can lead, in contrast to hydromorphological and physico-chemical QEs, to a downgrading of the ecological status. Exceedance of even just one EQS for RBSPs in a specific water body means that the ecological status (or ecological potential) can only be "moderate" (Class 3), even if the biological quality elements are all "good" or above. Monitoring frequencies are increased if considered necessary for a reliable and accurate statement on status. A QE can be exempt from assessment if it proves impossible to define reliable reference conditions due to the high degree of natural variability in a specific water body type. For RBSPs that are emitted in significant quantities, sampling should be carried out at least once every three months, and for pollutants relevant to chemical status at least once a month, unless higher frequencies are required for a reliable and precise assessment of status. If the EQS of a RBSP is exceeded, the substance will remain in the monitoring program until the monitoring results show that this substance is no more relevant in the water body concerned.

5. Modifications in Water Monitoring due to the WFD

Due to the specifications of the WFD, water monitoring has strongly changed over the past 15 years. In particular, this has affected

- the change monitoring focus (more biology and fewer individual substances);
- the temporal rate and areal scope of biological water monitoring;
- the range of BQEs monitored;
- the development, improvement and addition of biological assessment methods;
- and the level of standardization and harmonization of biological assessment methods.

Due to the WFD specifications, European water monitoring no longer focuses on the primary monitoring of saprobity, nutrients and pollutants, but concentrates on a comprehensive, integrative assessment concept which gives priority to biological QEs being indicators for the overall assessment of influences on the aquatic environment. Furthermore, the assessment includes both physical and chemical factors serving as supporting QEs and RBSPs as well as the hydrology and hydromorphology. Thus, monitoring and assessment of the water status under the WFD pursue a noticeably more holistic approach than in the past [10].

6. Changes in the Temporal Rate and Areal Scope of Biological Water Monitoring

Analyses by Beck *et al.* [11], which are based on data by the European Environment Agency (EEA) between 1965 and 2005, strongly indicated that water monitoring intensity, scope and extent in Europe distinctly increased over the past 40 years. In the course of the implementation of the WFD, the temporal rate and areal scope of German biological water monitoring continued to noticeably rise in the past 13 years. However, individual MS often feature considerable differences regarding monitoring density and sampling frequency [12]. Nevertheless, it is difficult to exactly quantify the degree to which European water monitoring has intensified. This is because the monitoring and assessment of the environmental state of European waters are performed by regional and national authorities, and the results are summarized for the state of the environment (SOE) assessments by the EEA at river basin district scale. SOE reporting covers many but not all monitoring sites of the MS resulting in an underestimation of monitoring activities. However, to our knowledge the EEA database forms the most comprehensive overview of freshwater monitoring activities in Europe. As of May 2012, the EEA database [13] included information on the water quality from

- more than 10,000 monitoring sites along running waters in 37 European countries,
- 3500 monitoring sites in lakes in 35 European countries, and
- 5000 monitoring sites in coastal waters in 28 European countries.

In the context of implementing the WFD, MS designated more than 127,000 surface water bodies. Eighty-two percent of these surface water bodies are running waters, 15% are lakes and 3% are coastal and transitional waters. Thus, the EEA database includes information on the 2009 water status of each of these water bodies, representing 1.1 million kilometers of running waters, about 19,000 lakes, approximately 370,000 km^2 of coastal and transitional waters and 3.8 million km^2 of groundwater bodies (*cf.* [13]).

7. Changes in the Range of Monitored Biological Quality Elements

Traditional monitoring of inland waters in Europe focused on the investigation of physical and chemical parameters. Prior to the entry into force of the WFD, water management applied just a few differing biological methods to assess the pressures on the water bodies (e.g., saprobic index). In the early 1990s, only half of the European MS used biological assessment methods to complement the monitoring of physical and chemical elements [14]. Today, monitoring under the WFD comprises many more biological quality elements (aquatic flora, invertebrate fauna and fish fauna) than formerly. Moreover, the monitoring frequency of biological quality elements noticeably increased due to the strict specifications of the WFD. Meanwhile, the volume of collected data available for the assessment of the water status in Europe is nowhere near as comprehensive as that available hitherto. Such a comprehensive picture of the flora and fauna of European waters had never been gained before. At the same time, biological data collected under the WFD monitoring may considerably support the implementation of other EU Directives (e.g., Habitats Directive [15], Marine Strategy Framework Directive [16]). In this context, the parallel and increasing collection of biological, physical and chemical data is extremely important since it facilitates the further statistical analysis of these data. In future, such data and analyses will increasingly be used to allow the assessment of the relative importance of different pressures on water systems for reaching good ecological status, which is an essential step towards prioritizing management measures in multiple pressure situations and also contributes to a better justification of their necessity.

8. Data Availability and Participation of Applied Sciences

Nowadays, many more German and European monitoring data are publicly available than was the case in the past. This is due to strikingly increased transparency regarding the publication of water monitoring results and to the reporting obligations to the European Commission. Such increased transparency might be considered a success of the WFD. The data collected within the framework of the WFD are of great interest for applied water ecology as well. This interest finds expression in an enormous number of scientific publications over the past years (*cf.* Figure 1).

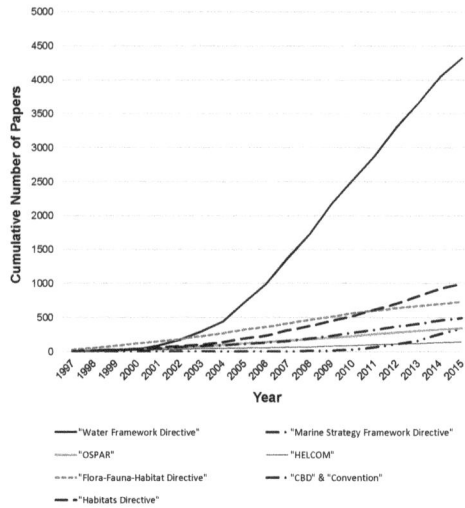

Figure 1. Comparison of the cumulative number of scientific publications referring to the Water Framework Directive (title, keywords, abstract) and other EU Directives, International Conventions on the Protection of the Marine Environment for the North Sea (OSPAR) and the Baltic Sea (HELCOM), the Habitats Directive, and the Convention on Biological Diversity (CBD) (Source: www.scopus.com, queried in December 2015).

9. Development, Improvement and Addition of Biological Assessment Methods

Numerous new biological assessment methods had to be developed to ensure that the biological assessment is compliant with the WFD specifications (cf. [1] Annex V, 1.2 *Normative definitions of ecological status classifications*). Hundreds of scientists were involved and helped to develop, at European and national levels, complex new sampling and assessment methods. In many MS, the development of national assessment methods was delayed. Although the attributes and properties of the "good ecological status" and the "very good ecological status" were precisely defined as standards in the WFD annexes, the scientific community had to specify these definitions, translating them into applicable methods in terms of indices and metrics [12]. Many of the newly developed assessment methods have a modular and multimetric design in common. It is intended to guarantee a pressure-specific assessment and analysis by using trait-based metrics of different BQEs as indicators of different pressures. The MS methods often differ with respect to the taxonomic resolution of the assessment methods, the way reference conditions are defined, interpreted or specified, and the indices and metrics applied with the method. In all German assessment systems, the organisms are identified to the lowest taxonomic level possible. Analyses of gathered WFD monitoring data sets have shown that the resolution of indication of trait-based metrics for lakes often work well, probably because "eutrophication" is the most dominant pressure in these systems. In river systems, with often multiple overlapping pressures acting, the resolution of pressure indication on the basis of trait-based approaches in many cases showed rather mixed results, making pressure prioritization much more difficult. One very important requirement for the development of the biological assessment systems was the differentiation of water categories into different types based on their natural biocenosis. In addition to these biological characteristics, geological, geographical, morphological and hydrological characteristics of the water categories were also used in Germany to distinguish twenty-five stream and river types, fourteen lake types, nine coastal water types and two types of transitional waters. As a result of the WFD, water type-specific and pressure-specific biological assessment systems are now the standard in Europe.

10. Level of Standardization and Harmonization of Biological Assessment Methods

Whereas many sampling and analysis methods for physical and chemical QEs were standardized long before the entry into force of the WFD and others became standard all over Europe in the course of the implementation of the WFD (e.g., CEN), formerly none or only national standards (e.g., DIN, Deutsches Institut für Normung/German Institute for Standardization) existed for biological assessment methods in the EU. The WFD addressed requirements of the MS to develop national assessment methods. By 2012, almost 300 assessment methods for biological quality elements had been developed in Europe to implement the WFD [17]. This is the reason why the WFD with its Annex V, 1.4.1 introduced a comprehensive harmonization process called "intercalibration". Its purpose is the establishment of consistent ecological status thresholds for the good-moderate and very good-good boundaries of the biological assessment systems by harmonizing the strengths of the different national approaches to biological assessment, rendering the assessment results comparable [18]. The biological assessment methods within the framework of the WFD have been intercalibrated by means of comprehensive statistical and numerical approaches. Class boundaries of the "very good" ecological status class and the "good" ecological status class are compared and harmonized through the intercalibration exercise. The three principal intercalibration methods developed are described by Birk *et al.* [17]. Although leading not to complete comparability of assessment systems in a mathematical-quantitative sense, allowing still considerable variability, the whole intercalibration process highly encouraged, at the international level, the exchange between experts on biological assessment methods, and thus brought about increased levels of harmonization of ecological standards (but for the good-moderate and very good-good boundaries only) relevant for water management at the European level. Major doubts on the level of comparability attained by the biological intercalibration approach are for instance:

- The three different methods (direct comparison, indirect comparison, common boundary setting with subsequent comparison; see Poikane *et al.* [19]) on which intercalibration was carried out were never compared as to whether they produce similar results.
- The so-called "intercalibration harmonization band", which marks the upper and lower limits of variability accepted by the intercalibration approach between good-moderate boundaries of MS's assessment methods, allows differences of up to 0.5 classes, which is considerable when only five status classes are being used.
- The "boundary bias", which was used as the major criterion for the evaluation of comparability, is defined as the deviation of a class boundary relative to the common view of the MS (*cf.* Poikane *et al.* [19]). Because the MS "common view" is defined via common metrics and by the global mean of all national assessment methods, the results of the intercalibration completely depend on the common or pseudo-common metrics selected, meaning that different metrics can lead to completely different overall results.
- In many instances, intercalibration was carried out by using so-called "common intercalibration metrics" as yardsticks for the comparison of different assessment methods. This leads to an intercalibration of different methods along only one dimension. This approach takes no account of the multidimensionality of the (multimetric) assessment methods. As a result, the decision as to which of the MS need to adjust their good-moderate boundary completely depends on the "common intercalibration metric" used. Changing this "metric" can completely change the result of the analysis.
- The intercalibration was carried out based on EQR values, which are defined as a ratio between observed assessment results and the expected value under reference conditions. An EQR = 1 marks the references status and EQR = 0 the most degraded one. Within the intercalibration approach, no method is integrated to compare whether the MS have a truly similar or comparable understanding of this "reference" and "most degraded" status. Without a common and quantitatively comparable understanding of this reference status no real comparability could be expected and the EQR scale used will mask differences of this issue rather than allowing for transparency.
- In many cases, only one metric or module of the multimetric assessment systems was harmonized instead of the whole system.
- Intercalibration was performed only at the BQE-level, but no comparison was made between the good-moderate boundaries among different BQEs.

- Intercalibration was carried out within so called geographic intercalibration groups (GIGs) and specific national typologies were forced into these much coarser groups (intercalibration types), which increased the variability within these groups and decreased the possibility of harmonization. In many cases, only the assessment methods of a few national types were officially intercalibrated. The grouping of specific national types, each with different reference communities, into broader types inevitable led to variability which is not controlled by the intercalibration methodology. The effects of this variability on the intercalibration results have not yet been quantified. Intercalibration of the same BQEs across geographic intercalibration groups (GIGs) and intercalibration across different BQEs was not performed.
- Regression and correlation analyses used to describe the relationships between "common or pseudocommon intercalibration metrics" and the national methods take only the variation explained by the models (R^2) into account and $R^2 \geqslant 0.5$ are deemed acceptable for the comparison of different assessment methods via intercalibration, but variation around the mean trend (uncertainty level of national methods) was not taken into account.
- Checks for feasibility of intercalibration included tests of comparability of water body types applied within each of the geographic group (GIG). In a considerable number of cases, the comparability tests lead to different assumptions as to what constitutes comparable national types within different GIGs.

However, the assessment strength of almost all German assessment methods was comparable to those of our neighbor states (*cf.* Commission Decision of 20 September 2013 [20]). A third period until the end of 2016 will complete the intercalibration of biological assessment methods of natural waters. To date, the harmonization of the assessment of the ecological potential of heavily modified water bodies at the European level remains to be completed. Within this activity, comparisons are being made on the basis of measures implemented by the MS for defined pressure categories (land drainage, water storage and others). Consequently, future scientific improvements of biological assessment methods must to be adjusted to the fixed and legally valid Commission Decision results. This may result in a never-ending intercalibration process. On the other hand, continuous improvements of biological assessment methods conflict with their role as decision-making tools for status and therefore for measures as part of the management process. In extreme cases, changes in the original assessment results could lead to results in the light of which measures taken and investments made might appear unnecessary. This would substantially lower their public acceptance. It is, therefore, important to balance improvements of the assessment systems with the needs of administration and management. Our opinion is that administration and management need reliable assessment results, and repeated changes of assessment systems within short time-frames will decrease the chances of successful WFD implementation and have large potential to lower public acceptance of the overall process. Consequently, an important issue within an adaptive management strategy is to explore timescales that are optimal for revisions and adjustments of the assessments systems.

11. Comparison of Supporting Physical and Chemical Quality Elements' Monitoring at European Level

Good-moderate boundaries for physical and chemical quality elements were set by the MS in order to reach a good ecological status of the biological communities. Analyses based on data from 20 MS (*cf.* [21]) of the first reporting to the European Commission on the river basin management plans in 2010 revealed that these MS monitor many different physical and chemical QEs potentially as a result of different national monitoring traditions. For rivers, 20 MS reported to monitor 32 different variables; for lakes, 16 MS reported to monitor 28 different variables; for coastal waters, 12 MS monitored 24 different variables, and for transitional waters 11 MS monitored 22 different variables (*cf.* Table 3 & Claussen *et al.* [21]). Within one water category, MS monitored a differing number of supporting physical and chemical variables (e.g., rivers: Slovakia and Poland monitored 16 variables, Finland monitored just three variables, and Germany was in the mid-range with nine variables). Total phosphorus, dissolved oxygen and total nitrogen were the three physical and chemical QEs that were monitored most in Europe. Only very few MS monitored hydromorphological variables. Good-moderate boundaries specified by the MS often showed significant differences for many supporting physical and chemical QEs. Figure 2 shows exemplary the high variability of the good-moderate boundaries specified by the MS for total phosphorous and orthophosphate in streams and rivers. In Figure 2, the order of MS from left to right on the x-axis is based on maximum values observed and shall not be interpreted as a ranking, because the differences among the MS may either be attributed to different river typologies due to river sizes, climate, geology and geographical position or to a different statistical information content or to different assumptions on reference conditions (e.g., background concentrations) but also on different assumptions on good-moderate boundaries which might reflect different ambitions of the MS on the "good status". Alternatively, the observed differences might also reflect different interpretations by the MS on the levels of nutrients that are acceptable in order to reach good ecological status or potential. To differentiate between reasons of variation was not part of the analysis, but is part of the ongoing work at European level within the CIS Working Group A on Ecological Status ("Ecostat"). Harmonization of rules to set good-moderate boundaries for supporting physical and chemical QEs is still necessary.

Table 3. Overview of supporting physical and chemical quality elements used by the EU Member States (MS). Source: modified after Claussen *et al.* 2011 [21]. Numbers indicate the total number of MS (a total of 20 MS were included in the analysis) that reported good-moderate boundary values for variables for different surface water categories. Missing numbers imply that good-moderate boundary values were not defined (the variable was not used in monitoring) or not reported for this variable.

Supporting Physical and Chemical Variables	Rivers & Streams (20 MS)	Lakes (17 MS)	Transitional Waters (11 MS)	Coastal Waters (12 MS)	Totals
Total phosphorus or total dissolved phosphorus or soluble reactive phosphorus or orthophosphate (mg·L^{-1})	19	17	7	7	50
Dissolved oxygen concentration (mg·L^{-1})	13	7	7	6	33
Total nitrogen (mg·L^{-1})	9	9	5	7	30
Total ammonium (mg·L^{-1})	16	5	3	2	26
pH or delta pH	14	7	3	2	26
Secchi depth (m) & transparency (m)	2	9	6	9	26
Nitrate (mg·L^{-1})	13	3	4	4	24
Oxygen saturation (%)	7	5	4	5	21
BOD5 (mg·L^{-1})	12	2	3	2	19
Water temperature (°C) or delta temperature	9	3	1	2	15
Nitrite (mg·L^{-1})	7	2	2	2	13
Chloride (mg·L^{-1})	8	4			12
Total inorganic nitrogen or dissolved inorganic nitrogen (mg·L^{-1})		1	5	6	12
Sulfate (mg·L^{-1})	5	2			7
Conductivity (µS·cm^{-1})	3	4			7

Table 3. *Cont.*

Supporting Physical and Chemical Variables	Rivers & Streams (20 MS)	Lakes (17 MS)	Transitional Waters (11 MS)	Coastal Waters (12 MS)	Totals
Morphological and/or hydromorphological conditions	2	2	1	1	6
CODCr (mg· L^{-1})	4	1			5
Total organic carbon (mg· L^{-1})	2	1		2	5
River continuity	4				4
Non ionized ammonia, NH4-N (mg· L^{-1})	1	1	1	1	4
Structure of riparian zone	3				3
Hydrological regime	2	1			3
COD (mg· L^{-1})	1	2			3
Salinity (‰)			1	2	3
Organic nitrogen (mg· L^{-1})			2	1	3
Kjeldahl nitrogen (mg N· L^{-1})	2				2
Structure and substrate of river bed	2				2
River depth and width variation	2				2
Alkalinity (mmol· L^{-1})	1	1			2
Hypolimnion oxygen (%)		2			2
Lake depth variation		2			2
Silicate (mg· L^{-1})			1	1	2
N total soluble (mg/kg settled sediment)			1	1	2
P total (mg/kg settled sediment)			1	1	2
DIN, winter, November–February, 0–10 m (EQR)			1	1	2
Turbidity (NTU)			1	1	2
Dissolved organic carbon (DOC) (mg· L^{-1})	1				1

26

Table 3. *Cont.*

Supporting Physical and Chemical Variables	Rivers & Streams (20 MS)	Lakes (17 MS)	Transitional Waters (11 MS)	Coastal Waters (12 MS)	Totals
BOD7 (mg·L^{-1})	1				1
CODMn (mg·L^{-1})	1				1
Connection to groundwater	1				1
Water flow	1				1
Acid neutralizing capacity (μ equivalents L^{-1})		1			1
Hardness (dH)		1			1
Nitrate/orthophosphate ratio		1			1
Wave exposure (after CIS Guidance 2.4)				1	1

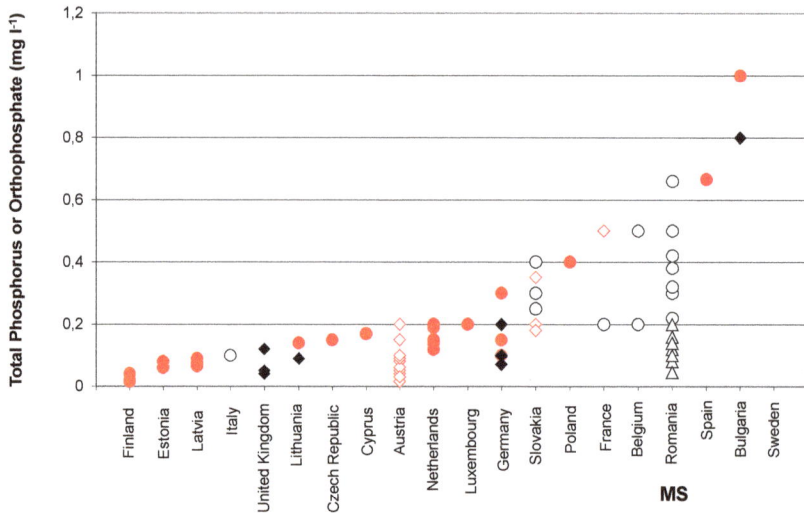

Figure 2. Total phosphorus or orthophosphate values for good-moderate boundaries of ecological status in European streams and rivers. Filled circles = Total phosphorus as annual mean values or mean values for defined seasons or no entry. Open circles = Total phosphorus as 90 percentile values of annual measurements. Triangles = Total phosphorus median values. Filled diamonds = Orthophosphate/soluble reactive phosphorus as annual mean values. Open diamonds = Orthophosphate as 90 percentile values of annual measurements. Values provided by CZ, FR, BG, PL, LT, LU, CYP and IT are good-moderate boundaries set for "All rivers" (without typological differences). Multiple points for other MS denote good/-moderate boundaries set for different river and stream types. Sweden reported good-moderate boundaries as EQR (not shown). Spain reported a value with unknown unit of measurement (mg/L^{-1} or EQR). UK values refer to soluble reactive phosphorus.

12. Comparison of RBSP Monitoring at the European Level

At present, EU MS classify the ecological status under the WFD on the basis of a highly varying number of RBSPs ([22,23], Figure 3). Within the first management plan period, the Czech Republic and the Netherlands reported the highest number of RBSPs, *i.e.*, 169 and 162 respectively. Germany came in third place, reporting 152 pollutants. Cyprus (three pollutants) implemented the lowest number of RBSPs (*cf.* Arle *et al.* [22]). At the European level, the mean of RBSPs amounted to 47 RBSPs. The EQSs were comparable between the MS in part, but there were also significant differences for the same pollutants. Irmer *et al.* 2014 [23] extended the comparisons made by Arle *et al.* [22] and found a total number of 452 substances regulated as RBSPs in the European Union. The average number of RBSPs per MS increased to 55 substances. Seven MS regulated more than the mean number of 55 substances (Figure 3). One hundred and eighty nine substances were regulated by only one MS each. EQSs for 263 substances were regulated by at least two MS and thus enabled a comparison of EQSs. Copper and zinc were both regulated by 22 MS and thus were the most often monitored RBSPs in the EU and should be the first candidates for EU-wide regulation as priority substances within chemical status.

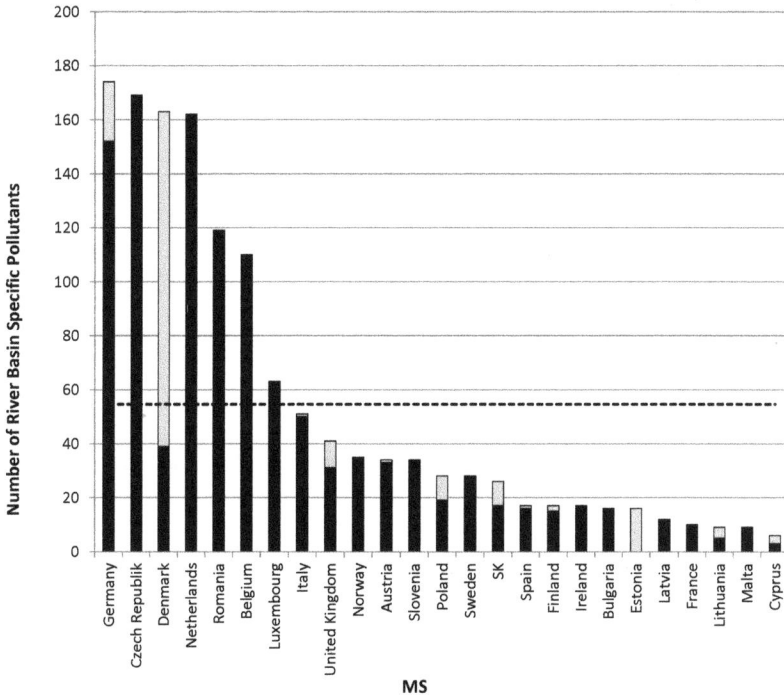

Figure 3. Number of river basin specific pollutants of the Member States as reported to the EU Commission within the first management plan period (black bars after Arle *et al.*, 2012 [22]) and updated numbers based on MS requests (grey bars after Irmer *et al.*, 2014 [23]). Dotted line = mean number of substances used by the MS as RBSPs, after Irmer *et al.*, 2014 [23].

The divergence between highest and lowest EQS for the same substance is exemplary for some substances visualized in Figure 4. All MS in question indicated identical EQSs for 40 substances (about 15%). Minimum and maximum EQS values for one third of the listed substances differ up to tenfold from each other. The values of a little more than half (53%) of all substances differ from more than tenfold to 10^5-fold from each other. Most of the differences observed seemed to not be caused by differences in the safety factors used by the MS. Rather the differences are indicative of differing approaches used by the MS to derive the EQSs. Under the WFD, EQSs for RBSPs are derived according to rules set out in Annex V, 1.2.6 which should eliminate flexibility of standard values in future. In many cases, the results for other water categories (lakes, transitional waters and coastal waters) also showed similar discrepancies between the highest and lowest EQSs for the same substance. The number of RBSPs regulated by the MS in lakes, transitional waters and coastal waters was considerably lower than those regulated in rivers. Comparisons of the MS EQSs for different pollutants with $PNEC_{freshwater}$—values available at the official webpage of the European Chemicals Agency (ECHA) (*cf.* [24]) and the National Recommended Water Quality Criteria—Aquatic Life Criteria Table of the United States Environmental Protection Agency (*cf.* [25]) indicate a generally high international variability in the threshold levels for different pollutants that are assumed to be protective of aquatic life. To ensure that the assessments of the ecological status in the EU may be comparable and that programs of measures aiming at reducing pollutant inputs into surface waters may coherently be initiated under the WFD, harmonization, at the European level, of EQSs for RBSPs is needed.

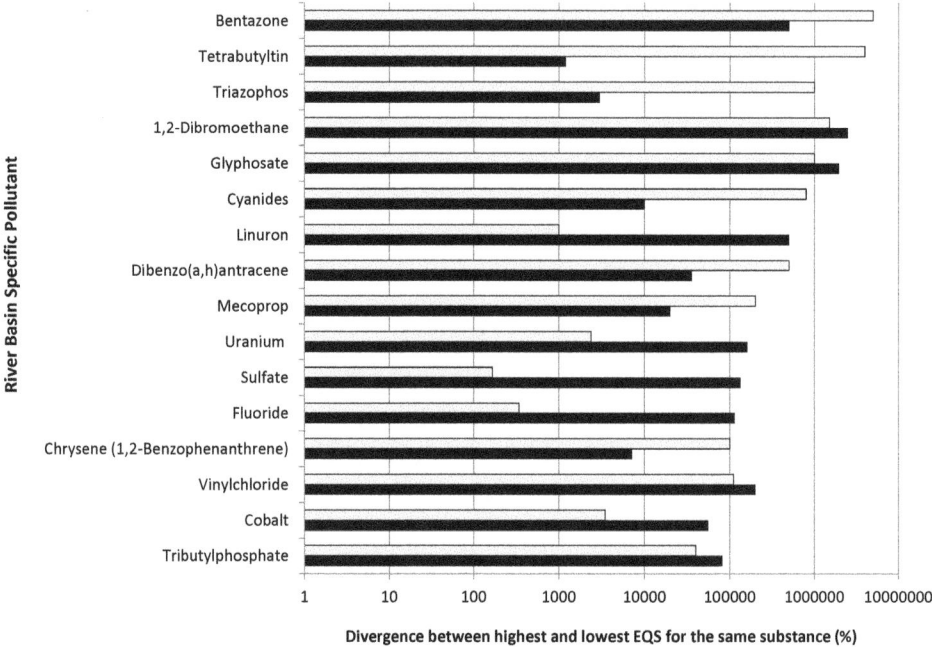

Figure 4. Divergence (in %) between minimum values and maximum values of EQSs for RBSPs reported by different MS. 16 substances are shown exemplary. Medium of measurement for the substances = water. Grey bars = values reported by Arle *et al.*, 2012 [22], Black bars = values reported by Irmer *et al.*, 2014 [23]. Under the WFD, EQSs for RBSPs are derived according to fixed rules which should eliminate flexibility in the determination of standard values. Similar to the quality standards for substances under the chemical status classification, the EQSs for RBSPs should show identical values.

13. Results of WFD Monitoring in Germany and the EU

Figure 5 shows the assessment results of ecological status/ecological potential of surface water bodies in Germany based on the preliminary drafts of the second river basin management plan. The results are within the range observed during the first management cycle. Of more than 9900 surface water bodies (including rivers, lakes, transitional and coastal waters) only 8.2% actually reach good or very good ecological status or potential, whereas >90% of all water bodies are classified to be at a moderate, poor or bad status. Slight improvements of the ecological status or potential in comparison to the assessment results of the first river basin management cycle (2010, as reported to the Commission) occurred in the lower classes. The number of water bodies at moderate status or potential increased, whereas the number of water bodies at poor or bad status slightly decreased. The final assessment results of the second river basin management cycle were submitted to the Commission in 2016. The most common reasons for failing to achieve a "good ecological status" are changes in hydromorphology of streams and rivers and high levels of nutrient loads originating from agricultural land use. In 2014, of the about 9000 streams and river water bodies, only 6.7% were in "very good" or "good" ecological status or potential. For the individual river biological quality elements higher percentages were at "very good" or "good" ecological status or potential: 20.2% of monitoring sites for fish ($n = 5918$), 27.5% of monitoring sites for macroinvertebrates ($n = 8105$), 21.3% of monitoring sites for macrophytes and phytobenthos ($n = 5647$) and 58% of monitoring sites for phytoplankton ($n = 241$, monitored in very large rivers only). These results show that the application of the "one-out-all-out" principle has major effects on the overall assessment result. The chemical status of all German water bodies is currently "not good". This is because of the ubiquitous distributed substance Mercury. It was found in all water bodies. In addition the priority substances Cadmium, Nickel, Polycyclic Aromatic Hydrocarbons (PAH), Tributyltin, Fluoranthene, Diruron, Isoproturon measured in concentrations higher than the EQS in more than the half of the ten river basin districts in Germany. Lead, Brominated Diphenylether, 1.2 Dichloorethane, Anthracene, Bis(2-ethylhexyl)phthalate, Hexachlorobenzene, Hexachlorobutadiene, Naphthalene, Nonylphenol, Octylphenol, Pentachlorobenzene, Tetrachloroethylene, Trichlorobenzene, Trichloroethene, Trichloromethane, Hexachlorocyclohexane, DDT and Chlorpyrifos exceeded the EQS in five or less river basin districts.

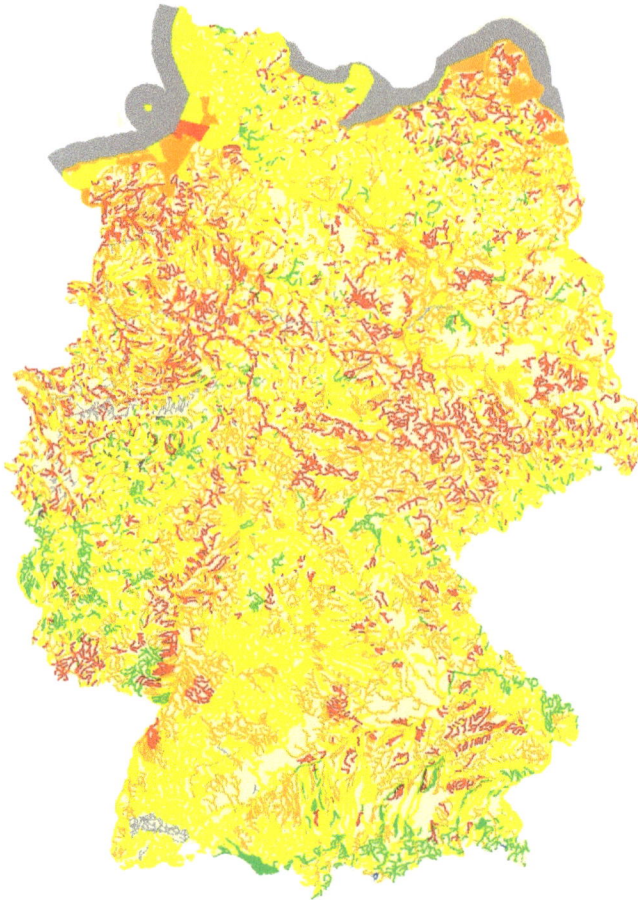

Figure 5. Ecological status/ecological potential of surface water bodies in Germany. Blue color = very good status; Green color = good status/potential; Yellow color = moderate status/potential; Orange color = poor status/potential; Red color = bad status/potential; Grey color = status/potential unknown. Source: Federal Environment Agency after data from the German reporting portal WasserBLIcK/BfG 2015.

In 2012, the European Environment Agency (EEA) presented several comprehensive reports on the European monitoring results according to the WFD (*cf.* [13]). These reports are based on assessments of both data that the MS had conveyed to the EU Commission as part of official WFD reporting (WISE-WFD database) and data the MS make available to the EEA on a one-year-cycle. According to the evaluation of the first river management plans by the EEA, more than half of the European surface waters are not at a "good ecological status" or do not reach a "good ecological potential" at present. On average, European rivers and transitional waters show a considerably worse ecological status and are subject to noticeably higher pressures than lakes and coastal waters [13]. As in Germany, many MS of the EU identified diffuse nutrient inputs and hydromorphological changes as the main pressures on European rivers. For the first river basin management cycle the EEA has criticized the insufficient knowledge of the status of waters in Europe [13]. The chemical status, for example, of 51% of rivers and of 54% of the lakes was not known in the EU due to the lack of data and assessments by some MS during the first river basin management plan cycle. In sum, about 90% of the surface waters assessed in Europe were at a good chemical status. Polycyclic aromatic hydrocarbons as well as mercury, for example, are considered to be the main reason why rivers in the EU fail to reach a good chemical status.

14. Using Biological Monitoring and Assessment Results to Determine Most Important Pressures

According to the WFD, the BQEs (fish, macroinvertebrates, phytobenthos, phytoplankton, macrophytes) and their assessment results are used in Germany as indicators for a "coarse" identification of the potentially most important pressures in a specific water body. For instance, fish is used as an indicator for hydromorphological degradation and disruptions of the longitudinal continuity in rivers, macroinvertebrates are used as indicators for organic pollution, whereas phytoplankton and phytobenthos are used as the primary indicators for eutrophication in lakes and rivers. The way to identify the relevant pressures in a specific water body does not follow a legally binding approach and so each stream manager can use his/her own knowledge. The identification of the most important pressures via the BQEs can be validated by the assessment of physical and chemical QEs (for instance pH, chloride, sulfate, phosphorous, nitrogen, oxygen, TOC, variables for hydromorphological degradation, variables for the extent of longitudinal discontinuity, and others). This often provides more detailed information for the identification of potential measures for a specific water body, which then should be subjected to economic analysis in order to check for feasibility and proportionality (cost-benefit ratio). The experiences gained from WFD implementation show that although the method described above is very useful to determine "major pressures" at larger scales and key measures, at the local level of the "water body" it sometimes is much more difficult to determine the most important pressures. This is especially the case in river systems where multiple natural factors and anthropogenic pressures act and interact. The resulting complexity does in many cases strongly limit our ability to draw explicit conclusions on the "relative importance of different pressures and natural factors" in a water body. Because of this, a prioritization of measures to be taken against different pressures is also difficult, though required due to financial constraints for measures. Many problems also arise from the questions of "how much of a measure is necessary?", "to what extent must a measure be taken in order to achieve good ecological status?", the effectiveness of a specific measure, and the selection of the measures with the best cost-benefit ratios. This is especially the case for pressures resulting from hydromorphological degradation, often because of uncertain correlative relationships with the ecological status. For many large-scale pressures, which cannot be tested in the laboratory for their effect on the biological community, pressure-response relationships are used to determine "management thresholds" for the WFD based on correlation analysis. However, because correlation does not necessarily imply causality, the derivation of measures can be difficult. For most of the chemical pollutants the derivation of EQSs is seen as much more reliable, because they are derived by extrapolating results to the real world from laboratory toxicity tests on common standard organisms for single pollutants. Although this is an international accepted strategy, some new results based on data analyses of WFD monitoring data have partly called this strategy into question [26]. The variability of EQSs for RBSPs presented in this review should not be seen as a contradiction to this ecotoxicological approach. Rather, it appears to be a result of inaccurate usage of the approach. A fundamental review of the way of deriving EQSs for RBSPs listed by at least three MS could noticeably reduce the divergence between the values, which in many cases differ by several degrees of magnitude. An EQS database by the European CIS Working Group on Chemicals (WG CHEM) could also contribute to reducing divergences, as could actions such as the harmonization of EQSs of particular substances or groups of substances such as herbicides, if effect values exist that are based on European-level certification. Since only three MS reported to have used the CIS Guidance No. 27 of 2011 [5] to select and derive EQSs for RBSPs, it will hopefully be harmonized in the course of its further application. For this purpose, the developments should continuously be monitored and recorded because a harmonized list of EQSs for RBSPs is important to achieving the same "good ecological status" for all European surface waters and to the coherent implementation of the WFD.

15. Summary and Conclusions

Biological water monitoring in Europe has noticeably intensified due to the specifications of the WFD. Never before has a comparably comprehensive view of the flora and fauna of European waters existed. The data collected under WFD monitoring are a solid basis for future water management due to their large quantity and high quality. The harmonization of assessment methods and management approaches has notably progressed within the European Union, but it needs further attention. German water quality has improved over the past 50 years. However, further efforts and measures are necessary to reach the ambitious aims of the

WFD by e.g., reducing nutrient losses from agriculture and by supporting natural hydromorphology, but also by addressing other anthropogenic pressures.

To further improve freshwater management practice under the WFD, the following topics need to be addressed by basic and applied research:

- The international harmonization of EQSs for pollutants,
- the interactions of multiple factors (natural factors and anthropogenic pressures) and their effects on freshwater communities,
- effects of "land use" as a large-scale pressure and attempts to disentangle the "land use—pressure bundle",
- the estimation of the relative importance of different pressures in their effect on the biological communities, in order to prioritize specific measures against different pressures,
- the role of biological interactions on the ecological status (e.g., non-native species),
- the determination of the efficiency of measures against different pressures, and
- the question as to whether biological systems of freshwaters "can be managed" or "can be restored" to a defined "ecological state", which is determined by the WFD as "slight deviation" from past reference conditions.

Acknowledgments: We are grateful to Ulrich Irmer for his many years of support and advice on multiple aspects of WFD implementation in Germany. Dagmar Larws (Federal Environment Agency) and Ute Bohnsack helped to improve the English language of the manuscript. We thank Katrin Blondzik (Federal Environment Agency) for the preparation of Figure 5. We like to thank Soon-Jin Hwang for inviting us to contribute to the special issue "Ecological Monitoring, Assessment, and Management in Freshwater Systems".

Author Contributions: The main author is Jens Arle. Volker Mohaupt supervised the work. Ingo Kirst contributed most of the content on the chemical status assessments.

Conflicts of Interest: The authors declare no conflict of interest. The views presented and conclusions drawn by the authors in this paper do not necessarily represent the official view of the Federal Environmental Agency.

References

1. Water Framework Directive, European Union (WFD E.U.). *Establishing a Framework for Community Action in the Field of Water Policy*; Directive 2000/60/EC of the European Parliament and of the Council of 23 October 2000; EU: Brussels, Belgium, 2000.
2. Federal Law Gazette (BGBl.). *Ordinance on the Protection of Surface Waters (Surface Waters Ordinance) of 20 July 2011*; BGBI: Bonn, Germany, 2011; pp. 1429–1469.
3. Federal Environment Agency (UBA). *Water Resource Management in Germany. Part 1: Fundamentals*; Irmer, U., Huber, D., Walter, A., Eds.; UBA: Berlin, Germany, 2014; Available online: https://www.umweltbundesamt.de/publikationen/water-resource-management-in-germany-part-1 (accessed on 1 January 2016).

4. Federal Environment Agency (UBA). *Water Resource Management in Germany. Part 2: Water Quality*; Irmer, U., Blondzik, K., Eds.; UBA: Berlin, Germany, 2014; Available online: https://www.umweltbundesamt.de/publikationen/water-resource-management-in-germany-part-2 (accessed on 1 January2016).

5. European Commission. *Common Implementation Strategy for the Water Framework Directive-Guidance Document No. 27-Technical Guidance for Deriving Environmental Quality Standards*; European Commission Technical Report; EU: Brussels, Belgium, 2011; p. 204.

6. Carvalho, R.; Marinov, D.; Ceriani, L.; Loos, R.; Lettieri, T. *STE Methodology for Monitoring-Based Prioritisation of Chemical Substances in Support to EU Water Policy*; Joint Research Centre (JRC): Ispra, Italy, 2015; p. 13.

7. Chirico, N.; Carvalho, R.; Ceriani, L.; Lettieri, T. *Modelling-Based Strategy for the Prioritisation Exercise under the Water Framework Directive*; Joint Research Centre (JRC): Ispra, Italy, 2015; p. 35.

8. Irmer, U.; Mathan, C.; Mohaupt, V.; Naumann, S.; Rechenberg, J.; Wolter, R. Auf dem Weg vom Monitoringprogramm zu Bewirtschaftungsplan und Maßnahmeprogramm. In *Gewässerschutz—Wasser— Abwasser, Band 217: 42. Essener Tagung für Wasser-und Abwasserwirtschaft, 18–20 März 2009*; Pinnekamp, J., Ed.; Institut für Siedlungswasserwirtschaft der Rhein.-Westf. Techn. Hochschule: Aachen, Germany, 2009; pp. 52/1–52/16.

9. Mohaupt, V.; Hoffmann, A.; Naumann, S.; Richter, S.; Völker, J. Monitoring an Fließgewässern: Konzepte, Ergebnisse, Bewertung. In *DWA-WasserWirtschafts-Kurse, N7, Fließgewässer: Moderne Gewässerentwicklung im Zeichen von Ökologie und Hochwasserschutz*; Deutsche Vereinigung für Wasserwirtschaft, Abwasser und Abfall e.V.: Hennef, Germany, 2012; pp. 6–17.

10. Borja, A.; Bricker, S.B.; Dauer, D.M.; Demetriades, N.T.; Ferreira, J.G.; Forbes, A.T.; Hutchings, P.; Jia, X.; Kenchington, R.; Marques, J.C.; *et al.* Overview of integrative tools and methods in assessing ecological integrity in estuarine and coastal systems worldwide. *Mar. Pollut. Bull.* **2008**, *56*, 1519–1537.

11. Beck, L.; Bernauer, T.; Kalbhenn, A. Environmental, political, and economic determinants of water quality monitoring in Europe. *Water Resour. Res.* **2010**, *46*, W11543.

12. Hering, D.; Borja, A.; Carstensen, J.; Carvalho, L.; Elliott, M.; Feld, C.K.; Heiskanen, A.S.; Johnson, R.K.; Moe, J.; Pont, D.; *et al.* The European water framework directive at the age of 10: A critical review of the achievements with recommendations for the future. *Sci. Total Environ.* **2010**, *408*, 4007–4019.

13. European Environment Agency. *European Waters—Assessment of Status and Pressures*; EEA Report No 8/2012; EEA: Kopenhagen, Denmark, 2012; Available online: http://www.eea.europa.eu/themes/water/water-assessments-2012 (accessed on 1 January 2016).

14. Hering, D.; Buffagni, A.; Moog, O.; Sandin, L.; Sommerhäuser, M.; Stubauer, I.; Feld, C.K.; Johnson, R.; Pinto, P.; Skoulikidis, N.; *et al.* The development of a system to assess the ecological quality of streams based on macroinvertebrates—Design of the sampling program within the AQEM project. *Int. Rev. Hydrobiol.* **2003**, *88*, 345–361.

15. Council of the European Communities. *Council Directive 92/43/EEC of 21 May 1992 on the Conservation of Natural Habitats and of Wild Fauna and Flora*; European Commission: Brussels, Belgium, 1992; Volume 35, pp. 7–50.

16. Marine Strategy Framework Directive, European Union (MSFD E.U.). *Establishing a Framework for Community Action in the Field of Marine Environmental Policy (Marine Strategy Framework Directive)*; Directive 2008/56/EC of the European Parliament and the Council of 17 June 2008; EU: Brussels, Belgium, 2008.

17. Birk, S.; Bonne, W.; Borja, A.; Brucet, S.; Courrat, A.; Poikane, S.; Solimini, A.; van de Bund, W.; Zampoukas, N.; Hering, D. Three hundred ways to assess Europe's surface waters: An almost complete overview of biological methods to implement the Water Framework Directive. *Ecol. Indic.* **2012**, *18*, 31–41.

18. Birk, S.; Bellack, E.; Böhmer, J.; Mischke, U.; Schaumburg, J.; Schütz, C.; Witt, J. The EU WFD Intercalibration Exercise—New Results and Résumé. *Wasserwirtschaft* **2013**, *1/2*, 52–55.

19. Poikane, S.; Zampoukas, N.; Borja, A.; Davies, S.P.; van de Bund, W.; Birk, S. Intercalibration of aquatic ecological assessment methods in the European Union: Lessons learned and way forward. *Environ. Sci. Policy* **2014**, *44*, 237–246.

20. European Commission. *Commission Decision of 20 September 2013, Establishing, Pursuant to Directive 2000/60/EC of the European Parliament and of the Council, the Values of the Member State Monitoring System Classifications as a Result of the Intercalibration Exercise and Repealing Decision 2008/915/EC (Notified under Document Number C(2013) 5915)*; European Commission: Brussels, Belgium, 2013; Available online: http://eurlex. europa.eu/LexUriServ/LexUriServ.do?uri=OJ:L:2013:266:0001:0047:EN:PDF (accessed on 1 December 2015).

21. Claussen, U.; Müller, P.; Arle, J. *Comparison of Environmental Quality Objectives, Threshold Values or Water Quality Targets Set for the Demands of European Water Framework Directive*; Report for the CIS Working Group A "ECOSTAT"; 2011; pp. 1–14. Available online: https://circabc.europa.eu/sd/a/e73efaa5-4819-4af2-81f7-24e0ae9de230/4a% 20-%20Nutrient_standards_ECOSTAT_WGA_2011.doc (accessed on 1 December 2015).

22. Arle, J.; Claussen, U.; Irmer, U. Ecological Environmental Standards of "River-basin specific pollutants" in Surface Waters—A Europe-wide Comparison. *Korresp. Wasserwirtsch.* **2012**, *10/12*, 556–558.

23. Irmer, U.; Rau, F.; Arle, J.; Claussen, U.; Mohaupt, V. Ecological Environmental Quality Standards of "River Basin Specific Pollutants". In *Surface Waters—Update and Development Analysis of a Comparison between EU Member States*; Working Group Ecological Status; Report for the WFD CIS Working Group A Ecological Status (ECOSTAT); Umweltbundesamt (UBA): Dessau-Roßlau, Germany, 2014; Available online: https://circabc.europa.eu/sd/a/93fc95d2-3afe-485f-93e0-be1aa6e1e432/ 9%20-%20Environmental%20Quality%20Standards_28012014.docx (accessed on 1 December 2015).

24. Official webpage of the European Chemicals Agency (ECHA). Available online: http://echa.europa.eu/de/information-on-chemicals/registered-substances (accessed on 1 November 2015).
25. The United States Environmental Protection Agency. National Recommended Water Quality Criteria—Aquatic Life Criteria Table. Available online: http://www.epa.gov/wqc/national-recommended-water-quality-criteria-aquatic-life-criteria-table (accessed on 11 December 2015).
26. Berger, E.; Haase, P.; Oetken, M.; Sundermann, A. Field data reveal low critical chemical concentrations for river benthic invertebrates. *Sci. Total Environ.* **2016**, *544*, 864–873.

Open, Sharable, and Extensible Data Management for the Korea National Aquatic Ecological Monitoring and Assessment Program: A RESTful API-Based Approach

Meilan Jiang, Karpjoo Jeong, Jung-Hwan Park, Nan-Young Kim,
Soon-Jin Hwang and Sang-Hun Kim

Abstract: Implemented by a national law, the National Aquatic Ecological Monitoring Program (NAEMP) has been assessing the ecological health status of surface waters, focusing on streams and rivers, in Korea since 2007. The program involves ecological monitoring of multiple aquatic biota such as benthic diatoms, macroinvertebrates, fish, and plants as well as water quality and habitat parameters. Taking advantage of the national scale of long-term aquatic ecological monitoring and the standardization of protocols and methods, the datasets in NAEMP provide many opportunities for various advanced comparative and synthetic studies, policy-making, and ecological management. In order to realize these potentials and opportunities, we have developed a RESTful API-based data management system called OSAEM (the Open, Sharable and Extensible Data Management System for Aquatic Ecological Monitoring), which is designed to be open, sharable, and extensible. In this paper, we introduce the RESTful API-based data management approach, present the RESTful API for the OSAEM system, and discuss its applicability. An OSAEM prototype system is currently available on a commercial cloud service (Amazon EC2) but the system remains under active development.

Reprinted from *Water*. Cite as: Jiang, M.; Jeong, K.; Park, J.-H.; Kim, N.-Y.; Hwang, S.-J.; Kim, S.-H. Open, Sharable, and Extensible Data Management for the Korea National Aquatic Ecological Monitoring and Assessment Program: A RESTful API-Based Approach. *Water* **2016**, *8*, 201.

1. Introduction

Implemented by a national law, the National Aquatic Ecological Monitoring Program (NAEMP) aims to assess the ecological health status of lotic environments in Korea by means of biological health indices focusing on benthic diatoms, macroinvertebrates, fish and aquatic plants [1]. Every six months, NAEMP has been conducting ecological monitoring on a national scale at 960 sites (whose number has grown from 540 in the beginning and will reach 3000 by 2018) in major rivers and their tributaries since 2007. Protocols and methods are standardized for monitoring and the same set of those standardized protocols and methods is used for collecting data at all the sites in NAEMP. The datasets from NAEMP are very different from those produced by individual or group projects because of both the national scale of the long-term aquatic ecological monitoring and the standardization of the protocols and methods. The NAEMP datasets are so standardized that they are highly comparable and easy to integrate. Therefore, the NAEMP datasets provide many opportunities for various advanced comparative and synthetic studies, policy-making, and ecological management, in addition to the ecological health assessment of lotic water.

In order to realize such potentials and opportunities fully, effective management and sharing of the ecological monitoring data is crucial [2–6]. However, the development of a data management system for ecological monitoring is really challenging because not only aquatic ecosystems *per se* but also approaches to ecological studies are very complicated and diverse. Furthermore, the integration of datasets from different ecological data management systems is even more challenging due to the exponential combination of such complexity and diversity [7–9].

In this paper, we present a data management system called OSAEM (the Open, Sharable and Extensible Data Management System for Aquatic Ecological Monitoring) designed to facilitate and promote the sharing of NAEMP datasets. Furthermore, the design of the OSAEM system focuses on enabling a variety of software systems, rather than human users, to access the NAEMP datasets as easily and flexibly as possible. Data management in the OSAEM system is based on the web technology called the RESTful API [10,11]. There are a number of active efforts to apply the RESTful API technology to the management of scientific and engineering data [12–15]. In these efforts, the RESTful APIs are used to support the efficient management of software services. As such, the conventional RESTful APIs are service-oriented.

In this paper, we present a novel, data model–oriented approach to the RESTful API. The data model–oriented RESTful API intends to make the management of ecological monitoring data:

- *Open*. From the OSAEM system, potential users can easily find information about what kind of data is available, what data models are applied to data, and what services are provided for data, without in-depth understanding of the internal design of the data management system. Such information is crucial for ecological analyses and syntheses.
- *Sharable*. New programs (e.g., internal analysis code or external information systems in other projects) can easily be developed to access datasets in the OSAEM system. The interface to data services in the OSAEM system is well defined and easy to use. Such service interface is crucial for integrating datasets from different projects for synthetic analyses.

- *Extensible.* Extensions or changes to the datasets in the OSAEM system are easily and efficiently supported. For example, new categories of living organisms, methods, or sites are easily added for the extension of monitoring and ecological studies.

For the last three decades, there have been active research efforts to develop data management systems for the sharing of ecological monitoring datasets [2,16–21]. In general, there are two kinds of approaches to data sharing: metadata-level sharing and data-level sharing. In the metadata-level data sharing, metadata that satisfies standards such as EML (Ecological Metadata Language) or Darwin Core are created and shared [16,17,22–25]. Such metadata describes basic information (e.g., keywords, locations, investigators, and related documents) about real monitoring data, but usually fails to provide sufficient information about the specific data structures of monitoring data. Therefore, this approach does not allow for the development of software programs (e.g., analysis code) to process monitoring datasets automatically. DataONE is a most notable example of this approach [16,17,21].

In data-level sharing, the data models (e.g., database schemas) for monitoring data are defined and monitoring datasets are organized and stored in data management systems according to the models. Since the data models are available, this approach allows for the development of software programs to analyze datasets automatically. CUAHSI is a most notable example of this approach [18,19].

However, this approach requires us to understand complicated data models, software tools, and service interfaces in such data management systems. For example, the CUASHI relational data model for ecological data called ODM (Observations Data Model) consists of a large number of database tables. In addition, a large number of various data services and software tools are provided and need to be understood for the development of data analysis software [26].

As opposed to these conventional approaches, the RESTful API-based data management approach in the OSAEM system combines data models and service interfaces both logically and physically. This feature makes the RESTful API-based approach really simple and intuitive to understand and to use. Furthermore, this approach does not require special software tools to access the datasets in the OSAEM. Generic software libraries or tools can be used. Therefore, the OSAEM system significantly facilitates the sharing of ecological data.

In addition, there have recently been active research efforts to manage ecological data as linked data in the global ecological research community [27,28]. However, the OSAEM system does not yet support linked data because we believe the ecological research community in Korea currently prefers ecological data in a traditional database style. In spite of the current design, the OSAEM system can be easily extended to support linked data because every data entity in the OSAEM system is represented as a web resource with a unique URI (Uniform Resource Identifier).

In this paper, we do not address issues in the design of databases in the OSAEM system because data management is based on the RESTful API and the design of databases is intentionally made transparent to the user or client software. This paper is structured as follows. Section 2 provides a brief overview of NAEMP. In this section, we discuss the potential opportunities and values of the NAEMP datasets as reference data for various ecological studies and management. In addition, we raise challenging issues about data management for NAEMP. In Section 3, we introduce the RESTful API-based approach to data management. In Section 4, we present the main features of the OSAEM system, focusing on the RESTful API: URIs, representations, and services. The RESTful API is intended for programming, not for the human user. In Section 5, we explain the web portal of the OSAEM to allow the user to access the NAEMP datasets in a GUI-based, user-friendly manner. In Section 6, we conclude this paper with discussions and future work.

2. Brief Overview of NAEMP

NAEMP monitors five major rivers and their tributaries in Korea due to a national law. The program was established in 2007, and since then, the number of sampling sites has grown from 540 to 960, covering the entire nation [1]. The total number of monitoring sites will gradually increase to 3000 by 2018. Figure 1 shows the spatial distribution of monitoring sites in NAEMP. All the datasets collected with the same standardized protocols and methods are comparatively and synthetically analyzed in order to assess the ecological health status of the national lotic ecosystem by means of developed biological health indices including benthic diatoms, macroinvertebrates, fish, and aquatic plants [29–34].

Every six months, NAEMP monitors the physio-chemical and biological parameters shown in Table 1. However, biological and habitat parameters (currently fish, benthic diatoms, and macroinvertebrates) are the main criteria used to assess the ecological health of the rivers and streams. At each monitoring site, the same set of parameters is measured with the same set of standardized protocols and methods during almost the same period of time. The locations of monitoring sites are chosen in order to allow the datasets to be scientifically representative of lotic ecosystems in Korea. The monitoring results at each site are quality-controlled during the lab analyses. Some parameters are observed and recorded at the sites on standardized forms (*i.e.*, field survey forms).

The lab analysis results and field survey paper sheets are sent to the NAEMP central data management team and organized into standardized spreadsheet files. The NAEMP data management team manages those spreadsheet files in a centralized manner. Figure 2 shows the data collection process for NAEMP.

Due to the standardized protocols and methods applied to every monitoring site, the datasets in NAEMP have great potential and value as reference datasets for various ecological studies and management. However, NAEMP currently lacks effective dataset management and data services for other ecological studies. Currently, the datasets are managed as spreadsheet files (*i.e.*, Microsoft Office Excel files) in NAEMP. Excel Macro programs embedded in the dataset files are used to generate biological health indices. Such spreadsheet file–based management of datasets is very simple, easy, and flexible, but inevitably *ad hoc* and error-prone. With this data management approach, it is very difficult to search for or locate data items of interest in the datasets, to share only some parts of the datasets (e.g., excepting sensitive or private data) with other users or information systems outside the NAEMP, and to support extensions to ecological monitoring such as the addition of new living organisms, sites, or methods.

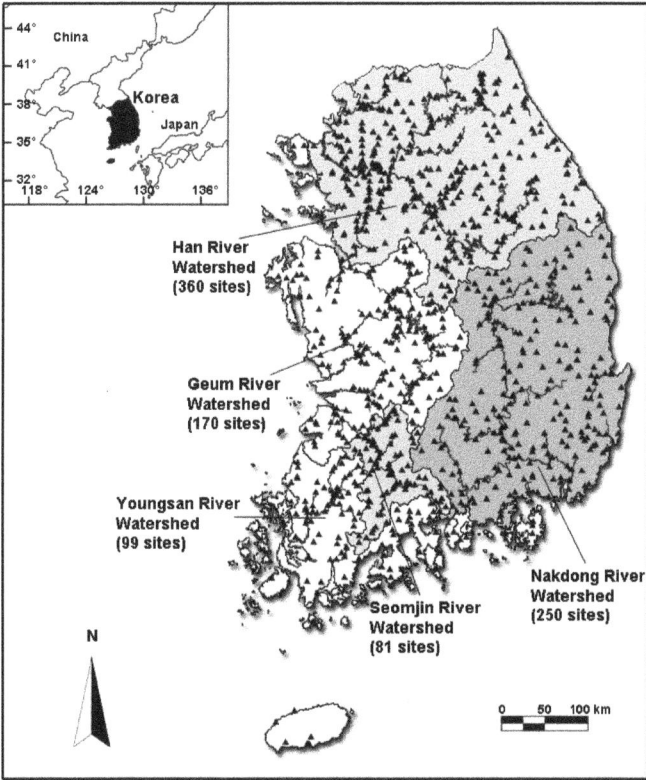

Figure 1. The monitoring sites in the five major rivers for the National Aquatic Ecological Monitoring Program in Korea. The triangles indicate 960 sites that are currently monitored.

Table 1. The parameters measured for fish.

Category	Parameter	Type	Unit	Method
Site Information	Water Basin	Text		
	River	Text		
	GPS/Address	Text		
	Stream Type	Text		
	Sampling Method	Text		
	Weather	Text		Observe
Biological Factor	Species	Text		
	Number of Species	Numeric Data		Observe
	Tolerance Guild	Text		
	Trophic Guild	Text		
	Habitat Guild	Text		
	Protected Species	Text		
	Exotic Species	Text		
	Abnormal Individuals	Text		
	Number of Abnormal individuals	Numeric Data		Observe
Environmental Factor	Substrata Structure	Numeric Data	%	Observe
	Type of Water Flows	Numeric Data	%	Observe
	Canopy	Numeric Data	%	Observe
	Vegetation Cover	Numeric Data	%	Observe
	Odor	Text		Observe
	Plant Structure	Numeric Data	%	Observe
	Land Use	Numeric Data	%	Observe
	Depth	Numeric Data	cm	Measure
	Stream Width	Numeric Data	m	Measure
	Water Velocity	Numeric Data	cm/sec	Measure
	Water Temperature	Numeric Data	°C	Measure
	Conductivity	Numeric Data	μS/cm	Measure
	Turbidity	Numeric Data	NTU	Measure
	pH	Numeric Data		Measure
	DO	Numeric Data	mg/L	Measure

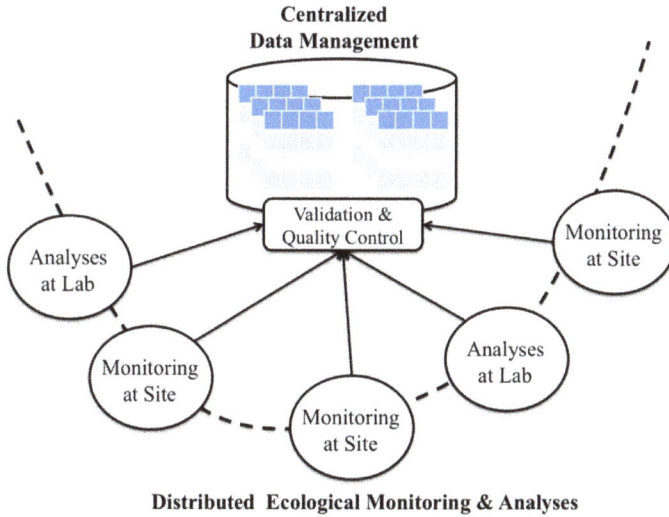

Figure 2. NAEMP data collection and management structure.

NAEMP must address serious challenges with respect to data management, in addition to converting the datasets in spreadsheet files to well-defined tables in a database system. First, there are no universally acceptable standards for such a national-scale, index-based ecological health assessment. Therefore, NAEMP has actually been developing effective health indices over the years and may need to change them for optimization (e.g., adding new living target organisms and parameters, changing methods, or adding new sites) in the future. These changes generally require substantial redesign of data models, data storage, and data services. However, the amount of such redesign on the other components of the information system (e.g., analyses, syntheses, or reporting) must be minimized.

Second, taking into consideration the fact that NAEMP continues to evolve or advance in terms of not only the total size of the data but also science and technology, there will be great potential or requirements for new analyses and syntheses for the assessment of ecological health. Such new analyses or syntheses must be easily testable, addable and supportable.

Finally, if the datasets in NAEMP are used as a kind of reference data for other aquatic ecological studies in Korea, other users or information systems for other projects outside NAEMP will need to access the datasets in NAEMP. For example, there may be new synthetic analyses to integrate data from NAEMP and those from other projects. Such data integration must be supported in an easy and efficient manner.

The OSAEM data management system is developed to address these challenging issues. For the rest of this paper, we explain the main features of the OSAEM system, focusing on the RESTful API and the web portal.

3. RESTful API-based Approach to Data Management

3.1. RESTful API

In this section, we briefly introduce REST (Representational State Transfer), a set of architectural and design styles for software services on the World Wide Web, and then present a novel data model–oriented RESTful API for data management [9]. The REST styles are based on technical concepts such as web resources, URIs (Uniform Resource Identifier), representations and services. Software services available on the Web that satisfy these REST principles are called RESTful web services. The set of resources, URIs (e.g., https://en.wikipedia.org/wiki/Fish), representations, and RESTful web services is called a RESTful API [10].

More specifically, major REST principles relevant to data management are summarized as follows:

- *Resource-oriented*. Everything that needs to be referenced or managed is treated as a logical entity called a *web resource* (hereafter, just *resource*). A resource can be either a real document (e.g., an HTML file or an image file) or a logical entity associated with software services.
- *Universally addressable*. Every resource that is either a real document or a logical entity has a URI assumed to be unique on the Internet. This URI allows the user or the application software to refer and access the resource from any computer on the Internet.
- *Representation-based*. Every resource including a logical entity associated with a software *service* is presented and accessed as if it were a document (or a data object). The document dynamically created by the associated software service for the logical entity is called a representation. In REST, the data structure and format of the representation for each logical entity must be well defined and available to the public. The representation of a resource can be text data, binary data (*i.e.*, data understandable only to particular machines or software), an image, an audio file, and a video. In REST, the most widely used syntax for representations is JSON (JavaScript Object Notation) [35].
- *Serviceable*. The RESTful API assumes the HTTP protocol for communication [36]. In the HTTP protocol, the URI for a resource can be considered to be the Internet address of software services associated with the resource. In the HTTP protocol, four types of software services can be assigned to each URI: POST, GET, PUT, and DELETE. The POST, GET, PUT, and DELETE types are intended for the Create, Read, Update, and Delete operations, respectively.

There are also some other architectural principles but since they are system-specific, they are not covered in this paper which is intended to address data management issues. Please refer to Roy Fielding's article for a complete explanation of REST principles [10].

We present Algorithm 1 and Figure 3 to illustrate how a RESTful web service works. Algorithm 1 shows an example of a representation in JSON. A data object in the JSON format is basically a list of key and value pairs enclosed by curly braces. A key plays the role of a data field in a relational database. A value can be a number, a Boolean value, a text string, another JSON object, or an array of JSON objects. Arrays of JSON objects are enclosed by square brackets. Having a JSON object or an array of JSON objects as a value allows JSON to support a hierarchical data model. In Algorithm 1, the JSON data object consists of two key-value pairs where the keys are "Name" and "Job". The value for the key "Name" is another JSON object.

Algorithm 1 A representation in JSON.

```
{
    "Name": {
        "firstName": "John",
        "lastName": "Smith"
    },
    "Job": "Teacher"
}
```

Figure 3 shows how a RESTful web service works on the Internet. In this example, a logical data entity or document (called a web resource, or just a resource in REST) is assumed to exist for a certain person on a web server where the entity has a URL (Uniform Resource Locator) or URI (Uniform Resource Identifier): http://www.rest.org/person/1. In fact, the logical entity does not exist physically on the server. However, if the URI of the logical entity is referenced on a web browser (e.g., entered into the URL bar window), a new real document is dynamically created from the database and returned by the web server as if the logical entity were a real document. The dynamically created real document is called a representation in REST.

In this example, instead of simply returning a real document file in the file system for the URI, the web server invokes the readPerson software service associated with the URI, then readPerson generates a document dynamically, and finally the web server returns the newly created document. In fact, readPerson obtains and merges three data records from three DB (Database) tables (one record from each DB table, respectively) into a new document (called a representation in REST) for the URI. The syntax for data in this example will be explained in Section 4. In REST, software services, therefore, appear as if they were documents. Such software services on web servers are called RESTful web services (hereafter, just web services).

47

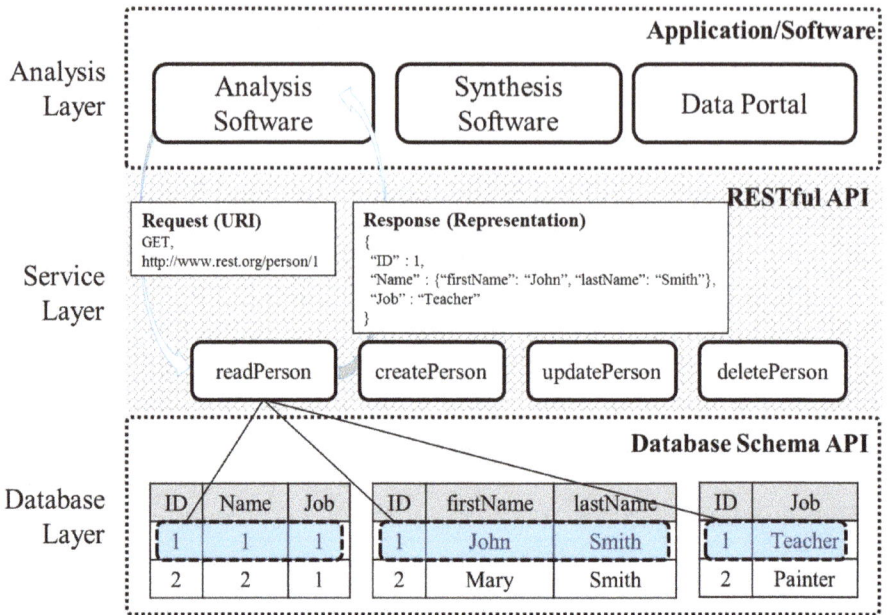

Figure 3. Web architecture for RESTful web services.

3.2. Data Model-Oriented RESTful API

Conventionally, the REST principles and APIs are used to model and define the interfaces for web services, although they have great potential for other kinds of modeling. In this paper, we propose a novel approach to the RESTful API where the REST principles are applied to the design of data models for data management. The approach is called the *data model–oriented RESTful API* where services are not explicitly defined but implicitly assumed.

The data model–oriented RESTful API is designed to support the following resource model:

- *Resource classification.* Every resource is classified as either an instance resource or a collection resource.
- *Instance resource.* Each individual resource is considered an instance resource. Each instance resource has a unique URI and is accessed as a representation.
- *Collection resource.* A collection resource represents the set of all the instance resources of the same type. Each collection resource also has a unique URI.
- *Representation schema.* The representation schema is defined to be the data model or structure for the representation of a resource. The representations of all the instance resources in a certain collection resource must be based on the same representation schema. Syntactically, the representation schema is defined to

be a list of key and data-type pairs enclosed by angle braces. However, the representation schema (created only as a document) is currently intended to help the user to understand the data models for resources, but is not used in system implementation.

- *Implicit CRUD (Create, Read, Update and Delete) services.* According to the HTTP protocol, each resource (its URI) is assumed to be associated with four types of services: Create, Read, Update and Delete. The Read service for a collection resource is designed to be a database query. The HTTP-based communication allows each resource to be independently accessed by those services.

This resource model has several advantages over other conventional RESTful APIs and data management systems.

- First, the resource model makes the RESTful API simple and intuitive to understand and use. Although the REST principles allow a great amount of generality and flexibility, they effectively do not enforce simplicity and consistency. This resource model combines and unifies data models and services logically and physically into only URIs and representations. The user can assume the HTTP CRUD services for each resource.

- Second, the resource model makes the RESTful API look like a database system. In fact, the resource model is logically similar to the database model. The instance and collection resource are analogous to the database record and tables, respectively. The representation schema is similar to the database schema. The URI for a resource can be considered to be the key for a database record. Therefore, this resource model can substantially help the software developer to understand the design of data management and to develop analysis software.

- Third, the resource model is managed independently of underlying database schemas that are usually much more complicated than the resource model. Figure 3 illustrates the independence between the resource model and underlying database schemas. Although the resource model looks like a database model, it is in fact the model for service interfaces (*i.e.*, RESTful APIs). Both the resource model and the underlying database design can be changed without affecting the other. This feature facilitates the support for hosting and portability significantly.

- Finally, if the resource model is publicly open and available to any user or information systems, no special custom-built software is required to access datasets. There are generic HTTP protocol-based client libraries widely available that can be used to invoke software services associated with the resources. Therefore, when any research group wants to develop analysis software, they do not need any special software from the NAEMP project and can easily develop their own analysis software only with publicly available generic HTTP protocol-based libraries

Because of these advantages, *the data model-oriented RESTful API* can effectively address the challenging issues in the sharing of ecological monitoring data discussed in Sections 1 and 2.

4. OSAEM: The Open, Sharable, and Extensible Data Management System for Aquatic Ecological Monitoring

In this section, we present a data management system for NAEMP datasets called OSAEM (the Open, Sharable, and Extensible Data Management System for Aquatic Ecological Monitoring) that is designed to support the data model–oriented RESTful API in Section 3.

4.1. Resource Model

First, major data components of the NAEMP are modeled and managed as instance and collection resources:

- *Living organisms.* Each individual species of fish, benthic diatoms and macroinvertebrates is modeled as a resource. For example, fish species such as "*Lethenteron camtschaticum* Tilesius" and "*Cyprinus carpio* Linnaeus" are treated as separate resources in OSAEM. Since 150 fish species are currently monitored in NAEMP, there are 150 resources in OSAEM. Currently, the OSAEM system does not support datasets for benthic diatoms and macroinvertebrates.
- *Sites* and *rivers.* The individual sites where ecological monitoring is carried out are modeled as resources. The rivers are also handled as resources. In NAEMP, each of the 960 sites is currently considered a separate resource.
- *Parameters* and *methods for observation.* The individual parameters are treated as resources. Therefore, this design requires all observation parameters to be explicitly defined and registered. Similarly, the individual measurement and analysis methods are managed as resources.
- *Observation data.* Each individual instance of observation is modeled as a separate resource. For example, if the water temperature at a certain site is measured 10 times, the 10 measurement results are considered separate resources.

The instance and collection resources are managed as follows:

- *Instance resources.* A single data entity (e.g., an individual fish species or an individual site) is managed as an instance resource. An instance resource is logically analogous to a database record in relational databases.
- *Collection resources.* All instance resources of the same type are managed as one collection resource. For example, the Fish collection resource is defined

as the set of all instance resources for fish species. Currently, there are eight collection resources in OSAEM: Fish, Site, River, Variable, Method, Source, Unit and Observation. Collection resources are logically analogous to database tables in relational database systems.

The organization of resources is illustrated in Figure 4 where two collection resources (Fish and Site) and their instance resources are shown. In the example, URI_BASE in URIs denotes the common prefix for every URI in OSAEM and is explained in Section 4.2. URI_BASE/fishes and URI_BASE/sites are the URIs of the Fish and Site collection resources, respectively. URI_BASE/fishes/1 and URI_BASE/sites/1 are the URIs of instance resources for specific fish species and site, respectively. The URI patterns are also explained in Section 4.2.

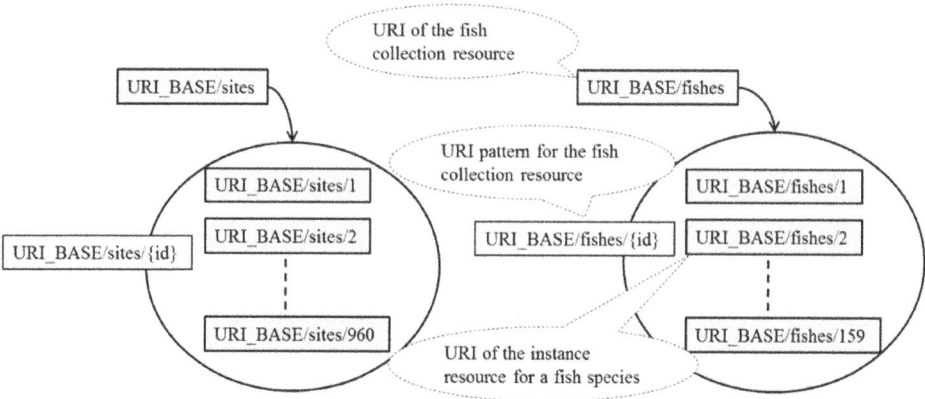

Figure 4. Organization of instance and collection resources. URI_BASE is the common prefix for every URI in OSAEM and is explained in Section 4.2. URI_BASE = http://db.cilaboratory.org:8080/naemp.

4.2. URI Scheme

In OSAEM, we support a simple scheme for assigning URIs to the instance and collection resources. This URI scheme intends to make the management of URIs intuitive and predictable in spite of a large number of instance resources. The URI scheme consists of:

- URI structure.
- URIs and URI patterns for the collection resources.
- System-assigned numeric IDs for the instance resources.

First, the structure of every URI consists of three components in OSAEM: the URI base, the name of the collection resource, and, optionally, the system-assigned

51

resource ID only for an instance resource. The URI base is the common prefix for every URI in OSAEM. The URI base is composed of the Internet network address of the web server (e.g., http://db.cilaboratory.org:8080/) and a web application name (currently, just /naemp) added to indicate the NAEMP project. Scheme 1 illustrates an example of the URI structure. Currently, the OSAEM prototype system is on a web server whose Internet network address is http://db.cilaboratory.org:8080, but the address will be changed to a permanent web server later. For the rest of this paper, we use the symbolic keyword URI_BASE in every URI to denote the URI base.

Scheme 1. URIs and URI pattern.

URI for the Fish collection resource

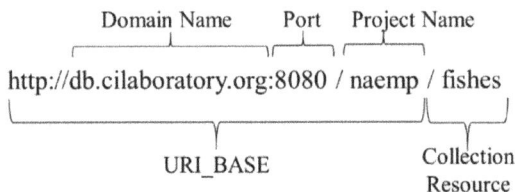

Domain Name   Port   Project Name

http://db.cilaboratory.org:8080 / naemp / fishes

URI_BASE   Collection Resource

URI for an instance resource in the Fish collection resource

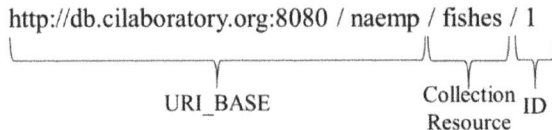

http://db.cilaboratory.org:8080 / naemp / fishes / 1

URI_BASE   Collection Resource   ID

URI Pattern for the Fish collection resource

http://db.cilaboratory.org:8080 / naemp / fishes / {id}

Second, the URI for every instance resource in the same collection resource is based on the same pattern called the URI pattern. Therefore, each collection resource is associated with the URI pattern shared by the instance resources. The URI pattern for a collection resource consists of the URI for the collection resource and a variable (*i.e.*, a placeholder) for a system-assigned numeric ID for a specific instance resource. In Scheme 1, the URI pattern for the Fish collection resource is URI_BASE/fishes/{id}, where {id} denotes a numeric ID for an instance resource. The URI for an instance resource (*i.e.*, a fish species) in the Fish collection resource is URI_BASE/fishes/1.

Together with their URIs and URI patterns, all eight collection resources are listed in Table 2.

Table 2. URIs and URI patterns for the collection resources.

Collection Resources	URI	URI Pattern	Number of Instance Resources
Fish	URI_BASE/fishes	URI_BASE/fishes/{id}	159
Site	URI_BASE/sites	URI_BASE/sites/{id}	960
River	URI_BASE/rivers	URI_BASE/rivers/{id}	760
Variable	URI_BASE/variables	URI_BASE/variables/{id}	12
Unit	URI_BASE/units	URI_BASE/units/{id}	11
Method	URI_BASE/methods	URI_BASE/methods/{id}	5
Source	URI_BASE/sources	URI_BASE/sources/{id}	17
Observation	URI_BASE/values	URI_BASE/values/{id}	139567

4.3. Representation Schema

In OSAEM, a resource (either a real document file or a logical entity associated with a software service) is handled as a data object (or a document) at run time. Therefore, every resource must be defined and handled as if it were a real document. For example, if a human user enters the URI of the resource for the fish species Arctic lamprey (in fact, URI_BASE/fishes/1) into the URL bar of a web browser, then the user will receive a document (*i.e.*, data) for the resource. Figure 5 gives a snapshot of a web browser window displaying the data for the fish species Arctic lamprey. Web browsers do not print the data in a readable format with proper indentations and line breaks; the same data is well formatted in Figure 6. Such a formatting service (e.g., http://codebeautify.org/jsonviewer) is available online.

The realization of a resource as a document or data object is called a representation. In RESTful APIs, resource representations can be based on various formats: JSON, XML, HTML, or multimedia formats (e.g., PNG, JPEG, MP3, and MP4). In OSAEM, JSON is used as the main data format for representations [35]. JSON is a simple text format in the JavaScript programming language but is widely used for a variety of Internet services.

In OSAEM, the representation of every instance resource in the same collection resource (e.g., the Fish collection resource) is based on the same data model or structure called the *representation schema*, whose name was chosen to show its resemblance to the database schema. The representation schema facilitates the management of representations. In OSAEM, the representation schema is similar to the JSON syntax but contains data types (enclosed by a pair of angled brackets) instead of values.

In OSAEM, there are 14 representation schemas. Appendix A shows the list of all the representation schemas. Algorithm 2 shows the representation schema for the Fish collection resource. In fact, the representation schema simply shows the keys and their data types. Figure 5 shows the representation for the fish species Arctic lamprey based on the representation schema.

{"family":"칠성장어과","order":"칠성장어목","invasiveSpecies":"","trophicGuild":{"term":"O","definition":"잡식종"},"habitatGuild":{"term":"-","definition":" "},"toleranceGuild":{"term":"IS","definition":"중간종"},"fishID":1,"fishClass":"두갑강","scientificName":"Lethenteron camtschaticum Tilesius","species":"칠성장어","endemicSpecies":" ","endangeredSpecies":"멸종위기 2급","naturalMonument":null,"imageLink":null,"feature":{"featureID":1,"featureName":"칠성장어","featureType":"Fish"},"description":null}

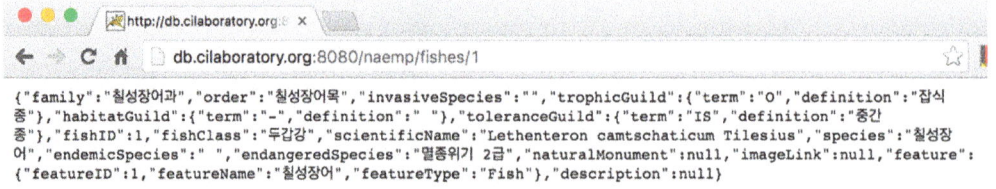

Figure 5. Snapshot of a web browser displaying the representation data for Arctic lamprey.

Result : Beautify Json

```
1 ▾ {
2        "family": "칠성장어과",
3        "order": "칠성장어목",
4        "invasiveSpecies": "",
5 ▾      "trophicGuild": {
6            "term": "O",
7            "definition": "잡식종"
8        },
9 ▾      "habitatGuild": {
10           "term": "-",
11           "definition": " "
12       },
13 ▾     "toleranceGuild": {
14           "term": "IS",
15           "definition": "중간종"
16       },
17       "fishID": 1,
18       "fishClass": "두갑강",
19       "scientificName": "Lethenteron camtschaticum Tilesius",
20       "species": "칠성장어",
21       "endemicSpecies": " ",
22       "endangeredSpecies": "멸종위기 2급",
23       "naturalMonument": null,
24       "imageLink": null,
25 ▾     "feature": {
26           "featureID": 1,
27           "featureName": "칠성장어",
28           "featureType": "Fish"
29       },
30       "description": null
31 }
```

Figure 6. Well-formatted representation data for the fish species Arctic lamprey generated by an online formatting service (http://codebeautify.org/jsonviewer).

Algorithm 2 Representation schema for the Fish collection resource.

```
{
        "fishID": <integer>,
        "class": <string>,
        "order": <string>,
        "family": <string>,
        "species": <string>,
        "scientificName": <string>,
        "toleranceGuild": <string>,
        "trophicGuild": <string>,
        "habitatGuild": <string>,
        "invasiveSpecies ": <string>,
        "endemicSpecies ": <string>,
        "endangeredSpecies ": <string>,
        "naturalMonument ": <string>,
        "imageLink": <string>,
        "description": < string>
}
```

The representation of a collection resource is simply the entire list of the representations of all its instance resources. All the representations in the list must conform to the representation schema for the collection resource. Algorithm 3 shows a representation (many details are omitted) for the Fish collection resource that a user can easily retrieve by entering its URI (URI_BASE/fishes) into the URL bar of any web browser. Thus, it is really simple for a user to determine which instance resources exist in a certain collection resource, as long as the user knows the URI for the collection resource. However, the OSAEM does not allow the user or other software to retrieve the representation of the Observation collection resource because the number of instance resources in the collection resource is very large.

Algorithm 4 shows the representation schema for the Observation collection resource. An instance resource of the Observation collection resource contains the data collected from a single monitoring activity. The data includes a value (*i.e.*, what value was collected), a site ID (*i.e.*, where monitoring occurred), an entity ID (*i.e.*, what species was monitored), a variable ID (*i.e.*, what parameter was measured), and a method ID (*i.e.*, what method was used). Each of these (site ID, entity ID, variable ID, and method ID) references an instance resource in one of the Site, Fish, Variable, and Method collection resources, respectively. The real URI for each instance resource can be automatically generated with such a system ID and the corresponding URI pattern.

Algorithm 3 Representation for the Fish collection resource.

```
[

{ "fishID": 1, "scientificName": "Lethenteron camtschaticum Tilesius" ... },
{ "fishID": 2, "scientificName": "Lethenteron reissneri Dybowski" ... },
{ "fishID": 3, "scientificName": "Anguilla japonica Temminck and Schlegel" ... }, ...
...

]
```

Algorithm 4 Representation schema for the Observation collection resource.

```
{
        "valueID": <integer>,
        "dateTime": <string>,
        "surveyYear": <integer>,
        "surveyTerm": <integer>,
        "dataValue": <double>,
        "site": <integer (siteID)>,
        "latitude": <string>,
        "longitude": <string>,
        "source": <integer (sourceID)>,
        "entity": <integer (speciesID)>,
        "variable": <integer (variableID)>,
        "method": <integer (methodID)>
}
```

4.4. RESTful Web Services

In OSAEM, a resource is serviceable because its URI is associated with software services. That is, the URI of each resource is used as the Internet network address for those software services. In OSAEM, the following software services are provided:

- *Four CRUD services* (explained in Section 3.2) for every instance resource. The CRUD services include the Create, Read, Update and Delete services. They are used to manage instance resources.
- *Query services* for collection resources. Query services are used to search instance resources of interest from collection resources.

4.4.1. CRUD Services for the Instance Resources

In OSAEM, every instance resource (*i.e.*, each URI) is assumed to have a set of four CRUD software services. Therefore, when instance resources must be managed (e.g., create, read, update, or delete them), these services can be invoked. However, the collection resources do not have their own CRUD services, with the exception of the Read operation, because management of the collection resources is really critical and is supported only through a separate admin tool available to the human system administrator.

In the system implementation, a set of CRUD services is in fact assigned to the URI pattern for each collection resource, instead of the URIs of individual instance resources, because the CRUD services are the same for every instance resource in the same collection resource. In other words, every individual resource in a collection resource shares the same set of CRUD services in the OSAEM system.

As explained in Section 3.2, software services associated with URIs are invoked according to the HTTP protocol, which is the URI-based request-response communication model. In the protocol, a client sends a URI-targeted request message to a server; then the server carries out the request on the URI and returns the result as a response to the client. In the protocol, there are four major request types: POST, GET, PUT, and DELETE.

Figure 7 illustrates how CRUD services are associated with the resources for Fish and invoked at runtime:

- First, the four CRUD services are associated with the URI pattern for the Fish collection resource: URI_BASE/fishes/{id}. The CRUD services are createFishSpecies, readFishSpecies, updateFishSpecies, and DeleteFishSpecies.
- Second, a client system generates CRUD service request messages for a certain fish species according to the HTTP protocol and sends them to the web server. The request message contains three components: the request type (*i.e.*, POST, GET, PUT, or DELETE for the create, read, update or delete operations, respectively), the URI for the target resource, the header information for attributes, and, optionally, the message body (usually a resource representation). In this example, for the fish species Arctic lamprey, whose URI is URI_BASE/fishes/1, HTTP requests are shown. The GET type of request message consists of the string text for the request type (*i.e.*, "GET") and the URI for Arctic lamprey. However, the POST type of request message contains a special URI (that is URI_BASE/fishes/new) instead of URI_BASE/fishes/1 because the URI does not exist at the time of the creation request.
- Third, when it receives a HTTP request message for a URI, the web server invokes a software service (in fact, a RESTful web service) according to the request type.
- Finally, the web service (*i.e.*, createFishSpecies for the POST request and readFishSpecies for the GET request) performs the request operation and generates a response message as a result. The web server then returns the result to the client system. For the GET type of request message, the response contains the representation of the target resource (*i.e.*, Arctic lamprey).

57

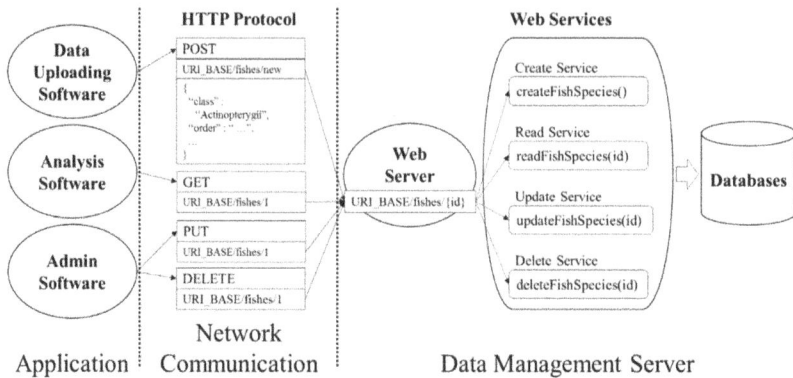

Figure 7. CRUD services for the fish instance resource.

4.4.2. Query Services for the Collection Resources

In addition to the CRUD services for the individual management of each instance resource, the OSAEM system also supports query services to search for instance resources of interest through collection resources. The OSAEM system does not support an official query language but, rather, a set of query services based on the representation schema. These query services are associated with the URIs of the collection resources because a query is a collection resource–level operation.

The logical model for a query is defined as follows:

- A query specifies how to select instance resources of interest from a target collection resource.
- Syntactically, a query is logically defined to be a series of query conditions on a representation schema where each query condition consists of a key and a list of matchable values for the key (in fact, a JSON array object). Therefore, a query itself is a JSON data object.
- The query result is a list of representations that match the query. The result is generated as a JSON array.

Figure 8 illustrates the logical model for the query in OSAEM. In the figure, the representation schema for the Site collection resource is shown on the left. Two queries about the collection resource are given on the right. The top query consists of only one query condition whose key is "siteName" and whose matchable values are the list of two site names. The bottom query is composed of two query conditions and those sites whose representations satisfy both queries are selected.

58

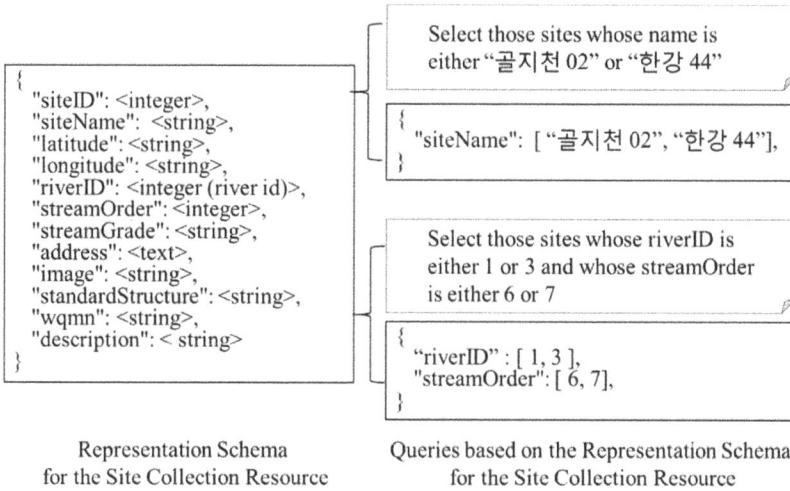

| Representation Schema for the Site Collection Resource | Queries based on the Representation Schema for the Site Collection Resource |

Figure 8. Logical model for the query in OSAEM.

The query model is supported in two ways:

- *GET requests with URIs.* In the HTTP protocol, a URI can have a number of additional query conditions. In OSAEM (in fact, in conventional RESTful web services), each query condition consists of a key and a value (any string).

- *POST requests with queries in JSON.* A query in JSON is directly submitted as a POST request. Queries with complicated query conditions are difficult to express in URIs. In this case, POST requests with queries in JSON are used instead of GET requests with URIs.

First, queries based on the GET request take advantage of the query feature of the URI syntax in the HTTP protocol. The HTTP protocol allows the URI to consist of the address and the query parts. The question mark "?" is used as the delimiter to separate the address and query parts in the URI. The query part consists of a list of query conditions separated by the ampersand sign "&". Each query condition is a pair of a query parameter (in fact, a key in the OSAEM query model) and a value combined with the equal sign "=".

Algorithm 5 shows both an example of a query expressed as a URI for the Site collection resource and the HTTP GET request message with the URI. The query URI consists of the address part (*i.e.*, URI_BASE/sites) and the query part ("?riverID=1&riverID=3&streamOrder=6&streamOrder=7"). The query part consists of two query conditions: "riverID=1, riverID=3" and "streamOrder=6, and streamOrder = 7". The riverID and streamOrder parameters are keys of the representation schema for the Site collection resource and the repeated parameters are recognized as an array for each parameter.

Algorithm 5 GET request with URIs as a query for the Site collection resource.

---

http:/db.cilaboratory.org/naemp/sites?riverID=1&riverID=3&streamOrder=6&streamOrder=7

GET /naemp/sites? riverID=1&riverID=3&streamOrder=6&streamOrder=7
HTTP/1.1
Host: db.cilaboratory.org:8080

In comparison with queries based on the GET request, queries based on the POST request are simple to create, but cannot be used in a web browser. They require the HTTP software library to invoke queries. A query (in fact, a JSON data object) is simply included as the message body of the POST request message. The use of the POST request for querying is not exactly consistent with the conventional semantics (*i.e.*, the creation of a resource) for the POST operation recommended in the HTTP protocol. However, since it is strongly recommended that the GET request not have a message body, the POST request is often used for complicated queries.

4.4.3. Query Services for the Observation Collection Resource

In this section, we explain the query services for the Observation collection resource in detail because those queries are the most important for the ecological studies. Therefore, the collection resource requires a kind of relational database join operation to compare and merge data from multiple collection resources. However, the other collection resources in the project datasets have no direct relationships with each other.

Figure 9 shows the relationships between the Observation and other collection resources. For example, the join operation is needed for a query such as 'select observation data that is collected at the site whose name is "골치천 02". In the current design of the OSAEM system, such a join operation is not fully supported yet. However, such a join operation can be easily programmed in OSAEM. Currently, only some primitive join operations are available for the Observation collection resource:

- The query services for every Collection resource are extended to enable the join mode. The join mode is enabled only for the Observation collection resource. That is, no new join operation is added.
- When the join operation is enabled, the representation of an instance resource in the Observation collection resource is automatically expanded to replace the system IDs for instance resources in the other collection resources such as Site and Fish and become their real representations.

Algorithm 6 illustrates the join mode for the Observation collection resource. The only difference between queries with and without the join mode is whether the query parameter "join" is added and set to "on". The algorithm shows the result of a

60

query with the join mode enabled. If the join mode is disabled in the query, then the representation shown in Figure 9 is retrieved.

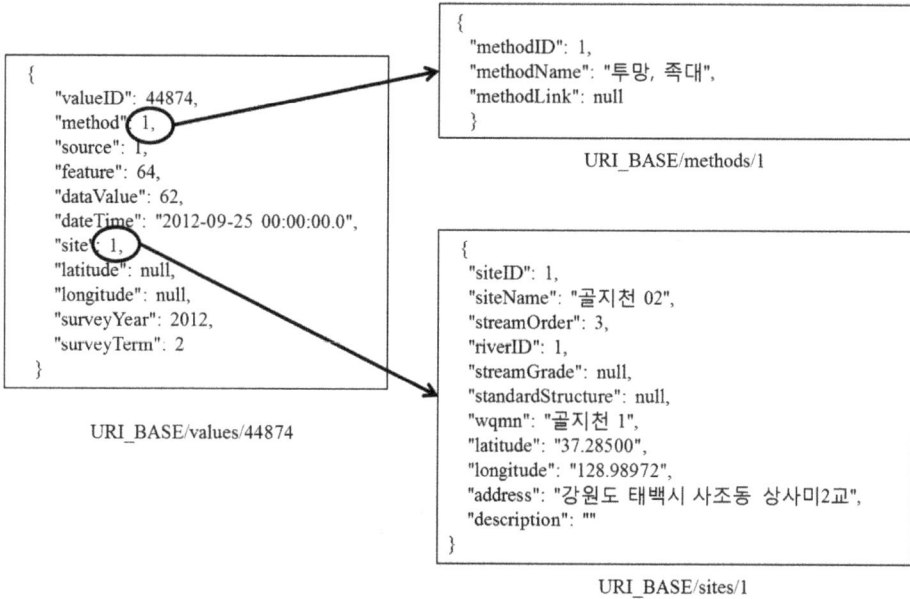

Figure 9. Relationships between the Observation and other collection resources.

Algorithm 6 Query with the join mode enabled: URI_BASE/values?valueID=44874&join=on.

```
URI_BASE/values?valueID=44874&join=on
{
    "valueID": 44874,
    "method": {
        "methodID": 1,
        "methodName": "투망, 족대",
        "methodLink": null
    },
    ... ....
    "dataValue": 62,
    "dateTime": "2012-09-25 00:00:00.0",
    "site": {
        "siteID": 1,
        "siteName": "골지천 02",
        ... .... .
    },
    ... .....
}
```

5. Web Portal

In OSAEM, the RESTful API is intended for software (e.g., programming synthetic analysis code), not for the human user, although the user can test most of the services with only a web browser. The API will be used by not only a variety of analysis, visualization, and reporting software but also by information systems in other projects for synthetic research. For the end user, OSAEM provides a web portal, a user-friendly GUI system based on the RESTful API. The main goal of the OSAEM web portal is to allow the user to access NAEMP datasets through an intuitive and convenient GUI, instead of the programming-oriented RESTful API. However, the OSAEM web portal uses the RESTful API explained in Section 4.

Currently, the OSAEM web portal consists of two major components:

- GUI services to manage the project datasets: the Fish, Site, River, Variable, Method, and Source collection resources
- GUI services to search through the observation datasets (*i.e.*, only the Observation collection resource).

5.1. CRUD Services and Query Services for the Project Datasets

The web portal allows a human user to manage the project datasets via an intuitive and interactive GUI as follows:

- First, the user selects a collection resource such as Fish or Site in the web portal. Then, the web portal displays all the instance resources in the selected collection resource on the web browser.
- Second, the user chooses instance resources of interest from the selected collection resource.
- Finally, the user requests CRUD operations for those selected instance resources in the collection resource. In the collection resource, the user can read, update, or delete an existing instance resource (e.g., a particular fish species or site). In addition, the user can create a new instance resource in the collection resource.

Figure 10 shows the GUI window of the web portal for the management of project datasets. In the window, when the user selects the Project Management button in the top command menu bar, the web portal shows the list of collection resources for the project datasets in a pop-up window. If the user selects a collection resource (e.g., Site) in the pop-up window, then the web portal displays all the instance resources in the collection resource in a tabular form in the window. Since the number of instance resources in the collection resources for project datasets is small, the web portal displays all instance resources in the selected collection resource in the GUI window at once.

Figure 10 shows the snapshot of the GUI window displaying all the instance resources in the Site collection resource. In this case, the web portal gets the data from the OSAEM server by sending a GET request with "URI_BASE/sites" as the URI. Then, the web portal in fact receives the representation of the Site collection resource from the OSAEM server as the response shown in Algorithm 7.

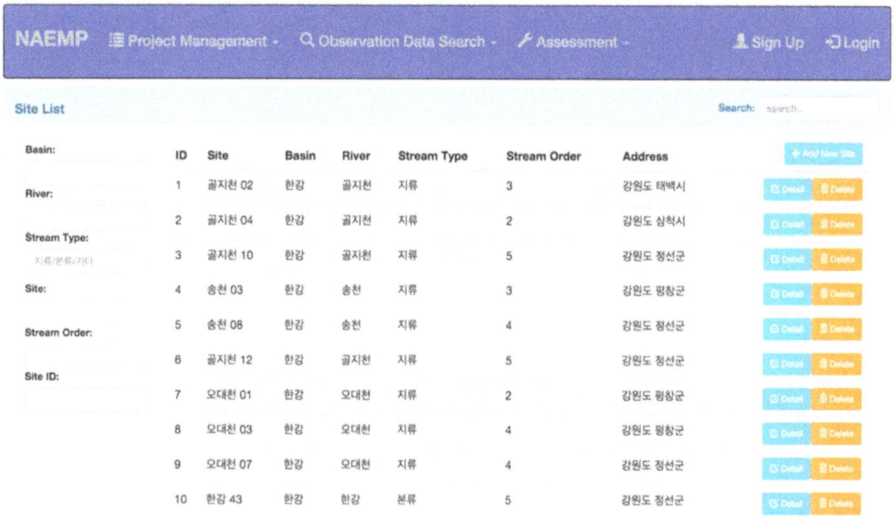

Figure 10. GUI window for the Site collection resource.

Once the instance resources are displayed in the GUI window, the user can read, update or delete a particular instance resource in the window by clicking on the corresponding button (*i.e.*, the Detail button for Read and Update, and the Delete button for Delete). In addition, the user can also create a new instance resource in the collection resource by clicking on the add button. Figure 11 shows the detailed view provided for a specific site when the detail button for the site resource is clicked.

Currently, the web portal does not support query services for project datasets because their sizes are small. In other words, it always retrieves all the instance resources from the OSAEM server and displays all of them in the window at once.

Algorithm 7 Representation for the Site collection resource.

```
[
  {
    siteID": 1,
    "siteName": "골지천 02",
    "streamOrder": 3,
    "wqmn": "골지천 1",
    address": "강원도 태백시 사조동  상사미 2 교",
    ......
  },
  {
    siteID": 2,
    "siteName": "골지천 04",
    "streamOrder": 2,
    "wqmn": "번천",
    "address": "강원도 삼척시 하장면 숙암리 번천리측 지류",
    ......
  }
  ......
]
```

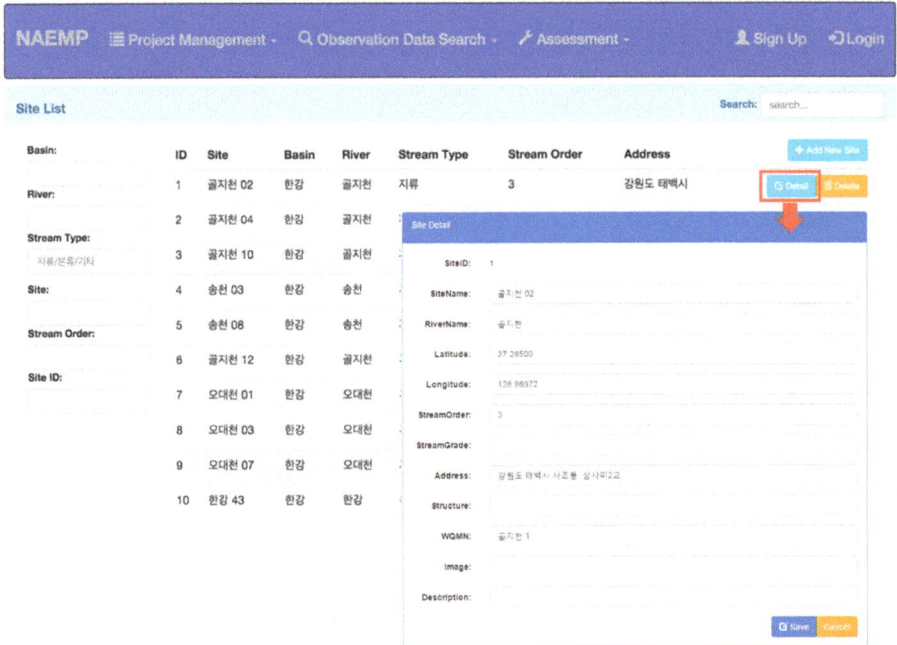

Figure 11. GUI window for the detailed view and update for a site.

5.2. Query Services for the Observation Datasets

Although the RESTful API supports a variety of ways to search for observation data, the OSAEM web portal currently allows the user to search for observation data through the Observation collection resource in two ways:

- *Site-based search.* The site-based search is intended to support site-specific data search and analyses. The user selects sites of interest in the GUI window. Then, the web portal retrieves all the observation data collected at the selected sites and displays them in a tabular form in the web browser.
- *Species-based search.* The species-based search is intended to support comparative and synthetic data search and analyses. The user selects observation targets of interest (e.g., fish species) in the GUI window. Then, the web portal shows all of the observation data for those targets collected at all of the sites.

Selection of the Observation Data Search button in the top command menu bar creates a pop-up window with two choices: Site and Species. If the Site menu is chosen, the web portal displays all the sites and allows the user to filter out those sites not of interest by specifying matching conditions about site attributes. In addition, the user can specify times of interest. Then, the web portal retrieves the observation data collected at those selected sites during the specified times.

Figures 12 and 13 show the GUI windows for the site selection and the display of observation data collected at the selected sites during specified times. In the web portal, the species-based search also works in a similar way. Those retrieved observation datasets are displayed in a tabular form by default. In addition, they can be graphically displayed in charts. Figure 14 shows observation datasets in charts.

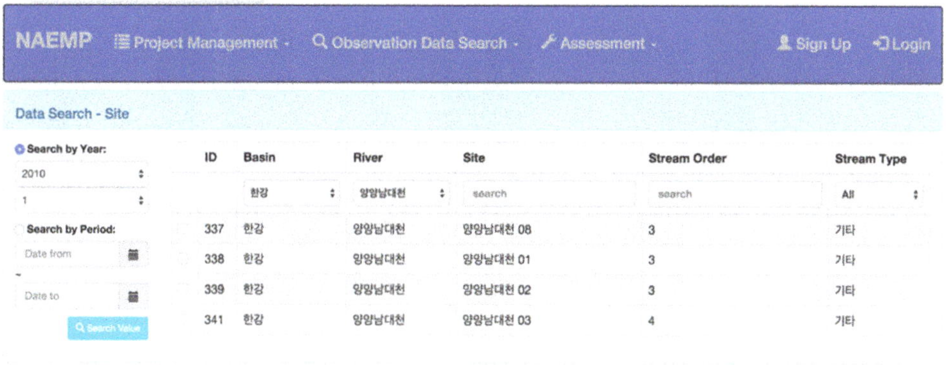

Figure 12. GUI window for the site-based observation data search.

Site	Domain	Target	Variable	Result
				> Value >
search	search	search	search variable	
양양남대천 01	Water	수질	T-N	1.603
양양남대천 01	Water	수질	Chl-a	0
양양남대천 01	Water	수질	NO3-N	1.23
양양남대천 01	Water	수질	BOD	1.2
양양남대천 01	Water	수질	NH3-N	0.001
양양남대천 01	Water	수질	T-P	0.014
양양남대천 01	Water	수질	PO4-P	0.002
양양남대천 02	Water	수질	BOD	1.1
양양남대천 02	Water	수질	Chl-a	0.9
양양남대천 02	Water	수질	NO3-N	0.71

Figure 13. Observation data collected at selected sites.

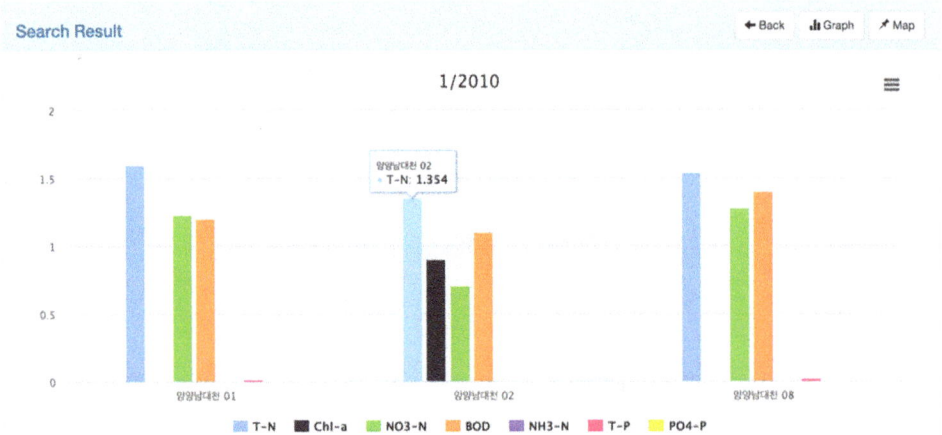

Figure 14. Observation data in the chart window.

6. Discussion and Conclusions

In this paper, we presented the data management system called OSAEM (the Open, Sharable, and Extensible Data Management System for Aquatic Ecological Monitoring), developed to manage datasets from the Korea National Aquatic Ecological Monitoring and Assessment Program (NAEMP) [11,29–34]. In NAEMP, the datasets are collected every six months from 960 sites according to standardized protocols and methods (the number of sites will be increased to 3000 by 2018). Therefore, they are highly comparable and easy to integrate with respect to both species and sites, and provide great potential and many opportunities for advanced comparative and synthetic analyses in various ecological studies and management. However, the realization of such potential and opportunities requires effective data management and sharing of the datasets from NAEMP.

The OSAEM system is designed to facilitate and promote data sharing of the NAEMP datasets. In OSAEM, the management and sharing of the datasets are based on web technology called the RESTful API (the set of resources, URIs, representations and RESTful web services) [10]. In this paper, we explained how the NAEMP datasets are modeled and managed as the RESTful API. The RESTful API for the NAEMP datasets unifies the data models and service interfaces both logically and physically. This feature significantly facilitates the design and implementation of the application software (e.g., analysis code) to access the datasets.

In this paper, we presented two main features of the OSAEM system: the RESTful API and web portal. In OSAEM, the RESTful API is mainly intended for software rather than the human user, although it is very logical and easy for the user to understand and test with only a web browser. In addition to the RESTful API, a web portal is provided to help the human user access the datasets in the OSAEM system in an easy and interactive manner. The web portal uses the RESTful API to access the datasets in the OSAEM server.

With respect to data sharing, this RESTful API-based data management approach has important advantages over the traditional database-based approach. First, the data models in the publicly available RESTful API (*i.e.*, instance resources, collection resources, URIs, URI patterns, representation schemas, and web services) are simple and intuitive to understand, and significantly facilitate the development of software (e.g., analysis code and visualization code) to access the datasets. In the data models, every entity or object (e.g., a fish species, a site, a parameter, or a method) in ecological monitoring is explicitly referenced and can be individually managed with URIs and representation schemas.

Second, the URIs for data objects (*i.e.*, resources) in the RESTful API can be directly invoked from any computer on the Internet without the installation of any special client software. In other words, the same URIs are used not only to identify data objects (e.g., a specific fish species) but also to invoke corresponding software services. Furthermore, datasets can be easily accessed with simple, generic HTTP clients including web browsers. In addition, the representation schema for the fish species is both its logical data organization and the data format of the response message from the service invocation. Therefore, data sharing can be significantly simplified.

For example, the URI for a fish species (e.g., URI_BASE/fishes/1) is both its unique ID and the Internet network address of the software service used to access the data for the fish species. With an ordinary web browser such as Google Chrome or Microsoft Internet Explorer, any user can access most datasets easily by entering URIs into the URL address bar of the web browser.

Finally, the data models in the RESTful API are not actual database schemas but separate logical models supported by software web services, independent of the underlying database schemas. This two-layered design of the data models makes data analysis or visualization (*i.e.*, using client or application software) independent of the underlying database schemas, and therefore allows the underlying databases to be extended or modified without affecting the software. For example, during the development of the OSAEM prototype system, we have been able to modify the underlying database schemas without changing the RESTful API at all, and *vice versa*.

However, the RESTful API-based data management in OSAEM currently has some drawbacks and limitations that need to be further addressed in future work. First, there is no explicit query language or engine available, although HTTP-based CRUD services are provided for important and frequent types of queries. Therefore, only a predefined set of queries can be supported and complex queries such as the relational join operation are not available. If a new type of query is needed, then a new web service must be developed.

Second, query processing requires more performance overhead because web services for query processing cause additional performance overheads and are not as optimized as database query engines. However, the size of the NAEMP datasets (*i.e.*, hundreds of thousands of data objects) is, fortunately, not large enough to cause serious performance overhead on conventional server computers. These performance issues will be addressed together with query engines in future work.

Third, the web portal currently allows the user to access a very limited subset of services from the current RESTful API. Furthermore, a variety of analysis or visualization code can be added to the web portal. Such extensions to the web portal are planned for 2016.

Currently, there are no data sharing policies established in NAEMP. Since the datasets are generated from the government program, they are generally assumed to be open to everyone. However, effective data sharing requires well-specified policies.

Finally, there is currently insufficient support for management tools, and especially little support for security such as authentication and authorization. For example, the catalogs of resources and services are created and made available on web pages in an *ad hoc* way in the current OSAEM system.

Currently, the OSAEM system is under active development to address and improve the above issues and limitations. In addition, a prototype system is installed and is being tested on the Amazon EC2 cloud service. The current prototype for the web portal is available at http://db.cilaboratory.org:8080/naemp. The current prototype system supports the monitoring datasets for fish and water quality. The addition of monitoring datasets for benthic diatoms and macroinvertebrates is planned for 2016.

Acknowledgments: This study was performed under the "National Aquatic Ecosystem Health Survey and Assessment" project in Korea and supported by the Ministry of Environment and the National Institute of Environmental Research, Korea. The authors are grateful to survey members involved in the project. The authors also thank the anonymous reviewers for their help in improving the scientific content of the manuscript.

Author Contributions: Meilan Jiang designed and implemented the OSAEM system. Karpjoo Jeong designed the RESTful API and led the overall system development work. Nan-Young Kim managed the datasets and contributed to the design of the data models. Jung-Hwan Park and Sang-Hun Kim worked on the analysis of requirements for services and contributed to the design of services. Soon-Jin Hwang planned and coordinated the entire development project for the OSAEM system, and also contributed to the overall system design.

Conflicts of Interest: The authors declare no conflict of interest.

Appendix A. Representation Schemas.

Collection Resources	Representation Scheme

Fish	{
	"fishID": <integer>,
	"class": <string>,
	"order": <string>,
	"family": <string>,
	"species": <string>,
	"scientificName": <string>,
	"toleranceGuild": <string>,
	"trophicGuild": <string>,
	"habitatGuild": <string>,
	"invasiveSpecies ": <string>,
	"endemicSpecies ": <string>,
	"endangeredSpecies ": <string>,
	"naturalMonument ": <string>,
	"imageLink": <string>,
	"description": < string>
	}

Site	{
	"siteID": <integer>,
	"siteName": <string>,
	"latitude": <string>,
	"longitude": <string>,
	"river": <integer (riverID) >,
	"streamOrder": <integer>,
	"streamGrade": <string>,
	"address": <string>,
	"image": <string>,
	"standardStructure": <string>,
	"wqmn ": <string>,
	"description": < string>
	}

Collection Resources	Representation Scheme

River	{
	"riverID": \<integer\>,
	"riverName": \<string\>,
	"basin": \<string\>,
	"waterSystem": \<string\>,
	"midWatershed ": \<string\>,
	"subWatershed ": \<string\>,
	"classification": \<string\>,
	"image": \<string\>,
	"description": \< string\>
	}

Variable	{
	"variableID": \<integer\>,
	"variableName": \<string\>,
	"valueType": \<string\>,
	"unit": \<integer (unitID)\>,
	"description": \<string\>
	}

Unit	{
	"unitID": \<integer\>,
	"unitName": \<string\>,
	"unitNameLong": \< string\>
	}

Method	{
	"methodID": \<integer\>,
	"methodName": \<string\>,
	"methodLink": \< string\>
	}

Source	{
	"sourceID": \<integer\>,
	"insttitution": \<string\>,
	"investigator": \<string\>,
	"phone": \<string\>,
	"email": \<string\>,
	"description": \<string\>
	}

Observation	{
	"valueID": \<integer\>,
	"dateTime": \<string\>,,
	"surveyYear": \<integer\>,
	"surveyTerm": \<integer\>,
	"dataValue": \<double\>,
	"site": \<integer (siteID)\>,
	"latitude": \<string\>,
	"longitude": \<string\>,
	"source": \<integer (sourceID)\>,
	"entity": \<integer (fishID)\>,
	"variable": \<integer (variableID)\>,
	"method": \<integer (methodID)\>
	}

References

1. Lee, S.-W.; Hwang, S.-J.; Lee, J.-K.; Jung, D.-I.; Park, Y.-J.; Kim, J.-T. Overview and application of the National Aquatic Ecological Monitoring Program (NAEMP) in Korea. *Ann. Limnol. Int. J. Lim.* **2011**, *47*, S3–S14.

2. Michener, W.K. Ecological data sharing. *Ecol. Inform.* **2015**, *29*, 33–44.

3. Michener, W.K.; Beach, J.H.; Jones, M.B.; Ludäscher, B.; Pennington, D.D.; Pereira, R.S.; Rajasekar, A.; Schildhauer, M. A knowledge environment for the biodiversity and ecological sciences. *J. Intell. Inf. Syst.* **2007**, *29*, 111–126.

4. Michener, W.K.; Jones, M.B. Ecoinformatics: Supporting ecology as a data-intensive science. *Trends Ecol. Evol.* **2012**, *27*, 85–93.

5. Hampton, S.E.; Strasser, C.A.; Tewksbury, J.J.; Gram, W.K.; Budden, A.E.; Batcheller, A.L.; Duke, C.S.; Porter, J.H. Big data and the future of ecology. *Front. Ecol. Environ.* **2013**, *11*, 156–162.

6. Molloy, J.C. The open knowledge foundation: Open data means better science. *PLoS Biol.* **2011**, *9*, e1001195.

7. Madin, J.S.; Bowers, S.; Schildhauer, M.P.; Jones, M.B. Advancing ecological research with ontologies. *Trends Ecol. Evol.* **2008**, *23*, 159–168.

8. Madin, J.; Bowers, S.; Schildhauer, M.; Krivov, S.; Pennington, D.; Villa, F. An ontology for describing and synthesizing ecological observation data. *Ecol. Inform.* **2007**, *2*, 279–296.

9. Reichman, O.J.; Jones, M.B.; Schildhauer, M.P. Challenges and opportunities of open data in ecology. *Science* **2011**, *331*, 703–705.

10. Fielding, R.T.; Taylor, R.N. Principled design of the modern Web architecture. *ACM Trans. Internet Technol. TOIT* **2002**, *2*, 115–150.

11. Guinard, D.; Trifa, V.; Wilde, E. A resource oriented architecture for the Web of Things. In Proceedings of the Internet of Things (IOT), Tokyo, Japan, 29 November–1 December 2010; pp. 1–8.

12. Barry, M.G.; Purcell, M.E.; Eck, B.J.; Hayes, J.; Arandia, E. Web services for water systems: The iWIDGET REST API. *Procedia Eng.* **2014**, *89*, 1120–1127.

13. Bianchi, L.; Paganelli, F.; Pettenati, M.C.; Turchi, S.; Ciofi, L.; Iadanza, E.; Giuli, D. Design of a RESTful Web information system for drug prescription and administration. *Biomed. Health Inform. IEEE J.* **2014**, *18*, 885–895.

14. Paganelli, F.; Turchi, S.; Bianchi, L.; Giuli, D. An information-centric and REST-based approach for EPC Information Services. *J. Commun. Softw. Syst.* **2013**, *9*, 14–23.

15. Li, L.; Wu, C. Design and describe REST API without violating REST: A Petri net based approach. In Proceedings of the 2011 IEEE International Conference on Web Services (ICWS), Washington, DC, USA, 4–9 July 2011; pp. 508–515.

16. Allard, S. DataONE: Facilitating eScience through collaboration. *J. eScience Librariansh.* **2012**, *1*, e1004.

17. Michener, W.K.; Allard, S.; Budden, A.; Cook, R.B.; Douglass, K.; Frame, M.; Kelling, S.; Koskela, R.; Tenopir, C.; Vieglais, D.A. Participatory design of DataONE—Enabling cyberinfrastructure for the biological and environmental sciences. *Ecol. Inform.* **2012**, *11*, 5–15.

18. Horsburgh, J.S.; Tarboton, D.G.; Maidment, D.R.; Zaslavsky, I. A relational model for environmental and water resources data. *Water Resour. Res.* **2008**, *44*, W05406.

19. Horsburgh, J.S.; Tarboton, D.G.; Piasecki, M.; Maidment, D.R.; Zaslavsky, I.; Valentine, D.; Whitenack, T. An integrated system for publishing environmental observations data. *Environ. Model. Softw.* **2009**, *24*, 879–888.

20. Whitlock, M.C. Data archiving in ecology and evolution: Best practices. *Trends Ecol. Evol.* **2011**, *26*, 61–65.

21. Berkley, C.; Bowers, S.; Jones, M.B.; Madin, J.S.; Schildhauer, M. Improving Data Discovery for Metadata Repositories through Semantic Search. In Proceedings of the 2009 International Conference on Complex, Intelligent and Software Intensive Systems, Fukuoka, Japan, 16–19 March 2009.

22. Jones, M.B.; Berkley, C.; Bojilova, J.; Schildhauer, M. Managing Scientific Metadata. *J. IEEE Internet Comput.* **2001**, *5*, 59–68.

23. Greenberg, J.; White, H.C.; Carrier, S.; Scherle, R. A Metadata Best Practice for a Scientific Data Repository. *J. Libr. Metadata* **2009**, *9*, 194–212.

24. Michener, W.; Vieglais, D.; Vision, T.; Kunze, J.; Cruse, P.; Janée, G. DataONE: Data observation network for earth-preserving data and enabling innovation in the biological and environmental sciences. *D-lib Mag.* **2011**, *17*, 1–12.

25. Wieczorek, J.; Bloom, D.; Guralnick, R.; Blum, S.; Döring, M.; Giovanni, R.; Robertson, T.; Vieglais, D. Darwin Core: An evolving community-developed biodiversity data standard. *PLoS ONE* **2012**, *7*, e29715.

26. Whiteaker, T.; Ernest, T. CUAHSI Web Services for Ground Water Data Retrieval. *Ground Water* **2008**, *46*, 6–9.

27. Patton, E.W.; Seyed, P.; Wang, P.; Fu, L.; Dein, J.; Bristol, S.; McGuinness, D.L. SemantEco: A semantically powered modular architecture for integrating distributed environmental and ecological data. *Future Gener. Comput. Syst.* **2014**, *36*, 430–440.

28. Moura, A.D.C.; Porto, F.; Poltosi, M.; Palazzi, D.C.; Magalhães, R.P.; Vidal, V. Integrating ecological data using linked data principles. In Proceedings of Joint V Seminar on Ontology Research in Brazil (ONTOBRAS) and VII International Workshop on Metamodels, Ontologies and Semantic Technologies (MOST), Recife, Brazil, 19–21 September 2012; pp. 156–167.

29. Stoddard, J.L.; Herlihy, A.T.; Peck, D.V.; Hughes, R.M.; Whittier, T.R.; Tarquinio, E. A process for creating multimetric indices for large-scale aquatic surveys. *J. N. Am. Benthol. Soc.* **2008**, *27*, 878–891.

30. Karr, J.R. Assessment of biotic integrity using fish communities. *Fisheries* **1981**, *6*, 21–27.

31. Hering, D.; Johnson, R.K.; Kramm, S.; Schmutz, S.; Szoszkiewicz, K.; Verdonschot, P.F.M. Assessment of European streams with diatoms, macrophytes, macroinvertebrates and fish: A comparative metric-based analysis of organism response to stress. *Freshw. Biol.* **2006**, *51*, 1757–1785.

32. Johnson, R.K.; Furse, M.T.; Hering, D.; Sandin, L. Ecological relationships between stream communities and spatial scale: Implications for designing catchment-level monitoring programmes. *Freshw. Biol.* **2007**, *52*, 939–958.

33. Johnson, R.K.; Goedkoop, W.; Sandin, L. Spatial scale and ecological relationships between the macroinvertebrate communities of stony habitats of streams and lakes. *Freshw. Biol.* **2004**, *49*, 1179–1194.

34. Kelly, M.G.; Whitton, B.A. The Trophic Diatom Index: A new index for monitoring eutrophication in rivers. *J. Appl. Phycol.* **1995**, *7*, 433–444.

35. Crockford, D. The application/json media type for JavaScript Object Notation (JSON). 2006. Available online: http://tools.ietf.org/html/rfc4627 (accessed on 27 November 2015).

36. Fielding, R.T.; Gettys, J.; Mogul, J.C.; Frystyk, H.F.; Masinter, L.; Leach, P.; Berners-Lee, T. *Hypertext Transfer Protocol—HTTP/1.1. Internet RFC 2616*; The Internet Society: Reston, VA, USA; June; 1999.

37. Murugesan, S. Understanding Web 2.0. *IT Prof.* **2007**, *9*, 34–41.

38. Seo, D.; Lee, J. Web_2.0 and five years since: How the combination of technological and organizational initiatives influences an organization's long-term Web_2.0 performance. *TELE* **2016**, *33*, 232–246.

Transferability of Monitoring Data from Neighboring Streams in a Physical Habitat Simulation

Byungwoong Choi, Sung-Uk Choi and Hojeong Kang

Abstract: Habitat simulation models heavily rely on monitoring data, which can have serious effects on the success of a physical habitat simulation. However, if data monitored in a study reach are not available or insufficient, then data from neighboring streams are commonly used. The problem is that the impact of using data from neighboring streams has rarely been studied before. Motivated by this, we report herein on an investigation of the transferability of data from neighboring streams in a physical habitat simulation. The study area is a 2.5 km long reach located downstream from a dam in the Dal River, Korea. *Zacco platypus* was selected as the target fish for the physical habitat simulation. Monitoring data for the Dal River and three neighboring streams were obtained. First, similarities in the data related to channel geometry and in the observed distribution of the target species were examined. Principal Component Analysis (PCA) was also carried out to see the characteristics of the habitat use of the target species. Habitat Suitability Curves (HSCs) were constructed using the Gene Expression Programming (GEP) model, and improved Generalized Habitat Suitability Curves (GHSCs) were proposed. The physical habitat simulations were then performed. The Composite Suitability Index (CSI) distributions were predicted, and the impact of using data from the neighboring streams was investigated. The results indicated that the use of data from a neighboring stream even in the same watershed can result in large errors in the prediction of CSI. The physical habitat simulation with the improved GHSCs was found to best predict the CSI.

Reprinted from *Water*. Cite as: Choi, B.; Choi, S.-U.; Kang, H. Transferability of Monitoring Data from Neighboring Streams in a Physical Habitat Simulation. *Water* **2015**, *7*, 4537–4551.

1. Introduction

A physical habitat simulation is a numerical tool that quantifies physical habitat in terms of the flow depth, velocity, and substrate at a particular discharge for a given stream [1]. Thus, it is capable of predicting the impact of a change in flow on habitat availability for target species. Physical habitat simulations have been successfully used to estimate the environmental flows in rivers [2–5], in designing a river restoration [6–8], to evaluate river health [9–11], and to assess the impact of river works or river development [12–14].

For the success of a physical habitat simulation, acquiring relevant monitoring data is important. This is because most habitat simulation models are heavily dependent on the monitoring data [15–19]. However, a significant portion of previous physical habitat simulations have used monitoring data from neighboring streams due to the lack of sufficient data. The monitoring data include such physical habitat variables as flow depth, velocity, substrate, and related issues. Obtaining monitoring data is, in general, costly and time-consuming. For example, more than one hundred physical habitat simulations have been carried out for streams in Korea. However, only about 15% of these studies used data obtained for the actual stream being studied [20,21]. The situation is not much better in the US [18,22–25].

The use of monitoring data from neighboring streams involves an implicit hypothesis that the knowledge-based or data-based models constructed using data from a neighboring stream are applicable to the stream being studied. The similarity of HSCs between the study stream and neighboring streams has been studied by many researchers [18,23,25–32]. However, the impact of using such data from neighboring streams has rarely been investigated and a general and efficient solution to this problem has never been proposed. This motivated the present study.

The goal of this study was to assess the impact of using data from neighboring streams in a physical habitat simulation and to propose a generalized and efficient Habitat Suitability Index (HSI) model using data from neighboring streams. For this, a 2.5 km long reach in the Dal River, Korea was selected. This study reach is a gravel-bed stream located downstream from a dam. For the physical habitat simulation, *Zacco platypus* was selected for the target fish. Monitoring data from three neighboring streams were obtained for the physical habitat simulations of the study reach. Similarities of data for the channel geometry and for the observed distribution of the target fish against physical habitat variables were studied. Physical habitat simulations were carried out using the CCHE2D model and the HSI model for hydraulic and habitat simulations, respectively. For the HSI model, HSCs were constructed using the GEP model, and improved GHSCs were proposed using the suitable range concept. First, the impact of using data from neighboring streams in physical habitat simulations was examined quantitatively. Then, the improved GHSCs were used to predict the CSI distribution, and simulated results were compared.

2. Materials and Methods

2.1. Study Area

Figure 1 shows the study area in the Dal River, Korea and its neighboring streams. The Dal River is a mid-sized stream, a tributary of the Han River, and the basin area is 682.41 km^2. The study reach is 2.5 km in length and extends from the Sujeon Bridge to the Daesu Weir. The Goesan Dam, located 0.92 km upstream from the Sujeon Bridge, regulates the flow in the study reach. The Goesan dam discharges water irregularly only for hydropower generation. As a result, the role of the Daesu Weir is to maintain a constant flow during periods when the dam is not discharging water. For the study reach, the discharges for drought flow, low flow, normal flow, and averaged-wet flow are 1.82, 4.02, 7.23, and 17.13 m^3/s, respectively [33].

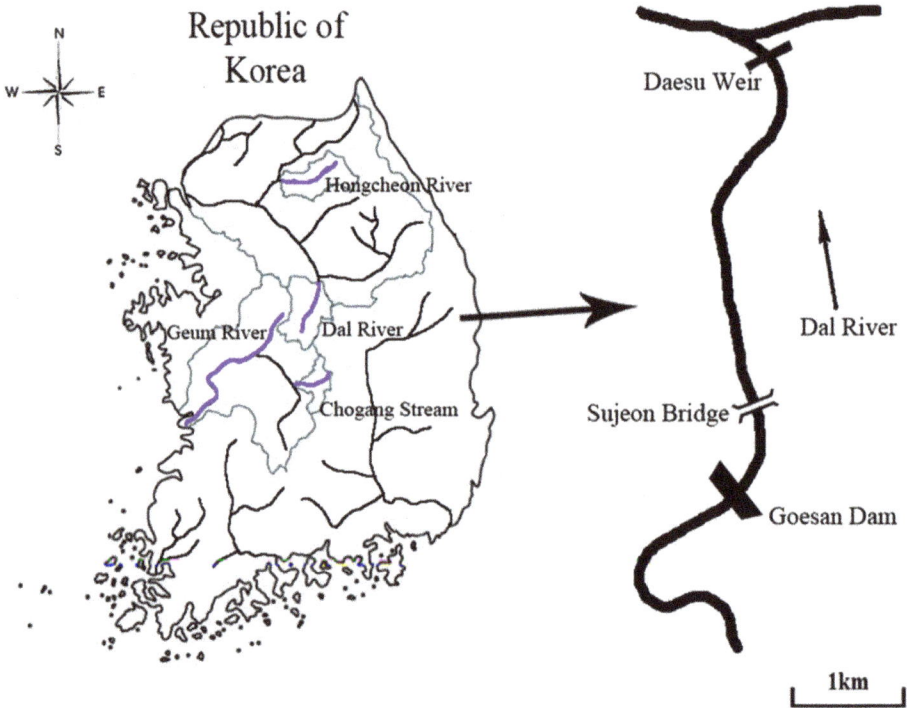

Figure 1. The Dal River and its neighboring streams.

2.2. Monitoring Data

The three neighboring streams include the Hongcheon River, the Geum River, and the Chogang Stream, which are shown in Figure 1. For the Dal River and three neighboring streams, hydrologic, water quality, and fish monitoring data were collected for the period of 2007–2010 through government R&D projects [34,35]. To measure the water level and velocity, a radar water gauge and a price current meter were used, respectively, at the Sujeon Bridge. Dissolved oxygen and pH were measured using the handheld dissolved oxygen meter and pH meter, respectively. Turbidity was measured by turbidity meter (PT-200). Fish monitoring was carried out using cast nets and kick nets, revealing that dominant species in the study area is a minnow (*Zacco platypus*), followed by dark chubs (*Zacco temmincki*) and swiri (*Coreoleuciscus splendidus*). They account for 27%, 15% and 15%, respectively. In the present study, the adult minnow was selected as the target fish. Since the monitoring data includes the number of individuals, flow depth, velocity, and substrate, they are habitat use data of Category II based on the criterion by Bovee [15].

2.3. Habitat Suitability Curves

In the present study, the GEP was used to construct HSCs. GEP takes advantage of both the Genetic Algorithm (GA) and Genetic Programming (GP). The GEP uses linear chromosomes with a fixed length and nonlinear parse trees with varied sizes and shapes. The former is obtained from the GA and the latter from the GP. Therefore, in the GEP, individuals are encoded as chromosomes, which are then expressed as expression trees. The combination of these separate entities, chromosomes and expression trees, enables the GEP to perform with a high degree of efficiency compared to GA and GP.

For constructing HSCs, GeneXpro-Tools 5.0, a GEP software program developed by Ferreira [36,37] was used. Four basic arithmetic operators (+, −, ×, ÷) and some mathematical functions (power, abs, cos, tan, arctan, min, max) were used with a combination of all genetic operators, including mutation, transposition, and crossover.

2.4. Improved Generalized Habitat Suitability Curves

Maki-Petays *et al.* [25] introduced GHSCs for physical habitat simulations of juvenile salmon in four rivers in Finland. They constructed GHSCs for each habitat variable using the arithmetic means of the habitat suitability indices for the four neighboring rivers. To smooth the curves, Maki-Petays *et al.* [25] used the distance weighted least square method.

In the present study, a new method for constructing GHSCs by using the suitable range concept is proposed. The improved GHSCs use the arithmetic means of the HSCs of neighboring streams constructed by using data in a suitable range. The suitable range, the concept of which was proposed by Thomas and Bovee [26], is the range containing the central 95% of the occupied locations in the HSC.

The ranges of data for the Dal River and three neighboring streams are presented in Figure 2 where the total range, suitable range, and optimum range are denoted by dotted, black and red bold arrows, respectively. The optimum range is the interval containing the central 50% of the occupied locations in the HSC [26]. The suitable and optimum ranges of the Dal River data are shadowed by black and red, respectively, in the figure. It can be seen that the suitable and optimum ranges of the Chogang Stream data are the most similar to those of the Dal River data.

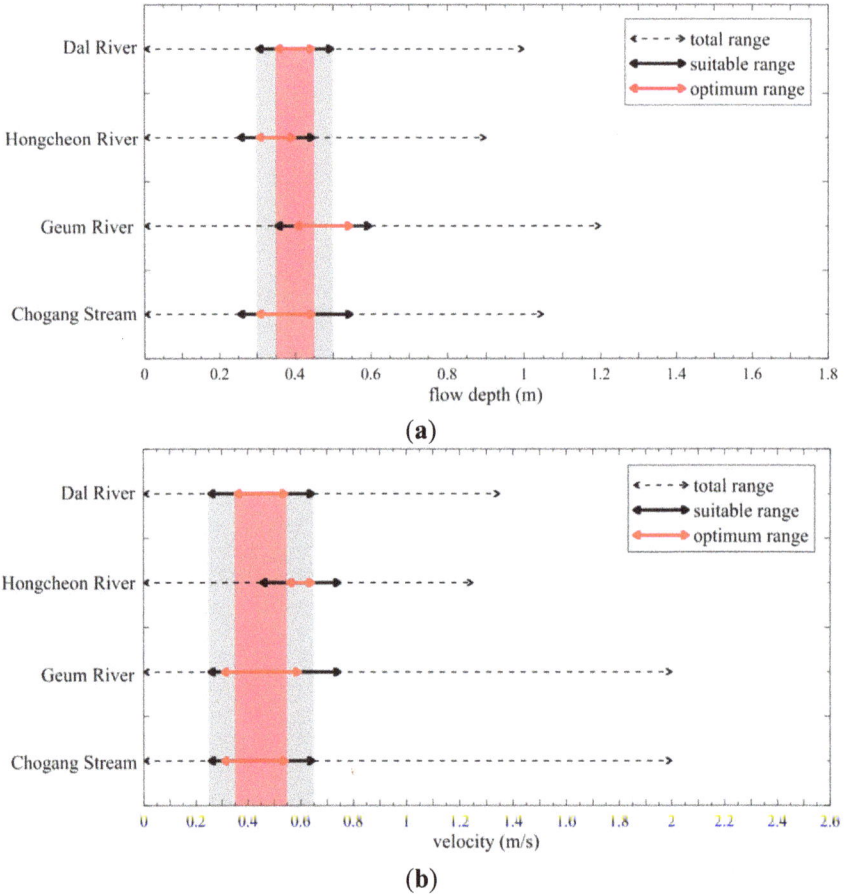

Figure 2. (a) Flow depth; (b) velocity. Suitable range and optimum range of data.

2.5. Hydraulic Simulation

CCHE2D, a numerical model for analyzing unsteady turbulent flows in an open-channel, was developed by the National Center for Computational Hydrosciences and Engineering at the University of Mississippi, US. The CCHE2D solves two-dimensional depth-averaged hydrodynamic equations using the efficient element method [38]. The continuity and longitudinal (x) and lateral (y) components of momentum equations are, respectively, given by:

$$\frac{\partial H}{\partial t} + \frac{\partial q_x}{\partial x} + \frac{\partial q_y}{\partial y} = 0 \tag{1}$$

$$\frac{\partial q_x}{\partial t} + \frac{\partial}{\partial x}\left(Uq_x\right) + \frac{\partial}{\partial y}\left(Vq_x\right) + \frac{g}{2}$$
$$= gH\left(S_{0x} - S_{fx}\right) + \frac{1}{\rho}\left\{\frac{\partial}{\partial x}\left(H\tau_{xx}\right)\right\} + \frac{1}{\rho}\left\{\frac{\partial}{\partial y}\left(H\tau_{xy}\right)\right\} \tag{2}$$

$$\frac{\partial q_y}{\partial t} + \frac{\partial}{\partial x}\left(Uq_y\right) + \frac{\partial}{\partial y}\left(Vq_y\right) + \frac{g}{2}\frac{\partial}{\partial x}H^2$$
$$= gH\left(S_{0x} - S_{fx}\right) + \frac{1}{\rho}\left\{\frac{\partial}{\partial x}\left(H\tau_{yx}\right)\right\} + \frac{1}{\rho}\left\{\frac{\partial}{\partial y}\left(H\tau_{yy}\right)\right\} \tag{3}$$

where H is the flow depth, U and V are the depth-averaged velocities in the x- and y-directions, respectively, q_x and q_y are respective discharges per unit width ($q_x = HU$, $q_y = HV$), g is the gravitational acceleration, ρ is the water density, S_{0i} and S_{fi} are the river bed slope and friction slope in the i-direction, and τ_{ij} is the horizontal turbulent stress tensor.

2.6. Method of Comparing Data and Results

For quantitative comparisons of geometric data of the streams, relative errors of each component defined below were computed.

$$RE_i = \frac{\left|\Phi_{i\_NS} - \Phi_{i\_Dal}\right|}{\Phi_{i\_Dal}} \times 100 \tag{4}$$

where Φ_{i_Dal} denotes the i-th geometric component of the Dal River and Φ_{i_NS} is the same component of the neighboring streams. In addition, to compare the observed and predicted CSIs, the following MAPE (Mean Average Percentage Error) is used:

$$MAPE = \frac{1}{n}\sum \frac{\left|CSI^o - CSI^p\right|}{CSI^o} \times 100 \ (\%) \tag{5}$$

in which n is the number of data and CSI^o and CSI^p are the observed and predicted CSIs, respectively.

3. Results and Discussions

3.1. Data Variability between Target and Source Streams

The Hongcheon River belongs to the Han River basin, the same as the Dal River, whereas the other two belong to the Geum River basin. Detailed characteristics of channel geometry and the substrate of the Dal River and neighboring streams are given in Table 1. The shape factor, defined by the basin area divided by the length of the stream, in the table denotes the average width of the basin. It should be noted in the table that the Geum River is a large-sized sand-bed river whose average slope is mild compared to the other streams. The geometric components include basin area, stream length, average width, mean elevation, and mean slope. Average values of the relative errors of five geometric components in Table 1 are 51%, 357% and 25% for the Hongcheon River, the Geum River, and the Chogang Stream, respectively. This indicates that the Chogang Stream is the most similar to the Dal River in terms of channel geometry.

Table 1. Characteristics of the data used.

Streams	Number of Datasets	basin Area (km^2)	Length (km)	Average Width (km)	Shape Factor	Mean Elevation (EL.m)	Mean Slope	Mean Diameter of Substrate (mm)
Dal River	468	682.41	82.8	8.30	0.10	381.00	1/650	102.3–165.9
Hongcheon River	138	1566.2	140.2	11.17	0.08	423.00	1/550	35.1–148.0
Geum River	291	9912.15	397.8	24.92	0.063	217.03	1/2500	0.0097–1.70
Chogang Stream	689	665.2	66.3	10.03	0.15	289.00	1/770	46.8–111.8

3.2. Habitat Use Characteristics of Freshwater Minnow and Constructed HSCs

The distribution of the *Zacco platypus* in the Chogang Stream is the most similar to that in the Dal River (Figure 3). However, the target species in the Hongcheon River and Geum River appears to be distributed differently. In the figure, the classification scheme by Wentworth [39] was used for the substrate. Specifically, the ranges for the high population of *Zacco platypus* in the Dal River are a velocity of 0.4–0.65 m/s, a flow depth of 0.4–0.7 m, and a substrate of 4–6. However, in the Hongcheon River, the target species are densely populated in the ranges of a velocity of 0.6–1.0 m/s, a flow depth of 0.5–0.8 m, and a substrate of 2–4. In the Geum River, the ranges for a high population are a velocity of 0.1–0.6 m/s, a flow depth of 0.2–0.6 m, and a substrate of 2–5. The figure implies that monitoring data for the Chogang Stream can be used acceptably for a physical habitat simulation of the target fish in the study reach if data for the Dal River are not available.

In order to investigate the habitat use difference between the Dal River and the three neighboring streams, Principal Component Analysis (PCA) was carried out with seven variables, namely observed number of individuals, flow depth, velocity, substrate, pH, DO, and turbidity. The total number of data is 1586. A two-dimensional plot of PC2 *versus* PC1 is shown in Figure 4. It can be seen that the data can be grouped into four, which is the number of streams in this study. For PC1, the data for the Dal River lie in the range of −60–5. Ranges of data for PC1 are 10–60, 5–60, and −45−−10 for the Hongcheon River, the Geum River, and the Chogang Stream, respectively. For PC2, the data for the Dal River, the Hongcheon

River, the Geum River, and the Chogang Stream range from -30–40, -60–0, 0–70, and 0–70, respectively. The results from PCA indicate that the pattern of the Chogang Stream data is the most similar to that of the Dal River data, followed by the Geum River data and the Hongcheon River data.

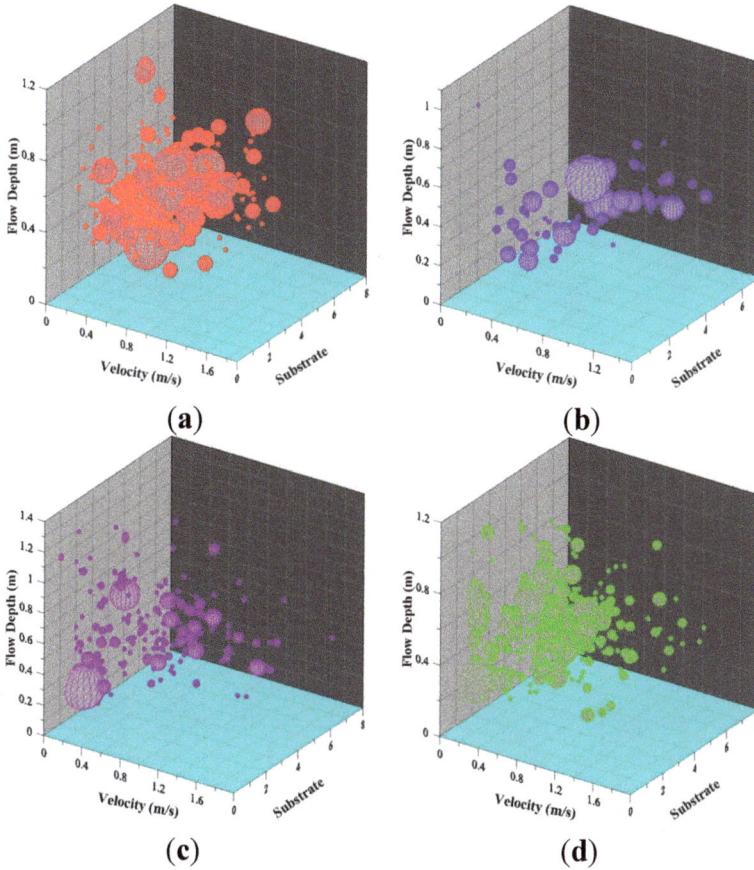

Figure 3. Observed distribution of *Zacco platypus* against velocity, flow depth, and substrate for (**a**) Dal River; (**b**) Hongcheon River; (**c**) Geum River; (**d**) Chogang Stream.

Before performing the physical habitat simulations, HSCs were constructed using the GEP model. For three physical habitat variables, HSCs were constructed with the three monitoring datasets from the neighboring streams and are compared with those constructed using the Dal River data. Figure 5 shows the resulting HSCs. The grey bars in the figure indicate that the number of observed individuals for each physical habitat variables. Choi and Choi [40] found that the HSCs by the GEP model are very similar to those by the method of Gosse [41] and the GEP model predicts HSCs better than the Adaptive Neuro Fuzzy Inference System (ANFIS) model. Furthermore, Choi and Choi [40] indicated that the GEP model is robust and non-subjective compared the method of Gosse [41].

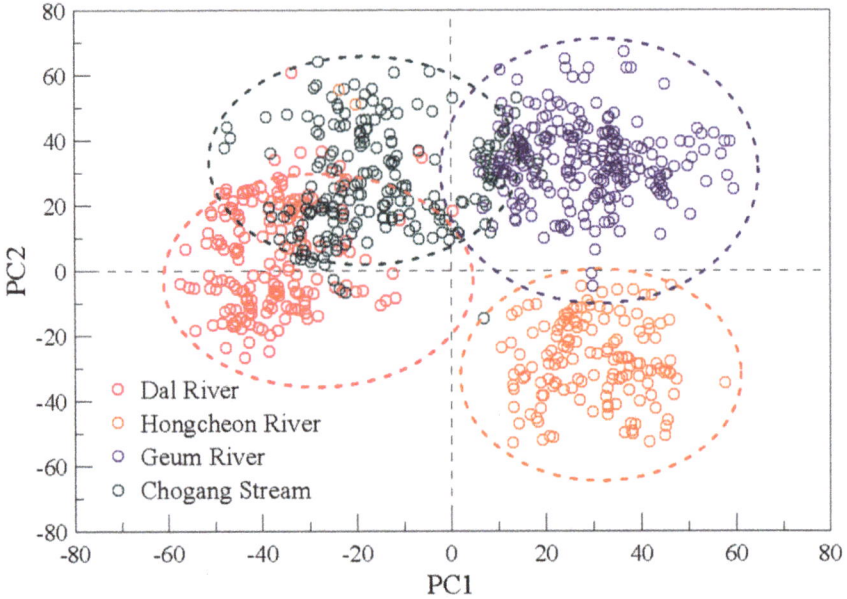

Figure 4. Result of a principal component analysis.

For the flow depth and velocity, the HSCs for the Chogang Stream appear to be the most similar to those of the Dal River (Figure 5). However, the HSCs for the Hongcheon River are substantially different from those for the Dal River even though the two streams are part of the same watershed. The level of similarity can be expected from prior investigations in Table 1 and Figures 3 and 4.

Regarding the substrate, all HSCs differ seriously in Figure 5c. This is because the target fish, *Zacco platypus*, does not have a substrate preference [42,43]. That is, *Zacco platypus* lives in both sand-bed and gravel-bed streams. Thus, hereafter, the substrate will not be considered in the physical habitat simulations.

The improved GHSCs are constructed and compared with the GHSCs constructed by the method proposed by Maki-Petays *et al.* [25] (Figure 6). It can be seen that both curves appear to be very similar except that the GHSCs have long tails in case where the values for the flow depth and velocity are large.

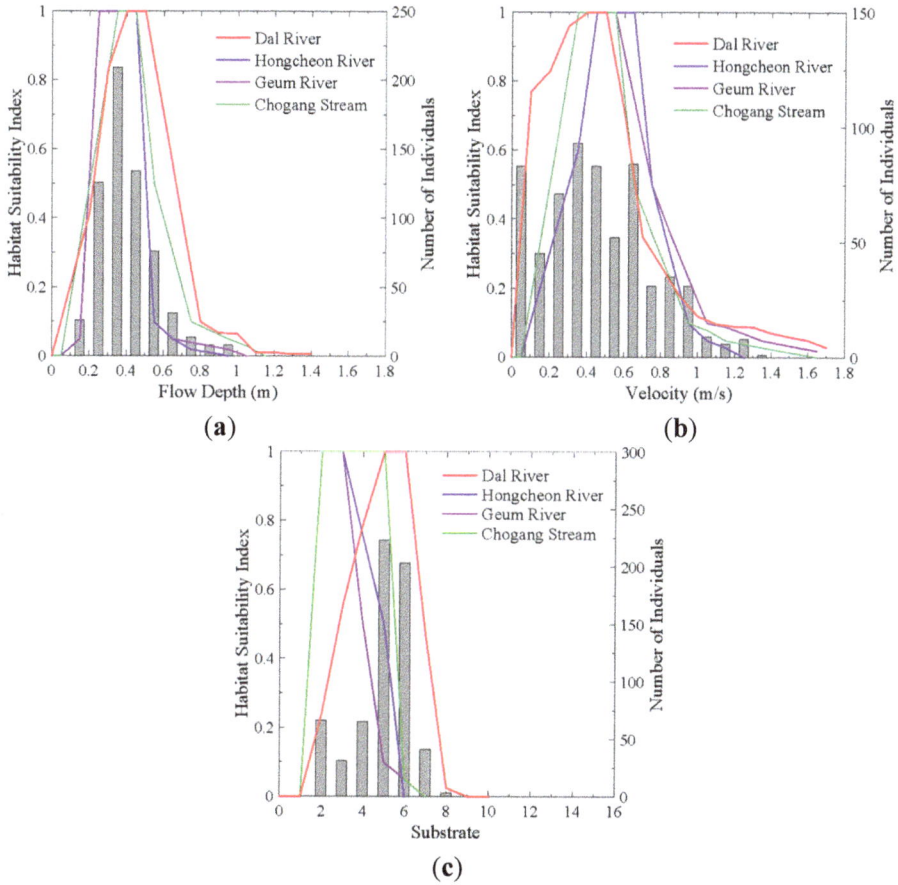

Figure 5. (**a**) Flow depth; (**b**) velocity; (**c**) substrate. Habitat suitability curves.

3.3. Transferability of HSCs and Validation of Improved GHSCs

In this section, CSI distributions of the Dal River are presented based on the physical habitat simulations. First, various HSCs in Figure 5, constructed using the study reach and the neighboring streams, were used for the HSI model. Then, GHSCs in Figure 6 were used to compute the CSI distributions. The former is to investigate the transferability of HSCs of the neighboring streams and the latter is to show the v alidity of the proposed GHSCs.

Using the HSCs in Figure 5, physical habitat simulations were carried out for *Zacco platypus*, and the resulting CSI distributions in the study reach are shown in Figure 7. The CSI was calculated using the multiplicative aggregation method. It can be seen that the CSI distribution computed using Chogang Stream data is the most similar to that of the Dal River. The CSI distribution computed using Geum River data is also similar, but the CSI distribution predicted with the use of Hongcheon River data is substantially different from the CSI distribution for the Dal River.

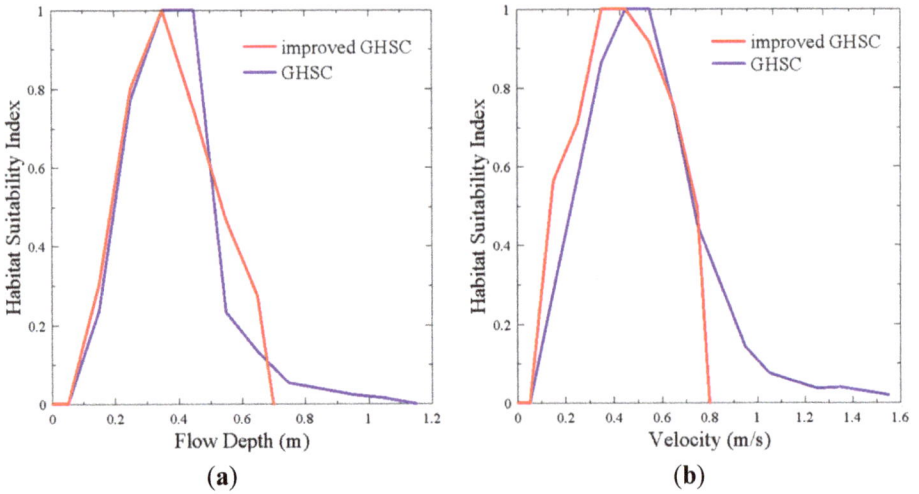

(a)

(b)

Figure 6. (a) Flow depth; (b) velocity. Generalized habitat suitability curves.

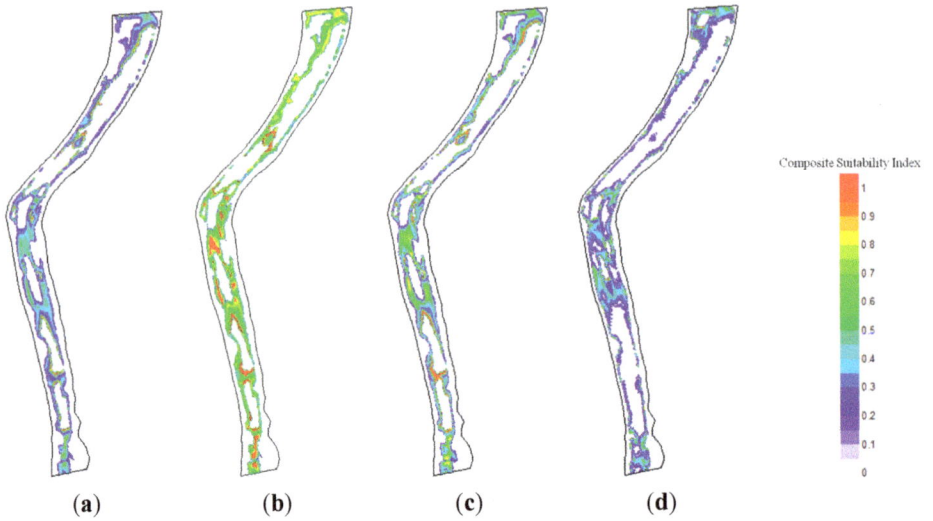

(a)   (b)   (c)   (d)

Figure 7. CSI distribution for Zacco platypus (a) with Dal River data; (b) with Hongcheon River data; (c) with Geum River data; (d) with Chogang Stream data.

In order to evaluate the impact of using data from neighboring streams, the CSI of the Dal River *versus* the CSI predicted using data from neighboring streams are plotted in Figure 8. Only non-zero values of CSI are plotted in the figure, where the 45 degree line indicates a perfect match. It can be seen that values of CSI predicted using Chogang Stream data provide the match best with those of the Dal River. However, the use of Hongcheon River data and Geum River data results in slight and serious over-predictions of CSI, respectively. The values for MAPE were also computed, and 67.5%, 39.2% and 25.1% were obtained for predictions using Hongcheon River data, Geum River Data, and Chogang Stream data, respectively. The results indicate that the use of data from a neighboring stream even in the same watershed results in larger errors in the prediction of the CSI. Prior investigations in Table 1 and Figures 3 and 4 can be useful for selecting appropriate datasets for a physical habitat simulation.

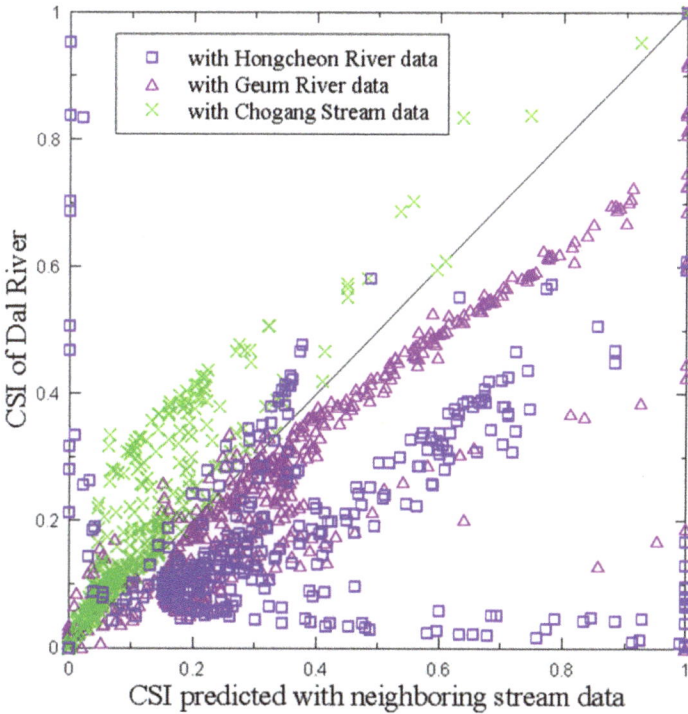

Figure 8. Comparison of predicted CSIs.

Figure 9 shows the CSI distributions for *Zacco platypus* in the Dal River. The GHSCs and improved GHSCs were used for CSI distributions in Figure 9a,b, respectively. It appears that the use of GHSCs substantially improves the CSI distribution compared to the CSI distributions in Figure 7a. Quantitatively, the uses of HSC constructed with the Chogang Stream data, GHSC, and improved GHSC result in MAPE values of 17.18%, 19.35% and 15.46%, respectively. This indicates that the use of the improved GHSC leads to a CSI distribution that is most similar to that of the Dal River.

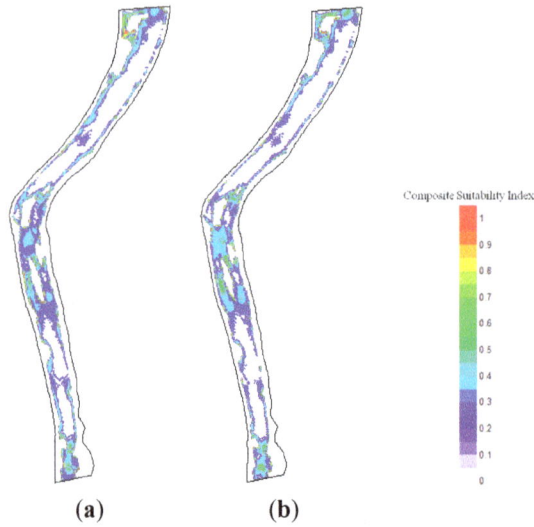

Figure 9. CSI distribution for Zacco platypus computed with GHSCs (**a**) with GHSC; (**b**) with improved GHSC.

4. Conclusions

This paper investigated the impact of using monitoring data from neighboring streams for a physical habitat simulation and proposed a new method for constructing GHSCs using monitoring data from neighboring streams. The present study showed that great attention should be paid to the use of data from neighboring streams when monitoring data are not available for the physical habitat simulation. That is, the data from a neighboring stream whose geometrical properties are similar to those of the study reach should be used. Even the data from a stream that shares the same watershed as the study reach can result in large errors in the prediction of the CSI.

In addition, for a general strategy, the present study proposed an improved GHSC, which can be used for the physical habitat simulation more confidentially with data from neighboring streams. The new method used the arithmetic means of the HSCs that were constructed with data only in a suitable range. The predicted CSI distribution was compared with that computed using the conventional HSI model, revealing that the prediction made using the improved GHSCs was better. However, the applicability of the proposed GHSCs can be investigated further by applying the methodology to various target streams that have their own monitoring data as well as that from neighboring streams.

Acknowledgments: This work was supported by the National Research Foundation of Korea (NRF) grant funded by the Korea Government (NRF-2014R1A2A1A11054236).

Author Contributions: All of the authors contributed extensively to the work. Byungwoong Choi and Sung-Uk Choi performed the calculation, result analysis, and wrote the research article. Hojeong Kang contributed to the article's discussion and edition.

Conflicts of Interest: The authors declare no conflict of interest.

References

1. Booker, D.J.; Dunbar, M.J. Application of physical habitat simulation (PHABSIM) modelling to modified urban river channels. *River Res. Appl.* **2004**, *20*, 167–183.
2. Freeman, M.C.; Bowen, Z.H.; Bovee, K.D.; Irwin, E.R. Flow and habitat effects on juvenile fish abundance in natural and altered flow regimes. *Ecol. Appl.* **2001**, *11*, 179–190.
3. Tharme, R.E. A global perspective on environmental flow assessment: Emerging trends in the development and application of environmental flow methodologies for rivers. *River Res. Appl.* **2003**, *19*, 397–441.
4. Acreman, M.C.; Dunbar, M.J. Defining environmental river flow requirements? A review. *Hydrol. Earth Syst. Sci. Discuss.* **2004**, *8*, 861–876.
5. Arthington, A.H.; Bunn, S.E.; Poff, N.L.; Naiman, R.J. The challenge of providing environmental flow rules to sustain river ecosystems. *Ecol. Appl.* **2006**, *16*, 1311–1318.
6. Frissell, C.A.; Liss, W.J.; Warren, C.E.; Hurley, M.D. A hierarchical framework for stream habitat classification: Viewing streams in a watershed context. *Environ. Manag.* **1986**, *10*, 199–214.
7. Downs, P.W.; Kondolf, G.M. Post-project appraisals in adaptive management of river channel restoration. *Environ. Manag.* **2002**, *29*, 477–496.
8. Im, D.; Kang, H.; Kim, K.H.; Choi, S.U. Changes of river morphology and physical fish habitat following weir removal. *Ecol. Eng.* **2011**, *37*, 883–892.
9. Maddock, I. The importance of physical habitat assessment for evaluating river health. *Freshw. Biol.* **1999**, *41*, 373–391.
10. Oberdorff, T.; Pont, D.; Hugueny, B.; Porcher, J.P. Development and validation of a fish-based index for the assessment of "river health" in France. *Freshw. Biol.* **2002**, *47*, 1720–1734.
11. Mouton, A.; Meixner, H.; Goethals, P.L.; De Pauw, N.; Mader, H. Concept and application of the usable volume for modelling the physical habitat of riverine organisms. *River Res. Appl.* **2007**, *23*, 545–558.
12. Richter, B.D.; Thomas, G.A. Restoring environmental flows by modifying dam operations. *Ecol. Soc.* **2007**, *12*, 12.
13. Tomsic, C.A.; Granata, T.C.; Murphy, R.P.; Livchak, C.J. Using a coupled eco-hydrodynamic model to predict habitat for target species following dam removal. *Ecol. Eng.* **2007**, *30*, 215–230.
14. Yi, Y.; Wang, Z.; Yang, Z. Two-dimensional habitat modeling of Chinese sturgeon spawning sites. *Ecol. Model.* **2010**, *221*, 864–875.

15. Bovee, K.D. *Development and Evaluation of Habitat Suitability Criteria for Use in the Instream Flow Incremental Methodology*; National Ecology Center, Division of Wildlife and Contaminant Research, Fish and Wildlife Service, US Department of the Interior: Fort Collins, CO, USA, 1986; Volume 86.

16. Beecher, M.D. Signaling systems for individual recognition: An information theory approach. *Anim. Behav.* **1989**, *38*, 248–261.

17. Conklin, D.J., Jr.; Canton, S.P.; Chadwick, J.W.; Miller, W.J. Habitat suitability curves for selected fish species in the central Platte River, Nebraska. *Rivers* **1995**, *5*, 250–266.

18. Moir, H.J.; Gibbins, C.N.; Soulsby, C.; Youngson, A.F. PHABSIM modelling of Atlantic salmon spawning habitat in an upland stream: Testing the influence of habitat suitability indices on model output. *River Res. Appl.* **2005**, *21*, 1021–1034.

19. Conallin, J.; Boegh, E.; Jensen, J.K. Instream physical habitat modelling types: An analysis as stream hydromorphological modelling tools for EU water resource managers. *Int. J. River Basin Manag.* **2010**, *8*, 93–107.

20. Kang, H.; Im, D.; Hur, J.W.; Kim, K.H. Estimation of Habitat Suitability Index of Fish Species in the Geum River Watershed. *J. Korean Soc. Civil Eng.* **2011**, *31*, 193–203. (In Korean)

21. Kang, H. Comparison of Physical Habitat Suitability Index for Fishes in the Rivers of Han and Geum River Watersheds. *J. Korean Soc. Civil Eng.* **2012**, *32*, 71–78. (In Korean)

22. Scott, D.; Shirvell, C.S. A critique of the instream flow incremental methodology and observations on flow determination in New Zealand. In *Regulated Streams*; Springer: New York, NY, USA, 1987; pp. 27–43.

23. Groshens, T.P.; Orth, D.J. Transferability of habitat suitability criteria for smallmouth bass, *Micropterus dolomieu*. *Rivers* **1994**, *4*, 194–212.

24. Freeman, M.C.; Bowen, Z.H.; Crance, J.H. Transferability of habitat suitability criteria for fishes in warmwater streams. *North Am. J. Fish. Manag.* **1997**, *17*, 20–31.

25. Mäki-Petäys, A.; Huusko, A.; Erkinaro, J.; Muotka, T. Transferability of habitat suitability criteria of juvenile Atlantic salmon (*Salmo salar*). *Can. J. Fish. Aquat. Sci.* **2002**, *59*, 218–228.

26. Thomas, J.A.; Bovee, K.D. Application and testing of a procedure to evaluate transferability of habitat suitability criteria. *Regul. Rivers Res. Manag.* **1993**, *8*, 285–294.

27. Leftwich, K.N.; Angermeier, P.L.; Dolloff, C.A. Factors influencing behavior and transferability of habitat models for a benthic stream fish. *Trans. Am. Fish. Soc.* **1997**, *126*, 725–734.

28. Glozier, N.E.; Culp, J.M.; Scrimgeour, G.J. Transferability of habitat suitability curves for a benthic minnow, Rhinichthys cataractae. *J. Freshw. Ecol.* **1997**, *12*, 379–393.

29. Lamouroux, N.; Capra, H.; Pouilly, M.; Souchon, Y. Fish habitat preferences in large streams of southern France. *Freshw. Boil.* **1999**, *42*, 673–687.

30. Mäki-Petäys, A.; Muotka, T.; Huusko, A. Densities of juvenile brown trout (*Salmo trutta*) in two subarctic rivers: Assessing the predictive capability of habitat preference indices. *Can. J. Fish. Aquat. Sci.* **1999**, *56*, 1420–1427.

31. Guay, J.C.; Boisclair, D.; Leclerc, M.; Lapointe, M. Assessment of the transferability of biological habitat models for Atlantic salmon parr (*Salmo salar*). *Can. J. Fish. Aquat. Sci.* **2003**, *60*, 1398–1408.

32. Nykänen, M.; Huusko, A. Transferability of habitat preference criteria for larval European grayling (*Thymallus thymallus*). *Can. J. Fish. Aquat. Sci.* **2004**, *61*, 185–192.

33. Ministry of Construction and Transportation. *The Basic Plan for Dal River Maintenance*; Ministry of Construction and Transportation: Seoul, Korea, 1995. (In Korean)

34. Ministry of Science and Technology. *Technology for Surface Water Resources Investigation*; Ministry of Science and Technology: Seoul, Korea, 2007. (In Korean)

35. Ministry of Land, Transport and Maritime Affairs. *Development of Techniques for Creation of Wildlife Habitat Environment*; Ministry of Land, Transport and Maritime Affairs: Seoul, Korea, 2011. (In Korean)

36. Ferreira, C. Gene expression programming in problem solving. In Proceedings of the 6th Online World Conference on Soft Computing in Industrial Applications (Invited Tutorial), Santiago, Chile, 10–24 September 2001.

37. Ferreira, C. Gene expression programming: A new adaptive algorithm for solving problems. *Complex Syst.* **2001**, *13*, 87–129.

38. Jia, Y.; Wang, S.S. *CCHE2D: Two-Dimensional Hydrodynamic and Sediment Transport Model for Unsteady Open Channel Flows over Loose Bed*; Technical Report No. NCCHE-TR-2001-1; National Center for Computational Hydroscience and Engineering: Oxford, MS, USA, 2001.

39. Wentworth, C.K. A scale of grade and class terms for clastic sediments. *J. Geol.* **1922**, *30*, 377–392.

40. Choi, B.; Choi, S.U. Physical Habitat Simulations of the Dal River in Korea using the GEP Model. *Ecol. Eng.* **2015**, *83*, 456–465.

41. Gosse, J.C. *Microhabitat of Rainbow and Cutthroat Trout in the Green River below Flaming Gorge Dam*; Final Report, Contract 81 5049; Utah Division of Wildlife Resources: Salt Lake City, UT, USA, 1982.

42. Yeom, D.H.; Lee, S.A.; Kang, G.S.; Seo, J.; Lee, S.K. Stressor identification and health assessment of fish exposed to wastewater effluents in Miho Stream, South Korea. *Chemosphere* **2007**, *67*, 2282–2292.

43. Page, L.M.; Burr, B.M. *Peterson Field Guide to Fresh-Water Fishes of North America North of Mexico*, 2nd ed.; Houghton Mifflin Harcourt: Boston, MA, USA, 2011.

Identification of Outlier Loci Responding to Anthropogenic and Natural Selection Pressure in Stream Insects Based on a Self-Organizing Map

Bin Li, Kozo Watanabe, Dong-Hwan Kim, Sang-Bin Lee, Muyoung Heo, Heui-Soo Kim and Tae-Soo Chon

Abstract: Water quality maintenance should be considered from an ecological perspective since water is a substrate ingredient in the biogeochemical cycle and is closely linked with ecosystem functioning and services. Addressing the status of live organisms in aquatic ecosystems is a critical issue for appropriate prediction and water quality management. Recently, genetic changes in biological organisms have garnered more attention due to their in-depth expression of environmental stress on aquatic ecosystems in an integrative manner. We demonstrate that genetic diversity would adaptively respond to environmental constraints in this study. We applied a self-organizing map (SOM) to characterize complex Amplified Fragment Length Polymorphisms (AFLP) of aquatic insects in six streams in Japan with natural and anthropogenic variability. After SOM training, the loci compositions of aquatic insects effectively responded to environmental selection pressure. To measure how important the role of loci compositions was in the population division, we altered the AFLP data by flipping the existence of given loci individual by individual. Subsequently we recognized the cluster change of the individuals with altered data using the trained SOM. Based on SOM recognition of these altered data, we determined the outlier loci (over 90th percentile) that showed drastic changes in their belonging clusters (D). Subsequently environmental responsiveness (E_k') was also calculated to address relationships with outliers in different species. Outlier loci were sensitive to slightly polluted conditions including Chl-a, NH_4-N, NO_X-N, PO_4-P, and SS, and the food material, epilithon. Natural environmental factors such as altitude and sediment additionally showed relationships with outliers in somewhat lower levels. Poly-loci like responsiveness was detected in adapting to environmental constraints. SOM training followed by recognition shed light on developing algorithms *de novo* to characterize loci information without *a priori* knowledge of population genetics.

Reprinted from *Water*. Cite as: Li, B.; Watanabe, K.; Kim, D.-H.; Lee, S.-B.; Heo, M.; Kim, H.-S.; Chon, T.-S. Identification of Outlier Loci Responding to Anthropogenic and Natural Selection Pressure in Stream Insects Based on a Self-Organizing Map. *Water* **2016**, *8*, 188.

1. Introduction

Water, a highly sensitive substrate environment in the biogeochemical cycle, is extremely vulnerable to various anthropogenic effects since disturbing agents are highly diffusive in aquatic ecosystems and their impacts are critical to the survival of living organisms (e.g., drinking resource) as well [1,2]. In addition to the hydrological aspect, water quality maintenance should also be considered from ecological and systematical viewpoints. Water is an essential element in aquatic ecosystems. Consequently water quality affects the status of biological organisms (*i.e.*, distribution and abundance).

Whereas physico-chemical indicators may only present non-biological aspects of water quality, monitoring biological organisms garners special attention since their status would reveal how water quality directly affects the survival of living organisms. Among various taxa, benthic macroinvertebrates respond characteristically to pollution sources and are suitable for monitoring aquatic ecosystems [3–5] since they are taxonomically diverse and sedentary in a habitat range with a reasonably long life span [4,6].

Moreover, biological organisms present another dimension in responding to environmental stress; gene information would correspondingly adapt to environmental constraints. Here, we report genetic information of aquatic insects adapting to natural and anthropogenic selection pressures in streams. Detecting adaptive genes under selection is a critical issue to infer how environmental heterogeneity can drive genetic divergence [7,8]. Studies on the genetic basis of adaptation are often based on candidate genes [9–11] and quantitative trait loci (QTL) approaches [12–14]. However, these methods are limited to model organisms and well-characterized genes, making it difficult to apply the approach in non-model organisms and anonymous genes.

An alternative approach is genome scanning for identifying large numbers of candidate loci including Amplified Fragment Length Polymorphism (AFLP) with statistical tests to detect outlier loci (candidate adaptive gene) under direct or indirect selection pressure [15–18]. A locus (plural loci) is a unique chromosomal location defining the position of a gene or DNA sequence in genetics [19]. In our study, it was simply defined as a gene but its position and DNA sequence are unknown due to AFLP analysis [15]. Conventionally, the two statistical methods Dfdist and BayeScan are widely used for outlier loci detection. Dfdist (adapted from Fdist [15]) uses coalescent simulations to generate thousands of loci evolving under a neutral model of islands with a mean global F_{ST} close to the observed global F_{ST}. Empirical loci with F_{ST} values significantly higher or lower than the simulated distribution were considered to be outlier loci under divergent (*i.e.*, high F_{ST}s) or balancing (*i.e.*, low F_{ST}s) selection [15]. BayeScan, based on a hierarchical Bayesian model, detects locus-specific (e.g., selection) and population-specific (e.g., immigration rates, local effective size) effects based on F_{ST} variability [16,18]. In practice, however, these statistical methods are prone to be affected by various factors such as dispersal, genetic drift, sample size, and nonselective evolutionary forces [20]. Assumptions used for data analyses regarding population structure, history, migration, and mutation rates may introduce a bias in the results if the genetic data violate the assumptions [7].

An alternative means could be considered to infer the complex relationships between genetic data and environmental impacts based on information processing. A separate line of research has been conducted with information processing in the field of water quality maintenance and ecosystem management since the 1990s [21], including the BayeScan model applied to prediction of harmful algal bloom [22] and an empirical modeling approach including neural networks and extra trees in phytoplankton dynamics to improve water-quantity-oriented management in reservoirs [23]. In this study we focus on implementation of information processing to gene information in association with environmental effects based on a self-organizing map (SOM).

SOM performs dimension compression of complex multivariable data while keeping the topology of the original data structure by training [24], and has been efficiently used in various fields including biology and ecology [25,26]. By adjusting weights between input (matching to variables in sample units) and output nodes adaptively through lateral inhibition among the nodes (*i.e.*, winner taking the chance of updating weights), the sample units sharing similar variables will be clustered together on the map of the reduced dimension. In molecular biology, SOM has been used in clustering of high-dimension molecular composition since the early 1990s [27], and earned great popularity recently in various fields including gene expression [28,29], genetic structure [30], and elucidating the effects of selection pressures [31].

SOM has been mainly used for patterning and visualization of complex datasets. Recently, however, a sensitivity test using SOM garnered additional attention for two purposes: characterizing network architecture and revealing the output sensitivity of the SOM responding to input data variability (e.g., data alteration). First, regarding network architecture, SOM performance was evaluated in response to model parameters including connection radius (neighbor size) and lateral inhibition (*i.e.*, proportion of weight contribution of the target variable comparing with competing neighbor variables) in modeling muscle groups in the proprioceptive cortex [32], whereas SOM sensitivity was also examined according to the number of clusters, sample size, and neighboring function parameters including initial neighbor size and reduction rate in image analysis [33].

Whereas sensitivity of network architecture focuses on model development, the second aspect of sensitivity analysis puts an emphasis on addressing output response to the input data variability in determining important variables, especially in ecosystem management for practical purposes. In revealing the input–output relationships, Paini *et al.* [34] conducted a SOM sensitivity test by altering presence/absence data of species occurrence (variables) in different sampling sites (sample units) in pest management. They revealed that small changes in a limited range of data supported the SOM's robust predictions of pest invasion risks. However, there has been no extensive study focusing on how to quantify the importance of each variable, specifically the SOM clustering results.

In this study, we implemented SOM to detect outlier loci in AFLP band presence/absence data. Being partly inspired by Paini *et al.* [34], we adopted a recognition process instead of retraining. We intended to investigate the local effect of altered data on trained clusters specifically for each individual instead of checking overall output variability in addressing stability of SOM performance. After initial training we altered the presence/absence data in each variable (loci) and recognized the cluster change on the trained map individual by individual (sample unit) across all the loci present in each species. This would allow a local alteration effect for each locus on the SOM, while not causing global cluster conformation changes due to retraining. The AFLP datasets of aquatic insects collected at various sampling sites across different streams [8] were used in this study (1) to characterize overall loci composition patterns by SOM training; (2) to reveal environmental responsiveness according to altered input data (presence/absence of loci) based on SOM recognition, and (3) to address associations between outlier loci and environmental variables.

2. Materials and Methods

2.1. Ecological and Genetic Data

The AFLP datasets were collected from three aquatic insect species in Trichoptera in six adjacent stream catchments in Miyagi Prefecture, Japan in July and November (summer and autumn) of the year 2006 [8] (Figure 1a). The surveyed region is largely mountainous, and streams in the area are characterized by high environmental heterogeneity with short and steep corridors. While the highland areas are generally clean, the lowland areas are slightly polluted due to agriculture or development of residential and commercial areas [8].

DNA band presence (1) or absence (0) (binary values) data for loci (DNA fragments) in AFLP were used for three aquatic insect species in Trichoptera: *Hydropsyche albicephela*, *Stenopsyche marmorata*, and *Hydropsyche orientalis*. In addition, 15 environmental factors including water quality indicators were concurrently recorded: "altitude", "stream order", "width", "velocity", "mean gravel size", "sediment", "epilithon", "Benthic Coarse Particulate Organic Matter (BCPOM)", "Suspended Fine Particulate Organic Matter (SFPOM)", "Chlorophyll-a (Chl-*a*)",

"Biochemical Oxygen Demand (BOD)", "Suspended Solid (SS)", "nitrite/nitrate nitrogen (NO$_X$-N)", "ammonia nitrogen (NH$_4$-N)", and "orthophosphate phosphorus (PO$_4$-P)". Since anthropogenic variables were not in a full scope (*i.e.*, less polluted), we characterized the habitats of the sampling sites using SOM according to nine natural variables (Table 1). Anthropogenic variables were separately dealt with regarding the effect of pollution on loci composition of surveyed species (see Section 3.3).

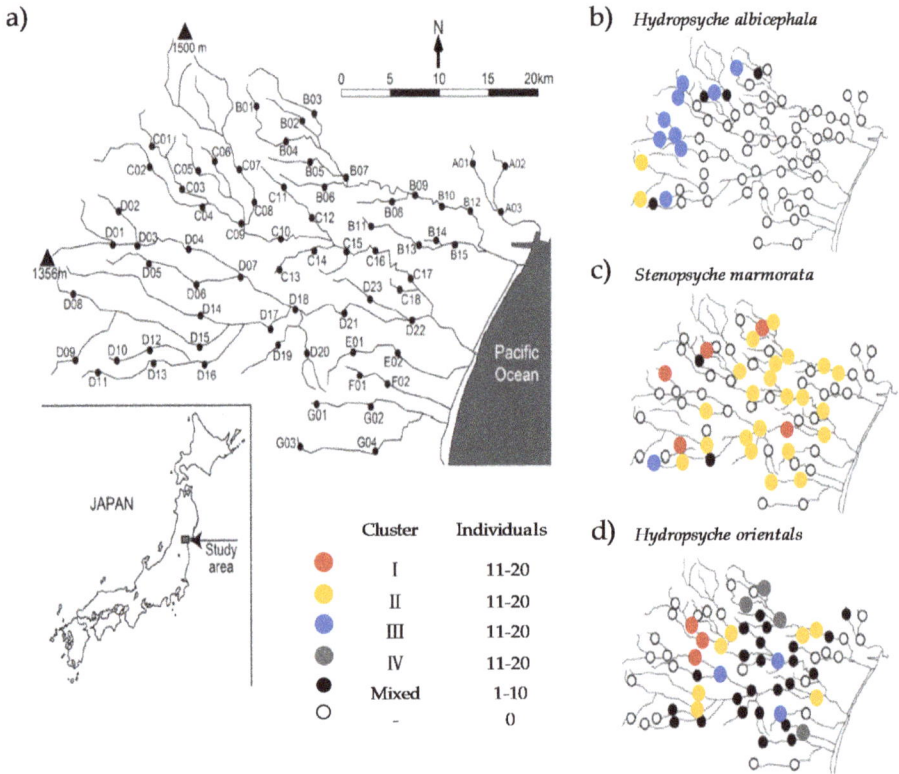

Figure 1. (a) Map of sampling sites; (b) group size of individuals collected within same cluster at each sampling site according to SOM (Figure 4) for *Hydropsyche albicephala*; (c) *Stenopsyche marmorata*, (d) *Hydropsyche orientals*. The color of the solid circles indicates a cluster in which 11 or more individuals (the total number of individuals in one site was 20) were grouped together, whereas the black circles mean the individuals are equal to or less than 10. The white circle stands for no individual collected at the sampling sites. The symbols, I, II, III and IV, in the legend indicate cluster names shown in Figure 4. Modified from Watanabe *et al.* [8].

2.2. SOM Applied to Environmental and AFLP Data

In order to characterize AFLP data and habitat features, a self-organizing map (SOM) [24] was applied to both environmental and AFLP data collected from the sampling sites. SOM performs dimension compression of complex multivariable data while keeping the topology of the original data structure by training. By adjusting weights adaptively between input nodes (matching variables of sample units) and output nodes through lateral inhibition, the sample units sharing similar variables will be clustered together on the map of reduced dimension. The SOM consists of two layers of artificial neurons (or nodes) in the input and output layers. Input nodes comprise the same number of variables with each neuron connected to all nodes in the output layer, on which learning procedures are projected through dimension compression. Starting with a randomly projected weight vector that connects input and output layers, the distance between input data and weight vector was calculated. The neuron that has the minimum distance was defined as the best matching neuron and selected as the winner. For the best matching neuron and its neighborhood neurons, the new weight vectors are adaptively updated (See [32,35,36] for details on SOM training in the ecological sciences).

Table 1. Summary of natural environmental variables measured at sampling sites (n = 57) during the survey period. When precise values were measured the higher number of significant digits are listed with the symbol "a".

Variables	Mean $\pm$ SD	Minimum	Maximum
Altitude (m)	180.33 $\pm$ 142.36	2	590
Stream order	2.44 $\pm$ 1.17	1	5
Width (m)	7.83 $\pm$ 7.87	1.24	38.33
Velocity (m$\cdot$s^{-1})	0.55 $\pm$ 0.23	0.03	1.28
Mean gravel size (cm)	12.71 $\pm$ 4.21	4.81	23.03
Sediment (mm)	10.43 $\pm$ 4.45	0.50	19
Epilithon (mg Chl-$a\cdot$cm^{-2})	0.0016 $\pm$ 0.0024 [a]	0.0001 [a]	0.01
BCPOM (mg AFDM$\cdot$m^{-2})	10.20 $\pm$ 9.58	0.89	54.07
SFPOM (mg AFDM$\cdot$L^{-1})	0.0069 $\pm$ 0.0052 [a]	0.0009 [a]	0.02

According to Vesanto's rule [37], the number of map units (m) could be approximately determined as $m = 5\sqrt{n}$, where n is the number of sample units. Starting from the initial value proposed by this rule, quantization error (QE) and topographic error (TE) [25] were obtained by SOM training by slightly adjusting the map size. We chose the map size with minimum QE and TE.

SOMs were used separately for both habitat and loci patterning. For habitat patterning, input data consisted of 57 sampling sites (sample units) and nine natural environmental factors (variables), and the number of nodes were

7 (vertical) × 6 (horizontal). For training AFLP, the number of nodes used for training was different since input data matrix (individuals (sample units) × loci (variables)) for each species varied according to the process of determining the number of nodes for training as stated above: 251, 571, and 753 individuals with 128, 220, and 129 loci were provided to the SOM consisting of 10 × 8, 14 × 10, and 14 × 10 output nodes for species *H. albicephela*, *S. marmorata*, and *H. orientalis*, respectively.

The SOM learning process was conducted under a Matlab environment (The Mathworks, R2009) using the SOM Toolbox [37] developed by the Laboratory of Information and Computer Science at the Helsinki University of Technology (http://www.cis.hut.fi/). The training was performed following suggestions made by the SOM Toolbox and Park *et al.* [36]. To reveal the degree of association between the SOM units, Ward's linkage method [38,39] was used to cluster the sample units according to the Euclidean distance.

2.3. Screening Outlier Loci and Environmental Responsiveness

In order to address associations between outlier loci and environmental variables, we adopted three processes: (1) initial training; (2) sensitivity analysis with altered datasets through recognition; and (3) calculating environmental responsiveness due to outlier loci. The following procedure was conducted for screening outlier loci and checking environmental responsiveness (Figure 2):

1. SOM is trained with the AFLP data and the trained SOM output units are classified to clusters (I, II, ... , N) (see Section 2.2).
2. A vector, **B** (list of clusters with training data for each individual) is produced (e.g., **B** = (I, II, I, III) with the first, second, third, and fourth individual matching cluster I, II, I, and III, respectively). Euclidian distance is calculated between clusters.
3. Each locus is altered by flipping over (switching either "presence to absence" or "absence to presence") separately for each individual.
4. Sensitivity analysis is conducted with altered datasets through recognition (See Figure A1).
5. A vector, **G** (list of clusters with altered data) is produced (e.g., **G** = (I, II, I, I) with each individual sequentially belonging to I, II, I, and I, similar to the case of **B** in process 2).
6. Mean cluster distance for each locus (*D*) is defined to determine the overall differences between training and recognition for individuals. *D* is calculated as average of the summed Euclidian distance according to **B** compared with **G**. If the change crossed over clusters with higher distance, higher values of distance would be given to this individual.

96

7. According to D outlier loci are determined. Loci with D value higher than the 90th percentile [40] were considered as outliers under selection in this study.
8. Once outlier loci were identified, we examined their relationships with each environmental variable. Indices $E_{k,i}$ and E_k were devised to present responsiveness of each environmental variable (k) in each cluster (i) and overall responsiveness across clusters, respectively, after SOM recognition as follows:

$$E_{k,i} = \left| \frac{e_{k,i} - h_{k,i}}{e_{k,i}} \right| \qquad (1)$$

$$E_k = \frac{1}{N} \sum_{1}^{N} E_{k,i}, \qquad (2)$$

where $e_{k,i}$ is the mean of environmental variable k for all individuals belonging to cluster i before data alteration, $h_{k,i}$ is the mean of the same environmental variable k for all individuals belonging to cluster i after recognition, and N is the total number of clusters. Higher $E_{k,i}$ value indicates a higher potential of variation in specific environmental factor, k, due to loci alteration between training and recognition in cluster i, whereas a higher E_k value represents overall responsiveness across all clusters on the SOM.

Take B = (I, II, I, III) and G = (I, II, I, I) in Figure 2 for the case of altitude matching to a certain locus as an example. In B, two individuals belonged to cluster I, corresponding with two values of altitude (100 m and 150 m). Subsequently, the average value was expressed as $e_{k,i}$ (=125 m). In G, cluster I has three individuals (corresponding with altitudes of 100 m, 150 m, and 200 m) after recognition, and the average of altitude was calculated as 150 m ($h_{k,i}$). In order to check the variability in environmental responsiveness, absolute values were used in this study. In the example case, the ratio of change in the mean altitude (absolute value) was 0.2 ($E_{k,1}$) between training and recognition in cluster I according to Equation (1). Similarly, we have $E_{k,2}$ and $E_{k,3}$ values of 0.0 and 1.0 for cluster II and III, respectively. Subsequently the E_k, value for altitude was calculated as the mean value of the three clusters, or 0.4. This indicates the degree of responsiveness of altitude due to data alteration in the locus as 0.4.

In order to present the relative importance of specific environmental factor to the maximum environmental responsiveness, the E_k value was further normalized as E_k' in relation to maximum value of E_k across all environmental variables (Figure 9) as shown below:

$$E_k' = \frac{E_k}{\max(E_k)} \qquad (3)$$

Suppose that there are only two environmental variables, altitude and Chl-*a* with E_k values of 0.4 and 0.8, respectively, for example. Then E_k' for altitude would be 0.5 whereas E_k' for Chl-*a* would be 1.0.

9. Based on outliers according to D (process 7), a locus-specific pattern showing degree of associations between outlier loci and environmental factors was determined according to the level of E_k' (process 8) (see Figure 9 for details).

The obtained environmental factors corresponding to outlier loci were cross-checked with the Kolmogorov-Smirnoff test (K-S test) to see whether environmental variables were indeed different across sampling sites [41,42]. In our study, the environmental values were divided into five classes according to different levels. Presence/absence of locus was counted for each individual belonging to the same environmental class and combined to give the overall frequency of loci presence for the individuals collected across different environmental classes. The hypothesis of uniform frequency of loci (*i.e.*, equal presence of loci in different environmental classes) was tested according to the K-S test. For comparing outlier loci, two conventional analyses, the Dfdist and BayeScan methods, were additionally conducted with the same datasets according to Watanabe *et al.* [8].

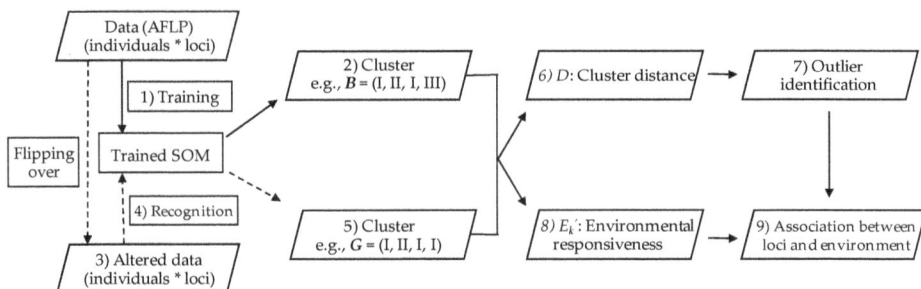

Figure 2. Flowchart for determining outlier loci and environmental responsiveness based on SOM sensitivity test through training and recognition. The numbers 1) to 9) show the processes described in Section 2.

3. Results

3.1. Habitat Specialization

Based on the SOM trained with nine natural environmental variables (57 sample sites in six streams), grouping was observed according to cluster analysis (Ward linkage method) (Figure 3a). Five clusters were formed based on the dendrogram. Clusters 1, 3, and 4 were observed at the upper area, corresponding with somewhat higher levels in altitude, sediment, BCPOM, and SFPOM according to profiles on the component plane (Figure 3c). Clusters 2 and 5 were placed at the lower area matching higher levels of order, width, mean gravel size, and SFPOM. However, the degree of matching varied according to clusters and environmental variables in both upper and lower groups. For instance, SFPOM was high in both upper and lower groups (Figure 3c). The sampling sites appeared to be mixed in all clusters (Figure 3b). There seemed to be no major environmental factors in determining the overall gradients at the surveyed area according to a visualization of the component profiles (Figure 3c). Habitats are variably characterized by heterogeneous environmental conditions at the sampling sites.

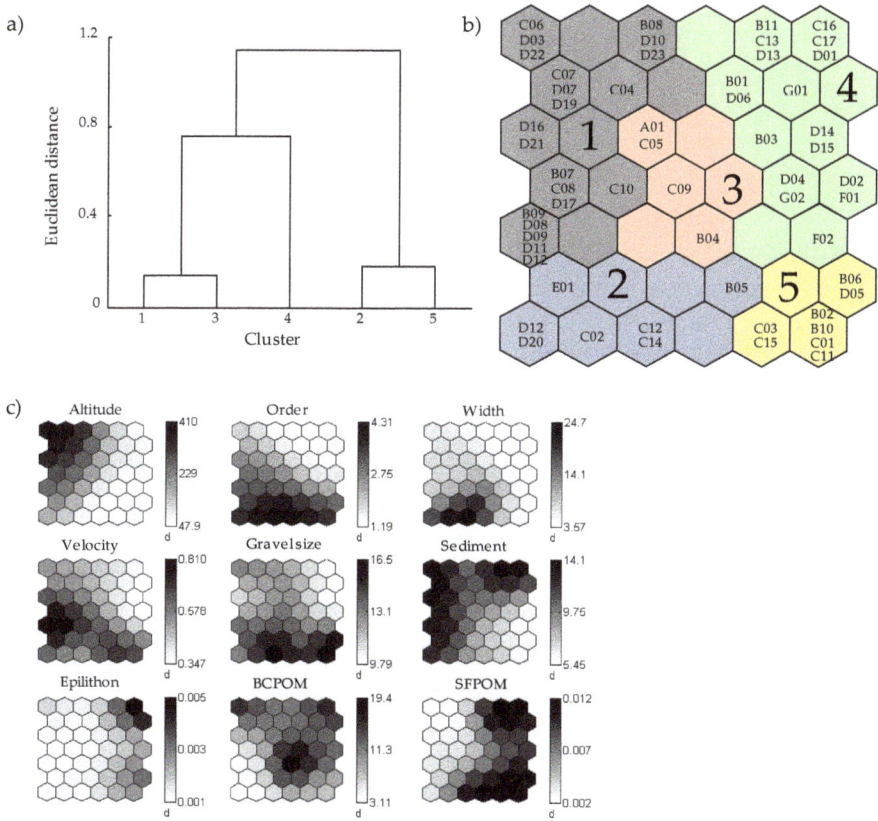

Figure 3. Patterning of sampling sites with SOM according to natural environmental factors. Dendrogram (**a**); ordination map (**b**); and component plane (**c**). The code in each node of map (**b**) refers to the name of sampling sites shown in Figure 1. Gray levels in (**c**) indicate the mean value of each variable, corresponding with sampling units presented on output nodes of SOM. Mean gravel size was calculated as the mean of the longest diameter among 36 grid points sampled in a 1 m^2 area of the stream bottom in riffles [8].

3.2. Patterning of Loci Data

Loci presence/absence data for all individuals were additionally trained with the SOM in each species (see Section 2.2). The number of clusters was determined according to conformation patterns on the U-matrix as shown in Figure A2. Three clusters were chosen for *H. albicephela* and *S. marmorata* (Figure 4a,b). In *H. orientalis*, however, diverse grouping was observed in the U-matrix (Figure A2). After initial training with various numbers of clusters in preliminary studies, four clusters were chosen for grouping the sample units since further division into more clusters did not provide feasible information in characterizing overall loci composition patterns in *H. orientalis* in this study (Figure 4c).

For the sake of convenience, I was assigned for the most strongly grouped cluster according to the U-Matrix (Figure A2, Figure 4), and the numbers II, III, and IV were subsequently given to the remaining clusters based on the Euclidean distance from cluster II, in ascending order. Clusters I and II occupied a small area and were strongly separated from clusters III and IV, which, by contrast, spanned over a broad area on the SOM (bottom panels, Figure 4).

Figure 1 additionally shows the degree of grouping in different clusters in each site for each species according to the SOM training in Figure 4. The size of solid circles indicates the groups of individuals within the same sampling site (the total number in one site was 20 individuals) in each cluster. The group size of individuals varied according to species. Overall, *H. albicephela* (Figure 1b) was collected in narrow upstream areas whereas *S. marmorata* (Figure 1c) and *H. orientalis* (Figure 1d) were broadly presented over the surveyed streams. In *S. marmorata*, the large groups (≥ 11 individuals) mostly belonged to intermediately (II) and strongly (I) grouped clusters (Figure 1c). The large groups were similarly found at the intermediately grouped cluster II in *H. orientalis* (Figure 1d). However, large groups were also found in all other clusters to a somewhat lesser degree in this species. It was also noteworthy that small groups (≤ 10 individuals) without dominant clusters were abundantly observed in *H. orientalis*. In *H. albicephela*, which was narrowly distributed upstream only, large groups were mainly observed in the weakly grouped cluster III (Figure 1b).

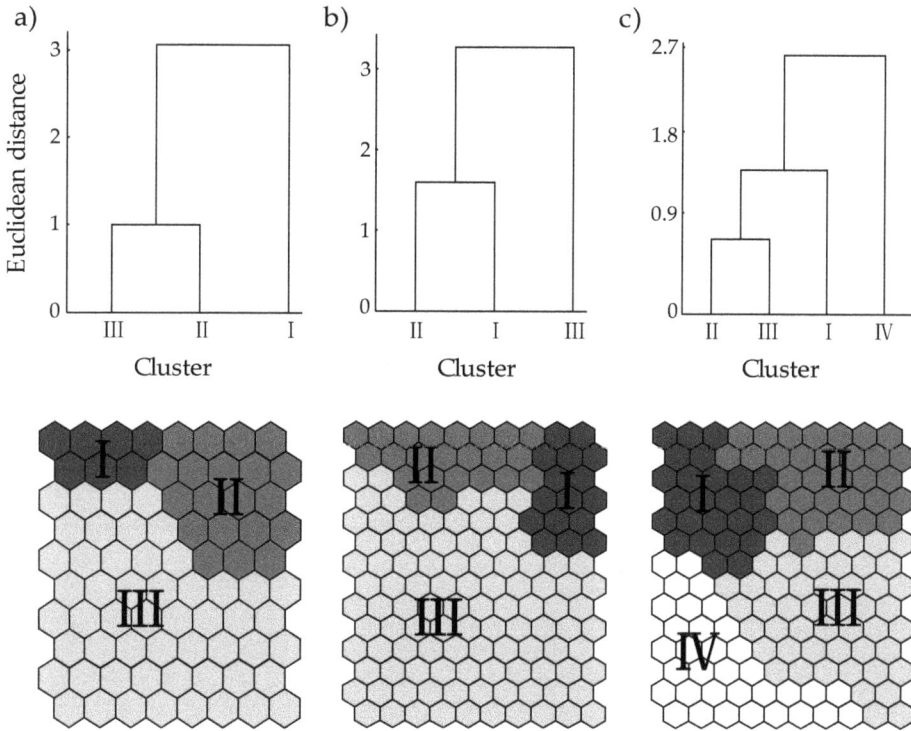

Figure 4. Dendrogram (top panels) and clustering (bottom panels) after SOM training on individual loci compositions (presence and absence) in *Hydropsyche albicephala* (**a**); *Stenopsyche marmorata* (**b**); and *Hydropsyche orientals* (**c**).

3.3. Identifying Outlier Loci and Environmental Responsiveness

The importance of variables (*i.e.*, loci) was checked by recognizing the altered data with the trained SOM in a similar sensitivity test (see Section 2.3 and Processes 3–9 in Figure 2). Cluster distances according to presence of loci (*i.e.*, higher presence, higher rank on *x* axis) are presented in Figure 5. The patterns of distance distribution varied according to species. The *D* values were variably observed in *H. albicephela* and *H. orientalis* (Figure 5a,c). The highest range, including peaks, was commonly observed at the ranks of 20–35 for both species. In *S. marmorata*, however, we found several fixed values of *D*, including 0.006 and 0.011 (Figure 5b). This indicated that a fixed number of individuals were commonly selected to experience cluster changes in this species. It was noteworthy that the two loci with the maximum *D* value (0.023) in *S. marmorata* were observed at loci 39 and 95, respectively (arrows in Figure 5b).

101

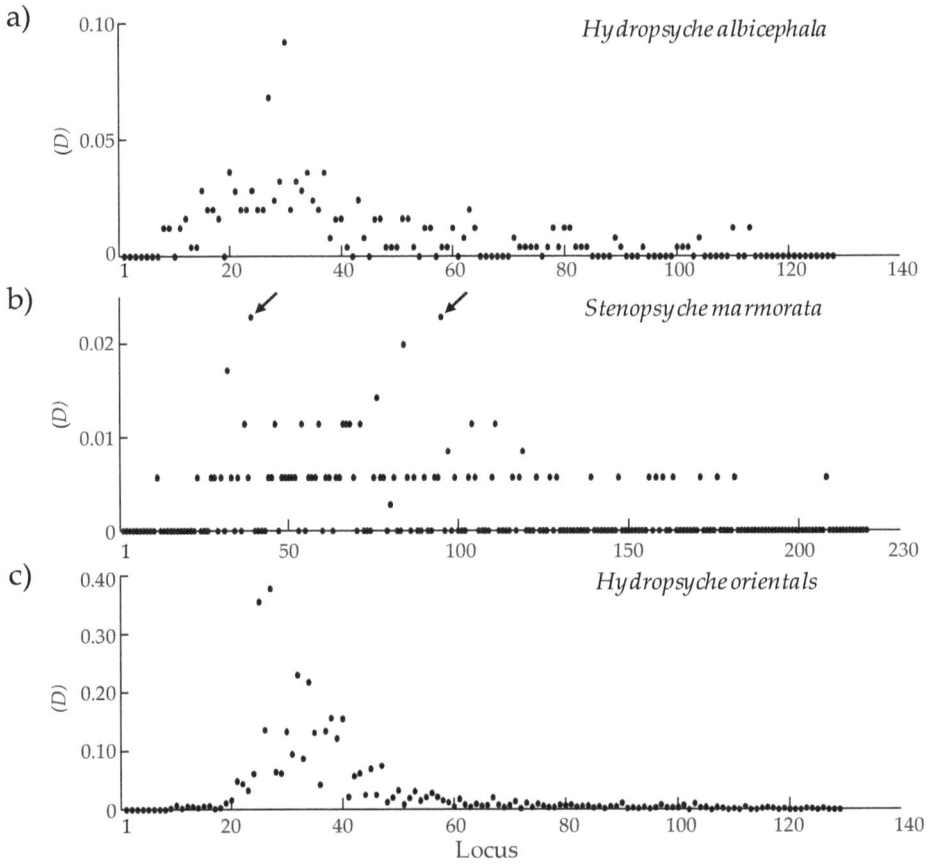

Figure 5. Cluster distance (D) in the order of loci presence (rank on x axis) after recognition in *Hydropsyche albicephala* (**a**); *Stenopsyche marmorata* (**b**); and *Hydropsyche orientals* (**c**).

Profiles of cluster distance are presented according to the order of D values, from low to high, in each species (Figure 6). Although SOMs could not be directly compared between species due to the separate SOM trainings for each species (see Section 2.2), overall trends and conformation of D values were separable and characteristic according to species. Usually a sharp drop of D was observed immediately after the peak by one or two-top ranked loci (Figure 6). The maximum D value (0.378) was observed in *H. orientalis* (Figure 6c), whereas *S. marmorata* had the minimum value (0.023) for the top rank locus (Figure 6b). Ranges in the number of individuals experiencing cluster changes (the gray shades in Figure 6) varied, with 0%–7.6%, 0%–0.7%, and 0%–21.5% for *H. albicephela*, *S. marmorata*, and *H. orientalis*, respectively. *S. marmorata* had the minimum proportion of individuals experiencing cluster changes, whereas the total number of individuals (571) used for training was intermediate compared with *H. albicephela* (251) and *H. orientalis* (753). Cluster change patterns were overall similar in the three species in accordance with the rank of loci, decreasing as the rank decreased (Figure 6).

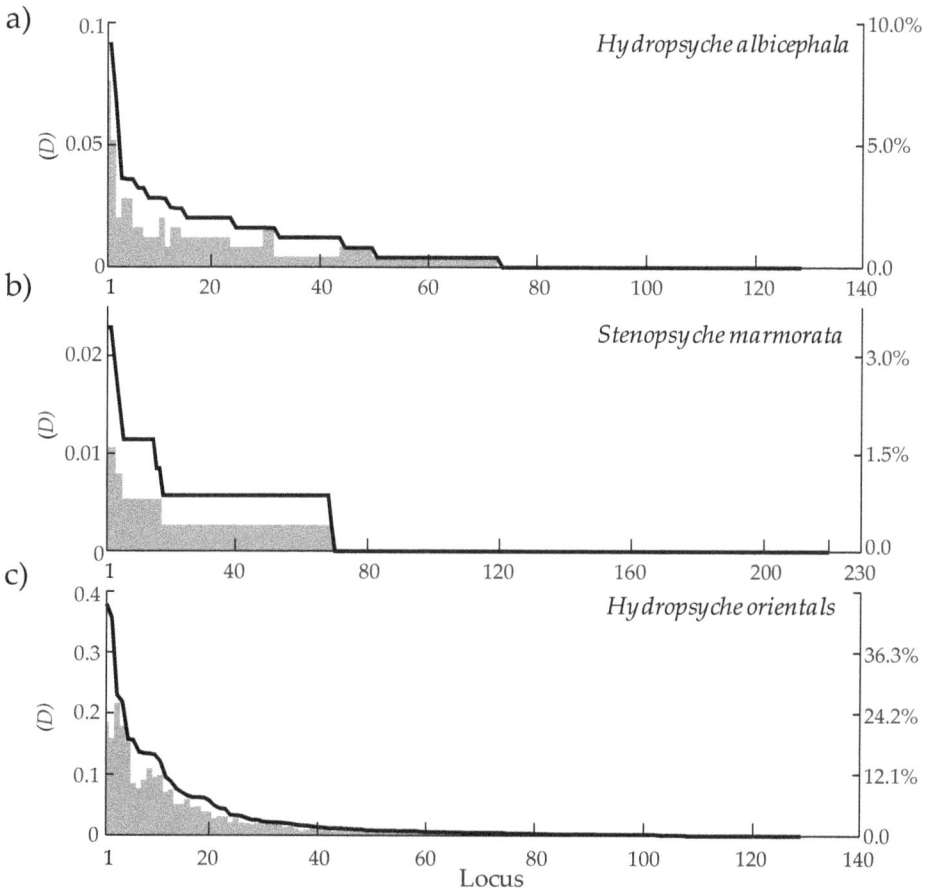

Figure 6. Cluster distance (D: solid line matching the scale in the left side of y axis) and proportion of individuals (gray shade matching the scale in the right side of y axis) experiencing cluster change in relation to the order of D (x axis) after recognition in *Hydropsyche albicephala* (**a**); *Stenopsyche marmorata* (**b**); and *Hydropsyche orientals* (**c**).

When frequencies were presented across different levels of D in the histogram, the maximum value was observed with the minimum level of D in different species (Figure 7). Frequencies sharply decreased in all species as D increased along the x axis. According to the frequency distribution shown in Figure 7, the outlier loci that responded strongly to data alteration were identified. In order to determine outliers, the 90th percentile was set as the threshold for each species (vertical lines in Figure 7) (see Section 2.3); the loci showing higher levels of D than the threshold were chosen as outlier loci in this study. Since many individuals showed the same value of D at the boundary of the 90th percentile line for *S. marmorata*, the grouped individuals that crossed the boundary of the percentile line were not included as outliers for *S. marmorata*. The numbers of outlier loci were 14, 17, and 12 for *H. albicephela*, *S. marmorata*, and *H. orientalis*, respectively (Table 2), and the outlier loci were listed in Table A1.

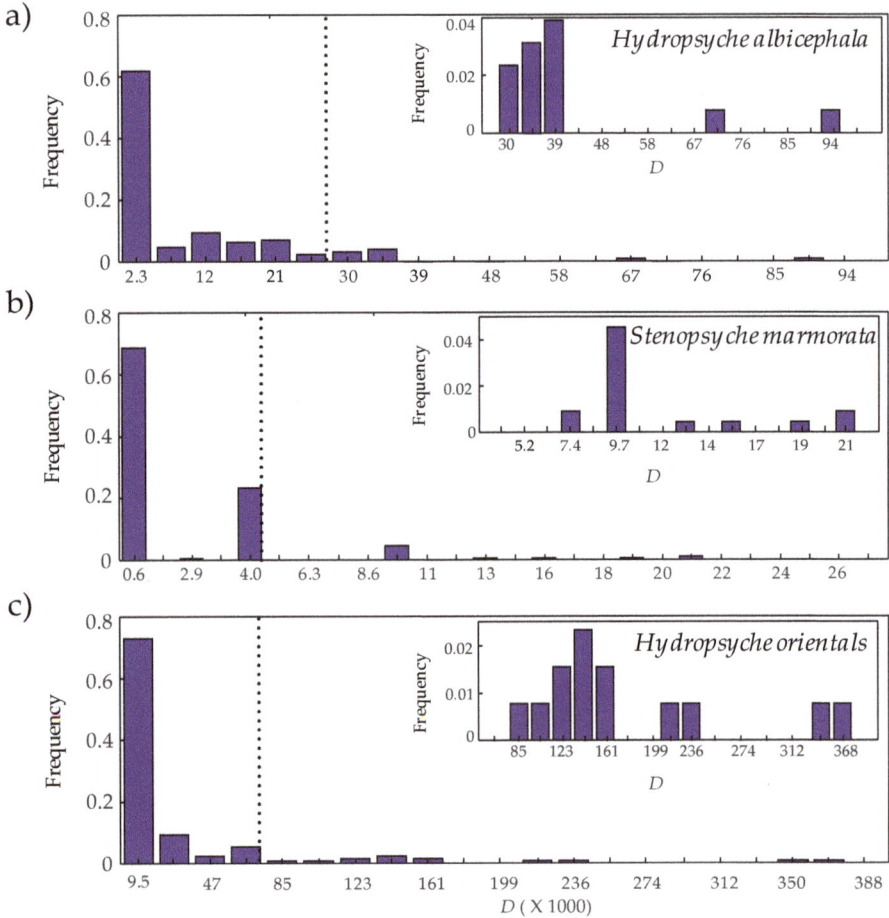

Figure 7. Histogram of cluster distance (D) in *Hydropsyche albicephala* (**a**); *Stenopsyche marmorata* (**b**); and *Hydropsyche orientals* (**c**) after recognition. The unit was expressed as 1000 times the actual values. The dotted vertical lines indicate the threshold for determining outliers (90th percentile). Insets show the upper range above the threshold.

Table 2. Similarity indices and number of common outliers between Dfdist, BayeScan, and SOM.

Species	No. of Outlier Loci			Common Outliers	Dfdist *vs.* BayeScan		Dfdist *vs.* SOM		BayeScan *vs.* SOM	
	BayeScan	Dfdist	SOM		Similarity Index	No. of Common Outliers	Similarity Index	No. of Common Outliers	Similarity Index	No. of Common Outliers
Hydropsyche albicephala	16	7	14	3	0.35	6	0.17	3	0.2	5
Stenopsyche marmorata	56	23	17	10	0.36	21	0.34	10	0.28	16
Hydropsyche orientalis	31	9	12	4	0.29	9	0.22	4	0.16	6

Subsequently we checked what clusters on the SOM responded to data alteration more sensitively. Figure 8 shows how clusters were affected when cluster change occurred for individuals in each species after SOM recognition (see Section 2.3). The cluster difference was evaluated according to the Jaccard similarity index [43] ($J = c/(a + b - c)$, where c is the commonly found individuals and a and b are the total number of individuals found in the cluster before and after recognition, respectively). Subsequently the value of $F = 1 - J$ was used to represent cluster difference. For convenience of visualization, all values were normalized to 0 to 1 based on the maximum F value obtained for each species, as shown by the vertical bars in Figure 8. The patterns of cluster change were different according to species. In *H. albicephela*, cluster II followed by cluster I were mainly affected by the altered data (Figure 8a). One locus in cluster II (locus 35) showed an especially high level of cluster difference. *S. marmorata* similarly showed a stronger response in clusters I and II (Figure 8b). In this case cluster I presented the highest level of F in selected loci (e.g., loci 71, 97, and 111). In *H. orientalis*, by contrast, cluster II and the weakly grouped clusters IV and III were more affected by the altered data (Figure 8c). The overall results indicated that sensitivity in loci composition varied according to different species.

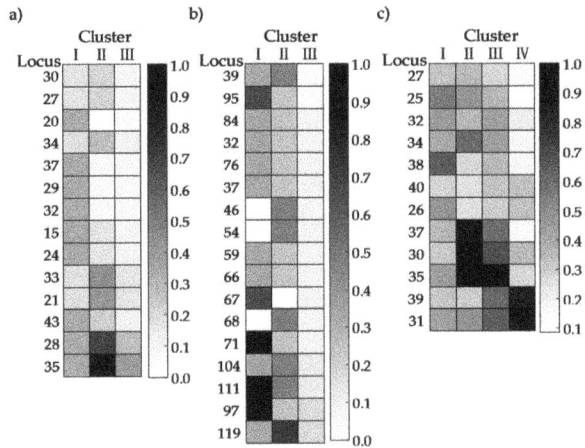

Figure 8. Cluster differences after recognition of altered data in *Hydropsyche albicephala* (**a**); *Stenopsyche marmorata* (**b**); and *Hydropsyche orientals* (**c**). The gray levels in the cells and the vertical bar indicate the degree of cluster difference (F value normalized to 0–1 according to the maximum in each species). The order of outlier loci was according to the D value from top down.

By calculating environmental responsiveness E_k' (Equation (3), see Section 2.3), each environmental factor was presented to associate specifically with outlier loci (Figure 9). For convenience of visualization, all E_k' values were normalized in the range of 0.00–1.00 based on the maximum E_k value across environmental variables in each species, as stated above (vertical bars, Figure 9). Maximum E_k values were found with Chl-a (1.28), NH$_4$-N (0.76), and Chl-a (0.22) for species $H.$ $albicephela$, $S.$ $marmorata$, and $H.$ $orientalis$ respectively. E_k values were relatively small for $H.$ $orientalis$ although the proportion of individuals experiencing cluster change was high (Figure 6c). The environmental responsiveness was characteristically observed according to species. In $H.$ $albicephela$, Chl-a (0.94–1.00) showed outstandingly high E values, followed by epilithon (0.43–0.44) (Figure 9a). $H.$ $orientalis$ similarly showed the highest responsiveness to Chl-a (0.75–1.00) and epilithon (0.64–0.72). In addition, PO$_4$-P (0.56–0.67), NO$_X$–N (0.54–0.56), and altitude (0.43–0.60) presented higher values in this species (Figure 9c). $S.$ $marmorata$ was characteristically sensitive to various anthropogenic stresses of NH$_4$–N (0.97–1.00), followed by PO$_4$–P (0.45–0.46), SS (0.32–0.33), and somewhat to sediment (0.24–0.25) (Figure 9b). Species overall responded highly to various anthropogenic factors and epilithon, and to a lower degree to natural factors (*i.e.*, altitude and sediment).

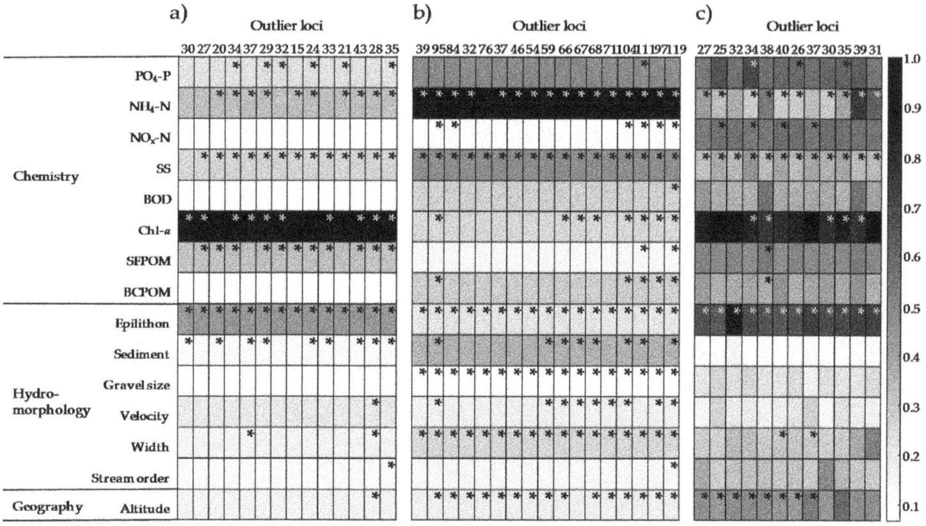

Figure 9. Environmental responsiveness in association with outlier loci in *Hydropsyche albicephala* (**a**); *Stenopsyche marmorata* (**b**); and *Hydropsyche orientals* (**c**). The gray levels indicate the $E_k{}'$ value according to the maximum across environmental variables in each species (*H. albicephela*: Chl-a (1.28), *S. marmorata*: NH$_4$–N (0.76), and *H. orientalis*: Chl-a (0.22)). The asterisk at the upper right corner of each cell indicates the significance ($p < 0.05$) of environmental variables according to the Kolmogorov-Smirnov test. For convenience of visualization, white asterisks are listed when $E_k{}'$ is equal to or greater than 0.6; otherwise there is black asterisk.

Two points can be summarized in terms of environment-outlier relationships. First, many loci were concurrently involved in responding to specific environmental factors, suggesting poly-loci-like responsiveness to environmental constraints (Figure 9). Secondly, outliers were mainly sensitive to either pollution agents (e.g., Chl-a, NH_4–N, NO_X–N, PO_4–P, and SS) or feeding material (*i.e.*, epilithon).

We further checked whether the results from SOM sensitivity would be in accordance with the difference in environmental factors measured at different sampling sites. A K-S test was conducted to check the significant difference between different levels of environmental variables (see Section 2.3). With all the environmental variables, the significance was only observed at $p < 0.05$ without a lower level of alpha error (e.g., $p < 0.01$). The symbol "*" was superimposed on the cells (presenting each combination of locus and environmental variable) shown in Figure 9. Significance in the K-S test corresponded with a high overall environmental responsiveness, indicating that the sensitivity in loci was in accordance with differences in environmental variables. There were some mismatches, however. PO_4–P in *S. marmorata* and *H. orientalis*, for instance, did not show overall significance in the K-S test, although the environmental responsiveness was high (Figure 9b,c).

It was noteworthy that the maximum values of pollution at the sampling sites were not indicative of severe conditions but rather indicated a low degree of pollution (Table 3). The gradients of water quality indicators across all the sampling sites are presented in Figure A3. According to WHO standards [44], water quality indicators measured in this study were substantially low. When the average values of the observed data were compared with the maximum level in the WHO standards, PO_4–P and NH_4–N showed the minimal range with 0.007%–1.880%. BOD had a somewhat higher level of 5.057% whereas NO_X–N, SS, and Chl-a presented maximal levels of 15.817%, 12.016%, and 12.000%, respectively (Table 3). All observed measurements in this study, however, were substantially low compared with the WHO standards.

The outliers detected by SOM, Dfdist [15], and BayeScan [18] are compared in Table 2. Since the number of detected outliers varied according to different methods, similarity in outliers between different methods was measured by the Jaccard similarity index [43] ($J = c/(a+b–c)$, where c is the commonly detected loci and a and b are the total number of detected loci by each method, respectively). Between Dfdist and BayeScan, the indices were in the range of 0.18–0.36 for all species (Table 2). Indices of SOM in relation to Dfdist and BayeScan were in a comparable range, 0.16–0.34. Among the three species, *S. marmorata* showed the highest number of outliers with 17, 56, and 23 loci for SOM, BayeScan, and Dfdist, respectively (Table 2, Table A1). A substantial number of common loci between different methods were also observed in *S. marmorata*: 21 in Dfdist *vs.* BayeScan, followed by 16 in BayeScan *vs.* SOM, and 10 in Dfdist *vs.* SOM. In the other two species the number of common loci between different methods was in the range 3–9 (Table 2).

Table 3. Summary of water quality indicators measured at sampling sites ($n = 57$) compared with WHO standards. (Minimum value of suspended solid is not recorded in WHO standards). For Chl-*a*, the number of significant digits is four for precise measurement.

Variables	Unit	Mean ± SD	Observed Values		WHO Standards [a]		Comparison (%) [b]
			Minimum	Maximum	Minimum	Maximum	
Chl-*a*	mg·L^{-1}	0.000,6 ± 0.001,2	0.0001	0.0087	<0.0025	0.005,0–0.140,0 [c]	12.000,0
BOD	mg·L^{-1}	0.455 ± 0.430	0.035	2.394	2 or less than 2	9	5.057
SS	mg·L^{-1}	3.004 ± 2.653	0.318	11.756	–	25	12.016
NO$_X$–N	mg·L^{-1}	0.475 ± 0.313	0.055	1.513	<0.1	3	15.817
NH$_4$–N	mg·L^{-1}	0.019 ± 0.022	0.004	0.147	0.04	1	1.880
PO$_4$–P	mg·L^{-1}	0.013,4 ± 0.014	0.001	0.078	0.001	200	0.007

[a]: Significant digits are based on the original data; [b]: Mean of observed value divided by maximum value (WHO standards); [c]: Maximum concentration of Chl-*a* was chosen as 0.0050 mg·L^{-1} instead of 0.1400 mg·L^{-1}.

4. Discussion

The current study demonstrated that specific associations between outlier loci and environmental variables could be effectively mined based on SOM even though *a priori* knowledge regarding population genetics (e.g., coalescent theory) is not available. Complexity residing in the presence/absence of loci was effectively extracted when they were exposed to complex effects of environmental factors according to different sampling sites. Whereas conventional methods such as Dfdist and BayeScan are mainly based on population genetics theories and statistical parameters (e.g., correlation coefficient between Wright's fixation index (F_{ST}) and environmental variables) [8,45,46], the SOM approach reveals relationships between outlier loci and environmental variables through information processing (Figures 4–9). The topology of complex data structure in multiple dimensions was effectively preserved in a reduced dimension through a self-organizing process [24]. It was noteworthy that environmental variables were specifically identified in association with each outlier based on the SOM approach *de novo* (Figure 9).

Loci were responsive to anthropogenic environmental change although the pollution level was low at the sampling sites (Figure 9, Table 3, and Figure A3). Based on the conventional methods (Dfdist and BayeScan), Watanabe *et al.* [8] reported that differences in F_{ST} at non-neutral loci were strongly related with natural and anthropogenic sources of Chl-*a* in *H. albicephela*, BCPOM in *S. marmorata*, and altitude in both *S. marmorata* and *H. orientalis*. A somewhat weak response was observed with velocity, epilithon, and NH$_4$–N in *H. orientalis*. This study partly confirmed results by Watanabe *et al.* [8] including Chl-*a* in *H. albicephela*, and altitude and epilithon in *H. orientalis*. However, overall differences were also observed. Response to anthropogenic stress appeared to be more prevalent in this study. *H. orientalis* showed high responsiveness to Chl-*a* in addition to *H. albicephela* (Figure 9c). *S. marmorata* additionally showed diverse responsiveness to NH_4–N, followed by PO_4–P and SS (Figure 9b). It was noteworthy that BOD showed generally lower levels of environmental responsiveness than other indicators in this study, except in *H. orientalis* (E_k' ranging from 0.31 to 0.44) (Figure 9c). Response to natural environmental factors, however, was not strongly observed except for the feeding material, epilithon, followed by altitude and sediment to a lesser degree, as stated above. Future study is warranted in revealing these loci-environment relationships in natural and anthropogenic variability more precisely along with gene functioning and physiological network studies.

Although overall trends were in accordance between the environmental response and the K-S results, there were discrepancies between the two tests (Figure 9). Whereas the K-S test only reveals statistical differences in environmental factors separately (*i.e.*, factor by factor), SOM deals with all variables together, accommodating complex interrelationships among variables concurrently in a non-linear manner through information processing. For instance, PO_4-P values were significantly different according to the K-S test in *S. marmorata* and *H. orientalis*, although the E_k' values were substantially lower (Figure 9). However, the detailed mechanism whereby environmental responsiveness was specifically sensitive to outlier loci is unknown. More studies regarding genetic functioning and molecular ecology are needed in the future.

Clusters with a small number of individuals ($\leqslant$ 10 individuals) were characteristically found in *H. orientalis* (Figure 1d). In *S. marmorata*, however, not many small groups were found and large groups of individuals were observed mainly in the intermediately grouped cluster II (Figure 1c). Following cluster II, more large individual groups were additionally found in the strongly grouped cluster I than in the weakly grouped cluster III in this species. In *H. albicephela*, by contrast, a large number of individuals were mainly observed in cluster III (Figure 1b). This indicated that loci composition patterns varied according to species. The somewhat more closely related loci (*i.e.*, being more different from other clusters according to SOM) observed in *S. marmorata*, for instance, had a stronger susceptibility to selection pressure, and consequently a higher potential for cluster change in more strongly grouped clusters. In *H. albicephela*, by contrast, the selection pressure may be not strong enough to cause loci variability in the strongly grouped clusters.

Difference in cluster change (*F*) also reflected the group responses of individuals. For instance, the intermediately and strongly grouped clusters I and II were particularly affected in *S. marmorata* (Figure 8b); large groups of individuals experiencing cluster changes were found in these clusters (Figure 1c). In *H. orientalis*, by contrast, the cluster change effect was mainly observed in the intermediately and weakly grouped clusters II, III, and IV, whereas a minimum response was observed in the strongly grouped cluster I (Figures 1d and 8c).

The loci composition patterns according to the SOM may match the dispersal ability of a species. It was reported that dispersal potential is weak for *H. orientalis*, whereas it is intermediate for *S. marmorata* according to Watanabe *et al.* [47]. The patterns of individual groupings were different between the two species. Whereas the intermediately and strongly grouped clusters II and I had large groups of individuals for cluster change at the same sampling sites in *S. marmorata* (Figure 1c), the small-group individuals ($\leqslant 10$ individuals) within one cluster (*i.e.*, without dominating clusters) were mainly observed for cluster change in *H. orientalis* (Figure 1d), as stated above. This indicated that loci information is more fractional (*i.e.*, there is a higher chance of variation in loosely grouped loci) in species like *H. orientalis* with low dispersal potential. However, in *S. marmorata*, which has intermediate dispersal potential, selection pressure would more strongly occur in strongly-grouped loci compositions (Figure 1c). However, drawing any conclusion regarding the relationships between dispersal ability and genetic divergence is still premature; additional studies on molecular ecology and gene functioning are needed in this regard in the future.

Specific individuals causing cluster changes could also be identified. We checked the fixed levels of D observed in *S. marmorata* (Figure 5b). The discrete values shown in *S. marmorata* were due to specific individuals changing consistently across different loci either in groups or as individuals. Initially *S. marmorata* showed individuals experiencing cluster changes separately. For example, individual 96 experienced cluster changes in numerous loci including 11, 23, 27, 44, *etc.*, while individual 537 separately experienced cluster changes in loci including 30, 45, 49, 64, *etc.* (Figure A4). In addition, certain individuals (e.g., (48, 317, 557), (96, 537)) contributed to cluster changes together. Since a fixed number of individuals showed cluster changes, this was the reason why these species showed discreteness in D (Figure 5b). Two peaks were additionally observed in D values in relation to the frequency of loci presence in *S. marmorata* (arrows in Figure 5b). The reason for obtaining two peaks in this species is currently unavailable. The strong responsiveness may be due to different types of data alteration (e.g., difference in "presence to absence" and "absence to presence" of loci), but close examination is required regarding input–output data relationships and genetic/ecological functioning.

In this study binary values were used as input (presence/absence). If continuous variables are used as input data, a small amount of noise could be added to cause continuous variability in the input data over a small range, similar to a conventional sensitivity analysis. However, the binary values were changed totally (either "1 to 0" or "0 to 1") in this study for training and recognition. Subsequently recognition was conducted to address the local effect of data alteration; the cluster change due to an altered datum for each locus was examined for each individual on the trained map (see Section 2.3).

In Paini *et al.* [34] binary data (presence or absence of pest species in different regions) were similarly used for training pest occurrence in different sampling sites. A different number of species (variables) was selected for data alteration (flipping over) according to different proportion levels (0.05, 0.10, 0.20, and 0.30). Subsequently the SOM was newly generated after data alteration. The variability of SOM output due to retraining was used as a criterion to determine stability in risk assessment. Data alteration of 0.30 indicated significant changes compared to original groupings whereas 0.20 indicated stability in groupings. The sensitivity test performed by Paini *et al.* [34] was useful for confirming the robustness of our predictions of pest invasion risk.

In our study, however, we focused on investigating the specific contribution of each locus (variable) on the cluster patterns. We adopted a recognition process after training (Figure 2). By examining the result of recognition on the trained SOM locally, the patterns of cluster change were specifically observed locus by locus for each individual (Figure A1). Further associations between loci and environmental variable were revealed according to this type of SOM training followed by recognition (Figure 9). However, variability would still be obtained in initial training if random conditions were different in different trials. Since convergence is obtained adaptively through a random process, as stated above, a minor degree of variability would exist in either clustering or determining outliers, although the overall trends would be similar. Further research is required to optimize the variability caused by randomness in SOM training in relationship with outlier determination in the future.

Considering flexibility in data handling and information extraction processes, SOM could be extended to link with the models used in conventional methods based on population genetics. Either a hybrid model or SOM network modification could be considered for addressing the interplay between information processing and molecular ecology. In addition, SOM could be further applied to diverse taxa and experimental data related to ecological (e.g., functional feeding group) and genetic functioning.

5. Conclusions

Addressing ecological functioning garners special attention in association with aquatic ecosystem management since the status of live organisms in ecosystems would reveal the direct effects of water quality. SOM was effective in addressing relationships between loci and environmental factors through training and recognition, although specific gene information and/or population genetics (e.g., coalescent theory) are not available for loci data. By applying the SOMs to AFLP of aquatic insects collected in six streams in Japan, we proved that the genetic information of aquatic insects responds sensitively to anthropogenic and natural selection pressures. Outlier loci over the 90th percentile were associated with environmental factors pertaining to specific sampling sites and were comparable with the conventional methods (Dfdist and BayeScan). Some loci were sensitive to pollutants at low levels including Chl-a, NH_4–N, NO_X–N, PO_4–P, and SS. The feeding material, epilithon, also served as a source of selection pressure. Loci compositions further responded to natural factors including altitude and sediment, but to a lesser degree. In addition, poly-loci-like responsiveness was detected in respond to environmental constraints.

Gene information adapting to environmental stress would be accordingly reflected in information processing. SOM training combined with recognition shed a light on developing algorithms *de novo* to characterize loci without *a priori* knowledge of population genetics used for conventional methods. For further understanding of genetic diversity in adapting to environmental constraints, more studies are warranted on both informatics (e.g., the development of networks) and biological (e.g., ecological/genetic functioning) aspects in the future.

Acknowledgments: This work was partially supported by the Japan Society for the Promotion of Science (JSPS) (grant numbers: 25289172, 25241024 and 26630247).

Author Contributions: Bin Li and Tae-Soo Chon conducted the data analyses and wrote the manuscript. Kozo Watanabe provided the field data. Kozo Watanabe and Heui-Soo Kim evaluated the results in terms of ecological and genetic aspects. Sang-Bin Lee, Dong-Hwan Kim, and Muyoung Heo assisted in model application and visualization. All authors participated in discussions and approved the final manuscript.

Conflicts of Interest: The authors declare no conflict of interest.

Abbreviations

The following abbreviations are used in this manuscript:

SOM	self-organizing map
AFLP	Amplified Fragment Length Polymorphism
QTL	quantitative trait loci
BCPOM	Benthic Coarse Particulate Organic Matter
SFPOM	Suspended Fine Particulate Organic Matter
Chl-a	Chlorophyll-a
BOD	Biochemical Oxygen Demand
SS	Suspended Solid
NO_X-N	nitrite/nitrate nitrogen
NH_4-N	ammonia nitrogen
PO_4-P	orthophosphate phosphorus
K-S test	Kolmogorov-Smirnoff test
F_{ST}	Wright's fixation index

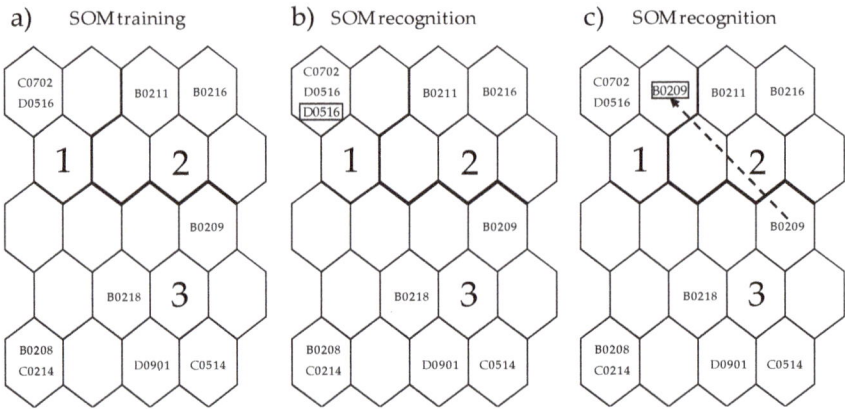

Figure A1. The process of SOM recognition. The SOM training results (**a**); individual (D0516) was recognized as having no cluster change (**b**); while individual (B0209) was recognized as having undergone a cluster change from cluster III to I (**c**).

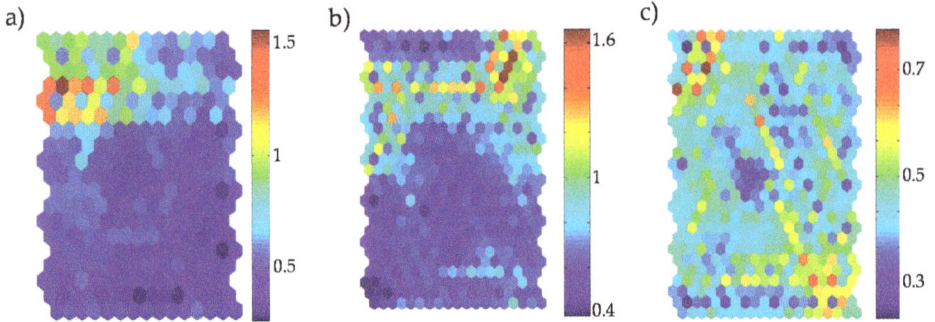

Figure A2. U-matrix based on SOM according to individual loci compositions in *Hydropsyche albicephala* (**a**); *Stenopsyche marmorata* (**b**); and *Hydropsyche orientals* (**c**). Color bars in the U-matrix indicate the mean Euclidean distance between the weights of nodes surrounding the U-matrix on the map.

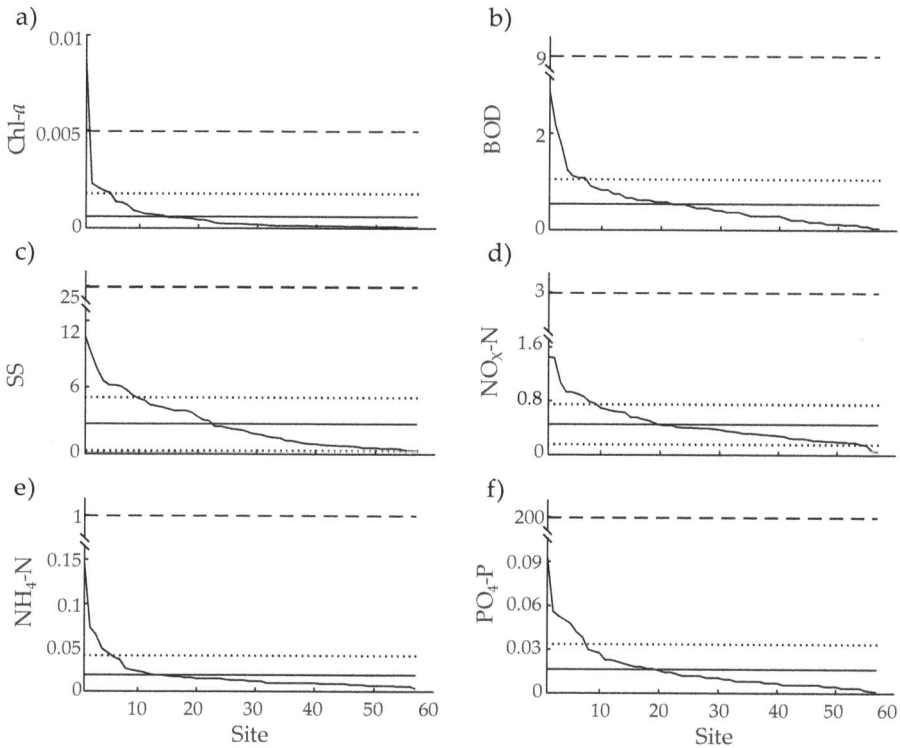

Figure A3. Gradients of water quality indicators across sampling sites. Dashed lines indicate the WHO standards, whereas solid and dotted lines stand for means and SDs of observed values, respectively. Chl-a (**a**); BOD (**b**); SS (**c**); NO_x-N (**d**); NH_4-N (**e**); and PO_4-P (**f**). (SD below mean was not listed in (**a**), (**b**), (**e**), and (**f**) because the value was either too close to 0 or negative).

Table A1. List of outlier loci identified by Dfdist, BayeScan, and SOM.

Species	Dfdist *vs.* BayeScan	Dfdist *vs.* SOM	BayeScan *vs.* SOM	Common Outliers
Hydropsyche albicephala	4, 5, 11, 15, 29, 43	15, 29, 43	15, 29, 33, 35, 43	15, 29, 43
Stenopsyche marmorata	32, 33, 34, 43, 45, 56, 66, 67, 68, 70, 76, 78, 80, 84, 85, 95, 97, 104, 106, 111, 116	32, 66, 67,68, 76,84, 95, 97, 104, 111	32, 37, 39, 46, 54, 59, 66, 67, 68, 71, 76, 84, 95, 97, 104, 111	32, 66, 67, 68, 76, 84, 95, 97, 104, 111
Hydropsyche orientals	20, 25, 27, 39, 40, 47, 49, 53, 67	25, 27, 39, 40	25, 26, 27, 32, 39, 40	25, 27, 39, 40

Outlier locus	Individuals experiencing cluster change											
	48	76	96	120	144	317	345	351	366	396	537	557
32	+					+						+
37				+			+					
39			+				+			+	+	
46	+		+									
54			+								+	
59			+								+	
66						+					+	
67						+			+			
68			+			+						
71			+								+	
76	+				+			+				
84		+	+		+						+	
95		+	+				+				+	
97			+		+							
104			+								+	
111			+								+	
119					+						+	

Figure A4. List of loci and individuals showing discrete values in cluster distance in *Stenopsyche marmorata*. Overlapping with four or more loci (or individuals) is represented by horizontal (or vertical) gray bars. The number of individuals is arbitrarily given.

References

1. Allan, J.D. Landscapes and riverscapes: The influence of land use on stream ecosystems. *Annu. Rev. Ecol. Evol. Syst.* **2004**, *35*, 257–284.
2. Walsh, C.J.; Roy, A.H.; Feminella, J.W.; Cottingham, P.D.; Groffman, P.M.; Morgan, R.P. The urban stream syndrome: Current knowledge and the search for a cure. *J. North Am. Benthol. Soc.* **2005**, *3*, 706–723.
3. Hellawell, J.M. Biological indicators. In *Biological Indicators of Freshwater Pollution and Environmental Management*, 1st ed.; Hellawell, J.M., Ed.; Elsevier Science Publishing Co., Inc: New York, NY, USA, 1986; pp. 45–63.
4. Rosenberg, D.M.; Resh, V.H. Introduction to Freshwater Biomonitoring and Benthic Macroinvertebrates. In *FreshwaterBiomonitoring and Benthic Macroinvertebrates*, 1st ed.; Rosenberg, D.M., Resh, V.H., Eds.; Chapman & Hall: New York, NY, USA, 1993; pp. 1–30.
5. Wright, J.F. An introduction to RIVPACS. In *Assessing the Biological Quality of Freshwaters, RIVPACS and Other Techniques*; Wright, J.F., David, W.S., Mike, T.F., Eds.; Freshwater Biological Association: Ambleside, UK, 2000; pp. 5–26.
6. Park, Y.S.; Song, M.Y.; Park, Y.C.; Oh, K.H.; Cho, E.C.; Chon, T.S. Community patterns of benthic macroinvertebrates collected on the national scale in Korea. *Ecol. Model.* **2007**, *203*, 26–33.

7. Storz, J.F. Invited Review: Using genome scans of DNA polymorphism to infer adaptive population divergence. *Mol. Ecol.* **2005**, *14*, 671–688.

8. Watanabe, K.; Kazama, S.; Omura, T.; Michael, T.M. Adaptive Genetic Divergence along Narrow Environmental Gradients in Four Stream Insects. *PLoS ONE* **2014**, *9*, e93055.

9. Kim, J.; Lee, T. An integrated approach of comparative genomics and heritability analysis of pig and human on obesity trait: Evidence for candidate genes on human chromosome 2. *BMC Genom.* **2012**, *13*, 711.

10. Tabor, H.K.; Risch, N.J.; Myers, R.M. Candidate-gene approaches for studying complex genetic traits: Practical considerations. *Nat. Rev. Genet.* **2002**, *3*, 391–397.

11. Carneiro, M.; Afonso, S.; Geraldes, A.; Garreau, H.; Bolet, G.; Boucher, S.; Tircazes, A.; Queney, G.; Nachman, M.W.; Ferrand, N. The genetic structure of domestic rabbits. *Mol. Biol. Evol.* **2011**, *28*, 1801–1816.

12. Daniels, S.E.; Bhattacharrya, S.; James, A.; Leaves, N.I.; Young, A.; Hill, M.R.; Faux, J.A.; Ryan, G.F.; le Souef, P.N.; Lathrop, G.M.; *et al.* A genome-wide search for quantitative trait loci underlying asthma. *Nature* **1996**, *383*, 247–250.

13. Verhoeven, K.J.F.; Vanhala, T.K.; Biere, A.; Nevo, E.; van Damme, J.M. The genetic basis of adaptive population differentiation: A quantitative trait locus analysis of fitness traits in two wild barley populations from contrasting habitats. *Evolution* **2004**, *58*, 270–283.

14. Rogers, S.M.; Bernatchez, L. Integrating QTL mapping and genome scans towards the characterization of candidate loci under parallel selection in the lake whitefish (Coregonus clupeaformis). *Mol Ecol.* **2005**, *14*, 351–361.

15. Beaumont, M.A.; Nichols, R.A. Evaluating loci for use in the genetic analysis of population structure. *Proc. Roy. Soc. B Biol. Sci.* **1996**, *263*, 1619–1626.

16. Beaumont, M.A.; Balding, D.J. Identifying adaptive genetic divergence among populations from genome scans. *Mol Ecol.* **2004**, *13*, 969–980.

17. Foll, M.; Gaggiotti, O.E. Identifying the environmental factors that determine the genetic structure of populations. *Genetics* **2006**, *74*, 875–891.

18. Foll, M.; Gaggiotti, O.E. Estimating selection with different markers and varying demographic scenarios: A Bayesian perspective. *Genetics* **2008**, *180*, 977–993.

19. Strachan, T.; Read, A. Genes in Pedigrees and Populations. In *Human Molecular Genetics*; Garland Science: New York, NY, USA, 2009.

20. Lotterhos, K.E.; Whitlock, M.C. Evaluation of demographic history and neutral parameterization on the performance of FST outlier tests. *Mol. Ecol.* **2014**, *23*, 2178–2192.

21. Straškraba, M. Ecotechnological models for reservoir water quality management. *Ecol. Model.* **1994**, *74*, 1–38.

22. Hamilton, G.; McVinish, R.; Mengersen, K. Bayesian model averaging for harmful algal bloom prediction. *Ecol. Appl.* **2009**, *19*, 1805–1814.

23. Fornarelli, R.; Galelli, S.; Castelletti, A.; Antenucci, J.P.; Marti, C.L. An empirical modeling approach to predict and understand phytoplankton dynamics in a reservoir affected by interbasin water transfers. *Water Resour. Res.* **2013**, *49*, 3626–3641.

24. Kohonen, T. The self-organizing map. *Proc. IEEE* **1990**, *78*, 1464–1480.

25. Park, Y.-S.; Tison, J.; Lek, S.; Coste, M.; Giraudel, J.; Delmas, F. Application of a self-organizing map in ecological informatics: Selection of representative species from large community dataset. *Ecol. Inform.* **2006**, *1*, 247–257.

26. Chon, T.-S. Self-Organizing Maps applied to ecological sciences. *Ecol. Inform.* **2011**, *6*, 50–61.

27. Ferrán, E.A.; Ferrara, P. Clustering proteins into families using artificial neural networks. *Bioinformatics* **1992**, *8*, 39–44.

28. Nikkilä, J.; Törönen, P.; Kaski, S.; Venna, J.; Castrén, E.; Wong, G. Analysis and visualization of gene expression data using Self-Organizing Maps. *Neural Netw.* **2002**, *15*, 953–966.

29. Chavez-Alvarez, R.; Chavoya, A.; Mendez-Vazquez, A. Discovery of Possible Gene Relationships through the Application of Self-Organizing Maps to DNA Microarray Databases. *PLoS ONE* **2014**, *9*, e93233.

30. Giraudel, J.L.; Aurelle, D.; Berrebi, P.; Lek, S. Application of the self-organizing mapping and fuzzy clustering to microsatellite data: How to detect genetic structure in brown trout (Salmo trutta) populations. In *Artificial Neuronal Networks*; Lek, S., Guégan, J.F., Eds.; Springer: New York, NY, USA, 2000; pp. 187–202.

31. Kontunen-Soppela, S.; Parviainen, J.; Ruhanen, H.; Brosché, M.; Keinäen, M.; Thakur, R.C.; Kolehmainen, M.; Kangasjävi, J.; Oksanen, E.; Karnosky, D.F.; *et al.* Gene expression responses of paper birch (Betula papyrifera) to elevated CO_2 and O_3 during leaf maturation and senescence. *Environ. Pollut.* **2010**, *158*, 959–968.

32. Cho, S.Z.; Jang, M.; Reggia, J.A. Effects of Varying Parameters on Properties of Self-Organizing Feature Maps. *Neural Process. Lett.* **1996**, *4*, 53–59.

33. Gonzalez, A.I.; Grana, M.; Anjou, A.D.; Albizuri, F.X.; Cottrell, M. A sensitivity analysis of the self-organizing maps as an adaptive one-pass non-stationary clustering algorithm: The case of color quantization of image sequences. *Neural Processing Lett.* **1997**, *6*, 77–89.

34. Paini, D.R.; Worner, S.P.; Cook, D.C.; De Barro, P.J.; Thomas, M.B. Using a self-organizing map to predict invasive species: Sensitivity to data errors and a comparison with expert opinion. *J. Appl. Ecol.* **2010**, *47*, 290–298.

35. Lek, S.; Guégan, J.F. *Artificial Neuronal Networks: Application to Ecology and Evolution*; Springer: Berlin, Germany, 2000.

36. Park, Y.-S.; Céréghino, R.; Compin, A.; Lek, S. Applications of artificial neural networks for patterning and predicting aquatic insect species richness in running waters. *Ecol. Model.* **2003**, *160*, 265–280.

37. Vesanto, J.; Alhoniemi, E. Clustering of the self-organizing map. *Neural Netw.* **2000**, *11*, 586–600.

38. Ward, J.H. Hierarchical Grouping to Optimize an Objective Function. *J. Am. Stat. Assoc.* **1963**, *58*, 236–244.

39. Wishart, D. An algorithm for hierarchical classifications. *Biometrics* **1969**, *25*, 165–170.

40. Kolmogorov, A. Sulla determinazione empirica di una legge di distribuzione. *Giornale dell'Istituto Italiano degli Attuari* **1933**, *4*, 83–91.

41. Marsaglia, G.; Tsang, W.W.; Wang, J. Evaluating Kolmogorov's Distribution. *J. Stat. Softw.* **2003**, *8*, 1–4.
42. Parkhomenko, E.; Tritchler, D.; Lemire, M.; Pingzhao, H.; Beyene, J. Using a higher criticism statistic to detect modest effects in a genome-wide study of rheumatoid arthritis. *BMC Proceed.* **2009**, *S40*, 1–4.
43. Jaccard, P. The distribution of the flora in the alpine zone. *New Phytol.* **1912**, *11*, 37–50.
44. Chapman, D.; Kimstach, V. Selection of water quality variables. In *Water Quality Assessments-A Guide to Use of Biota, Sediments and Water in Environmental Monitoring*, 2nd ed.; Chapman, D., Ed.; E&FN Spon: London, UK, 1996; pp. 90–100.
45. Bonin, A.; Miaud, C.; Taberlet, P.; Pompanon, F. Explorative genome scan to detect candidate loci for adaptation along a gradient of altitude in the common frog (Ranatemporaria). *Mol. Biol. Evol.* **2006**, *23*, 773–783.
46. Fischer, M.C.; Foll, M.; Excoffier, L.; Heckel, G. Enhanced AFLP genome scans detect local adaptation in high-altitude populations of a small rodent (Microtusarvalis). *Mol. Ecol.* **2011**, *20*, 1450–1462.
47. Watanabe, K.; Monaghan, M.T.; Takemon, Y.; Omura, T. Dispersal ability determines the genetic effects of habitat fragmentation caused by reservoirs in three species of aquatic insect. *Aquat. Conserv. Mar. Freshw. Ecosyst.* **2010**, *20*, 574–579.

Spatial Distribution of Benthic Macroinvertebrate Assemblages in Relation to Environmental Variables in Korean Nationwide Streams

Yung-Chul Jun, Nan-Young Kim, Sang-Hun Kim, Young-Seuk Park,
Dong-Soo Kong and Soon-Jin Hwang

Abstract: Conserving and enhancing freshwater biodiversity are global issues to ensure ecosystem integrity and sustainability. To meet this, it is critical to understand how the biological assemblages are determined by environmental gradients in different spatial scales. Nevertheless, information on their large-scale environmental relationships remains scarce in Korea. We aimed to understand nationwide spatial distribution patterns of benthic macroinvertebrates and important environmental factors affecting their distribution in 388 streams and rivers across Korea. A total of 340 taxa, belonging to 113 families in 23 orders of five phyla, were identified. Assemblage composition in most Korean streams included a few predominant colonizers and a majority of rare taxa. Cluster analysis based on benthic macroinvertebrates classified a total of 720 sampling sites into five clusters according to the pollution levels from fast-flowing less polluted streams with low electrical conductivity to moderately or severely polluted streams with high electrical conductivity and slow water velocity. Canonical correspondence analysis revealed that altitude, water velocity and streambed composition were the most important determinants, rather than watershed and water chemistry variables, for explaining the variation in macroinvertebrate assemblage patterns. The results provide basic information for establishing the conservation and restoration strategies of macroinvertebrate biodiversity against anthropogenic disturbances and developing more confident bio-assessment tools for diagnosing stream ecosystem integrity.

Reprinted from *Water*. Cite as: Jun, Y.-C.; Kim, N.-Y.; Kim, S.-H.; Park, Y.-S.; Kong, D.-S.; Hwang, S.-J. Spatial Distribution of Benthic Macroinvertebrate Assemblages in Relation to Environmental Variables in Korean Nationwide Streams. *Water* **2016**, *8*, 27.

1. Introduction

Freshwater ecosystems occupying a very tiny fraction of the Earth's surface support remarkable biodiversity and abundance of benthic macroinvertebrates [1,2]. However, growing human pressures have substantially deteriorated the overall ecological integrity and induced a biodiversity crisis. This problem now increasingly becomes a

120

global issue, and thereby calls attention to establish relevant conservation strategies and wise management practices to maintain sustainable freshwater environments.

Although Asian riverine ecosystems contain remarkable taxonomic diversity as well as high levels of endangerment and endemism, studies on biodiversity and ecology of these ecosystems have been poor [1,3]. Year-to-year variations in seasonality and its relationship to monsoonal climate, particularly in temperate Asian regions, can be major drivers that profoundly affect the hydrologic regime and geomorphology in stream environments, consequently determining the distribution and abundance of macroinvertebrates [4]. For example, a wet season low and a dry season high are expected for periodic seasonal patterns in abundance, depending on the frequency and intensity of summer monsoon rainfall. However, habitat destruction and water quality degradation in Asian river systems have become more dramatically epidemic than any other continents due to the fast growing human population and demands for economic development [5,6]. Because failure of conservation efforts has mainly resulted from scant evidence on river ecology, the priority over scientific and practical challenges to overcome this circumstance must be put on establishing species inventories and ecological information based on the accumulation of region-specific case studies, particularly at watershed or national scales [4,7].

Macroinvertebrate diversity and abundance are significant community attributes that are controlled by a variety of mechanisms at different spatial scales. A number of studies have documented how macroinvertebrate assemblages respond to environmental variables and which variables best explain their distribution and abundance. Some studies showed good relationships among macroinvertebrate assemblages, chemical variables [8], and the organic energy base [9], whereas habitat-related physical factors were widely demonstrated as primary contributors such as substrate composition [10], flow and current velocity [11], elevation and stream size [12] and temperature [13]. Vegetation, geology and human land use are also important for their spatial distribution [14,15]. While there remains a large number of documents focusing on such environmental relationships at finer spatial scales, studies on large-scale spatial patterns have been relatively small (e.g., at national scale [16,17] and at continental scale [18,19]).

Studies on the spatial distribution patterns of macroinvertebrate assemblages based on their environmental relationships are crucial. Firstly, those studies provide fundamental information for the conservation and restoration of biodiversity against anthropogenic disturbances. Secondly, knowledge of the species response to environmental gradients is important to separate the effects of pollution from the effects of natural variables on community structures [20]. Thirdly, it is possible to develop more confident bio-assessment tools for diagnosing stream ecosystem integrity because macroinvertebrate responds sensitively to environmental alterations that lead to changes in composition and community structures [21,22]. Thus, we performed a synoptic study of the large scale spatial distribution of benthic macroinvertebrates in relation to environmental variables. Specifically, the objectives of this study were: (i) to characterize the spatial distribution and assemblage structures of macroinvertebrates; (ii) to identify environmental distinction of Korean stream ecosystems based on their assemblages; (iii) to determine major environmental variables that affect their distribution; and (iv) to provide methodological considerations to improve biomonitoring programs, particularly focusing on sampling procedures. A relevant scientific basis would then be established with these data for developing sustainable management, conservation practices, and a reliable biomonitoring program. The results of this study could also contribute to a better understanding of the spatial distribution of freshwater macroinvertebrates in poorly explored Asian regions.

2. Materials and Methods

2.1. Study Area

Samples were collected at 720 sampling sites from 388 streams and rivers on a nationwide scale in South Korea. South Korea is located between 37°00′ N latitude and 127°30′ E longitude, encompassing the southern half of the Korean Peninsula with an entire area of approximately 100,033 km². The annual precipitation is 1308 mm, but there is substantial variation among seasons [23]: severe flooding events during the summer monsoon period, and base flow or even drought conditions in the other seasons.

The entire stream system throughout South Korea comprises five watersheds, including five large rivers, their tributaries, and other small independent streams (Figure 1). Land use types in each watershed generally display well-managed forests upstream, and agricultural and urban development from middle to lower regions. Most Korean streams have suffered from a variety of human activities that alter the physicochemical stream environments, particularly caused by channel modification and eutrophication [24].

Among 720 sampling sites, the largest number of sites belonged to the Han River watershed (HRW) (n = 320), followed by the Nakdong River watershed (NRW) (n = 130), the Geum River watershed (GRW) (n = 130), the Youngsan River watershed (YRW) (n = 76), and the Seomjin River watershed (SRW) (n = 64). It was assumed that such a nationwide survey covered the majority of stream types in Korea to understand how lotic macroinvertebrates are spatially distributed in relation to environmental factors. Field sampling was conducted during spring (May 2009) under base flow conditions because the highest benthic macroinvertebrate diversity was expected at that time. More information of the five major river watersheds in Korea can be found in Hwang *et al.* [25].

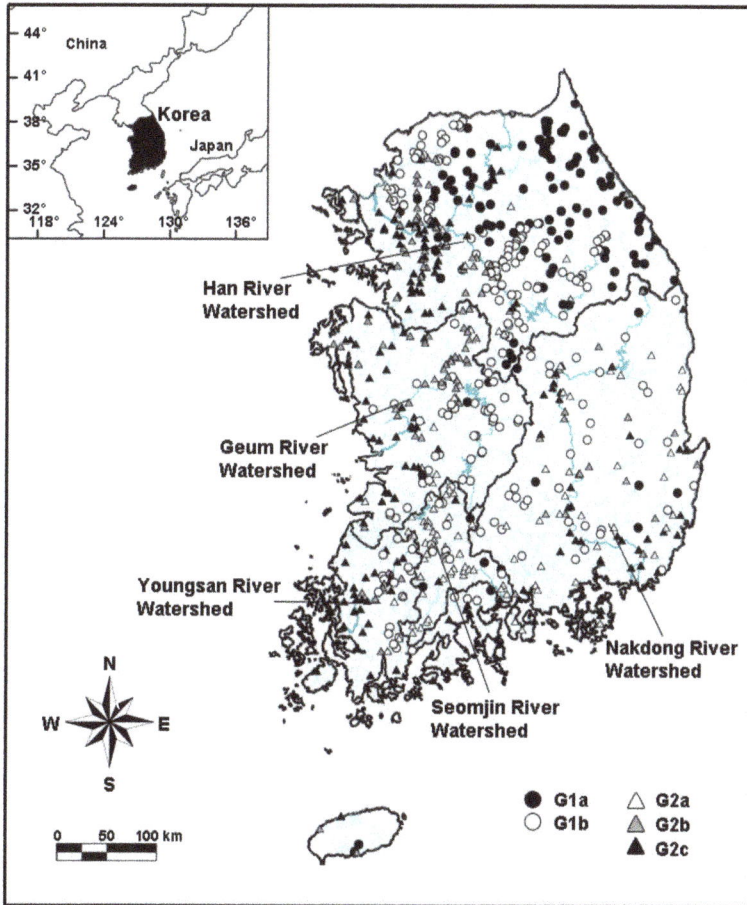

Figure 1. Geographic locations of the study sites. Five macroinvertebrate-based site groups were classified as G1a ($n = 126$, black circles), G1b ($n = 199$, white circles), G2a ($n = 106$, white triangles), G2b ($n = 118$, gray triangles), and G2c ($n = 171$, black triangles) based on Sørenson distance measure cluster analysis. Dark lines indicate the five large rivers and the light lines display their tributaries and small independent streams.

2.2. Measurement of Environmental Variables

Environmental variables were measured both in the field and in the laboratory to define their effects on benthic macroinvertebrate assemblages. Three categories of environmental variables were considered such as watershed-related regional variables, physical in-stream properties, and water quality elements.

Altitude and land-use types were considered as regional variables for watershed characteristics. The altitude at each sampling site was obtained using digital elevation data (Openmate Inc., Seoul, Korea). The proportion of prevalent land use types was determined for each sampling site using a topographic map (1:50,000).

Physical in-stream properties were measured during field surveys: (i) stream width for the distance from bank to bank at a transect representative of the stream channel; (ii) wetted stream width using a range finder (model LRM-1500M, Newcon Optik Inc., Toronto, ON, Canada); (iii) water depth of the vertical distance from the water surface to stream bottom; (iv) current velocity at riffles or gliding runs using a current meter (3000-LX, Swoffer Instruments, Inc., Tukwila, WA, USA), or calculated by the Craig method [26,27]; (v) percent substrate composition visually estimated as fine (<2 mm) and coarse-sized particles ($\geqslant$2 mm).

Water quality variables such as pH, water temperature, dissolved oxygen (DO), turbidity, and electrical conductivity (EC) were measured with a multi-probe portable meter (e.g., YSI 6920, YSI Inc., Yellow Springs, OH, USA or Horiba U-22XD, Kyoto, Japan) at the center of each sampling stretch. Two liters of water samples were collected in sterilized plastic bottles at each site, kept in a container with ice, and transported to the laboratory. Laboratory measurements were conducted to determine the biochemical oxygen demand (BOD), total nitrogen (TN), and total phosphorus (TP) following Standard methods [28].

2.3. Sampling of Benthic Macroinvertebrate

A Surber sampler (30 cm × 30 cm, 1 mm mesh) was employed to collect benthic macroinvertebrates. Samplings were conducted at fast-flowing riffle or gliding run habitats (in the case when suitable riffles were not available) within 100 m. Quantitative samples were taken from three randomly selected riffles at each site, pooled together in a 500 mL plastic bottle with 80% ethanol, and labeled. The sampling device and procedures followed the guidelines of the National Aquatic Ecological Monitoring Program (NAEMP), Korea [27].

After field sampling, all organisms were hand-sorted from detritus and inorganic material, and stored in 70% ethanol. Subsampling was permitted only for dominant taxonomic groups (e.g., Oligochaeta, Ephemerellidae, Chironomidae, and Hydropsychidae) with large numbers of specimens available in each sample. Each macroinvertebrate specimen was identified to the lowest possible taxonomic level (usually genus or species). However, several taxa with limited systematic information (e.g., Coleoptera and Diptera) were identified only to the family level. All individuals were counted and converted to individuals/m^2.

2.4. Data Analysis

Differences in environmental variables and macroinvertebrate assemblages were identified among the five major river watersheds by the Kruskal–Wallis test, followed by Dunn's nonparametric multiple comparison test if they are significantly different ($p < 0.05$). Spearman's rank correlation analysis was used to indicate the relationships between the most dominant taxa and environmental variables. A cluster analysis was conducted to classify the benthic macroinvertebrate communities using the flexible beta method (beta = -0.25) with the Sørenson distance measure [29]. Samples were classified into clusters based on similarities in the community composition. The multi-response permutation procedure (MRPP, [30]), which is a non-parametric method to test for differences in assemblage structure among *a priori* defined groups [31], was conducted to evaluate differences among clusters. Differences in environmental variables among clusters were evaluated using the Kruskal–Wallis test and Dunn's nonparametric multiple comparison test. The Kruskal–Wallis test, Dunn's test and Spearman correlation analysis were performed with STATISTICA software (StatSoft, Inc., version 7, Tulsa, OK, USA).

Indicator species analysis (IndVal, [32]) was performed to evaluate potential indicator species in each cluster defined in advance. The indicator value of a species is the product of its relative abundance and frequency ($\times$ 100), ranging from 0 (no indication) to 100 (perfect indication) [33]. A perfect indicator of a particular group should be faithful and exclusive to that group, never occurring in other groups [29]. A species in a cluster that had an indicator value greater than in any other cluster was defined as good indicator species for that cluster in this study. A Monte Carlo method was performed to test the significance of indicator values.

Canonical correspondence analysis (CCA) was used to relate macroinvertebrate assemblages to environmental variables and to identify which environmental variables could best differentiate among the clusters [34]. Monte-Carlo simulations were carried out to verify whether variables exerted a significant effect ($p < 0.05$) on macroinvertebrate distributions. For the cluster analysis and CCA, those taxa with less than 0.2% of total abundance were excluded to minimize the effects of rare taxa. Taxa abundance data were transformed to log ($x + 1$) in both analyses to down-weight the effects of dominant taxa. Square-root-transformation was used for the environmental parameters expressed in percentages (e.g., type of land use and substrate composition), and the other variables were transformed to log ($x + 1$) except pH. After these processes all environmental variables were rescaled in the range of 0 and 1 based on the minimum-maximum range normalization [25]. Spearman correlation coefficients between scores and environmental variables were calculated to assist interpretation of changes in community profile using STATISTICA software (StatSoft, Inc., version 7, Tulsa, OK, USA). Cluster analysis, IndVal, MRPP and CCA were conducted using PC-ORD software (ver. 4.25, MjM Software Design, Gleneden Beach, OR, USA) [34].

3. Results

3.1. Environmental Characteristics

Despite a large variation, physicochemical factors were significantly different among watersheds (Table 1). On average, the altitude, percent forest, and water velocity of the HRW and SRW were the higher than the other watersheds. The SRW streambeds were very heterogeneous with the highest percentage of coarse-sized particles, whereas those of the YRW contained a large amount of fine sediment. High altitude, fast water velocity and substrate complexity in both the HRW and SRW indicated good in-stream habitat conditions and potential to support high biodiversity and abundance of macroinvertebrates (Table 1a). The other three watersheds (NRW, GRW and YRW) were characterized by low altitude and a high proportion of agricultural land use.

Water quality parameters varied more prominently than physical variables among river watersheds (Table 1b). The average concentrations of DO, BOD, TN, TP, and turbidity were generally lower in the SRW than those in the other watersheds. EC varied widely depending on the sampling location, with the highest values near estuaries. Nutrient-related factors were particularly high in the GRW, YRW, and HRW.

3.2. Benthic Macroinvertebrate Assemblages

A total of 340 taxa, belonging to 113 families in 23 orders of five phyla, were identified in the five major river watersheds during the survey. Most of these were aquatic insects (272 species) including 144 Ephemeroptera Plecoptera Trichoptera (EPT; 62 mayflies, 24 stoneflies and 58 caddisflies) taxa and 35 Dipteran species. Taxa richness from the 720 sampling sites ranged from 0 to 49 with a mean of 14.4 ($\pm$ 9.1) with the highest diversity at HRW and the lowest in YRW (Table 2a). In contrast, homogeneous streambeds and nutrient enrichment attributed to relatively low Shannon diversity index and high dominance index in both the NRW and YRW. The proportion of EPT taxa richness ranged from 40.1% (GRW) to 54.8% (SRW). One-way ANOVA revealed that taxa abundance significantly differed among river watersheds at $p < 0.01$, whereas there were no great differences for the relative abundance of each taxonomic group among watersheds (Table 2b).

126

Table 1. Descriptive statistics of (**a**) regional and physical instream variables and (**b**) chemical variables in the five major river watersheds: the Han River (HRW), Nakdong River (NRW), Geum River (GRW), Youngsan River (YRW) and Seomjin River Watershed (SRW). Top and bottom lines of each variable indicate the average with standard deviation (in parenthesis) and the range, respectively. Kruskal–Wallis test (K–W) was performed with the environmental variables to compare the differences of each variable among river watersheds. The same small letters indicate no significant difference based on Dunn's multiple comparison tests. Abbreviations: DO, dissolved oxygen; BOD, biochemical oxygen demand; EC, electrical conductivity; TN, total nitrogen; TP, total phosphorus.

Variable	HRW (n = 320)	NRW (n = 130)	GRW (n = 130)	YRW (n = 76)	SRW (n = 64)	Total (n = 720)	p (K–W)
		(a) Regional and physical instream variables					
Altitude (m)	147.5 (151.6) [d]	89.6 (119.1) [b,c]	57.7 (61.2) [a,b]	32.5 (35.1) [a]	118.1 (79.3) [c,d]	106.1 (126.2)	0.000
	1.0–721.0	1.0–629.0	0.0–278.0	0.0–211.0	1.0–335.0	0.0–721.0	
% Urban	31.7 (32.0) [b]	23.0 (30.8) [a,b]	30.9 (35.8) [b]	31.4 (30.2) [b]	14.6 (12.4) [a]	28.4 (31.5)	0.000
	0–100	0–100	0–100	0–90	0–80	0–100	
% Agriculture	24.7 (27.5) [a]	42.5 (31.0) [b,c]	46.6 (38.9) [c]	45.4 (30.7) [c]	35.9 (20.7) [b]	35.0 (31.7)	0.000
	0–100	0–100	0–100	0–90	0–80	0–100	
% Forest	35.8 (33.2) [b]	31.4 (27.8) [b]	17.9 (30.1) [a]	22.6 (24.3) [a]	48.8 (24.6) [c]	31.5 (31.3)	0.000
	0–100	0–100	0–100	0–100	0–95	0–100	
Water velocity (cm/s)	53.8 (27.9) [c]	14.7 (14.0) [a]	38.1 (34.5) [b]	17.7 (13.1) [a]	41.9 (26.0) [b]	39.0 (30.4)	0.000
	0.0–140.0	0.0–67.0	0.0–137.7	0.4–47.5	1.4–98.2	0.0–140.0	
% Fine particles	31.1 (27.9) [b]	36.5 (36.8) [b,c]	43.6 (34.5) [c]	54.7 (31.3) [d]	10.8 (22.9) [a]	35.0 (32.6)	0.000
	0.0–100.0	0.0–100.0	0.0–100.0	0.0–100.0	0.0–100.0	0.0–100.0	
% Coarse particles	68.6 (28.1) [c]	63.5 (36.8) [b,c]	56.4 (34.5) [b]	45.3 (31.3) [a]	89.2 (22.9) [d]	64.8 (32.7)	0.000
	0.0–100.0	0.0–100.0	0.0–100.0	0.0–100.0	0.0–100.0	0.0–100.0	

Table 1. *Cont.*

Variable	HRW (n = 320)	NRW (n = 130)	GRW (n = 130)	YRW (n = 76)	SRW (n = 64)	Total (n = 720)	p (K–W)
(b) Chemical variables							
pH	8.1 (0.8) [b,c] 6.5–10.1	8.0 (0.8) [a,b] 6.2–10.6	8.3 (0.9) [c] 7.0–11.1	8.0 (0.7) [a,b] 6.7–10.1	7.8 (0.7) [a] 6.7–9.6	8.1 (0.8) 6.2–11.1	0.003
DO (mg/L)	9.78 (2.47) [a] 2.42–16.10	10.81 (2.36) [b] 2.55–17.34	10.86 (3.36) [b] 2.74–17.86	10.06 (1.69) [a] 6.19–15.40	9.57 (1.32) [a] 7.11–12.69	10.17 (2.54) 2.42–17.86	0.000
BOD (mg/L)	3.1 (3.7) [b,c] 0.3–37.5	1.9 (1.3) [a] 0.4–10.4	3.7 (2.0) [c] 0.8–9.1	3.5 (3.0) [b,c] 0.3–13.3	2.7 (1.7) [b] 0.3–12.3	3.0 (2.9) 0.3–37.5	0.000
EC (µS/cm)	299.8 (337.6) [a] 10.3–2729.0	1358.7 (6067.1) [b] 19.7–44000.0	404.8 (349.8) [a] 86.1–2780.0	270.7 (402.6) [a] 28.5–3082.0	1284.6 (5823.4) [b] 32.6–33360.0	594.4 (3141.6) 10.3–44000.0	0.000
TN (mg/L)	3.15 (2.97) [b] 0.38–23.78	2.10 (1.66) [a] 0.32–11.30	3.60 (2.42) [b] 0.29–14.27	3.36 (3.59) [b] 0.69–27.71	2.32 (1.39) [a] 0.44–5.77	2.99 (2.69) 0.29–27.71	0.000
TP (mg/L)	0.15 (0.41) [b,c] 0.00–5.59	0.07 (0.15) [a,b] 0.00–0.91	0.14 (0.15) [b,c] 0.01–1.01	0.21 (0.28) [c] 0.01–1.66	0.05 (0.06) [a] 0.00–0.32	0.13 (0.30) 0.00–5.59	0.000
Turbidity (NTU)	9.3 (16.6) [b] 0.0–152.0	8.6 (7.4) [b] 0.0–34.2	13.9 (22.3) [b] 0.4–182.4	7.8 (45.9) [b] 0.0–400.0	2.1 (3.3) [a] 0.0–16.9	9.2 (21.2) 0.0–400.0	0.000

Table 2. Assemblage attributes (**a**) and average abundance (individuals·m^{-2}) (**b**) of benthic macroinvertebrates with standard deviation (in parenthesis) in five major river watersheds: the Han River (HRW), Nakdong River (NRW), Geum River (GRW), Youngsan River (YRW), and Seomjin River Watershed (SRW). Relative abundance (RA) indicated as an average of total density for each taxonomic group. Kruskal–Wallis test (K–W) was performed for each taxonomic group. The same small letters indicate no significant difference based on Dunn's multiple comparison test.

Biological Attributes		HRW (n = 320)	NRW (n = 130)	GRW (n = 130)	YRW (n = 76)	SRW (n = 64)	Total (n = 720)	RA	p (K–W)
		(a) Assemblage attributes							
Taxa richness		15.7 (9.4) [a]	11.7 (7.3) [b]	15.5 (10.8) [a]	10.6 (6.3) [b]	15.1 (8.0) [a]	14.4 (9.1)	–	0.000
EPT richness		10.3 (8.0) [c]	5.3 (5.5) [a]	7.6 (8.2) [b]	4.5 (4.7) [a]	8.5 (5.4) [b]	8.1 (7.5)	–	0.000
Taxa abundance		2867.5 (9791.9) [b]	864.9 (1035.2) [a]	3944.4 (5964.7) [b]	947.0 (1054.1) [a]	612.0 (548.2) [a]	2297.1 (7121.0)	–	0.000
EPT abundance		1066.9 (1702.9) [b]	303.2 (726.5) [a]	1552.5 (2441.0) [c]	316.4 (545.0) [a]	278.6 (315.8) [a]	867.4 (1647.2)	–	0.000
Dominance index		0.65 (0.21) [b]	0.70 (0.20) [b,c]	0.71 (0.21) [b,c]	0.73 (0.20) [c]	0.52 (0.23) [a]	0.67 (0.22)	–	0.000
Shannon diversity index		2.33 (1.00) [b]	2.08 (0.91) [b]	2.09 (1.02) [b]	1.79 (0.94) [a]	2.63 (1.09) [c]	2.21 (1.01)	–	0.000
		(b) Taxa abundance of higher taxonomic group							
Non-Insecta	Platyhelminthes	18.7 (78.8)	28.7 (196.7)	25.5 (102.4)	9.9 (49.2)	17.2 (37.8)	20.7 (109.6)	0.01	0.447
	Nematomorpha	0.2 (0.9)	0.1 (0.6)	0.1 (0.5)	0.1 (0.8)	0.0 (0.0)	0.1 (0.7)	0.00	0.082
	Mollusca	10.7 (39.4) [a]	44.4 (151.5) [c]	30.4 (54.3) [b]	27.4 (49.5) [b]	21.2 (30.9) [a,b]	23.0 (76.5)	0.01	0.000
	Annelida	603.5 (6826.2) [b]	39.9 (90.8) [a]	548.3 (2997.6) [b]	21.0 (61.4) [a]	68.1 (91.7) [a]	382.7 (4731.0)	0.20	0.000
	Crustacea	1.9 (12.2) [a]	42.9 (214.3) [a]	41.2 (243.9) [a]	109.8 (560.6) [b]	1.4 (7.0) [a]	27.7 (230.6)	0.01	0.000
Insecta	Ephemeroptera	692.6 (1132.9) [b]	316.6 (594.6) [a]	565.1 (1015.1) [b]	199.0 (308.7) [a]	377.3 (389.0) [a]	521.5 (936.0)	0.27	0.000
	Odonata	3.9 (12.3) [a]	4.2 (12.4) [a]	14.5 (37.1) [b]	11.9 (29.2) [b]	0.7 (2.2) [a]	6.4 (21.3)	0.00	0.000
	Plecoptera	6.6 (21.6) [b]	5.6 (29.6) [b]	4.2 (21.5) [a,b]	0.1 (0.7) [a]	0.7 (4.2) [a]	4.8 (21.3)	0.00	0.000
	Hemiptera	1.1 (9.3) [a]	38.7 (207.4) [b]	25.0 (129.1) [a,b]	25.5 (213.5) [a,b]	0.0 (0.3) [a]	14.7 (125.7)	0.01	0.000
	Megaloptera	2.6 (10.4) [b]	1.2 (6.0) [a,b]	1.6 (6.0) [a,b]	0.5 (2.3) [a]	1.0 (9.8) [a,b]	1.8 (8.4)	0.00	0.000
	Coleoptera	15.3 (97.6) [a]	28.6 (80.6) [a]	80.9 (200.8) [c]	8.9 (31.5) [a]	56.9 (94.9) [b]	32.6 (119.3)	0.02	0.000
	Diptera	449.3 (780.2) [b]	309.8 (452.1) [b]	806.2 (1853.3) [c]	344.1 (712.1) [b]	105.0 (164.4) [a]	446.9 (1009.5)	0.23	0.000
	Trichoptera	549.2 (1029.0) [b]	306.6 (771.8) [a]	845.3 (1564.4) [c]	114.5 (296.2) [a]	126.4 (295.1) [a]	475.4 (1044.8)	0.24	0.000
	Lepidoptera	0.0 (0.1) [a]	–	0.1 (0.9) [b]	0.0 (0.6) [a,b]	–	0.0 (0.4)	0.00	0.000
	Neuroptera	–	–	–	0.0 (0.3)	–	0.0 (0.1)	0.00	0.076

Korean stream ecosystems were characterized by a few predominant colonizers and a majority of rare taxa in macroinvertebrate assemblage composition. The most abundant and widespread taxa were a worm (*Limnodrilus gotoi* Hatai) and midge larvae (Chironomidae spp.) with their relative abundance of approximately 50% of total density throughout the whole river watershed. Other dominant species were mayflies (*i.e.*, *Baetis fuscatus* (Linnaeus), *Epeorus pellucidus* (Brodsky) and *Uracanthella punctisetae* (Matsumura)) and netspinning caddisflies (*Cheumatopsyche brevilineata* Iwata, *Hydropsyche valvata* Martynov and *Hydropsyche kozhantschikovi* Martynov) among the different stream and river systems, most of which were dominant in somewhat nutrient-rich habitats at middle or lower streams. However, over 50% (195 taxa) of the fauna was present with low occurring frequency (less than 2% of all sites) and 90% with low abundance (less than 0.2%).

The environmental relationships of the dominant taxa were stronger with the physical variables (*i.e.*, altitude, water velocity and streambed conditions) than with the chemical variables, among which water velocity was the most significant parameter. Consequently, significant positive relationships existed in most dominant taxa with peak abundance at a moderate velocity of 50–100 cm/s, particularly for *E. pellucidus* ($r = 0.410$, $p < 0.001$) and *C. brevilineata* ($r = 0.435$, $p < 0.001$) (Figure 2). *Baetiella tuberculata* (Kazlauskas), *U. punctisetae* and *H. valvata* tended to occur in their highest densities at the fast velocity (120–140 cm/s). No clear tendency was observed for Chironomidae spp.

3.3. Macroinvertebrate-Based Site Classification

The cluster analysis, based on the similarity in the benthic macroinvertebrate composition, largely classified the 720 sampling sites into two clusters and subsequently sub-clustered them into five groups. As a result, 126 sampling sites were included in Group 1a; 199 sites in Groups 1b; 106 sites in Group 2a; 118 sites in Group 2b; and 171 sites in Group 2c (Figure 1). MRPP validated these five groups with significant differences ($A = 0.484$, $p < 0.001$). The differences in environmental variables among the clusters are shown in Figure 3. The most important indicator taxa for each cluster are shown with their indicator values in Table 3.

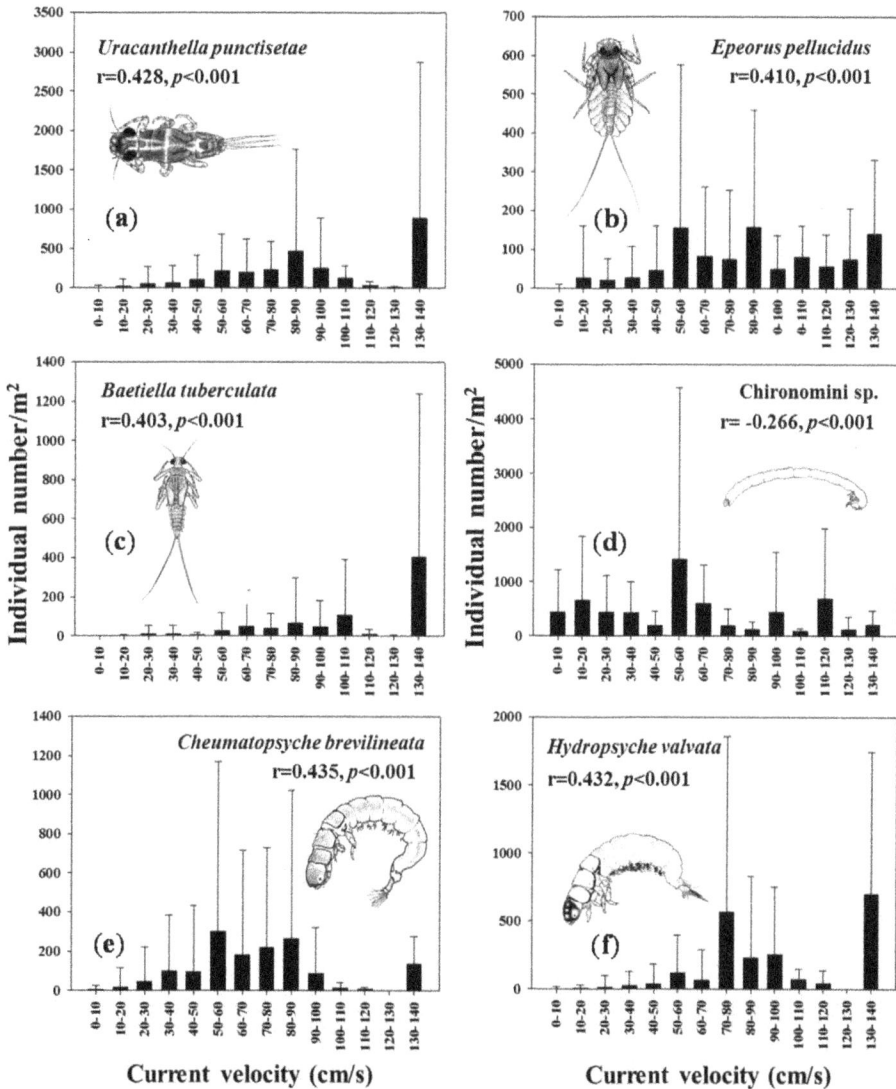

Figure 2. Distribution patterns of dominant macroinvertebrates along with water velocity. Data are given as arithmetic means with standard deviation. Spearman's rank correlation coefficients between water velocity and abundance of each dominant species are included. (**a**) *Uracanthella punctisetae*; (**b**) *Epeorus pellucidus*; (**c**) *Baetiella tuberculata*; (**d**) Chironomini sp.; (**e**) *Cheumatopsyche brevilineata*; (**f**) *Hydropsyche valvata*.

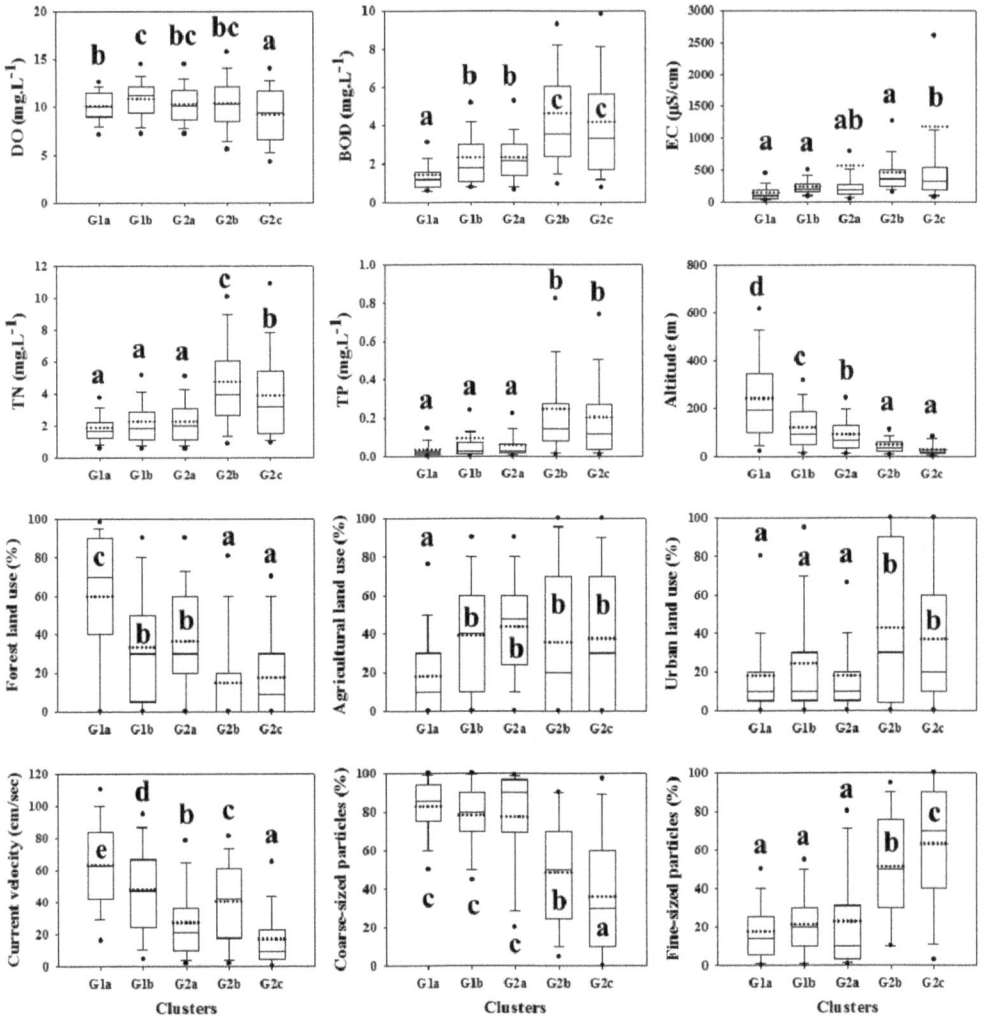

Figure 3. Distribution of selected environmental variables among the five clusters of the 720 sampling sites which were identified by cluster analysis with Sørenson distance measure. Box represents the 25th and 75th percentiles and whiskers indicate the 5th and 95th percentiles with standard deviations (error bar). The mean (horizontal dotted line) and median (horizontal solid line) are shown in each box. Different small letters indicate significant difference based on a Dunn's multiple comparison test at $p < 0.05$.

132

Table 3. Indicator values (%) for the most important species ($p < 0.05$) in each cluster group. Monte Carlo tests (999 permutations) were used to assess the significance of each species as an indicator for the respective group (G1a–G2c). In total 50 species whose contribution to total density was higher than 0.2% were analyzed. Less important species were not shown in this table.

Taxa	Cluster Group					p
	G1a	G1b	G2a	G2b	G2c	
Rhyacophila nigrocephala Iwata	47	16	0	0	0	0.001
Epeorus nipponicus (Uéno)	46	1	0	0	0	0.001
Glossosoma KUa	44	4	0	0	0	0.001
Drunella aculea (Allen)	44	1	0	0	0	0.001
Hydropsyche orientalis Martynov	40	7	1	1	0	0.001
Uracanthella punctisetae (Matsumura)	21	48	6	6	0	0.001
Hydropsyche valvata Martynov	6	46	2	9	0	0.001
Cheumatopsyche brevilineata Iwata	15	41	5	10	0	0.001
Hydropsyche kozhantschikovi Martynov	21	40	3	16	0	0.001
Psychomyia sp.	0	32	1	5	0	0.001
Ephemera orientalis McLachlan	2	20	41	5	2	0.001
Ecdyonurus levis (Navás)	5	28	40	1	0	0.001
Ecdyonurus joernensis Bengtsson	0	12	25	1	0	0.001
Mataeopsephus KUa	0	16	23	0	0	0.001
Asellus sp.	0	2	11	3	6	0.001
Hirudo nipponia Whitman	1	10	2	28	4	0.001
Chironomini sp.	1	7	9	27	21	0.001
Limnodrilus gotoi Hatai	5	12	18	23	9	0.001
Micronecta sedula Horváth	0	0	0	0	20	0.001
Physa acuta Draparnaud	2	3	2	12	16	0.001
Micronecta sp.	0	0	5	0	5	0.007
Total number of significant indicator species	12	21	8	3	3	-

Each cluster was clearly differentiated according to the differences of instream physicochemical conditions and geographical location of sampling sites. The cluster analysis discriminated less polluted streams with low EC and fast flowing water (Group 1) from moderately or severely polluted steams with high EC and slow water velocity (Group 2). Group 1a (G1a), congregating in the HRW, consisted specifically of mountainous upper streams although a tenth of this group was scattered over the other watersheds except the YRW. This group consisted of oligotrophic streams with distinguishing features of the highest altitude, the lowest BOD and nutrient concentrations, and the fastest water velocity. Additionally, the catchment was predominantly comprised of forested area. The best indicator species for this group were characterized as high sensitivity against organic pollution (e.g., *Rhyacophila nigrocephala* Iwata, *Epeorus nipponicus* (Uéno), *Drunella aculea* (Allen) and *Hydropsyche orientalis* Martynov). Group 1b (G1b) contained the largest number of sampling sites and was widely distributed throughout all river watersheds but mostly encompassing the agricultural and forested catchment. G1a and G1b closely resembled chemical environments, but steam sites belonging to G1b displayed mesotrophic condition with slightly higher BOD and nutrients than those of G1a. There existed the largest number of indicator species in G1b, among which *U. punctisetae* showed the highest indicator value.

G2a mostly included sites located in middle reaches of large rivers and their tributaries, particularly in the NRW and SRW. Although Group 2a (G2a) was similar in overall environmental characteristics to G1b, G2a was characterized with slightly lower altitude, higher EC and a lower water velocity when compared with G1b. Sampling sites suffering from poor water quality with organic degradation and nutrient enrichment were confined to Group 2b with the highest BOD, TN, and TP concentrations. These sites were influenced by the highest degree of agriculture and/or urbanization. The significant indicator species of this group were *Hirudo nipponia* Whitman, Chironomini sp. and *L. gotoi*, indicating high trophy and saprobity [35]. Water quality conditions in Group 2c (G2c) were as poor as those of G2b with the highest EC. Sites in this group were characterized with the lowest water velocity and the largest proportion of fine particles because G2c was gathered by streams adjacent to estuaries, large rivers, and dammed streams. The only three weak indicator species appeared in G2c.

3.4. Environmental Variables Affecting Macroinvertebrate Distributions

A CCA was performed to understand how macroinvertebrates were distributed along environmental gradients (Figure 4). Total variability explained in the species data was 15.9% (Table 4). The eigenvalues of the first CCA axis (0.281) and the second CCA axis (0.101) were significant ($p < 0.01$; 99 Monte Carlo permutation test). All three CCA results were significant based on a Monte Carlo permutation test ($p < 0.01$).

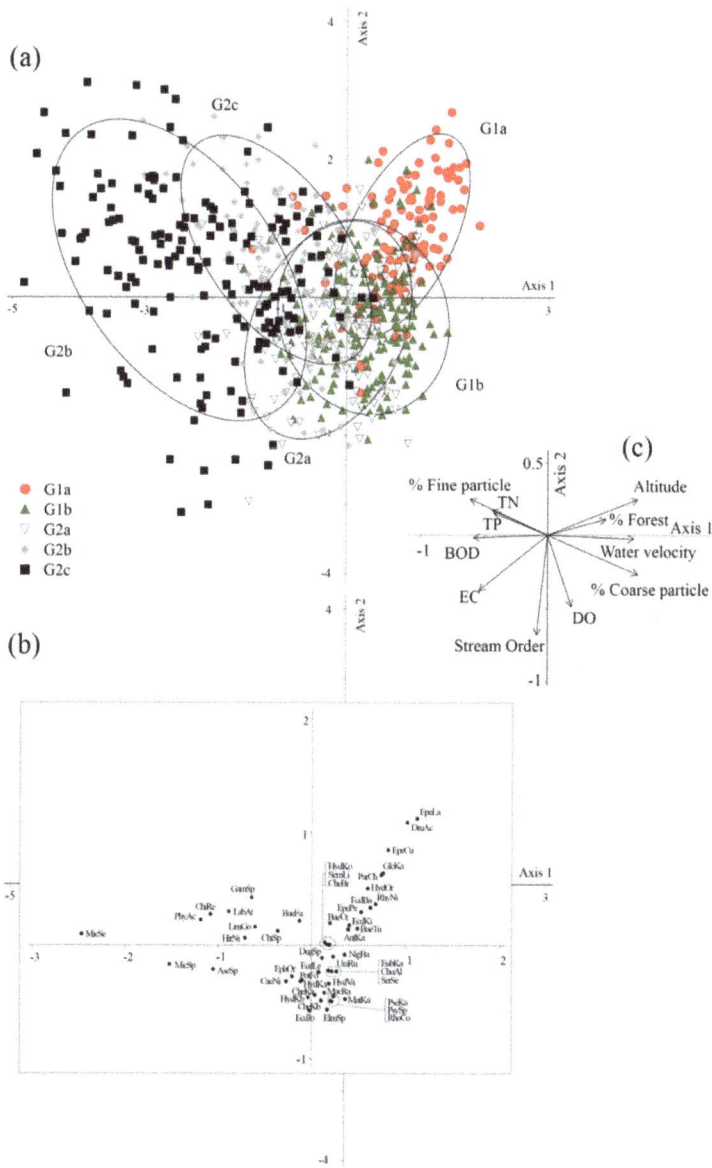

Figure 4. Ordination plots constructed by canonical correspondence analysis for all sampling sites ($n = 720$). The plots present the ordination of sampling sites (**a**) and macroinvertebrates (**b**) with relative contributions of environmental variables (**c**) for the first two axes. Only species contributing >0.2% to total abundance were included. The summary of ordination results is presented in Table 4 and the codes for each taxon are shown in Appendix Table A1.

Table 4. Spearman's rank correlation coefficients and probabilities of environmental variables and the CCA axes (** $p < 0.01$, * $p < 0.05$) (a), and summary of CCA results ($n = 720$) (b). All axes were significant based on Monte Carlo permutation procedures.

Variables	Axis 1	Axis 2	Axis 3
(a) Correlation coefficients			
Altitude	0.793 **	0.236 **	0.013
Stream order	−0.105 **	−0.658 **	0.099 **
% Urban	−0.218 **	0.134 **	−0.317 **
% Agriculture	−0.199 **	−0.273 **	0.133 **
% Forest	0.504 **	0.055	0.266 **
Water velocity	0.662 **	−0.054	−0.510 **
% Fine particles	−0.744 **	0.325 **	−0.076 *
% Coarse particles	0.747 **	−0.327 **	0.075 *
pH	0.011	−0.236 **	0.007
DO	0.304 **	−0.467 **	−0.166 **
BOD	−0.550 **	0.164 **	−0.370 **
EC	−0.196 **	−0.278 **	−0.116 **
TN	−0.463 **	0.266 **	−0.387 **
TP	−0.373 **	0.229 **	−0.314 **
(b) Summary of CCA results			
Eigenvalue	0.281	0.101	0.073
% variance explained in taxa data	9.8	3.5	2.5
Cumulative % variance explained	9.8	13.3	15.9
p value	0.010	0.010	0.010
Total variance	2.869	-	-

The five clusters were well separated in the CCA ordination plot (Figure 4a). Both the G1a and G1b groups with good physicochemical environments were positioned on the right side of the ordination plot, and G2b and G2c were on the opposite side. However, the G2a group straddled both sides. Eight environmental variables (*i.e.*, BOD, TN, TP, altitude, water velocity, % forest, % fine particles, and % coarse particles) had significant correlations with the first axis, among which altitude was the most significant contributor ($r = 0.793$, $p < 0.01$), followed by % coarse particles ($r = 0.747$, $p < 0.01$) and % fine particles ($r = −0.744$, $p < 0.01$) (Table 4). All physical factors except for % fine particles were in the opposite direction from the chemical variables (Figure 4c). This result suggests that the decline in altitude reflected deterioration in water quality, accompanied by increases in organic material and nutrients. The heterogeneity of the macroinvertebrate habitats also decreased with the longitudinal gradient toward downstream. The second axis was negatively related with stream order ($r = −0.658$, $p < 0.01$) and DO ($r = −0.467$, $p < 0.01$). High positive scores with the first axis were denoted for the rhithronic (*i.e.*, pertaining to the headwaters) and intolerant species (e.g., *Drunella aculea* (Allen), *E. nipponicus*,

Glossosoma sp., *R. nigrocephala* Iwata and *H. orientalis*), whereas negative scores were observed for the potamic (*i.e.*, pertaining to rivers) and tolerant taxa (e.g., *Micronecta sedula* Horváth, *Physa acuta* Draparnaud, *Labiobaetis atrebatinus* (Eaton) and *Asellus hilgendorfii* Bovalius) (Figure 4b).

4. Discussion

The Asian monsoon region is a global biodiversity hotspot suffering from increasing anthropogenic disturbances, but aquatic biodiversity and ecosystem integrity remain poorly explored [3]. This is the same situation in Korea, but recent establishment of the National Aquatic Ecological Monitoring Program [36] opened a new era to assess ecosystem health and biodiversity in Korea. Our work presented in this paper takes advantage of the opportunity of such a nationwide scale of survey. Although the five river watersheds in Korea exhibited differences in environmental conditions and macroinvertebrate taxa abundance, overall taxonomic composition at each watershed displayed little difference (Table 2). Instead, a considerable spatial variation in benthic macroinvertebrate assemblages accounted for the combination of both ultimate (e.g., altitude and the degree of land use) and proximate factors (e.g., flow, stream bed substrate, BOD, EC, and nutrients). This finding was in line with previous studies of benthic diatoms in Korea [25] and macroinvertebrates in other Asian countries [12,37], suggesting the importance of various multi-scale factors in structuring macroinvertebrate assemblages [38]. Our results provide basal information for the sustainable management and conservation practices of stream ecosystems.

4.1. Macroinvertebrate Taxonomic Composition

The macroinvertebrate assemblages in Korean stream ecosystems generally bear resemblance to those in tropical and other temperate streams at higher taxonomic levels [35,39]. However, temperate Korean streams were relatively rich in Ephemeroptera, Diptera, and Trichoptera when compared with tropical streams for the higher biodiversity of Gastropoda, Decapoda and other insect orders such as Odonata and Hemiptera [39,40] (Table 2). The taxonomic composition also included a large number of rhithronic fauna that prefer stony substrates. For example, Baetidae, Heptageniidae, and Ephemerellidae mainly dominated the Ephemeroptera, and Rhyacophilidae and Hydropsychidae composed the Trichoptera in this study.

We confirmed a total of 340 macroinvertebrate taxa in this study. Of the macroinvertebrate taxa, chironomid midge larvae and a small minnow mayfly (*B. fuscatus*) were extensively encountered throughout Korean stream environments with the highest occurring frequency with 94% and 76% of total sampling sites, respectively. Temperate Korean streams were also characterized with a few predominant colonizers and a majority of rare taxa in macroinvertebrate taxonomic composition. Moreover, only six cosmopolitan taxa were noticeably comprised of >60% of all macroinvertebrate samples, which were *L. gotoi*, Chironomidae spp., *U. punctisetae*, *C. brevilineata*, *H. valvata* and *H. kozhantschikovi*. These dominant taxa were found to be significant indicators of mesotrophic or polytrophic streams in our study (Table 3), consistent with a result that most Korean streams and rivers were degraded in both chemical and biological status [36]. On the other hand, most Korean streams were occupied by a great number of rare macroinvertebrate species with a small distribution range and/or low abundance, as was also demonstrated in other studies [41,42].

4.2. Environmental Relationships with Macroinvertebrate Distribution

Benthic habitats are complex, and a variety of environmental variables acting at multiple spatial scales regulate the composition and distribution patterns of stream macroinvertebrate assemblages in an exclusive or synergistic fashion [38,43]. We revealed that the variables associated with altitude and in-stream habitats best accounted for the largest amount of variability in our macroinvertebrate data set, supporting the more important determinants of local environmental factors than broad or regional parameters [16,44]. The importance and role of local environmental variables have also been highlighted in other aquatic communities: aquatic macrophytes [45], freshwater phytoplankton [46], benthic diatoms [44], intertidal macroinvertebrates [47], and fish [48]. However, a comprehensive understanding of multispatial scales is needed because of significant correlations between macrohabitat and microhabitat characteristics, depending on the relative size of the area studied [3].

In our study, the variability among the macroinvertebrate-based stream groups was more prominently explained by the altitudinal gradients together with streambed composition and water velocity than chemical variables (Table 3). First, the most widely accepted theory related to altitudinal changes is the river continuum concept (RCC), which displays structural and functional responses to the longitudinal gradient [49]. Altitude has also been well documented in other studies as a main descriptor to determine macroinvertebrate richness as well as other environmental variables such as temperature, hydrology, food availability, streambed condition, and water chemistry [3,50]. However, we revealed poor relationships of altitude with water chemistry in contrast with significant associations with physical variables and biological attributes such as taxa richness, EPT richness, EPT abundance, and the Shannon diversity index. This may be due to the fact that over half of the studied streams corresponded to lowland streams below 100 m a.s.l., which were characterized by moderate or slightly poor water quality with severe variations in BOD, TN, TP, and Shannon diversity-based saprobity [36,51] (Table 1). Harding *et al.* [52] demonstrated that increasing human land use intensity along a river continuum caused water quality degradation, consequently leading to changes in taxonomic composition from intolerant EPT dominated to tolerant taxa-dominated assemblages despite a gradual increase in taxa abundance by a few dominant species.

Riffle habitats are commonly characterized by shallow water depth, oxygenating fast-flowing water, and stony beds, and exhibit higher taxa richness and abundance than that at pools or habitats with fine sediment, as in our case [3,42,53]. Such hydraulic conditions are critical determinants for the distribution and species composition of benthic organisms [16,54] and are the driving forces for evolution of their morphologies and life history [11]. Rheophilous (*i.e.*, having an affinity for running waters) and limnophilous taxa were clearly discriminated based on the correlation analysis (Figure 2) and CCA (Figure 4) results. For example, well-known rhithronic species in two Ephemeropteran (Ephemerellidae and Heptageniidae) and one Trichopteran (Hydropsychidae) families mostly displayed high preference to the mid- or fast current conditions, which corresponds to the intolerant scraper or filtering collector groups abundant in upper and middle stream reaches [49]. Thus, biological monitoring and assessment programs using rheophilic macroinvertebrates would benefit from their ecological characteristics and their indication of good environmental quality (e.g., [42]) although seasonal fluctuations in hydraulic parameters by the Asian monsoon more critically cause a catastrophic drift and washout of benthic organisms.

Streambed composition is one of the most important factors to directly influence richness and abundance of macroinvertebrates on local scales [12,16]; thus, close correlations would be expected with variables related to longitudinal changes in stream ecosystems. Boulders and cobbles are typically the major structural elements in steep gradient upper streams, whereas sand and smaller sediments predominate in the lower reaches [55]. There have been a large number of studies on the macroinvertebrate-substrate relationship, most of which have revealed that macroinvertebrate diversity and density increase with higher heterogeneity due to the available stable and diverse microhabitats (e.g., [10,43]). Therefore, factors determining macroinvertebrate communities at local scales obviously put a priority on the streambed conditions rather than water quality or other physical variables. We found that the mainstreams of the Nakdong River, which contained about 60% fine particles, retained the lowest taxa richness (8.2 on average, $n = 18$) and abundance (747 individuals/m^2) among five major rivers despite its good water quality condition (mean BOD, 1.9 and TN, 1.8 mg/L) with the exception of the severely polluted Youngsan River (8.9 taxa richness and 7.3 BOD, $n = 13$). These results indicate that homogeneous streambeds with greater fine particles support lower diversity and abundance even when streams maintain good water quality [56]. Additionally, organic pollution could transform the coarse substrate into an organic rich soft bottom, altering the community structure from dominated by diverse and intolerant species to communities predominated by a few tolerant species [3,20], as in the case of the Youngsan River.

4.3. Considerations to Improve Macroinvertebrate Biomonitoring Programs

Not until 2006 did Korea adopt biological water quality criteria and the concept of ecological integrity in the water quality program [25,36]. Since then, Korean government has led a biological survey of streams and rivers (*i.e.*, NAEMP) every year to assess the current biological status of stream and river ecosystems, and to develop a strategy for the restoration and management of disturbed systems [25]. As a part of a nationwide survey benthic macroinvertebrates are also monitored based on the guidelines and assessment tools [27]. Notwithstanding their suitability for convenient and rapid bio-assessment, there remain debatable issues as to whether the methods for sampling and treating macroinvertebrates are effective to provide a reliable and accurate indication of the macroinvertebrate fauna throughout the country.

The Korean national biomonitoring program presents field sampling in a cost-effective and time-saving manner for rapid bio-assessment. To satisfy this purpose benthic macroinvertebrates are optimized to be collected at riffle and/or gliding run habitats using a Surber sampler with three replicates [27]. However, such sampling methods probably underestimate overall biodiversity in stream ecosystems, considering the whole stream environment. First, single-habitat sampling possibly produces incomplete taxa lists and includes no target habitats at certain sites despite its advantage that the influences of both water and habitat quality on macroinvertebrates are not confounded by instream habitat variation [57]. In this regard, a multihabitat approach would be more profitable for the estimation of taxonomic diversity due to its consistent application across stream types especially at large-scaled survey, comprehensive taxa lists, and effective assessment of ecological conditions [58]. Second, the Surber sampler is one of the most commonly used quantitative tools in lotic systems and provides high-precision information on the abundance and composition of macroinvertebrate assemblages [21,59]. This method is, on the other hand, usually more appropriate for riffle habitats of shallow streams, presumably underestimating overall biodiversity in a region. Recent studies on the comparison between sampling devices suggested that artificial substrates (e.g., leaf-bags) would be used as complementary tools due to their discriminative taxonomic composition [60–62]. Third, a majority of biomonitoring programs widely adopt three to five replicates for lotic ecosystems because one way to reduce monitoring costs is to decrease the sample size [21,63,64]. The previous studies also showed that such a small number of replicates rarely influenced ecological health assessment, particularly for the indices applying sensitivity/tolerance taxa, as the same for NAEMP [65–67]. Nevertheless, small sample sizes may influence the values of biological measures and the representativeness for a real benthic community based on the asymptotic relationship of the number of taxa with both the sampling area and the collected individuals [68]. Finally, sieve mesh size could also affect the accurate estimates of taxonomic diversity and environmental quality assessment [67]. Finer mesh sizes more accurately represent macroinvertebrate assemblages than coarser mesh sizes, whereas they need more efforts to handle specimens. On the contrary, coarser mesh sizes undervalue the original density and taxa richness of macroinvertebrate assemblages by passing small individuals through the sieves and result in higher diversity [69]. Considering the tradeoff between cost and effectiveness, many rapid bio-assessments determine 0.5 mm mesh size as a reasonable choice [21,64,70]. Therefore, future researches are required for identifying optimal sampling effort comprehensively considering cost-effectiveness, easy applicability, and well representative of resident macroinvertebrate assemblages.

5. Conclusions

Large-scale knowledge on environmental relationships of aquatic communities is crucial to conserve freshwater biodiversity and sustain ecological integrity. Korean stream macroinvertebrate assemblages were determined not only by regional and physical instream variables but also by pollution-related parameters at nation-wide scale. Macroinvertebrate-based site classification evidently provided an environmental characterization for different types of Korean streams, indicating that benthic macroinvertebrates are a valuable biomonitoring material. The results of this study provide important information and a bridge for further work such as the assemblage—specific responses to environmental disturbance. Our results also contribute to establishing effective management practices, implementing conservation measures, and developing more reliable biological monitoring tools for sustainable freshwater ecosystems.

Acknowledgments: This study was performed under the project of "National Aquatic Ecosystem Health Survey and Assessment" in Korea, and was supported by the Ministry of Environment and the National Institute of Environmental Research, Korea. The authors are grateful to survey members involved in the project. The authors also thank the anonymous reviewers for their help in improving the scientific content of the manuscript.

Author Contributions: Yung-Chul Jun developed the concept of the study, performed data analysis, and wrote the manuscript. Nan-Young Kim assisted data processing and figure construction. Sang-Hun Kim conducted quality assurance of the physico-chemical and biological data. Young-Seuk Park and Dong-Soo Kong assisted study design and statistical analysis. Soon-Jin Hwang supervised the research, and assisted data interpretation and manuscript preparation. All the authors contributed to the review of the manuscript.

Conflicts of Interest: The authors declare no conflict of interest.

Appendix A

Table A1. Codes for macroinvertebrates contributing to more than 0.2% of total abundance for canonical correspondence analysis in Figure 4.

Code	Taxon
	Phylum Platyhelminthes
DugSp	*Dugesia* sp.
	Phylum Mollusca
SemLi	*Semisulcospira libertina* (Gould)
PhyAc	*Physa acuta* Draparnaud
	Phylum Annelida
LimGo	*Limnodrilus gotoi* Hatai
HirNi	*Hirudo nipponia* Whitman
	Phylum Arthropoda
	Class Crustacea
AseSp	*Asellus* sp.
GamSp	*Gammarus* sp.
	Class Insecta
	Order Ephemeroptera
BaeTu	*Baetiella tuberculata* (Kazlauskas)
BaeFu	*Baetis fuscatus* (Linnaeus)
BaeUr	*Baetis ursinus* Kazlauskas
LabAt	*Labiobaetis atrebatinus* (Eaton)
NigBa	*Nigrobaetis bacillus* (Kluge)
EcdBa	*Ecdyonurus bajkovae* Kluge
EcdJo	*Ecdyonurus joernensis* Bengtsson
EcdKi	*Ecdyonurus kibunensis* Imanishi
EcdLe	*Ecdyonurus levis* (Navás)
EpeNi	*Epeorus nipponicus* (Uéno)
EpeLa	*Epeorus latifolium* (Uéno)
EpePe	*Epeorus pellucidus* (Brodsky)
ChoAl	*Choroterpes altioculus* Kluge
ParJa	*Paraleptophlebia japonica* (Matsumura)
PotFo	*Potamanthus formosus* Eaton
RhoCo	*Rhoenanthus coreanus* (Yoon and Bae)
DruAc	*Drunella aculea* (Allen)
EphOr	*Ephemera orientalis* McLachlan
SerSe	*Serratella setigera* (Bajkova)
UraPu	*Uracanthella punctisetae* (Matsumura)
CaeNi	*Caenis nishinoae* Malzacher
	Order Hemiptera
MicSe	*Micronecta sedula* Horváth
MicSp	*Micronecta* sp.
	Order Coleoptera
ElmSp	Elmidae sp.
EubKa	*Eubrianax* KUa
MatKa	*Mataeopsephus* KUa
PseKa	*Psephenoides* KUa

Table A1. *Cont.*

Code	Taxon
	Order Diptera
AntKa	*Antocha* KUa
CulSp	*Culex* sp.
ChiSp	Chironomidae spp. (non-red type)
ChiRe	Chironomini spp. (red-type)
	Order Trichoptera
RhyNi	*Rhyacophila nigrocephala* Iwata
HydKa	*Hydroptila* KUa
GloKa	*Glossosoma* KUa
CheBr	*Cheumatopsyche brevilineata* Iwata
CheKa	*Cheumatopsyche* KUa
CheKb	*Cheumatopsyche* KUb
HydKo	*Hydropsyche kozhantschikovi* Martynov
HydKb	*Hydropsyche* KUb
HydOr	*Hydropsyche orientalis* Martynov
HydVa	*Hydropsyche valvata* Martynov
MacRa	*Macrostemum radiatum* McLachlan
PsySp	*Psychomyia* sp.

References

1. Dudgeon, D.; Arthington, A.H.; Gessner, M.O.; Kawabata, Z.I.; Knowler, D.J.; Lévêque, C.; Naiman, R.J.; Prieur-Richard, A.H.; Soto, D.; Stiassny, M.L.J.; *et al.* Freshwater biodiversity: Importance, threats, status and conservation challenges. *Biol. Rev.* **2006**, *81*, 163–182.
2. Strayer, D.L.; Dudgeon, D. Freshwater biodiversity conservation: Recent progress and future challenges. *J. N. Am. Benthol. Soc.* **2010**, *29*, 344–358.
3. Li, F.; Chung, N.; Bae, M.-J.; Kwon, Y.-S.; Park, Y.-S. Relationships between stream macroinvertebrates and environmental variables at multiple spatial scales. *Freshwater Biol.* **2012**, *57*, 2107–2124.
4. Dudgeon, D. The ecology of tropical Asian rivers and streams in relation to biodiversity conservation. *Annu. Rev. Ecol. Syst.* **2000**, *31*, 239–263.
5. De Silva, S.S.; Abery, N.W.; Nguyen, T.T.T. Endemic freshwater finfish of Asia: Distribution and conservation status. *Divers. Distrib.* **2007**, *13*, 172–184.
6. Vörösmarty, C.J.; McIntyre, P.B.; Gessner, M.O.; Dudgeon, D.; Prusevich, A.; Green, P.; Glidden, S.; Bunn, S.E.; Sullivan, C.A.; Liermann, C.E.; *et al.* Global threats to human water security and river biodiversity. *Nature* **2010**, *467*, 555–561.
7. Strayer, D.L. Challenges for freshwater invertebrate conservation. *J. N. Am. Benthol. Soc.* **2006**, *25*, 271–287.
8. Buss, D.F.; Baptista, D.F.; Silveira, M.P.; Nessimian, J.L.; Dorvillé, L.F.M. Influence of water chemistry and environmental degradation on macroinvertebrate assemblages in a river basin in south-east Brazil. *Hydrobiologia* **2002**, *481*, 125–136.

9. Bott, T.L.; Brock, J.T.; Dunn, C.S.; Naiman, R.J.; Ovink, R.W.; Petersen, R.C. Benthic community metabolism in four temperate stream systems: An inter-biome comparison and evaluation of the river continuum concept. *Hydrobiologia* **1985**, *220*, 109–117.

10. Merz, J.R.; Ochikubo Chan, L.K. Effects of gravel augmentation on macroinvertebrate assemblages in a regulated California river. *River Res. Appl.* **2005**, *21*, 61–74.

11. Nelson, S.M.; Lieberman, D.M. The influence of flow and other environmental factors on benthic invertebrates in the Sacramento River, USA. *Hydrobiologia* **2002**, *489*, 117–129.

12. Jiang, X.M.; Xiong, J.; Qiu, J.W.; Wu, J.M.; Wang, J.W.; Xie, Z.C. Structure of macroinvertebrate communities in relation to environmental variables in a subtropical Asian river system. *Int. Rev. Hydrobiol.* **2010**, *95*, 42–57.

13. Vought, L.B.M.; Kullberg, A.; Petersen, R.C. Effect of riparian structure, temperature and channel morphometry on detritus processing in channelized and natural woodland streams in southern Sweden. *Aquat. Conserv. Mar. Freshw. Ecosyst.* **1998**, *8*, 273–285.

14. Townsend, C.R.; Hildrew, A.G.; Francis, J. Community structure in some southern English streams: The influence of physicochemical factors. *Freshw. Biol.* **1983**, *13*, 521–544.

15. Richards, C.; Haro, R.J.; Johnson, L.B.; Host, G.E. Catchment and reach-scale properties as indicator of macroinvertebrate species traits. *Freshw. Biol.* **1997**, *37*, 219–230.

16. Sandin, L. Benthic macroinvertebrates in Swedish streams: Community structure, taxon richness, and environmental relations. *Ecography* **2003**, *26*, 269–282.

17. Leps, M.; Tonkin, J.D.; Dahm, V.; Haase, P.; Sundermann, A. Disentangling environmental drivers of benthic invertebrate assemblages: The role of spatial scale and riverscape heterogeneity in a multiple stressor environment. *Sci. Total Environ.* **2015**, *536*, 546–556.

18. Heino, J. Biodiversity of aquatic insects: Spatial gradients and environmental correlates of assemblage-level measures at large scales. *Freshw. Rev.* **2009**, *2*, 1–29.

19. Shah, D.N.; Tonkin, J.D.; Haase, P.; Jähnig, S.C. Latitudinal patterns and large-scale environmental determinants of stream insect richness across Europe. *Limnologica* **2015**, *55*, 33–43.

20. Rossaro, B.; Pietrangelo, A. Macroinvertebrate distribution in streams: A comparison of CA ordination with biotic indices. *Hydrobiologia* **1993**, *263*, 109–118.

21. Rosenberg, D.M.; Resh, V.H. *Freshwater Biomonitoring and Benthic Macroinvertebrates*; Chapman and Hall: New York, NY, USA, 1993.

22. Charvet, S.; Statzner, B.; Usseglio-Polatera, P.; Dumont, B. Traits of benthic macroinvertebrates in semi-natural French streams: An initial application to biomonitoring in Europe. *Freshw. Biol.* **2000**, *43*, 277–296.

23. Korea Meteorological Administration (KMA). *The Meteorological Yearbook in Korea*; KMA: Seoul, Korea, 2010. (in Korean)

24. Jeong, K.S.; Hong, D.G.; Byeon, M.S.; Jeong, J.C.; Kim, H.G.; Kim, D.K.; Joo, G.J. Stream modification patterns in a river basin: Field survey and self-organizing map (SOM) application. *Ecol. Inform.* **2010**, *5*, 293–303.

25. Hwang, S.-J.; Kim, J.-Y.; Yoon, S.-A.; Kim, B.-H.; Park, M.-H.; You, K.-A.; Lee, H.-Y.; Kim, H.-S.; Kim, Y.-J.; Lee, J.; *et al.* Distribution of benthic diatoms in Korean rivers and streams in relation to environmental variables. *Ann. Limnol. Int. J. Limnol.* **2011**, *47*, S15–S33.

26. Craig, D.A. Some of what you should know about water. *Bull. N. Am. Benthol. Soc.* **1987**, *4*, 178–182.

27. The Ministry of Environment/National Institute of Environmental Research (MOE/NIER). *Survey and Evaluation of Aquatic Ecosystem Health in Korea*; MOE/NIER: Incheon, Korea, 2012.

28. American Public Health Association (APHA). *Standard Methods for the Examination of Water and Wastewater*, 21st ed.; APHA: Washington, DC, USA, 2001.

29. McCune, B.; Grace, J.B. *Analysis of Ecological Communities*; MjM Software Design: Gleneden Beach, OR, USA, 2002.

30. Mielke, P.W.; Berry, K.J.; Johnson, E.S. Multiresponse permutation procedures for a priori classifications. *Commun. Stat.* **1976**, *A5*, 1409–1424.

31. Zimmerman, G.M.; Goetz, H.; Mielke, P.W. Use of an improved statistical method for group comparisons to study effects of prairie fire. *Ecology* **1985**, *66*, 606–611.

32. Dufrene, M.; Legendre, P. Species assemblages and indicator species: The need for flexible asymmetrical approach. *Ecol. Monogr.* **1997**, *67*, 345–366.

33. Petersen, W.T.; Keister, J.E. Interannual variability in copepod community composition at a coastal station in the northern California Current: A multivariate approach. *Deep Sea Res.* **2003**, *50*, 2499–2517.

34. McCune, B.; Mefford, M.J. *Multivariate Analysis of Ecological Data (Version 4.25)*; MjM Software: Gleneden Beach, OR, USA, 1999.

35. Wang, X.; Cai, Q.; Tang, T.; Yang, S.; Li, F. Spatial distribution of benthic macroinvertebrates in the Erhai basin of southwestern China. *J. Freshw. Ecol.* **2012**, *27*, 89–96.

36. Lee, S.-W.; Hwang, S.-J.; Lee, J.-K.; Jung, D.-I.; Park, Y.-J.; Kim, J.-T. Overview and application of the National Aquatic Ecological Monitoring Program (NAEMP) in Korea. *Ann. Limnol. Int. J. Limnol.* **2011**, *47*, S3–S14.

37. Hoang, T.H.; Lock, K.; Chi Dang, K.; de Pauw, N.; Goethals, P.L.M. Spatial and temporal patterns of macroinvertebrate communities in the Du River Basin in Northern Vietnam. *J. Freshw. Ecol.* **2010**, *25*, 637–647.

38. Bae, M.-J.; Kwon, Y.; Hwang, S.-J.; Con, T.-S.; Yang, H.-J.; Kwak, I.-S.; Park, J.-H.; Ham, S.-A.; Park, Y.-S. Relationships between three major stream assemblages and their environmental factors in multiple spatial scales. *Ann. Limnol. Int. J. Limnol.* **2011**, *47*, S91–S105.

39. Jacobsen, D.; Cressa, C.; Mathooko, J.M.; Dudgeon, D. Macroinvertebrates: Composition, life histories and production. In *Aquatic Ecosystems: Tropical Stream Ecology*; Dudgeon, D., Ed.; Elsevier Science: London, UK, 2008; pp. 65–105.

40. Hoang, D.H.; Bae, Y.J. Aquatic insect diversity in a tropical Vietnamese stream in comparison with that in a temperate Korean stream. *Limnology* **2006**, *7*, 45–55.

41. Cao, Y.; Williams, D.D.; Williams, N.E. How important rare species in aquatic community ecology and bioassessment? *Limnol. Oceanogr.* **1998**, *43*, 1403–1409.

42. Nijboer, R.C.; Schmidt-Kloiber, A. The effect of excluding taxa with low abundances or taxa with small distribution ranges on ecological assessment. *Hydrobiologia* **2004**, *516*, 347–363.

43. Braccia, A.; Voshell, J.R. Environmental factors accounting for benthic macroinvertebrate assemblage structure at the sample scale in streams subjected to a gradient of cattle grazing. *Hydrobiologia* **2006**, *573*, 55–73.

44. Soininen, J.; Paavola, R.; Muotka, T. Benthic diatom communities in boreal streams: Community structure in relation to environmental and spatial gradients. *Ecography* **2004**, *27*, 330–342.

45. Mikulyuk, A.; Sharma, S.; Egeren, S.V.; Erdmann, E.; Nault, M.E.; Hauxwell, J. The relative role of environmental, spatial, and land-use patterns in explaining aquatic macrophyte community composition. *Can. J. Fish. Aquat. Sci.* **2011**, *68*, 1778–1789.

46. Stomp, M.; Huisman, J.; Mittelbach, G.G.; Litchman, E.; Klausmeier, C.A. Large-scale biodiversity patterns in freshwater phytoplankton. *Ecology* **2011**, *92*, 2096–2107.

47. Rodrigues, A.M.; Quintino, V.; Sampaio, L.; Freitas, R.; Neves, R. Benthic biodiversity patterns in Ria de Aveiro, Western Portugal: Environmental-biological relationships. *Estuar. Coast. Shelf Sci.* **2011**, *95*, 338–348.

48. Waite, I.R.; Carpenter, K.D. Associations among fish assemblage structure and environmental variables in Willamette Basin streams, Oregon. *Trans. Am. Fish. Soc.* **2000**, *129*, 754–770.

49. Vannote, R.L.; Minshall, G.W.; Cummins, K.W.; Sedell, J.R.; Cushing, C.E. The river continuum concept. *Can. J. Fish. Aquat. Sci.* **1980**, *37*, 130–137.

50. Beauchard, O.; Gagneur, J.; Brosse, S. Macroinvertebrate richness patterns in North African streams. *J. Biogeogr.* **2003**, *30*, 1821–1833.

51. Pan, Y.; Hill, B.H.; Husby, P.; Hall, R.K.; Kaufmann, P.R. Relationships between environmental variables and benthic diatom assemblages in California Central Valley streams (USA). *Hydrobiologia* **2006**, *561*, 119–130.

52. Harding, J.S.; Young, R.G.; Hayes, J.W.; Shearer, K.A.; Stark, J.D. Changes in agricultural intensity and river health along ariver continuum. *Freshw. Biol.* **1999**, *42*, 345–357.

53. Boyero, L.; Bosch, J. The effect of riffle-scale environmental variability on macroinvertebrate assemblages in a tropical stream. *Hydrobiologia* **2004**, *524*, 125–132.

54. Potapova, M.G.; Charles, D.F. Benthic diatoms in USA rivers: Distributions along spatial and environmental gradients. *J. Biogeogr.* **2002**, *29*, 167–187.

55. Halwas, K.; Church, M. Channel units in small, high gradient streams on Vancouver Island, British Columbia. *Geomorphology* **2002**, *43*, 243–256.

56. Jun, Y.-C.; Kim, N.-Y.; Kwon, S.-J.; Han, S.-C.; Hwang, I.-C.; Park, J.-H.; Won, D.-H.; Byun, M.-S.; Kong, H.-Y.; Lee, J.-E.; *et al.* Effects of land use on benthic macroinvertebrate communities: Comparison of two mountain streams in Korea. *Ann. Limnol. Int. J. Limnol.* **2011**, *47*, S35–S49.

57. Parsons, M.; Norris, R.H. The effect of habitat-specific sampling on biological assessment of water quality using a predictive model. *Freshw. Biol.* **1996**, *36*, 416–434.

58. Gerth, W.J.; Herlihy, A.T. Effect of sampling different habitat types in regional macroinvertebrate bioassessment surveys. *J. N. Am. Benthol. Soc.* **2006**, *25*, 501–512.

59. Gilles, C.L.; Hose, G.C.; Turak, E. What do qualitative rapid assessment collections of macroinvertebrates represent? A comparison with extensive quantitative sampling. *Environ. Monit. Assess.* **2009**, *149*, 99–112.

60. Pashkevich, A.; Pavluk, T.; de Vaate, A.B. Efficiency of a standardized artificial substrate for biological monitoring of river water quality. *Environ. Monit. Assess.* **1996**, *40*, 143–156.

61. Saliu, J.K.; Ovuorie, U.R. The artificial substrate preference of invertebrates in Ogbe Creek, Lagos, Nigeria. *Life Sci. J.* **2007**, *4*, 77–81.

62. Di Sabatino, A.; Cristiano, G.; Pinna, M.; Lombardo, P.; Miccoli, F.P.; Marini, G.; Vignini, P.; Cicolani, B. Structure, functional organization and biological traits of macroinvertebrate assemblages from leaf-bags and benthic samples in a third-order stream of Central Apennines (Italy). *Ecol. Indic.* **2014**, *46*, 84–91.

63. Vlek, H.E.; Šporka, F.; Krno, I. Influence of macroinvertebrate sample size on bioassessment of streams. *Hydrobiologia* **2006**, *566*, 523–542.

64. Carter, J.L.; Resh, V.H. After site selection and before data analysis: Sampling, sorting, and laboratory procedures used in stream benthic macroinvertebrate monitoring programs by USA state agencies. *J. N. Am. Benthol. Soc.* **2001**, *20*, 658–682.

65. Metzeling, L.; Miller, J. Evaluation of the sample size used for the rapid bioassessment of rivers using macroinvertebrates. *Hydrobiologia* **2001**, *444*, 159–170.

66. Kim, A.R.; Oh, M.W.; Kong, D.S. The influence of sample size on environmental assessment using benthic macroinvertebrates. *J. Korean. Soc. Water Environ.* **2013**, *29*, 790–798.

67. Pinna, M.; Marini, G.; Mancinelli, G.; Basset, A. Influence of sampling effort on ecological descriptors and indicators in perturbed and unperturbed conditions: A study case using benthic macroinvertebrates in Mediterranean transitional waters. *Ecol. Indic.* **2014**, *37*, 27–39.

68. Vinson, M.R.; Hawkins, C.P. Effects of sampling area and subsampling procedure on comparisons of taxa richness among streams. *J. N. Am. Benthol. Soc.* **1996**, *15*, 392–399.

69. Marini, G.; Pinna, M.; Basset, A.; Mancinelli, G. Estimation of benthic macroinvertebrate taxonomic diversity: Testing the role of sampling effort in a Mediterranean transitional water ecosystem. *Transit. Waters Bull.* **2013**, *7*, 28–40.

70. Buss, D.F.; Borges, E.L. Application of rapid bioassessment protocols (RBP) for benthic macroinvertebrates in Brazil: Comparison between sampling techniques and mesh sizes. *Neotrop. Entomol.* **2008**, *37*, 288–295.

Agricultural Rivers at Risk: Dredging Results in a Loss of Macroinvertebrates. Preliminary Observations from the Narew Catchment, Poland

Mateusz Grygoruk, Magdalena Frąk and Aron Chmielewski

Abstract: Ecosystem deterioration in small lowland agricultural rivers that results from river dredging entails a significant threat to the appropriate ecohydrological conditions of these water bodies, expressed as homogenization of habitats and loss of biodiversity. Our study was aimed at a comparison of abundance and taxonomic structure of bottom-dwelling macroinvertebrates in dredged and non-dredged stretches of small lowland rivers and tributaries of the middle Narew River, namely: Czaplinianka, Turośnianka, Dąb, and Ślina. The experimental setup was (1) to collect samples of the bottom material from the river stretches that either persisted in a non-modified state (dredging was not done there in the last few years) or had been subjected to river dredging in the year of sampling; and (2) to analyze the abundance and taxonomic structure of macroinvertebrates in the collected samples. The study revealed that at the high level of statistical significance (from $p = 0.025$ to $p = 0.001$), the total abundance of riverbed macroinvertebrates in the dredged stretches of the rivers analyzed was approximately 70% lower than in non-dredged areas. We state that the dredging of small rivers in agricultural landscapes seriously affects their ecological status by negatively influencing the concentrations and species richness of benthic macroinvertebrates.

Reprinted from *Water*. Cite as: Grygoruk, M.; Frąk, M.; Chmielewski, A. Agricultural Rivers at Risk: Dredging Results in a Loss of Macroinvertebrates. Preliminary Observations from the Narew Catchment, Poland. *Water* **2015**, *7*, 4511–4522.

1. Introduction

Although modified water bodies in agricultural landscapes may potentially play an important role as refuges for freshwater biodiversity [1–4], inappropriate management of these ecosystems vastly decreases the aquatic ecosystems' health [5–7]. Particularly, mechanic dredging of the river bed degrades the structure and composition of riverbanks and bottoms and negatively affects macroinvertebrate communities [6,8–12]. Considering the scale of river dredging in Poland in recent years, reported as critically affecting ecohydrological features of small and medium lowland rivers [13,14], and wishing to follow the Water Framework Directive's (WFD) call for European Union member states to conserve the status of their waters, we

148

believed that technical measures applied in a country-wide manner for the "reduction of flood risk in agricultural areas" had to be revisited to assess their compliance with the requirements of environmental conservation and to protect rivers. As the first step towards revealing the responses of aquatic ecosystems to bottom dredging, we intended to undertake comparative research on the bottom macroinvertebrates of selected dredged and non-dredged stretches of small lowland rivers. Due to the fact that a high diversity of bottom macroinvertebrates reflects the appropriate ecohydrological status of rivers (resulting from feedbacks of ecological, hydrological, and micro-habitat processes [15]), we focused on differences in the abundance and taxonomic composition of macroinvertebrates. Our research was performed in small lowland rivers located in northeastern Poland, known for its unique environmental features. The preliminary results, despite being based upon a small sample of collected data, indicate that the responses of the examined aquatic ecosystems to dredging tend to be critically negative and demand re-consideration in terms of fulfilling the requirements of the WFD in any similar cases on the European scale.

2. Materials and Methods

We investigated four rivers of which certain sections were subjected to dredging. These sandy lowland rivers and streams (Czaplinianka, Dąb, Ślina, and Turośnianka) are tributaries of the middle Narew, northeast Poland (Figure 1). The lowermost stretches of the Czaplinianka and Turośnianka rivers are located within the Natura 2000 sites, and the remaining rivers are situated within a few kilometers of the protected areas. The rivers sampled flow through the agricultural landscapes (hay meadows, pastures), and due to their morphological and hydrological features, they can be classified as small lowland rivers (Table 1). Catchments of the rivers analyzed are located in a temperate climate with strong continental influences. The average air temperature in the region is 7 °C, and the average annual precipitation equals 580 mm [16].

The shares of the analyzed river catchments covered with forests range from 19% to 36% for the catchments of Dąb and Czaplinianka, respectively, and the areas of agricultural lands range from 61% to 78% for the catchments of Czaplinianka and Dąb, respectively [17]. The research was conducted in September and October 2013 when, in summer and autumn, river dredging was implemented to keep the geometric shape of the river channel's cross-sections and to remove 0.3 m of the bottom sediments. This action is expected to eventually mitigate the flood risk. However, the probable influence of these actions on reduction of the floodwave was never examined nor proven by the river management authority and river dredging is being implemented with no particular pre-assessment of its probable efficiency.

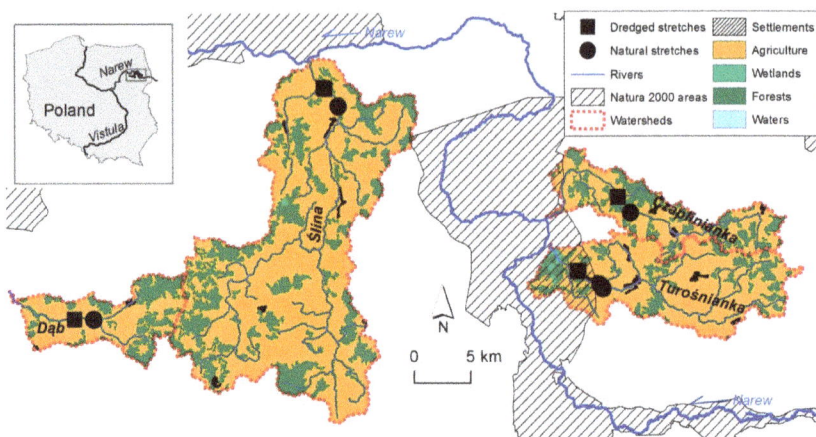

Figure 1. Study area—catchments of tributaries of the middle Narew river: Dąb, Czaplinianka, Ślina, and Turośnianka. Location of selected sampling stretches, land use, hydrography, and boundaries of protected areas (Natura 2000 and Narew National Park).

Table 1. Hydrological features of sampled rivers.

River	Ślina	Dąb	Czaplinianka	Turośnianka
Length [km]	39.3	16.5	31.3	31.4
Catchment area [km^2]	359.29	66.79	77.95	137.73
Average width in sampling locations [m]	7	2	3	4
Average depth in sampling locations [m]	1.2	0.3	0.8	0.8
Average flow velocity in sampling locations [m/s]	0.15	0.11	0.09	0.10
Average discharge in the confluence [m^3/s]	1.5	0.3	1.1	1.2
X Coordinate of the centroid of the non-dredged stretch of the river (GPS)	22.67306	22.32890	23.05540	22.98110
Y Coordinate of the centroid of the non-dredged stretch of the river (GPS)	53.16604	53.00912	53.07010	53.02692
X Coordinate of the centroid of the dredged stretch of the river (GPS)	22.65379	22.30228	23.03923	23.00352
Y Coordinate of the centroid of the dredged stretch of the river (GPS)	53.18109	53.01021	53.08307	53.01727

Macroinvertebrate sampling was done as follows: in each of the four rivers we selected two sampling stretches of 100 m; one stretch was located within the fragment of the river recently subjected to dredging, and the other was a reference stretch, where dredging had not been implemented recently. Dredged and non-dredged sampling stretches were located between 1.7 and 2.5 km from one another (Figure 1). Dredged stretches of the rivers sampled were cleared with an excavator, so the riparian vegetation was very poor. Non-dredged river stretches were vegetated. In Ślina and Czaplinianka, the dominant macrophytes were *Nuphar lutea* and *Saggittaria* sp., while the macrophytes in the sampled stretches of Dąb and Czaplinianka were poorly developed due to river size and shading from adjacent trees and shrubs. All of the rivers sampled have sandy banks. Water levels during the sampling were below the average annual water level and oscillated around the median of the lowest annual water levels recorded in the most recent multi-year period.

On each of the eight selected sampling stretches (Figure 1) we collected five samples of bottom sediments. Sampling locations were distributed every 20 m along the stretch. Samples were collected with the standard Eckman-Birge's bottom-sediment sampler that allows sampling of 225 cm^2 of the river bottom [18]. As the study was oriented at the analysis of bottom-dwelling macroinvertebrates, field research did not include drift measurements. Collection locations were selected to cover the most representative aquatic habitats of each stretch in various water depths, from an average distance of one-third of the river width. Samples were collected wading in the river. Forty samples of bottom sediments were examined in total (20 collected from dredged stretches of rivers and 20 collected from reference stretches that had not been dredged recently). Sampling was done from five days (Ślina river) to approximately one month after the river bed dredging (Turośnianka river). River flow velocities in the sampled stretches were approximated with surface flow measurements averaged to one value representative for the whole sampling stretch. Each sample was stored in a plastic bag immediately after collection and transferred to the laboratory where the abundance and species composition of macroinvertebrates were assessed. Macroinvertebrates were sorted after preservation in ethanol and counted by the naked eye. Obtained results of macroinvertebrate abundance and taxonomic composition were tested in order to reveal their statistical significance.

3. Results

Field research revealed the presence of 10 taxa of macroinvertebrates, namely *Amphipoda, Bivalia, Diptera, Ephemeroptera, Gastropoda, Hirudinea, Isopoda, Megaloptera, Oligochaeta,* and *Trichoptera*. It was generally observed that both the total abundance of macroinvertebrates and their taxonomic compositions were significantly higher in the natural stretches of rivers (Figure 2A) than in the freshly dredged ones (Figure 2B).

We tested the sampling results for statistical relevance. The Student's *t*-test for dependent variables (for n-2 degrees of freedom) and a Mann-Whitney-Wilcoxon test were applied in order to determine the statistical significance of observed differences in the abundance and species compositions of macroinvertebrates between the dredged and natural stretches of the rivers examined. The *t*-values for the total macroinvertebrates' abundance analysis reached 2.813; that gives a statistical significance at the level of $p = 0.023$. The *t*-values for the comparison of the taxonomic composition of macroinvertebrates of dredged and natural river stretches reached 4.420; that gives a statistical significance at the level of $p = 0.001$. The U-value of the Mann-Whitney-Wilcoxon test analyzing total macroinvertebrate abundance was 19 (giving $p = 0.025$), and for the taxonomic composition of macroinvertebrates of dredged and natural river stretches they were $U = 61$ and $p = 0.01$. The results of the statistical tests applied allow us to state that the recorded differences between the total abundances of macroinvertebrates and their taxonomic compositions are statistically significant (the lowest recorded level of statistical significance was 0.025, which we considered satisfactory).

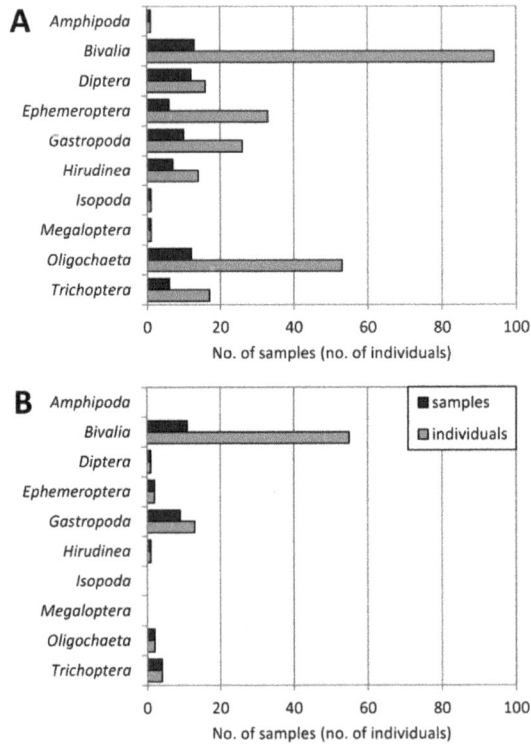

Figure 2. Comparison of the abundance of macroinvertebrate taxa *vs.* number of samples in which particular taxa were recorded in natural (**A**) and dredged (**B**) stretches of rivers.

Considering the total abundance of macroinvertebrates in all samples collected, we recorded approximately 70% lower concentrations of macroinvertebrates in the dredged stretches than in undredged ones (Figure 3A). The biggest differences between the total abundance of macroinvertebrates were found in the analyzed stretches of the Ślina and Dąb rivers (Figure 3B,C), which were nearly 91% and 98%, respectively. In the analyzed stretches of Czaplinianka and Turośnianka, differences in macroinvertebrate abundances were lower than in the cases of Ślina and Dąb, but still reached approximately 50% (Figure 3D,E). The most significant differences in the abundance of macroinvertebrate taxa were reported for the taxon of *Ephemeroptera*, whose numbers in dredged river segments were 83% to nearly 100% lower than in the undisturbed stretches of the rivers analyzed. Equally significant differences between dredged and non-dredged stretches were recorded for *Trichoptera, Gastropoda,* and *Diptera.* Differences in the abundance of *Amphipoda, Isopoda,* and *Megaloptera* between the dredged and non-dredged stretches of the rivers were insignificant. Of the rivers analyzed, the most critical disparities of macroinvertebrate abundance and composition were recorded in the smallest of the sampled rivers, the Dąb. Although individuals of *Amphipoda, Bivalia, Diptera, Ephemeroptera, Gastropoda, Hirudinea,* and *Oligochaeta* were recorded in the non-dredged stretch of this river, these taxa were not reported in dredged stretches. *Bivalia* was the taxon with the lowest disparities of abundance.

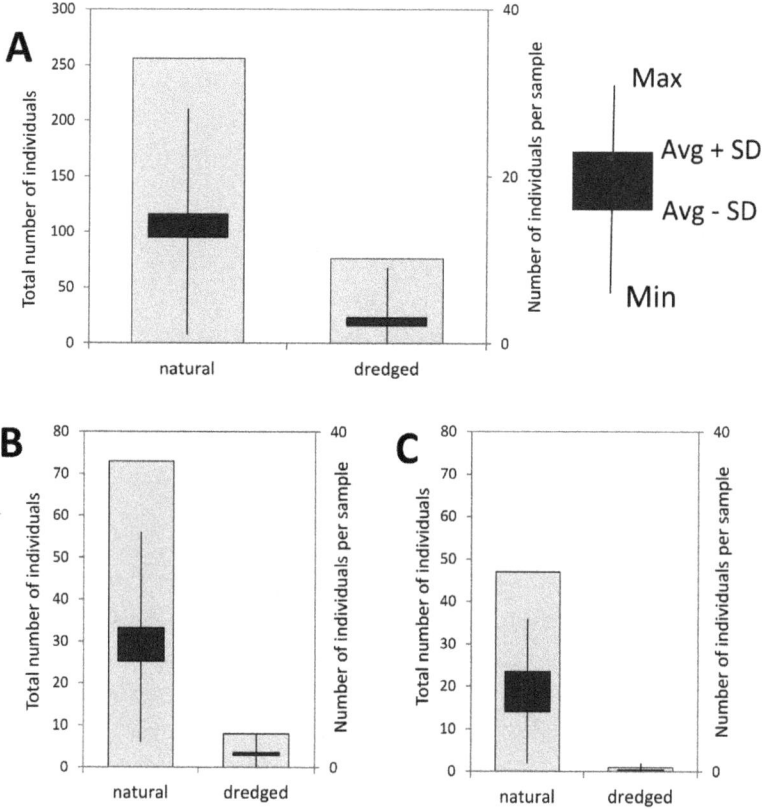

Figure 3. *Cont.*

153

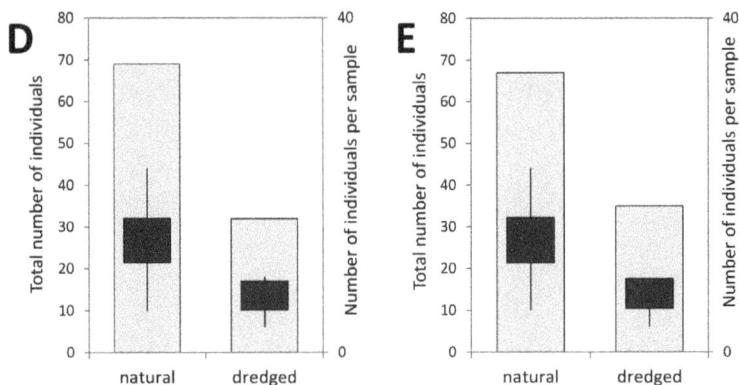

Figure 3. Comparison of abundance of macroinvertebrates between natural and dredged stretches: total number of sampled individuals and numbers of individuals per sample: (**A**) whole set of samples ($n = 20 + 20$); (**B**) Ślina river ($n = 5 + 5$); (**C**) Dąb river ($n = 5 + 5$); (**D**) Czaplinianka river ($n = 5 + 5$); and (**E**) Turośnianka river ($n = 5 + 5$).

4. Discussion

In subtle cases of river dredging where the response of macroinvertebrate communities (based on calculated macroinvertebrate indexes) to ecological disturbances might not be clearly linked to river regulation and maintenance works (dredging, macrophyte removal), additional mesohabitat assessments of river reaches or complex multivariate analyses may be required [8]. However, the strongly negative response of macroinvertebrate structure and abundance to river dredging found in our study tends to be clear and obvious. We see that, similarly to the studies of Armitage and Pardo [8], Bylak *et al.* [6], and Holmes *et al.* [15], the research we present should be extended to the other factors of the aquatic ecosystems examined (*i.e.*, the structure of bottom sediments, debris and microhabitat analysis, flow velocity distribution, and water quality assessment). However, in the cases of the Czaplinianka, Dąb, Ślina, and Turośnianka rivers, dredging to prevent floods involved mechanically removing (using excavators) some 0.3–0.5 m of the sediment material from the river bottom for the whole width of the river stretch and depositing the material on the river bank, causing the degradation of hydromorphology (Figure 4A). This type of river structure modification was proven as critically negative for freshwater mussels *Bivalvia* [9], but the negative impact of such river management measures can be extended to herein presented macroinvertebrates and other taxa (including amphibians, Figure 4B; Ukrainian brook lamprey *Eudontomyzon mariae*, Figure 4C; or numerous other taxa of macroinvertebrates, Figure 4D).

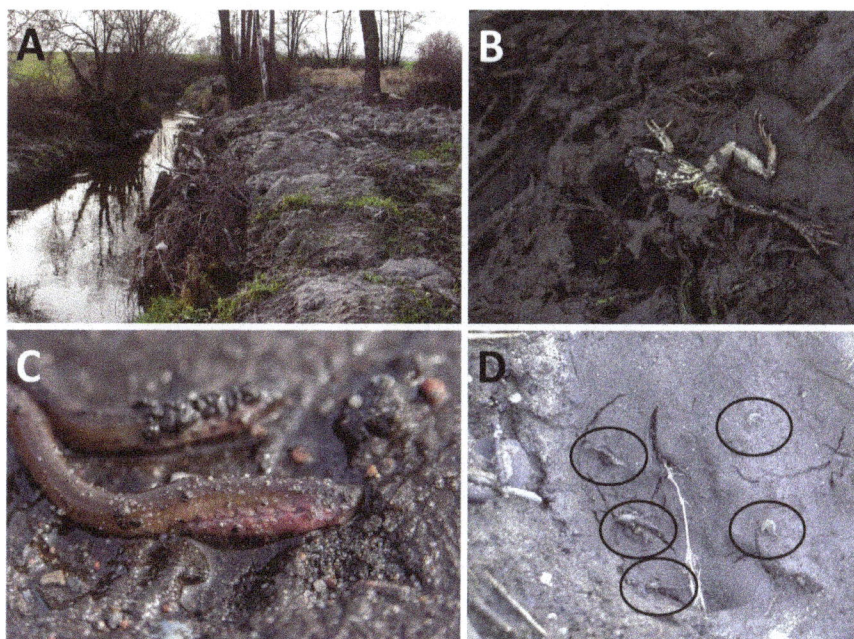

Figure 4. Environmental consequences of dredging agricultural rivers: (**A**) modified hydromorphology of the Czaplinianka River; (**B**) dead amphibians found in the excavated material (Ślina River); (**C**) dead Ukrainian brook lampreys *Eudontomyzon mariae* (Ślina River); (**D**) dead macroinvertebrates (*Gammaridae* and *Ephemeroptera*). Photos: courtesy of Paweł Fiedorczuk.

As the biggest differences in macroinvertebrate abundances between the dredged and non-dredged stretches were observed in the smallest river sampled (Dąb), we suspect that narrow and shallow (up to 2 m wide and up to 0.2 m deep) agricultural rivers may face the greatest risk of damage to their aquatic ecosystems as a result of inappropriate and devastating management.

It is likely that in the long run the self-restoration of the dredged river stretches analyzed may result in the re-establishment of hydromorphological conditions by allowing the re-development of macroinvertebrate abundance and species composition toward the reference values reported for non-dredged stretches [19,20]. However, if long headwater parts of these rivers remain under the pressure of dredging every 1–3 years (which is the case of analyzed and adjacent rivers), it is likely that the spontaneous restoration of macroinvertebrate populations of the whole river systems may be—if still possible—very slow. We rather expect the hydromorphological and ecological status of lowland agricultural rivers being dredged on up to 70% of their total length on an annual and bi-annual basis to deteriorate considerably. As it was observed that the structure of aquatic habitats and sediments, especially in the headwater streams (e.g., the small rivers examined in the presented research), promotes the abundance and taxonomic richness of bottom macroinvertebrates [15,21], it appears that once small lowland rivers' hydromorphology begins to suffer, the aquatic ecosystems subjected to dredging lose their resilience. Such speculation requires consideration in the catchment-scale river management. It should result in implementation of river dredging adaptation strategies by avoiding long stretches being dredged regularly and promoting dredging of selected short stretches only, if the flood risk was proven to be increasing due to the accumulation of sediments. This would require continuous monitoring of longitudinal profiles of the river bottom, which is predominantly not done in the case of small lowland agricultural rivers. As such actions are not anticipated in river management plans in northeastern Poland, we stress that only the reliable monitoring-based criteria of river dredging should be used to determine the relevance of this action for flood reduction.

In light of our analysis, and emphasizing the results obtained by Bylak *et al.* [6], we state that technical measures referred to as "river regulation" have a critically negative influence on the ecological status of rivers. For small lowland agricultural rivers, it is regular dredging that poses an equally significant challenge to the ecological status by deterioration of river hydromorphology and populations of macroinvertebrates. Preliminary results presented in this paper require replication, especially in terms of additional sampling of the same river stretches over time (months or years after dredging was implemented), and extension to other elements of the riverine environment in order to provide a comprehensive analysis of rivers' responses to dredging [22]. Extension of the research to, for example, the response of fish communities to dredging-induced changes in mesohabitat structure and macroinvertebrate composition would allow results to reveal the relevance of the maintenance of agricultural rivers to fishery management [23,24]. Research on these aspects is now ongoing. However, we stress that despite the strength of the correlations and covariance between the abundance of benthic macroinvertebrates and other elements of the environment, the "everyday" management of small lowland agricultural rivers requires revision, detailed environmental impact assessment, and the analysis of trade-offs between the potential (not certain) reduction of flood risk and the loss of resilient aquatic ecosystems. In some specific cases, when dredging is oriented at the removal of contaminated sediments which pose a decent risk of losing river ecosystems' quality, one should consider if the probable negative response of benthic macroinvertebrates to mechanic sediment removal is of lower importance than preventing the deterioration of water and sediment quality.

Facing the above facts and the potential changes in environmental legislation in Poland to permit the standard "maintenance" of river channels (including dredging) without any environmental impact assessments, the scale of dredging is expected to become even greater than it is now. It is likely that small agricultural rivers that are "maintained" by regular dredging, contra the examples given by Chester and Robson [1] and supported with the observations of Tonkin *et al.* [20], will never again serve as a refuge for freshwater biodiversity. Avoiding this scenario will require management approaches, especially in cases of small agricultural rivers, to be individualized so each of the complex river ecosystems could retain its unique and specific environmental values [25]. It is intended that the preliminary observations presented in this paper will allow us to reveal the appropriate meaning of "river maintenance" which, in presented examples, underpinned the vast deterioration of the biodiversity of the agricultural rivers analyzed. We also hope that learning from river management mistakes will allow implementing efficient restoration and maintenance strategies for the recreation and sustainability of resilient aquatic ecosystems in the future [26], where flood protection does not contradict keeping the good ecological status of agricultural rivers.

5. Conclusions

This study revealed that the species composition and abundance of macroinvertebrates is much lower in dredged stretches of the rivers analyzed than in the stretches where river dredging was not done, leaving the structure and thickness of bottom sediments untouched. The most significant differences in the abundance of macroinvertebrates' taxa were reported for *Ephemeroptera*, which were 83% to nearly 100% less abundant in the dredged areas. We revealed that in dredged stretches of rivers, the quantitative loss of macroinvertebrate populations (numbers of individuals) was much more significant than the quantitative loss of taxonomic composition (decreasing number of taxa). Based on the examples analyzed, we conclude that river dredging entails a potential significant threat to species diversity and the abundance of bottom macroinvertebrates and, potentially, to the whole river ecosystem. We stress that more research directed at before-after control-impact studies of river dredging's influence on aquatic ecosystems (especially small rivers located within protected areas) is needed in order to reveal the full extent of river ecosystem degradation. Although one could argue that presented results could have been expected and the reported negative response of biota and hydromorphology to dredging small agricultural rivers is obvious, they highlight that regular actions oriented at the so-called "maintenance" of rivers by mechanic sediment removal are negative to the environment of riverscapes. Our preliminary study only underlines this issue, which should be considered in the everyday management of rivers. Finally, regardless of the probable self-restoration dynamics in agricultural rivers, we stress that the implementation of technical measures aimed at mud removal and shaping the river channel may lead to the deterioration of the ecological status of rivers by affecting their ecohydrological features. Such actions contradict national and international (EU Water Framework Directive) environmental legislation.

Acknowledgments: This study was funded by the Stefan Batory Foundation through a project implemented by Stowarzyszenie Niezależnych Inicjatyw Nasza Natura, aimed at the control of legal and environmental effects of the technical maintenance of rivers in NE Poland. Paweł Fiedorczuk, Jędrzej Grygoruk, and Zofia Namyślak are acknowledged for help in data collection. Reviewers are kindly acknowledged for valuable comments that allowed us to improve the manuscript.

Author Contributions: Mateusz Grygoruk developed the concept of the study, planned field research, and wrote the manuscript. Magdalena Frąk and Aron Chmielewski performed field research, laboratory analysis of macroinvertebrates, and assisted in manuscript preparation.

Conflicts of Interest: The authors declare no conflict of interest.

References

1. Chester, E.T.; Robson, B.J. Anthropogenic refuges for freshwater biodiversity: Their ecological characteristics and management. *Biol. Conserv.* **2013**, *166*, 64–75.
2. Clarke, S.J. Conserving freshwater biodiversity: The value, status and management of high quality ditch systems. *J. Nat. Conserv.* **2015**, *24*, 93–100.
3. Herzon, I.; Helenius, J. Agricultural drainage ditches, their biological importance and functioning. *Biol. Conserv.* **2008**, *141*, 1171–1183.
4. Watson, A.M.; Ormerod, S.J. The distribution of three uncommon freshwater gastropods in the drainage ditches of British grazing marshes. *Biol. Conserv.* **2004**, *118*, 455–466.
5. Adynkiewicz-Piragas, M.; Drabiński, A. Wpływ inwestycji hydrotechnicznych Na ekosystem rzeki Smortawy. *Zesz. Nauk. Akad. Rol. Wroc.* **2001**, *417*, 7–28.
6. Bylak, A.; Kukuła, K.; Kukuła, E. Influence of regulation on ichtyofauna and benthos of the Różanka stream. *Ecohydrol. Hydrobiol.* **2009**, *9*, 211–223.
7. Hachoł, J.; Krzemińska, J. Influence of the regulation of the Smortawa River on the self-purification processes for oxygen indicators. *Infrastruct. Ecol. Rural Areas* **2008**, *9*, 207–216.
8. Armitage, P.D.; Pardo, I. Impact assessment of regulation at the reach level using macro invertebrate information from mesohabitats. *Regul. Rivers Res. Manag.* **1995**, *10*, 147–158.
9. Aldridge, D.C. The impacts of dredging and weed cutting on a population of freshwater mussels (Bivalvia: Unionidae). *Biol. Conserv.* **2000**, *95*, 247–257.
10. Rader, R.B.; Ward, J.V. Influence of regulation on environmental conditions and the macroinvertebrate community in the upper Colorado River. *Regul. Rivers Res. Manag.* **1988**, *2*, 597–618.
11. Sawa, K.; Popek, Z. Analysis of the connection between hydromorphological conditions and biocenic diversity on the example of the Zwoleńka River. *Ann. Warsaw. Univ. Life Sci. SGGW Land Reclam.* **2011**, *43*, 173–184.

12. Vermonden, K.; Brodersen, K.P.; Jacobsen, D.; van Kleef, H.; van der Velde, G.; Leuven, R.S.E.W. The influence of environmental factors and dredging on chironomid larval diversity in urban drainage systems in polders strongly influenced by seepage from large rivers. *J. North Am. Benthol. Soc.* **2011**, *30*, 1074–1092.

13. Jabłońska, E.; Kotkowicz, M.; Manewicz, M. *Summary and Interpretation of the Report, Inventory and Assessment of Environmental Consequences of River Maintenance in Małopolskie, Mazowieckie, Opolskie, Świętokrzyskie, Warmińsko-Mazurskie, Wielkopolskie and Zachodnipomorskie Provinces of Poland, in Years 2010–2012*; Evaluation on the Basis of Public Procurement Procedures Arranged by Regional Irrigation and Drainage Councils (Wojewódzkie Zarządy Melioracji I Urządzeń Wodnych) and Survey Analysis; WWF Poland: Warsaw, Poland, 2013.

14. Grygoruk, M.; Fiedorczuk, P.; Kasjaniuk, A.; Kostecka, A.; Grygoruk, J. Monitoring of river maintenance works and flood damages removal implemented by the Regional Irrigation and Drainage Council in Białystok (Wojewódzki Zarząd Melioracji i Urządzeń Wodnych w Białymstoku): Evaluation of the accordance of actions and intervetions in case of reporting inconsistences. Available online: http://naszanatura.com.pl/wp-content/uploads/2015/01/nasza_natura_raport.pdf (accessed on 18 May 2015).

15. Holmes, K.L.; Goebel, P.C.; Williams, L.R.; Schecengost, M. Environmental influences on macroinvertebrate assemblages in headwater streams of northeastern Ohio. *J. Freshw. Ecol.* **2011**, *26*, 409–422.

16. Górniak, A. *Klimat Województwa Podlaskiego*; IMGW: Białystok, Poland, 2000.

17. Chmielewski, A. Analysis of Abundance and Species Composition of Macroinvertebrates in the Bottom Sediments of Selected Tributaries of the Narew River in the Context of Dredging of River Channels. Bachelor Thesis, Department of Hydraulic Engineering, Warsaw University of Life Sciences—SGGW, Warsaw, Poland, 2014.

18. Klimaszyk, P.; Trawiński, A. *Ocena Stanu Rzek na Podstawie Makrobezkręgowców Bentosowych. DEKS BMWP-PL*; Department of Water Protection, Adam Mickiewicz University: Poznań, Poland, 2007; Available online: http://www.staff.amu.edu.pl/~zow/pobieranie/BMWP-PL.pdf (accessed on 18 July 2014).

19. Kiraga, M.; Popek, Z. Using the River Habitat Survey method in forecasting effects of river restoration. *Ann. Warsaw. Univ. Life Sci. SGGW Land Reclam.* **2014**, *46*, 125–138.

20. Tonkin, J.D.; Stoll, S.; Sundermann, A.; Haase, P. Dispersal distance and the pool of taxa, but not barriers, determine the colonization of restored river reaches by benthic invertebrates. *Freshw. Biol.* **2014**.

21. Larsen, S.; Vaughan, I.P.; Ormerod, S.J. Scale-dependent effect of fine sediments on temperate headwater invertebrates. *Freshw. Biol.* **2009**, *54*, 203–219.

22. Wyżga, B.; Amirowicz, A.; Oglęcki, P.; Hajdukiewicz, H.; Radecki-Pawlik, A.; Zawiejska, J.; Mikuś, P. Response of fish and benthic invertebrate communities to constrained channel conditions in a mountain river: Case study of the Biała, Polish Carpathians. *Limnologica* **2014**, *46*, 58–69.

23. Stranko, S.A.; Hildebrand, R.H.; Palmer, M.A. Comparing the fish and benthic macroinvertebrate diversity of restored urban streams to reference streams. *Restor. Ecol.* **2011**.

24. Wallace, J.B.; Webster, J.R. The role of macroinvertebrates in stream ecosystem function. *Ann. Rev. Entomol.* **1996**, *41*, 115–139.

25. Leps, M.; Tonkin, J.D.; Dahm, V.; Haase, P.; Sundermann, A. Disentangling environmental drivers of benthic invertebrate assemblages: The role of spatial scale and riverscape heterogeneity in a multiple stressor environment. *Sci. Total Environ.* **2015**, *536*, 546–556.

26. Grygoruk, M.; Acreman, M. Restoration and management of riparian and riverine ecosystems: Ecohydrological experiences, tools and perspectives. *Ecohydrol. Hydrobiol.* **2015**.

Spatio-Temporal Variability in Benthic Macroinvertebrate Communities in Headwater Streams in South Korea

Mi-Jung Bae, Jung Hwa Chun, Tae-Soo Chon and Young-Seuk Park

Abstract: Comprehensive research on the structural and functional variability of benthic macroinvertebrate communities within headwater streams is limited, despite the fact that the majority of streams within a watershed are headwater streams that form the primary link between terrestrial and aquatic ecosystems. Therefore, we investigated the structure and function of benthic macroinvertebrate communities in four headwater streams at two different spatial scales (*i.e.*, sampling sites (*i.e.*, reaches) >samples (*i.e.*, riffles)) over three seasons (*i.e.*, spring, summer and autumn) of the year. Community indices, functional feeding guilds and habit trait guilds varied significantly depending on the seasons rather than on sites in two-way ANOVA based on spatial (*i.e.*, sampling sites) and seasonal effects in each headwater stream. Non-metric multidimensional scaling analyses showed the differences between communities according to the considered spatial and temporal scales. At the individual stream scale, the differences between samples followed seasonal variation more than spatial differences. Site differences became more important when performing an ordination within a single season (*i.e.*, spring, summer, and autumn). Continued research and monitoring employing both multidisciplinary and multidimensional approaches are required to maintain macroinvertebrate diversity within headwater streams.

Reprinted from *Water*. Cite as: Bae, M.-J.; Chun, J.H.; Chon, T.-S.; Park, Y.-S. Spatio-Temporal Variability in Benthic Macroinvertebrate Communities in Headwater Streams in South Korea. *Water* **2016**, *8*, 99.

1. Introduction

Biodiversity has been declining at an increasing rate worldwide [1] as a result of anthropogenic habitat disruption. Although freshwater occupies less than 1% of the Earth's surface area, and rivers and streams represent only 0.006% of all freshwater resources [2], they exhibit high biodiversity, comprising approximately 10% of known species [3,4].

Headwater streams are extremely heterogeneous ecosystems with high spatial and temporal variation [5], comprising a significant proportion (*i.e.*, more than three-quarters) of the total stream channel length within a watershed [6]. Headwater streams are main sources of water, sediments, and organic materials that are transported downstream [7–10], and their small catchments couple terrestrial and aquatic ecosystems such as food web dynamics [11,12] including allochthonous input [13], inputs of terrestrial invertebrates [14], *etc.* (see Nakano *et al.* [15] for a detailed explanation). Furthermore, they are essential for sustaining the structure and function of watersheds [7,8,10,16]. Headwater streams provide valuable habitats for unique and diverse communities of aquatic flora and fauna [16–18]. Therefore, it has become increasingly clear that headwater streams are essential for maintaining biodiversity in both terrestrial and aquatic habitats [7,8,10,17,19,20].

Benthic macroinvertebrates perform central ecological roles in stream ecosystems [21], such as processing of detritus, participation in animal-microbial interactions and functioning as primary and secondary consumers through critical trophic interactions [22,23]. Headwater streams are characterized by diverse microhabitats (*i.e.*, refugia) that help protect macroinvertebrates from competition, predation and natural disturbances, and therefore support a rich regional biodiversity [20]. Research on the environmental and biological parameters that determine the structure and function of macroinvertebrate community in headwater streams is essential for the basic understanding of the ecology, biodiversity, and conservation of these important ecosystems [24,25].

The composition of the macroinvertebrate community can be differentiated by various factors, including latitudinal gradients [26], stream segmentation and microhabitat [27,28]. Heino *et al.* [26] suggested that local filters (e.g., water quality) in headwater streams were relatively weak whereas they showed the clear latitudinal gradients of macroinvertebrate community composition. Ligeiro *et al.* [27] found that the composition of macroinvertebrate community was differentiated according to stream segments and microhabitats in a tropical headwater catchment, and García-Roger *et al.* [28] reported that during the dry season, the species richness was decreased especially in the temporary headwater streams due to the reduction of available habitats. The diversity of different guilds (*i.e.*, functional feeding guilds and habit trait guilds) in headwater streams is affected by pH, stream width, moss cover, stream particle size, nitrogen, and water color [19]. Moreover, algae-scraping invertebrates represent longitudinal zonation patterns along the river systems whereas within riffles, algal abundance can determine the invertebrates in small-scales [29–31]. The distributions of leaf-shredding invertebrates often reflect longitudinal and among-stream variability in riparian conditions [32,33] as well as riffle-scale patchiness of leaf detritus on stream bottoms [34,35]. Chung *et al.* [36] reported that the variation in the trophic structure was affected by habitat characteristics in each channel reach, including channel morphology, proportion of habitat type, and benthic organic matter availability. However, there has been little research on aquatic biodiversity in headwater streams considering both seasonal and spatial differences.

162

Therefore, we examined the diversity of a benthic macroinvertebrate community in four different headwater streams at two different spatial scales (*i.e.*, sampling sites >samples (riffles)) in three different seasons (*i.e.*, spring, summer, autumn). We tested hypothesis that the composition of macroinvertebrate communities would be spatially and temporally heterogeneous at different spatial scales in headwater streams [37–39]. We considered only headwater streams free of anthropogenic disturbance to exclude interaction effects between anthropogenic and natural factors on macroinvertebrate communities.

2. Materials and Methods

2.1. Study Area

We studied benthic macroinvertebrate communities at the headwater streams in four different regions of the northern part (Gwangreng: GR and Hongcheon: HC) and southern part (Wando: WD and Geumsan: GS) of South Korea (Figure 1 and Table 1). All streams were in forested areas, free of anthropogenic disturbance (Table 1). For instance, GR and WD are in the National Arboretum and people have rarely visited HC and GS due to the accessibility. *Acer pseudosieboldianum*, *Quercus mongolica* and *Securinega suffruticosa* were dominant trees in riparian areas of GR, *Sambucus racemosa* L. ssp. sieboldiana and *Deutzia grandiflora* Bunge var. baroniana were dominant in HC. Meanwhile, the riparian vegetation of GS was mainly composed of *Pinus densiflora*, *Styrax obassia*, and *Phragmites japonica*, and *Eurya japonica*, *Camellia japonica* and *Quercus acuta* were mainly observed in the riparian vegetation of WD. There were no houses or farms in the stream catchments of study areas. All sampling sites were in the first or second order streams based on a geographical map (scale: 1:50,000). There were clear gradients of climate (*i.e.*, temperature and precipitation) according to the climate data from the Korea Meteorological Administration (KMA) [40]. Annual precipitation in the study areas was higher in the southern study area (WD: 1532.7 mm and GS: 1512.8 mm) than in the northern study area (GR: 1450.5 mm and HC: 1405.4 mm). Due to the monsoon climate, more than 50% of the precipitation was concentrated in summer (especially, June or July to August); whereas other periods (mainly from October to March) were dry [41]. Annual average temperature based on the data from 1980 to 2010 from KMA is the lowest in HC (10.8 °C) followed by KR (12.7 °C), KS (13.4 °C) and WD (14.3 °C). Monthly temperature range is the highest in HC from −11.5 °C to 30.2 °C followed by GS (−5.8 °C–30.3 °C), GR (−5.9 °C–29.6 °C) and WD (−0.4 °C–29.2 °C).

2.2. Ecological Data

Benthic macroinvertebrates were collected with a Surber sampler (30 × 30 cm, 300 um mesh) to a depth of 10 cm at 12 sampling sites in four different streams

(Figure 1). Sampling was conducted seasonally in spring, summer, and autumn in 2009 (GS), 2010 (GR), 2011 (HC), and 2014 (WD). Samples could not be collected in winter because the streams were frozen. In each stream, three riffle sites (e.g., GS1, GS2 and GS3 in GS stream) were selected at less than 0.5-km intervals between the adjacent sites. Within each riffle, three to five replicates were sampled on a longitudinal direction within 1- to 3-m distances between the adjacent sampling replicates (see [26,42]). Therefore, a total of 177 samples were collected (four streams × three sites × three–five replicates × three seasons). In the laboratory, macroinvertebrates were sorted and preserved in 70% ethanol. All the individuals were identified mainly to the species level except Chironomidae under a stereo microscope (SMZ800N) at 400× based on literature [43–48].

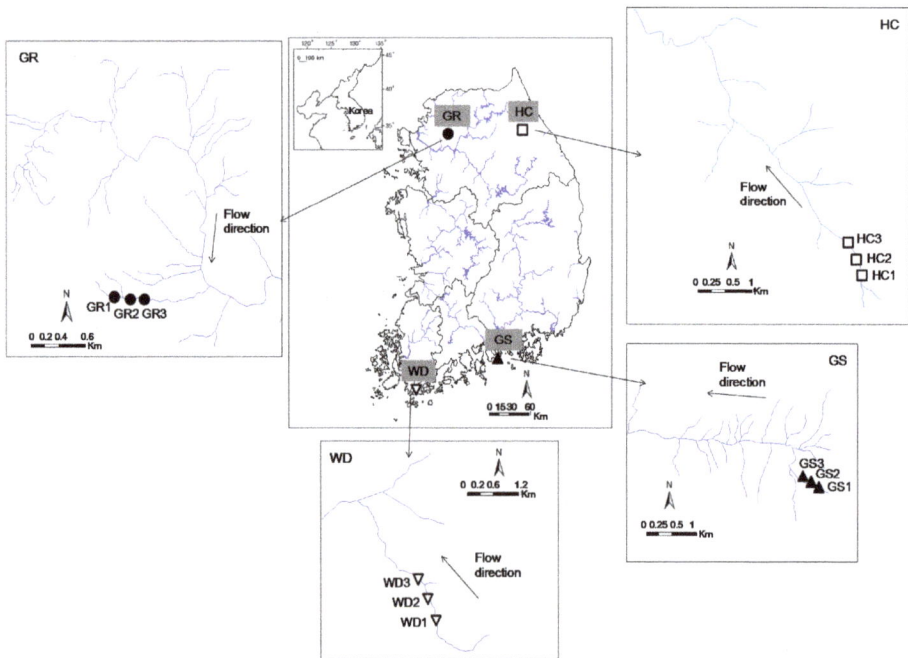

Figure 1. Locations of the sampling sites in four different headwater streams.

Table 1. Average (standard deviation) of physico-chemical characteristics of headwater streams.

Environmental Variable	GR (2010)			HC (2011)			GS (2009)			WD (2014)		
	GR1	GR2	GR3	HC1	HC2	HC3	GS1	GS2	GS3	WD1	WD2	WD3
Geography												
Altitude (m)	248	172	156	824	794	787	162	155	145	189	179	116
Stream order	1	2	2	2	2	2	1	1	1	1	1	2
Hydrology												
Velocity (cm/s)	33.3 (18)	34.9 (17.3)	52.2 (32.7)	42.8 (24.4)	29.4 (23)	35.3 (28.3)	30.6 (35.4)	27.1 (30.6)	34.7 (38)	21.6 (14.1)	24.9 (20.7)	32.9 (25.6)
Depth (cm)	7.5 (2.7)	13 (5.2)	12.3 (6.1)	23.6 (7)	21.7 (10.1)	22.6 (8.4)	12.2 (4.6)	9.6 (3.3)	9.9 (3.7)	14.3 (7)	26.9 (17.7)	32.9 (25.6)
Width (cm)	96 (12)	147 (49)	292 (112)	403 (108)	307 (153)	339 (100)	267 (129)	464 (50)	433 (272)	128 (27)	166 (23)	197 (26)
Substrate (%)												
<8mm	2.4 (1.4)	14.3 (23.5)	10.4 (18)	2.3 (3.2)	1.0 (2.1)	1.7 (2.4)	4.2 (4)	2.4 (1.9)	2.6 (2.2)	1.9 (1.0)	2.3 (1.4)	1.5 (0.7)
>8mm	5.9 (2.7)	3.9 (3.7)	5.4 (4.1)	5.7 (4.2)	4.3 (4.4)	3.7 (3.5)	7.6 (6.7)	6.0 (4.1)	9.0 (7.3)	4.4 (2.6)	5.1 (3.7)	3.1 (1.9)
>16mm	10.6 (4)	4.9 (3)	6.7 (5.4)	10.3 (6.7)	6.2 (5.1)	7.3 (4.2)	11.5 (7.1)	12.0 (6.7)	12.1 (7.1)	8.7 (5.3)	8.7 (4.4)	6.6 (3.4)
>32mm	17.3 (7.7)	10.1 (9.5)	8.7 (8.1)	14.3 (9.2)	12.5 (7.4)	10.3 (5.2)	20.7 (10.7)	24.1 (8.1)	17.0 (4.8)	14.0 (9.4)	16.2 (5.1)	10.9 (5.8)
>64mm	28.4 (20.9)	13.4 (14.7)	19.2 (11)	20.3 (9.7)	20.6 (12.7)	15.3 (7.7)	21.6 (7.9)	28.3 (14.5)	29.4 (16)	23.3 (8.6)	21.3 (10.2)	18.9 (9.6)
>128mm	13.0 (17.4)	21.0 (18.6)	29.5 (24.6)	22.3 (11.5)	27 (17.6)	34 (16.2)	23.7 (16.8)	18.3 (13.9)	26.5 (19.1)	33.7 (15.8)	29.0 (15.7)	31.0 (8.5)
>256mm	22.3 (29.4)	32.3 (32.4)	20.2 (26.2)	24.7 (30)	28.3 (37.5)	27.7 (26)	10.7 (18.9)	8.8 (17.9)	3.4 (13.3)	14.0 (15.9)	17.4 (19.5)	28.0 (19.2)
Water quality												
Conductivity (µS/cm)	72.3 (3.5)	58.4 (0.9)	60.9 (0.8)	45.8 (9.6)	44.8 (10.1)	48.7 (7.5)	37.9 (3.3)	46.3 (14.5)	42.9 (9.9)	77.3 (6.4)	77.6 (6.1)	79.6 (4.7)
Dissolved oxygen (mg/L)	9.9 (1.0)	10.6 (1.2)	9.8 (0.9)	10 (1.4)	9.9 (1.4)	9.8 (1.5)	10 (0.1)	8.9 (0.1)	9.5 (0.4)	9.5 (1.6)	9.5 (1.4)	9.7 (1.2)
pH	7.2 (0.4)	7 (0.2)	6.8 (0.1)	6.8 (0.1)	7.0 (0.1)	7.0 (0.5)	7.7 (0.0)	7.7 (0.2)	8.0 (0.3)	7.2 (0.0)	7.2 (0.0)	7.3 (0.1)

Values in parentheses for each headwater stream indicate the sampling year.

All specimens were categorized into both functional feeding guilds (FFGs, predators: PR, scrapers: SC, collector-gatherers: CG, collector-filterers: CF, and shredders: SH) and habit trait guilds (HTG, clinger: CL, burrower: BU, swimmer: SW, sprawler: SP, and climber: CM) based on Merrit and Cummins [34], except Chironomidae, because of the difficulties in taxonomic classification.

Physico-chemical environmental factors were also measured at each sampling site during the field sampling, including hydrological variables (stream depth, width, and discharge), substrates, and water quality variables. Substrate composition was measured based on substrate sizes (D): boulders (D $\geqslant$ 256 mm), coarse cobbles (128 mm $\leqslant$ D < 256 mm), fine cobbles (64 mm $\leqslant$ D < 128 mm), pebbles (16 mm $\leqslant$ D < 64 mm), gravel (2 mm $\leqslant$ D < 16 mm), and smaller substrates (D < 2 mm) [35] using each size of standard sieves (Testing sieve; Korea, Chung-gye). Water temperature, dissolved oxygen (DO), pH, and electric conductivity (conductivity) were measured using a multifunction meter (Orion®RA223). Altitude and stream order were extracted from a digital map using ArcGis (Ver. 10.1) [49].

2.3. Data Analysis

We conducted two steps of analyses to compare the differences between macroinvertebrate communities according to the spatial and temporal differences. First, variations of community indices (abundance, species richness, Shannon diversity index, Simpson diversity index, and Evenness) and proportions (%) of each class of FFGs and HTGs were analyzed using two-way analysis of variance (two-way ANOVA) to determine spatial and/or seasonal differences in each headwater stream. Second, we analyzed the abundance of macroinvertebrates using non-metric multidimensional scaling (NMS) and the Bray-Curtis distance to identify the relative differences between the sample units over multiple spatial scales and seasons. NMS is an indirect ordination analysis that compares the distribution of the macroinvertebrate community across all the sampling units without including any prior information about how the structure or taxa of macroinvertebrates could be altered or respond to environmental variables [50,51]. NMS was applied to the datasets at two different spatial scales: (1) each individual stream (three sites each) and (2) each site.

Prior to NMS analyses and statistical tests, we transformed the abundance of each taxon that showed large variations using the natural logarithm. Before transformation, the number one was added to the variables to avoid the logarithm of zero [52].

Two-way ANOVA were conducted with the package *stats* in R software [53], and NMS analyses were conducted with PC-ORD version 5 [54].

3. Results

Overall, 126 taxa with 53,002 individuals were collected (*i.e.*, GR: 77 taxa with 18,621 individuals, HC: 78 taxa with 16,981 individuals, GS: 53 taxa with 5247 individuals, and WD: 58 taxa with 11,973 individuals). At the site scale, species richness varied from 9 (WD1 in summer) to 50 (HC3 in spring) and abundance ranged from 267 (GS1 in summer) to 4854 (GR1 in summer) (Table 2). At the microhabitat scale, species richness ranged from 2 (WD1-4 in summer) to 36 (HC3-3 in spring) and abundance ranged from 21 (GS1-3 in spring) to 1705 (GR1-3 in summer).

The seasonal differences in community indices, FFGs and HTGs were mainly observed more frequently than the site differences except GR (Tables 3–5). For instance, their statistical differences (*i.e.*, community indices, FFGs and HTGs) were relatively larger among sites in GR (9 in 15 cases). Only scrapers and shredders showed seasonal differences or spatial differences in all cases (*i.e.*, sites, season and interaction between sites and season). In HC, species richness, Shannon diversity and scrapers showed seasonal differences. Only swimmers showed significant differences among sites. In GS and WD, the frequencies of seasonal differences were also higher (e.g., species richness, collector-gatherers, clingers, burrowers and swimmers in GS) than among sites (e.g., evenness, predators in GS).

In the NMS ordination for each stream, the distribution of the sampling units reflected seasonality rather than the differences among sites (first two stress values in GR: 22.4, HC: 14.9, GS: 22.3 and WD: 14.7) (Figure 2). For example, in HC, sampling units were clearly differentiated into three parts, indicating seasonal effects. The sampling units in the spring (green colored symbols in Figure 3) were located at the lower-left of the ordination map, the units in summer (sky-blue colored symbols) were located in the upper part and the units in autumn (plum colored symbols) were ordinated towards the lower right. In GS, seasonal effects in sampling units were shown according to axis 1. The units in autumn were mainly located in the right part of the NMS, the units in spring were in the middle and lastly, the units in summer were located in left part in the NMS. In the NMS ordination for each stream over different seasons, the sampling units were ordinated mainly according to site differences, especially in summer (Figure 3). For example, in WD in summer, the sampling units at WD1 were mainly located in the upper parts of the ordination, the units at WD2 were in the left part and the units at WD3 were in the right part. In GR, based on the axis 2, the units in GR1 were located in lower parts whereas the units in GR2 and 3 were ordinated in upper parts. In addition, based on axis 1, the units in GR2 were in the left parts whereas the units in GR3 were in the right parts of the NMS.

Table 2. Abundance, species richness (SR), evenness (E), Shannon diversity index (H') and Simpson diversity index (D') in four headwater streams.

Season	Site	Abundance	SR	E	H'	D'	Season	Site	Abundance	SR	E	H'	D'
Spring	GR1	2016	38	0.56	2.03	0.77	Spring	GS1	285	19	0.70	2.05	0.82
	GR2	1856	32	0.48	1.67	0.64		GS2	280	22	0.72	2.22	0.83
	GR3	1518	40	0.54	1.98	0.67		GS3	677	19	0.59	1.73	0.74
Summer	GR1	4854	40	0.25	0.91	0.34	Summer	GS1	267	23	0.69	2.16	0.80
	GR2	2269	46	0.35	1.32	0.50		GS2	288	20	0.62	1.87	0.76
	GR3	2076	41	0.65	2.42	0.84		GS3	1409	25	0.39	1.26	0.50
Autumn	GR1	1710	41	0.45	1.67	0.65	Autumn	GS1	686	32	0.67	2.32	0.83
	GR2	1441	42	0.52	1.93	0.73		GS2	894	31	0.62	2.14	0.80
	GR3	881	34	0.56	1.96	0.71		GS3	641	27	0.70	2.29	0.860
Spring	HC1	2888	41	0.56	2.07	0.76	Spring	WD1	1410	32	0.54	1.87	0.73
	HC2	2927	39	0.61	2.23	0.80		WD2	1055	30	0.60	2.05	0.77
	HC3	2187	50	0.60	2.36	0.80		WD3	467	28	0.63	2.10	0.75
Summer	HC1	632	35	0.70	2.48	0.87	Summer	WD1	4188	9	0.13	0.28	0.12
	HC2	617	34	0.69	2.42	0.84		WD2	1677	14	0.15	0.38	0.13
	HC3	388	30	0.70	2.39	0.86		WD3	828	11	0.29	0.70	0.38
Autumn	HC1	1851	34	0.63	2.22	0.78	Autumn	WD1	1020	15	0.16	0.44	0.16
	HC2	2469	41	0.53	1.95	0.70		WD2	587	22	0.43	1.33	0.55
	HC3	3022	42	0.55	2.04	0.70		WD3	741	19	0.28	0.81	0.31

Table 3. Summary of two-way ANOVAs for community indices at different sites and seasons.

Variable	Factor	GR				HC				GS				WD			
		Df	MS	F value	P	Df	MS	F value	P	Df	MS	F value	P	MS	F value	P	Df
Abundance	Sites	1	375910	2.718	0.107	1	1703	0.023	0.880	1	**71,108**	**5.416**	**0.025**	**753,667**	**7.752**	**0.008**	1
	Season	1	19458	0.141	0.710	1	14,520	0.197	0.660	1	23,595	1.797	0.187	2387	0.025	0.876	1
	Sites:Season	1	3080	0.022	0.882	1	175,219	2.373	0.131	1	19,911	1.517	0.225	8572	0.088	0.768	1
	Residuals	39	138318			40	73,840			41	13,129			97,220			41
Species richness	Sites	1	53.4	1.822	0.185	1	2.7	0.093	0.762	1	3.25	0.419	0.521	14.7	0.83	0.368	1
	Season	1	0.44	0.015	0.903	1	**140.83**	**4.866**	**0.033**	1	**237.67**	**30.678**	**<0.001**	**432.1**	**24.395**	**<0.001**	1
	Sites:Season	1	0.18	0.006	0.938	1	57.8	1.997	0.165	1	16.82	2.172	0.148	54.1	3.056	0.088	1
	Residuals	39	29.31			40	28.94			41	7.75			17.7			41
Evenness	Sites	1	**0.13153**	**6.32**	**0.016**	1	0.000145	0.012	0.912	1	**0.06476**	**6.069**	**0.018**	0.1599	3.631	0.064	1
	Season	1	0.00463	0.222	0.640	1	0.025579	2.154	0.150	1	0.00311	0.292	0.592	**0.8593**	**19.516**	**<0.001**	1
	Sites:Season	1	0.02039	0.98	0.328	1	0.012152	1.023	0.318	1	0.01255	1.176	0.285	0.0001	0.003	0.959	1
	Residuals	39	0.02081			40	0.011874			41	0.01067			0.044			41
Shannon diversity	Sites	1	**0.14666**	**4.952**	**0.032**	1	0.00273	0.335	0.566	1	0.03774	3.434	0.071	0.0812	1.527	0.224	1
	Season	1	0.00738	0.249	0.620	1	**0.06153**	**7.553**	**0.009**	1	0.04466	4.063	0.051	**1.3223**	**24.867**	**<0.01**	1
	Sites:Season	1	0.03464	1.169	0.286	1	0.01486	1.824	0.184	1	0.00107	0.097	0.757	0.0113	0.213	0.647	1
	Residuals	39	0.02962			40	0.00815			41	0.01099			0.0532			41
Simpson diversity	Sites	1	**1.4748**	**6.763**	**0.013**	1	0.0103	0.122	0.729	1	0.2774	3.021	0.090	0.32	1.028	0.317	1
	Season	1	0.0404	0.185	0.669	1	0.6195	7.313	0.010	1	1.1574	12.606	0.001	**8.145**	**26.212**	**<0.001**	1
	Sites:Season	1	0.1912	0.877	0.355	1	0.2627	3.101	0.086	1	0.0082	0.089	0.767	0.138	0.443	0.510	1
	Residuals	39	0.2181			40	0.0847			41	0.0918			0.311			41

Df: degree of freedom and MS: mean square.

Table 4. Summary of two-way ANOVAs for functional feeding groups at different sites and seasons.

Variable	Factors	GR				HC				GS				WD			
		Df	MS	F value	P	Df	MS	F value	P	MS	F value	P	Df	Df	MS	F value	P
Predator	Sites	1	1118.8	2.366	0.132	1	340	0.317	0.576	503.7	6.492	0.015	1	1	503.7	6.492	0.015
	Season	1	19.1	0.04	0.842	1	4	0.004	0.951	138.2	1.781	0.190	1	1	138.2	1.781	0.190
	Sites: Season	1	20.7	0.044	0.835	1	6266	5.849	0.020	7.1	0.092	0.763	1	1	7.1	0.092	0.763
	Residuals	39	472.8			41	1071			77.6			40	40	77.6		
Scraper	Sites	1	33.6	0.148	0.702	1	4225	1.365	0.249	964.3	3.561	0.066	1	1	24	0.132	0.718
	Season	1	2290.2	10.116	0.003	1	12,855	4.154	0.048	888.7	3.282	0.078	1	1	3852	20.891	<0.001
	Sites: Season	1	21.7	0.096	0.758	1	884	0.286	0.596	818.5	3.023	0.090	1	1	162	0.879	0.354
	Residuals	39	226.4			41	3094			270.7			41	41	184		
Collector-gatherer	Sites	1	448,017	4.262	0.046	1	145.2	0.082	0.777	853	1.15	0.290	1	1	648858	6.519	0.015
	Season	1	28,793	0.274	0.604	1	1080	0.607	0.441	18,302	24.669	<0.001	1	1	48386	0.486	0.490
	Sites: Season	1	8235	0.078	0.781	1	115.2	0.065	0.800	1887	2.543	0.119	1	1	704	0.007	0.933
	Residuals	39	105,131			41	1780.6			742			41	41	99533		
Shredder	Sites	1	1299.9	11.346	0.002	1	7	0.005	0.946	2.145	0.269	0.607	1	1	0.133	0.297	0.589
	Season	1	578.9	5.053	0.030	1	4713	3.295	0.077	25.964	3.259	0.079	1	1	4.929	10.974	0.002
	Sites: Season	1	1314.7	11.475	0.002	1	1862	1.302	0.261	0.347	0.044	0.836	1	1	0.166	0.369	0.547
	Residuals	39	114.6			41	1430			7.968			41	41	0.449		
Collector-filterer	Sites	1	2530.5	14.692	<0.001	1	16.1	0.053	0.819	0.2031	0.416	0.523	1	1	16.133	2.774	0.103
	Season	1	40.9	0.237	0.629	1	472	1.546	0.221	0.0019	0.004	0.950	1	1	0.215	0.037	0.849
	Sites: Season	1	31.2	0.181	0.673	1	6	0.02	0.889	0.0006	0.001	0.973	1	1	0.025	0.004	0.948
	Residuals	39	172.2			41	305.3			0.488			41	41	5.815		

Df: degree of freedom and MS: mean square.

Table 5. Summary of two-way ANOVAs for habit trait groups at different sites and seasons.

Variable	Factors	GS				HC				GS				WD			
		Df	MS	F value	P	Df	MS	F value	P	Df	MS	F value	P	Df	MS	F value	P
Clinger	Sites	1	22,552	8.083	0.007	1	101	0.008	0.931	1	1482	1.667	0.204	1	178	0.244	0.624
	Season	1	4444	1.593	0.214	1	26,049	1.98	0.167	1	20,344	22.891	<0.001	1	15,216	20.925	<0.001
	Sites:Season	1	139	0.05	0.825	1	20,930	1.591	0.214	1	3729	4.196	0.047	1	1428	1.963	0.169
	Residuals	39	2790			41	13,153			40	889			41	727		
Burrower	Sites	1	724.9	1.685	0.202	1	0	0	1	1	13.05	1.277	0.265	1	132.3	7.651	0.008
	Season	1	452.8	1.053	0.311	1	0	0	1	1	296.82	29.041	<0.001	1	108.02	6.247	0.016
	Sites:Season	1	224.6	0.522	0.474	1	7.2	1.538	0.222	1	2.47	0.241	0.626	1	23.89	1.382	0.247
	Residuals	39	430.2			41	4.683			40	10.22			41	17.29		
Swimmer	Sites	1	5.32	0.154	0.697	1	997.6	7.484	0.009	1	1191.7	4.004	0.052	1	80	3.538	0.067
	Season	1	52.31	1.511	0.226	1	537.6	4.033	0.051	1	1262.1	4.24	0.046	1	537.8	23.774	<0.001
	Sites:Season	1	0.27	0.008	0.930	1	281.2	2.11	0.154	1	11.8	0.04	0.843	1	129.6	5.728	0.021
	Residuals	39	34.61			41	133.3			40	297.6			41	22.6		
Sprawler	Sites	1	650,630	6.106	0.018	1	187	0.161	0.690	1	13.88	0.368	0.547	1	627,853	6.191	0.017
	Season	1	33,788	0.317	0.577	1	3245	2.789	0.103	1	117.05	3.106	0.086	1	66,881	0.66	0.421
	Sites:Season	1	12,351	0.116	0.735	1	884	0.76	0.388	1	90.46	2.4	0.129	1	104	0.001	0.975
	Residuals	39	106,549			41	1164			40	37.69			41	101,408		
Climber	Sites	1	129.1	9.62	0.004	1	187	0.161	0.690	1	73.3	3.445	0.071	1	0.03333	0.382	0.540
	Season	1	11	0.82	0.371	1	3245	2.789	0.103	1	7.12	0.335	0.566	1	0.02917	0.334	0.566
	Sites:Season	1	35.27	2.628	0.113	1	884	0.76	0.388	1	29.08	1.367	0.249	1	0.00292	0.033	0.856
	Residuals	39	13.42			41	1164			40	21.28			41	0.08729		

Df: degree of freedom and MS: mean square.

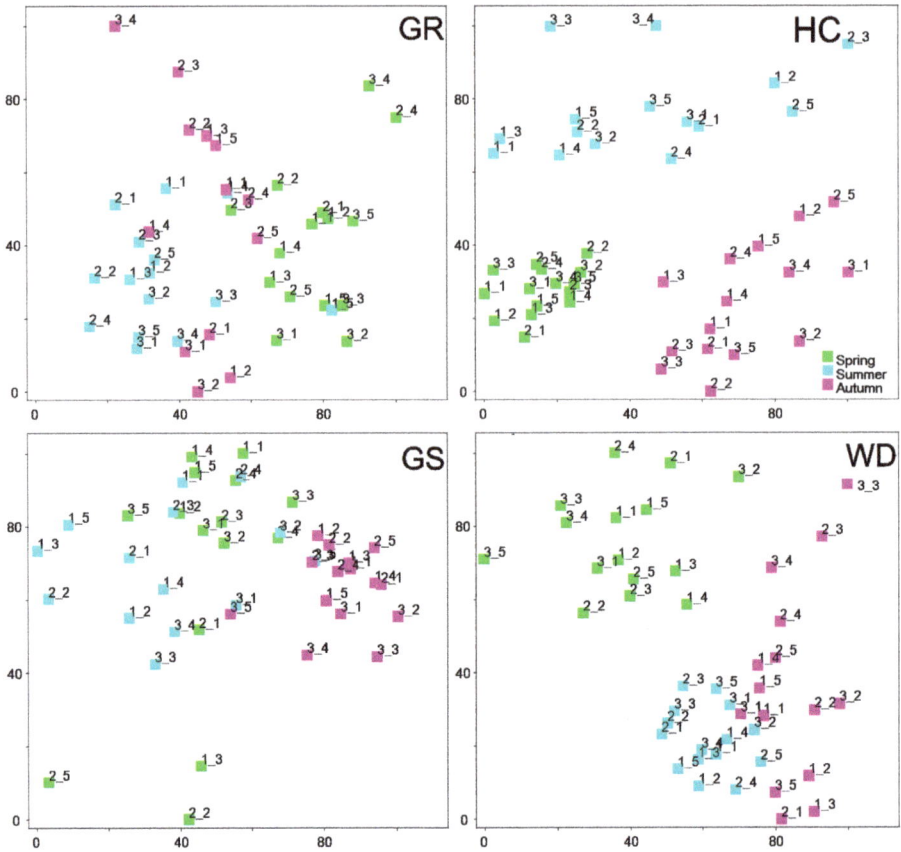

Figure 2. Spatial and/or temporal changes in macroinvertebrate communities using NMS ordination in four different headwaters. Acronyms in NMS units stand for the samples: the first numbers indicate sampling sites (*i.e.*, 1, 2 and 3) in each headwater and the last numbers represent replicates in each sampling site (1, 2, 3, 4 and 5). Each axis was rescaled on the 0–100 range based on the min-max scores of the NMS axes. (The stress values of the first two axes at GR: 22.4, HC: 14.9, GS: 22.3 and WD: 14.7).

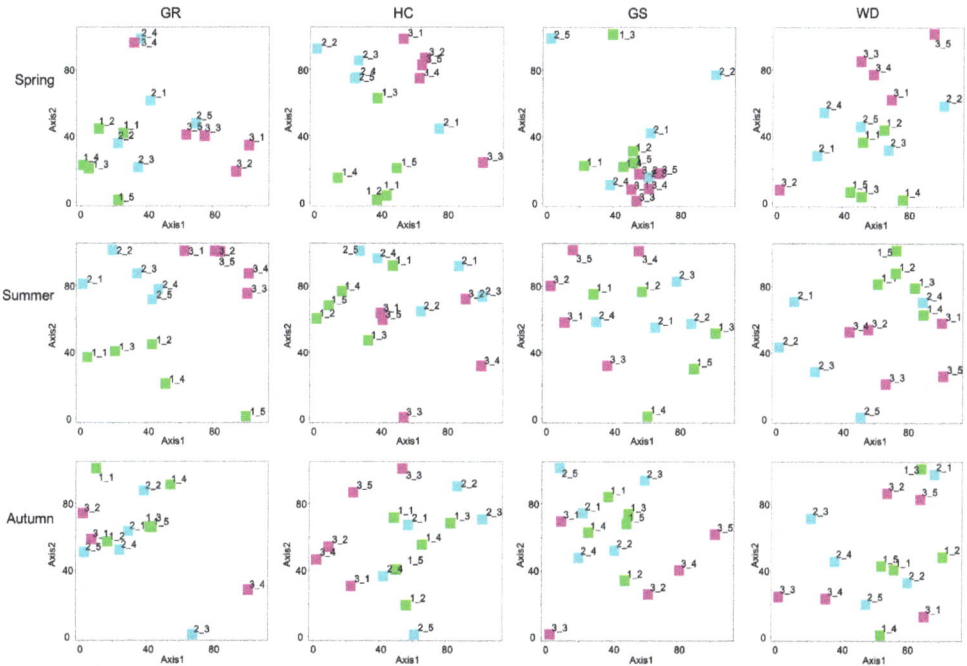

Figure 3. Spatial differences in macroinvertebrate communities using NMS ordination in four different headwaters in each season. Acronyms in NMS units stand for the samples: the first numbers indicate sampling sites (*i.e.*, 1, 2 and 3) in each headwater and the last numbers represent replicates in each sampling site (*i.e.*, 1, 2, 3, 4 and 5). Each axis was rescaled on the 0–100 range based on the min-max scores of NMS axes. (The stress values of the first two axes at GR: spring 14.8, summer 14.8, and autumn 10.3, at HC: spring 13.4, summer 18.4, and autumn 12.3, at GS: spring 10.5, summer 16.0, and autumn 16.4, and at WD: spring 16.8, summer 16.4, and autumn 104).

4. Discussion

Headwater streams are highly heterogeneous environments [9,10,26,55], supporting unique faunas that can differ from those in larger downstream areas [11]. Further, spatial and seasonal variations of various environmental factors create complex habitat conditions [56]. Upstream diversity influences the diversity of species found downstream and thus is important for the re-establishment of populations following local extinction events [57]. Despite the importance of headwater ecosystems for the resilience of species diversity upstream and downstream, little attention has been given to scale-dependent or multi-scale dependent variability in macroinvertebrate communities in headwater streams [58,59].

Our results showed that community indices were significantly different between seasons and sites that were closely located geographically (<500 m). The differences in species richness at the local scale could be caused by local processes such as habitat heterogeneity [60], biotic interactions [61], and biogeographical processes [62]. Moreover, because all the riparian zones were predominately forested, with no anthropogenic disturbance, the main factors differentiating the community composition at the stream and site scales likely relate to the natural variability of physical habitats and seasonal changes (e.g., canopy cover and the degree of autumn-shed leaves) [63]. For example, the differences in riparian vegetation, latitude, discharge rate and substrate composition prevailing among riffles and/or sites in each stream sections influence the distributions of macroinvertebrates. The amount, magnitude, and intensity of precipitation could also differ between headwater streams, reflecting regional differences (*i.e.*, southern and northern regions in Korea) [64]. In addition, each season can harbor unique habitats with interactions among differential environmental factors and organisms. Periphyton biomass can be limited by light in autumn and summer but not in spring, while nutrients can limit periphyton when light availability is higher [30]. Furthermore, seasonality in hydrology can be influential to structure macroinvertebrate composition [65]. During spring, snow-melting can be the main source of surface water supply as well as groundwater recharge. Particularly, in Korea, sequential floods (*i.e.*, summer) and droughts (*i.e.*, autumn) are main natural disturbances in headwater streams that affect the composition of benthic macroinvertebrate composition [66].

Differences within FFGs and HTGs were also observed among sites and seasons. Taxa associated with a particular habit category (*i.e.*, HTGs) exhibit certain morphological, physiological and behavioral adaptations to various microhabitats in freshwater ecosystems [67]. They can exist at low discharge rates compared with areas downstream because headwater streams are generally supplied by small catchment areas [64]. Clingers have morphological adaptations (e.g., curved tarsal claws, dorsoventral flattening, ventral gills arranged as a sucker, suction discs, and use of silk to construct attached retreats) that allow them to cling to substrate surfaces [68]. Therefore, in this study, the differences between hydrological variables as well as substrate compositions may have caused the significant differences in the abundance of clingers among streams. Furthermore, scrapers showed differences among streams and sites over time compared to other FFGs in this study. This was likely due to the differences in stream width and canopy cover. For example, the distribution of grazing invertebrates is directly influenced by the distribution of benthic algae, and therefore indirectly influenced by canopy cover [69,70]. Many researchers have suggested that scraper abundance tends to exhibit small-scale patchiness, resulting in localized variations depending on their algal food resources [29,30].

In our study, in NMS, samples were differentiated by seasons more than by spatial differences in each headwater stream. Within each season, the longitudinal differences in benthic macroinvertebrate communities were reflected in the NMS ordination. The units were clearly differentiated according to site differences even though the ordination patterns in each season were dissimilar. This indicated that in spite of their short distances between the adjacent sites in each stream (*i.e.*, less than 500 m) without anthropogenic disturbances, they have their own habitat characteristics among sites, which have different resilience and resistance in comparison to seasonal effects, reflecting complicated interactions among spatial and temporal cues.

5. Conclusions

Our study examined the structure and function of the macroinvertebrate community at two different spatial scales during three seasons. Community and functional diversity indices varied significantly within seasons and/or sites as well as by the category of FFGs or HTGs. In NMS, within a single headwater stream, samples were separated by seasonality rather than spatial differences. Within each season, sample ordination reflected site differences, suggesting that macroinvertebrate communities respond to multiple and interacting spatial and temporal cues. Therefore, continuous monitoring and research on the interactions between species diversity and spatio-temporal and physiochemical effects are fundamental to maintain catchment biodiversity and to provide strategies for watershed restoration of macroinvertebrate communities.

Acknowledgments: This study was supported by the Korea Forest Research Institute and by the Ministry of Environment and the National Institute of Environmental Research, Korea.

Author Contributions: Mi-Jung Bae and Young-Seuk Park conducted all the analyses and wrote the manuscript. Jung Hwa Chun and Tae-Soo Chon contributed materials. All authors participated in discussions as well as approved the final manuscript.

Conflicts of Interest: The authors declare no conflict of interest.

References

1. Heywood, V.H.; Watson, R.T.; Baste, I. *Global Biodiversity Assessment*; Heywood, V.H., Ed.; Cambridge University Press: Cambridge, UK, 1995.
2. Shiklomanov, I.A. World fresh water resources. In *Water in Crisis. A Guide to the World's Fresh Water Resources*; Gleick, P.H., Ed.; Oxford University Press: New York, NY, USA, 1993; pp. 13–24.
3. Allan, J.D.; Flecker, A.S. Biodiversity conservation in running waters. *BioScience* **1993**, *43*, 32–43.
4. Strayer, D.L.; Dudgeon, D. Freshwater biodiversity conservation: Recent progress and future challenges. *J. N. Am. Benthol. Soc.* **2010**, *29*, 344–358.

5. Townsend, C.R. The patch dynamics concept of stream community ecology. *J. N. Am. Benthol. Soc.* **1989**, *8*, 36–50.

6. Leopold, L.B.; Wolman, M.G.; Miller, J.P. *Fluvial Processes in Geomorphology*; W. H. Freeman and Co.: San Francisco, CA, USA, 1964.

7. Meyer, J.L.; Wallace, J.B. *Lost Linkages and Lotic Ecology: Rediscovering Small Streams*; Blackwell Scientific: Oxford, UK, 2001.

8. Gomi, T.; Sidle, R.C.; Richardson, J.S. Understanding processes and downstream linkages of headwater systems. *BioScience* **2002**, *52*, 905–916.

9. Clarke, A.; Mac Nally, R.; Bond, N.; Lake, P.S. Macroinvertebrate diversity in headwater streams: A review. *Freshw. Biol.* **2008**, *53*, 1707–1721.

10. Lowe, W.H.; Likens, G.E. Moving headwater streams to the head of the class. *BioScience* **2005**, *55*, 196–197.

11. Vannote, R.L.; Minshall, G.W.; Cummins, K.W.; Sedell, J.R.; Cushing, C.E. The river continuum concept. *Can. J. Fish. Aquat. Sci.* **1980**, *37*, 130–137.

12. Naiman, R.J.; Décamps, H. The ecology of interfaces: Riparian zones. *Annu. Rev. Ecol. Syst.* **1997**, 621–658.

13. Chloe, W., III; Garman, G.C. The energetic importance of terrestrial arthropod inputs to three warm-water streams. *Freshw. Biol.* **1996**, *36*, 104–114.

14. Allan, J.D.; Wipfli, M.S.; Caouette, J.P.; Prussian, A.; Rodgers, J. Influence of streamside vegetation on inputs of terrestrial invertebrates to salmonid food webs. *Can. J. Fish. Aquat. Sci.* **2003**, *60*, 309–320.

15. Nakano, S.; Miyasaka, H.; Kuhara, N. Terrestrial-aquatic linkages: Riparian arthropod inputs alter trophic cascades in a stream food web. *Ecology* **1999**, *80*, 2435–2441.

16. Wipfli, M.S.; Richardson, J.S.; Naiman, R.J. Ecological linkages between headwaters and downstream ecosystems: Transport of organic matter, invertebrates, and wood down headwater channels. *J. Am. Water Resour. Assoc.* **2007**, *43*, 72–85.

17. Dieterich, M.; Anderson, N.H. Life cycles and food habits of mayflies and stoneflies from temporary streams of western Oregon. *Freshw. Biol.* **1995**, *34*, 47–60.

18. Muchow, C.L.; Richardson, J.S. Unexplored diversity: Macroinvertebrates in coastal British Columbia headwater streams. In Proceedings of a Conference on the Biology and Management of Species and Habitats at Risk, Kamloops, BC, Canada, 15–19 February 1999; Darling, L.M., Ed.; British Columbia Ministry of Environment, Lands and Parks: Victoria, BC, Canada, 2000; pp. 503–506.

19. Heino, J. Functional biodiversity of macroinvertebrate assemblages along major ecological gradients of boreal headwater streams. *Freshw. Biol.* **2005**, *50*, 1578–1587.

20. Meyer, J.L.; Strayer, D.L.; Wallace, J.B.; Eggert, S.L.; Helfman, G.S.; Leonard, N.E. The contribution of headwater streams to biodiversity in river networks. *J. Am. Water Resour. Assoc.* **2007**, *43*, 86–103.

21. Boulton, A.J. Parallels and contrasts in the effects of drought on stream macroinvertebrate assemblages. *Freshw. Biol.* **2003**, *48*, 1173–1185.

22. Wallace, J.B.; Webster, J.R. The role of macroinvertebrates in stream ecosystem function. *Annu. Rev. Entomol.* **1996**, *41*, 115–139.

23. Covich, A.P.; Palmer, M.A.; Crowl, T.A. The role of benthic invertebrate species in freshwater ecosystems. *Bioscience* **1999**, *49*, 119–127.

24. Cairns, J., Jr.; Marshall, K.E.; Johnson, R.K.; Norris, R.H.; Reice, S.R.; Walker, I.R.; Buikema, A.L., Jr.; Cooper, S.D.; Brinkhurst, R.O.; Pratt, J.R.; *et al. Freshwater Biomonitoring and Benthic Macroinvertebrates*; Rosenberg, D.M., Resh, V.H., Eds.; Chapman and Hall: London, UK, 1993.

25. Palmer, M.A.; Ambrose, R.F.; Poff, N.L. Ecological theory and community restoration ecology. *Restor. Ecol.* **1997**, *5*, 291–300.

26. Heino, J.; Muotka, T.; Paavola, R. Determinants of macroinvertebrate diversity in headwater streams: Regional and local influences. *J. Anim. Ecol.* **2003**, *72*, 425–434.

27. Ligeiro, R.; Melo, A.S.; Callisto, M. Spatial scale and the diversity of macroinvertebrates in a Neotropical catchment. *Freshw. Biol.* **2010**, *55*, 424–435.

28. García-Roger, E.M.; del Mar Sánchez-Montoya, M.; Gómez, R.; Suárez, M.L.; Vidal-Abarca, M.R.; Latron, J.; Rieradevall, M.; Prat, N. Do seasonal changes in habitat features influence aquatic macroinvertebrate assemblages in perennial *versus* temporary Mediterranean streams? *Aquat. Sci.* **2011**, *73*, 567–579.

29. Kohler, S.D. Search mechanism of a stream grazer in patchy environments: The role of food abundance. *Oecologia* **1984**, *62*, 209–218.

30. Vaughn, C.C. The role of periphyton abundance and quality in the microdistribution of a stream grazer, *Helicopsyche borealis* (Trichoptera, Helicopsychidae). *Freshw. Biol.* **1986**, *16*, 485–493.

31. Richards, C.; Minshall, G.W. The influence of periphyton abundance on *Baetis bicaudatus* distribution and colonization in a small stream. *J. N. Am. Benthol. Soc.* **1988**, *7*, 77–86.

32. Molles, M.C. Trichopteran communities of streams associated with aspen and conifer forests: Long-term structural change. *Ecology* **1982**, *63*, 1–6.

33. Cummins, K.W.; Wilzbach, M.A.; Gates, D.M.; Perry, J.B.; Taliaferro, W.B. Shredders and riparian vegetation. *BioScience* **1989**, *39*, 24–30.

34. Dobson, M. Microhabitat as a determinant of diversity: Stream invertebrates colonizing leaf packs. *Freshw. Biol.* **1994**, *32*, 565–572.

35. Haapala, A.; Muotka, T.; Laasonen, P. Distribution of benthic macroinvertebrates and leaf litter in relation to streambed retentivity: Implications for headwater stream restoration. *Boreal Environ. Res.* **2003**, *8*, 19–30.

36. Chung, N.; Bae, M.-J.; Li, F.; Kwon, Y.-S.; Kwon, T.-S.; Kim, J.S.; Park, Y.-S. Habitat characteristics and trophic structure of benthic macroinvertebrates in a forested headwater stream. *J. Asia Pac. Entomol.* **2012**, *15*, 495–505.

37. Hart, D.D.; Finelli, C.M. Physical-biological coupling in streams: The pervasive effects of flow on benthic organisms. *Annu. Rev. Ecol. Evol. Syst.* **1999**, *30*, 363–395.

38. Palmer, M.A.; Covich, A.P.; Lake, S.A.M.; Biro, P.; Brooks, J.J.; Cole, J.; Dahm, C.; Gibert, J.; Geodkoop, W.; Martens, K.; *et al.* Linkages between aquatic sediment biota and life above sediments as potential drivers of biodiversity and ecological processes. *BioScience* **2000**, *50*, 1062–1075.

39. Townsend, A.R.; Howarth, R.W.; Bazzaz, F.A.; Booth, M.S.; Cleveland, C.C.; Collinge, S.K.; Dobson, A.P.; Epstein, P.R.; Holland, E.A.; Keeney, D.S.; *et al.* Human health effects of a changing global nitrogen cycle. *Front. Ecol. Environ.* **2003**, *1*, 240–246.

40. KMA. Available online: http://web.kma.go.kr/eng/index.jsp (accessed on 2 May 2015).

41. Bae, M.J.; Chon, T.S.; Park, Y.S. Characterizing differential responses of benthic macroinvertebrate communities to floods and droughts in three different stream types using a Self-Organizing Map. *Ecohydrology* **2014**, *7*, 115–126.

42. Schmera, D.; Erős, T. Sample size influences the variation of invertebrate diversity among different levels of a stream habitat hierarchy. *Int. Rev. Hydrobiol.* **2012**, *97*, 74–82.

43. Brinkhurst, R.O.; Jamieson, B.G.M. *Aquatic Oligochaeta of the World*; University of Toronto Press: Toronto, ON, Canada, 1971; p. 860.

44. Quigley, M. *Invertebrates of Streams and Rivers: A Key to Identification*; Edward Arnold: London, UK, 1977.

45. Pennak, R.W. *Freshwater Invertebrates of the United States*; John Wiley and Sons, Inc.: New York, NY, USA, 1978.

46. Brighnam, A.R.; Brighnam, W.U.; Gnika, A. *Aquatic Insects and Oligochaetes of North and South Carolina*; Midwest Aquatic Enterprise: Mahomet, IL, USA, 1982.

47. Yun, I.B. *Illustrated Encyclopedia of Fauna and Flora of Korea. Aquatic Insects*; Aquat. Insects, Ministry Education: Seoul, Korea, 1988; Volume 30, pp. 430–551.

48. Merritt, R.W.; Cummins, K.W. *An Introduction to the Aquatic Insects of North America*; Kendall Hunt: Dubuque, IA, USA, 2006.

49. Environmental Systems Research Incorporated ArcGIS 10.0. Environmental Systems Research Incorporated, Redlands, CA. USA, 2011.

50. Shepard, R.N. The analysis of proximities: Multidimensional scaling with an unknown distance function. II. *Psychometrika* **1962**, *27*, 219–246.

51. Mahecha, M.D.; Martinez, A.; Lischeida, G.; Beckc, E. Nonlinear dimensionality reduction: Alternative ordination approaches for extracting and visualizing biodiversity patterns in tropical montane forest vegetation data. *Ecol. Inform.* **2007**, *2*, 138–149.

52. Bae, M.J.; Kwon, Y.; Hwang, S.J.; Chon, T.S.; Yang, H.J.; Kwak, I.S.; Pakr, J.H.; Ham, S.A.; Park, Y.S. Relationships between three major stream assemblages and their environmental factors in multiple spatial scales. *Ann. Limnol. Int. J. Limnol.* **2011**, *47*, S91–S105.

53. R Development Core Team. R: A Language and Environment for Statistical Computing. R Foundation for Statistical Computing 2006, Vienna, Austria. ISBN 3-900051-07-0. Available online: http://www.R-project.org (accessed on 10 April 2015).

54. McCune, B.; Mefford, M.J. *PC-ORD. Multivariate Analysis of Ecological Data*; version 5.0; MjM Software Design: Gleneden Beach, OR, USA, 2006.

55. Furse, M.T. The application of RIVPACS procedures in headwater streams an extensive and important national resource. In *Assessing the Biological Quality of Freshwaters, RIVPACS and Other Techniques, 7991*; Wright, J.F., Sutcliffe, D.W., Furse, M.T., Eds.; The Freshwater Biological Association: Cumbria, UK, 2000.

56. Clinton, S.M.; Grimm, N.B.; Fisher, S.G. Response of a hyporheic invertebrate assemblage to drying disturbance in a desert stream. *J. N. Am. Benthol. Soc.* **1996**, *15*, 700–712.

57. Callanan, M.; Baars, J.R.; Kelly-Quinn, M. Macroinvertebrate communities of Irish headwater streams: Contribution to catchment biodiversity. *Biol. Environ.* **2014**, *114*, 143–162.

58. Downes, B.J.; Hindell, J.S.; Bond, N.R. What's in a site? Variation in lotic macroinvertebrate density and diversity in a spatially replicated experiment. *Austral Ecol.* **2000**, *25*, 128–139.

59. Parsons, M.; Thoms, M.C.; Norris, R.H. Scales of macroinvertebrate distribution in relation to the hierarchical organization of river systems. *J. N. Am. Benthol. Soc.* **2003**, *22*, 105–122.

60. Grönroos, M.; Heino, J. Species richness at the guild level: Effects of species pool and local environmental conditions on stream macroinvertebrate communities. *J. Anim. Ecol.* **2012**, *81*, 679–691.

61. Field, R.; Hawkins, B.A.; Cornell, H.V.; Currie, D.J.; Diniz-Filho, J.A.F.; Guégan, J.F.; Kaufman, D.M.; Kerr, J.T.; Mittelbach, G.G.; Oberdorff, T.; *et al.* Spatial species-richness gradients across scales: A meta-analysis. *J. Biogeogr.* **2009**, *36*, 132–147.

62. Ricklefs, R.E.; Schluter, D.S. *Species Diversity in Ecological Communities: Historical and Geographical Perspectives*; The University of Chicago Press: Chicago, IL, USA, 1993.

63. Mykrä, H.; Heino, J.; Muotka, T. Variability of lotic macroinvertebrate assemblages and stream habitat characteristics across hierarchical landscape classifications. *Environ. Manag.* **2004**, *34*, 341–352.

64. Heino, J.; Louhi, P.; Muotka, T. Identifying the scales of variability in stream macroinvertebrate abundance, functional composition and assemblage structure. *Freshw. Biol.* **2004**, *49*, 1230–1239.

65. Kanandjembo, A.N.; Platell, M.E.; Potter, I.C. The benthic macroinvertebrate community of the upper reaches of an Australian estuary that undergoes marked seasonal changes in hydrology. *Hydrol. Process.* **2001**, *15*, 2481–2501.

66. Lake, P.S. Disturbance, patchiness, and diversity in streams. *J. N. Am. Benthol. Soc.* **2000**, *19*, 573–592.

67. Lamouroux, N.; Dolédec, S.; Gayraud, S. Biological traits of stream macroinvertebrate communities: Effects of microhabitat, reach, and basin filters. *J. N. Am. Benthol. Soc.* **2004**, *23*, 449–466.

68. Cummins, K.W.; Merritt, R.W.; Berg, M.B. Ecology and distribution of aquatic insects. In *An Introduction to the Aquatic Insects of North America*; Merritt, R.W., Cummins, K.W., Berg, M.B., Eds.; Kendell/Hunt: Dubuque, IA, USA, 2008; pp. 105–122.

69. Behner, D.J.; Hawkins, C.P. Effects of overhead canopy on macroinvertebrate production in a Utah stream. *Freshw. Biol.* **1986**, *16*, 287–300.

70. Hill, W.R.; Ryon, M.G.; Schilling, E.M. Light limitation in a stream ecosystem: Responses by primary producers and consumers. *Ecology* **1995**, *76*, 1297–1309.

Multiple Time-Scale Monitoring to Address Dynamic Seasonality and Storm Pulses of Stream Water Quality in Mountainous Watersheds

Hyun-Ju Lee, Kun-Woo Chun, Christopher L. Shope and Ji-Hyung Park

Abstract: Rainfall variability and extreme events can amplify the seasonality and storm pulses of stream water chemistry in mountainous watersheds under monsoon climates. To establish a monitoring program optimized for identifying potential risks to stream water quality arising from rainfall variability and extremes, we examined water chemistry data collected on different timescales. At a small forested watershed, bi-weekly sampling lasted over two years, in comparison to three other biweekly sampling sites. In addition, high-frequency continuous measurements of pH, electrical conductivity, and turbidity were conducted in tandem with automatic water sampling at 2 h intervals during eight rainfall events. Biweekly monitoring showed that during the summer monsoon period, electrical conductivity (EC), dissolved oxygen (DO), and dissolved ion concentrations generally decreased, but total suspended solids (TSS) slightly increased. A noticeable variation from the usual seasonal pattern was that DO levels substantially decreased during an extended drought. Bi-hourly storm event samplings exhibited large changes in the concentrations of TSS and particulate and dissolved organic carbon (POC; DOC) during intense rainfall events. However, extreme fluctuations in sediment export during discharge peaks could be detected only by turbidity measurements at 5 min intervals. Concomitant measurements during rainfall events established empirical relationships between turbidity and TSS or POC. These results suggest that routine monitoring based on weekly to monthly sampling is valid only in addressing general seasonal patterns or long-lasting phenomena such as drought effects. We propose an "adaptive" monitoring scheme that combines routine monitoring for general seasonal patterns and high-frequency instrumental measurements of water quality components exhibiting rapid responses pulsing during intense rainfall events.

Reprinted from *Water*. Cite as: Lee, H.-J.; Chun, K.-W.; Shope, C.L.; Park, J.-H. Multiple Time-Scale Monitoring to Address Dynamic Seasonality and Storm Pulses of Stream Water Quality in Mountainous Watersheds. *Water* **2015**, *7*, 6117–6138.

1. Introduction

Headwater streams are a habitat for aquatic organisms, provide drinking water for downstream population centers, and contribute sources of organic matter, nutrients, and sediments to higher-order streams and rivers [1,2]. Mountainous watersheds are particularly important for providing water resources, as illustrated by the disproportionately high contribution of mountainous areas (32% of the world's land area) to the total discharge in the world's major river basins (63%) [3]. Land use changes on steep mountainous terrain, such as deforestation and agricultural expansion, have been associated with elevated flood risks [4,5] and water quality deterioration caused by suspended sediment eroded from disturbed soils [6,7]. More frequent intense monsoon rainfall events, which might occur as a consequence of climate change [8], can also have significant impacts on material transport and surface water quality in mountainous watersheds [9]. Instantaneous changes in discharge rates and stream chemistry in response to intense rainfalls cannot be adequately captured by conventional monitoring approaches and networks based on low frequency water sampling at weekly to monthly intervals. Novel monitoring approaches employing high frequency sampling and advanced sensor techniques have been proposed as tools to detect pulses of dissolved solutes, suspended sediments, or organic matter during important hydrologic events such as heavy rainfalls or snowmelts [10,11]. However, there have been few systematic assessments of sampling frequency and water quality components in the context of assessing climate-induced risks to water quality in mountainous headwater streams.

The strong coupling of headwater streams to hillslope processes often results in large sediment flows from episodic slope failures during extreme hydrologic events [1,7]. Forest management practices have been shown to significantly affect soil erosion on steep forested slopes through changes in forest conditions such as tree density, species composition, and soil compaction [12]. Recent studies have shown the effects of extreme climatic events such as typhoons on the flow of sediments and nutrients in mountainous watersheds in East Asia [13–15]. Suspended sediment is a major water quality problem in headwater streams, as it degrades drinking water quality and impacts aquatic organism habitats [6,16]. It also plays an important role in transporting nutrients [17], particulate organic carbon (POC) [11,18,19], and toxic metals [20].

Temporal variations in stream water chemistry reflect changes in both the forest ecosystem processes and the hydro-biogeochemical processes on the watershed level [21], offering a window through which we can monitor changes in environmental conditions that are important for regulating the production and release of materials within a forested watershed [22]. For example, long-term changes in stream water nitrogen and sulfur concentrations have been linked to the changing rates of acid deposition over recent decades [23,24]. It has also been suggested that changes in the amount and spatiotemporal distribution of precipitation, attributed to global climate change, has significant influences on the biogeochemical processes that regulate nutrient production and hydrologic export in forested watersheds [9,25,26]. Changes in rainfall amount and intensity can affect hydrologic flowpaths, along which nutrients released from soil sources are transported into streams [27]. Compared to the relatively well-established relationships between dissolved organic carbon (DOC) in forest streams and changing hydrologic conditions during storms [28,29], inconsistent patterns have been observed regarding nutrient releases in responses to storms (e.g., [30,31]). It has recently been reported that the concentrations and fluxes of DOC and POC in headwater streams respond differently to rainfall events of varying intensity and duration, with much stronger responses of POC occurring during extreme rainfall events [11,32].

Extreme rainfall events are often defined as events exceeding relative (*i.e.*, the upper first percentile of long-term data) or absolute (e.g., 100 mm per day) thresholds of precipitation [33]. The frequency and intensity of extreme rainfall events has been increasing over a large part of the Northern Hemisphere [8,34], including the Korean Peninsula [35]. There have been many reports on storm-induced water quality deterioration in large watersheds where storm responses might be more moderate compared to more rapid responses at smaller watersheds (e.g., [10,36,37]). Humic-like substances associated with aromatic and condensed structures have been shown to increase in forested watersheds during intense rainfall events, elevating the potential of disinfection byproduct formation upon chlorination in drinking water facilities [38,39]. These studies provide rare examples of climate-induced risks to water quality in headwater streams draining forested watersheds, yet more high-resolution monitoring data should be collected across a range of climatic conditions and watershed types to enhance our capacity to predict stream water quality changes during extreme rainfall events of varying duration and intensity.

The primary objective of this study was to compare different monitoring approaches based on different timescales and sampling methods in order to establish monitoring programs that can address large temporal variations of stream water quality associated with rainfall variability and extremes. Stream water chemistry was monitored on sub-hourly to bi-weekly timescales at small forested watersheds in the Haean Basin—a mountainous basin in northern South Korea. A specific goal was to identify major climate-induced risks to stream water quality in mountainous watersheds. Another important goal was to examine the potential of high-frequency, continuous measurements of turbidity to capture short-term changes in stream concentrations of suspended sediment and POC during rainfall events. For this purpose, high-frequency (5 min) instrumental measurements of pH, electrical conductivity (EC), and turbidity during several rainfall events of various duration and intensity were compared with bi-hourly measurements using an automatic water sampler.

2. Materials and Methods

2.1. Study Site

The study was conducted at four small forested watersheds in the Haean Basin (38°15′–38°20′ N; 128°05′–128°10′ E; 400 m–1304 m asl), a bowl-shaped, mountainous basin 1–2 km south of the demilitarized zone (DMZ) between South and North Korea (Figure 1) [19,20]. The bedrock below the center of the basin consists of highly weathered biotite granite surrounded by metamorphic rocks along the mountain ridges. Mixed deciduous forests were re-established through natural processes on the mountain slopes in the study area since the late 1970s, after recurrent forest fires during the three decades following the Korean War in 1950–1953. Rapid conversion of low-elevation forests to agricultural fields has occurred in recent decades, and natural forests now remain only on steep slopes (>20°) and along the mountain ridges, comprising 58% of the entire 60 km^2 area of the basin. The dominant tree species in the basin include Mongolian oak (*Quercus mongolica*), Daimyo oak (*Quercus dentata*), and Korean ash (*Fraxinus rhynchophylla*). Typical soils on the forested mountain slopes are dry to slightly moist, brown soils (acid Cambisols according to the FAO World Reference Base for Soil Resources), overlain by moder-like forest floors with a distinct Oi horizon and less distinct Oe/Oa horizons.

2.2. Sampling and In Situ Measurements

Routine bi-weekly water sampling was conducted at the four headwater streams from May 2008 to April 2010 (Figure 1). The four watersheds are all similar with regard to vegetation, soil, and topography, but their sizes vary (A: 38 ha; B: 56 ha; C: 190 ha; D: 171 ha). Grab water samples were collected at 10 cm below the stream surfaces near the centers of the streams using a 1 L Teflon bottle. The stream sampling procedure included measuring *in situ* water temperature, pH, EC, and dissolved oxygen (DO) using a portable pH meter (Orion 5 star, Thermo). In one forested watershed (A), throughfall ($n = 4$) and forest floor leachate (collected using custom-made zero-tension lysimeters; $n = 4$) were also sampled bi-weekly. Comparison of streamwater chemistry data with those for throughfall and forest floor leachate might provide some insights into the relative importance of atmospheric inputs and soil pools as potential sources of exported materials from the watershed.

Intensive storm event sampling at 2 h intervals was conducted eight times throughout two summer monsoon periods at the same stream location of watershed A previously described. During the eight rainfall events, stream samples were collected every 2 h using an autosampler (6712 Portable Sampler, ISCO Inc., Lincoln, NE, USA). In tandem with the automatic water sampling during rainfall events, in-stream, high-frequency measurements of pH, EC, and turbidity at 5 min intervals were conducted using a multi-parameter probe (6920 Water Quality Monitoring

System, YSI, Inc., Yellow Springs, OH, USA). The water quality probe was routinely checked with standard solutions and calibrated when necessary.

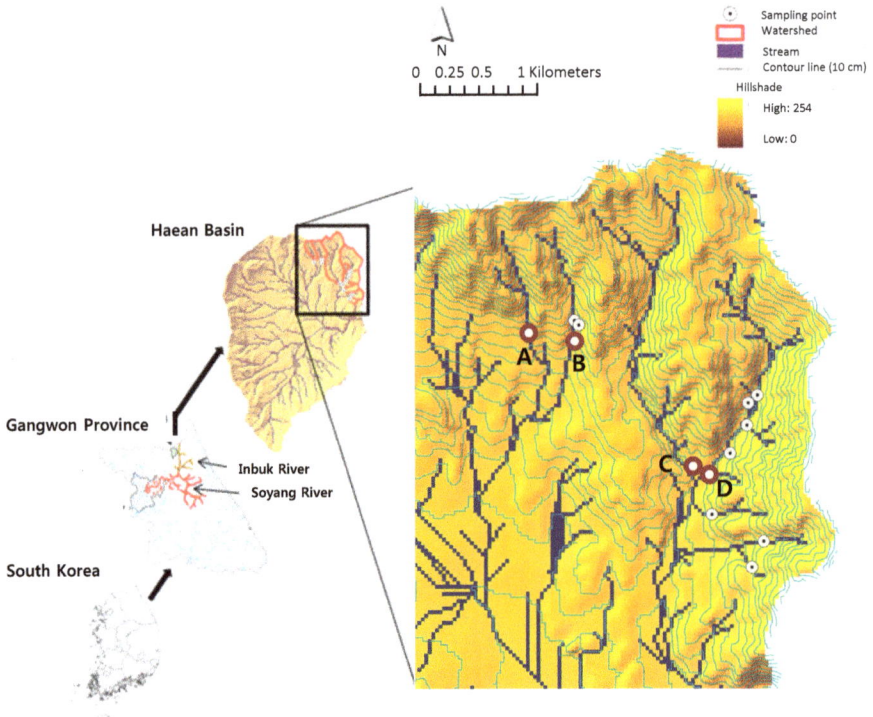

Figure 1. Stream water sampling locations for routine bi-weekly monitoring at the four streams (A–D). The location of the V-notch weir and the weather station within watershed A is indicated by the letter of X.

Continuous micrometeorological measurements were conducted at watershed A using a data logger (CR10X, Campbell Scientific, Logan, UT, USA) that was connected to various sensors, including precipitation, air and soil temperature, and volumetric soil water content at 10 and 30 cm depths (Figure 1). Discharge measurements were initiated at watershed A in July 2009, with the construction of a V-notch weir (Figure 1). A vented pressure transducer (Druck PDCR 830, Druck Ltd., Leicester, UK) was installed in a fully screened polyvinyl chloride (PVC) tube in a shallow pool upstream of the V-notch weir. Water level measurements were collected every 5 min. Discharge from the forested watershed during the period prior to the weir construction was estimated using a hydrologic model (HBV-Lite) [40]. The model was calibrated for the period from 16 July through 20 September 2009 on the basis of the observed soil moisture and stream discharge data. We then validated the model with the same parameters for two additional periods (30 June–30 August 2010; 5 August–5 October 2010). The validation period R^2 and efficiency values were 0.89 and 0.88 for the first period and 0.92 and 0.91 for the second period, respectively.

2.3. Chemical and Statistical Analyses

All water samples were transported in an ice box to the laboratory and then kept refrigerated at <4 °C. Within 24 h after sampling, a portion of the water sample (50–200 mL) was filtered through a pre-combusted glass fiber filter (GF/F, Whatman; nominal pore size of 0.7 μm, Whatman, Clifton, NJ, USA) after pre-filtering through a plastic sieve with 2 mm pores. The concentrations of total suspended solids (TSS) were measured gravimetrically as the difference in filter weight before and after filtering. Prior to filtering, the GF/F filters were combusted at 450 °C to remove any organic materials in the filters and then weighed. After filtering of the water samples, the filters were dried at 65 °C to a constant weight and re-weighed for the calculation of the TSS concentrations. For eight monsoon rainfall event samples, the dried filters were fumed with HCl in a sealed desiccator for 24 h to remove inorganic C prior to the analysis of POC. The concentrations of POC in the acidified filters were measured with an elemental analyzer (Vario MAX CN, Elementar, Germany).

Filtered water samples were analyzed for dissolved ions (Cl$^-$, SO$_4^{2-}$, NO$_3^-$, and NH$_4^+$) using an ion chromatograph (DX-320, Dionex, Sunnyvale, CA, USA). For the eight monsoon rainfall event samples, DOC in the filtered water samples was measured with a TOC analyzer using high-temperature combustion of organic matter followed by thermal detection of CO$_2$ (TOC 5000a, Shimadzu, Japan). As part of the quality control, a laboratory blank (Mili-Q ultrapure water) and a standard solution of a concentration that was typical of the concentration range of samples were analyzed for every batch of ten samples. Relative standard deviations (standard deviations divided by mean in percentage) from the repeated measurements of check standards were generally within 5%. Replicate analysis was conducted for approximately 10% of the total number of samples. Contamination from sampling and filtering was checked with field blank samples (Mili-Q ultrapure water).

Pearson correlation coefficients were used to characterize the correlations between bi-weekly water quality data and hydroclimatic variables. To reflect large changes in solute concentrations in response to rapidly changing discharge during rainfall events, event mean solute concentrations were obtained by weighting discharge to each corresponding solute concentration. Linear regression analyses were conducted to establish the relationships between the discharge-weighted event mean solute concentration (TSS, POC, and DOC) and the total rainfall amount or the mean rainfall intensity of each rainfall event. Most of POC and DOC data presented in this study (all events except events 5 and 6) have previously been used in two publications on differential storm responses of DOC and POC [11,19]. We included the published data in this study to provide a more complete view of watershed biogeochemical responses to storm events, incorporating both the traditional water quality and novel carbon cycle perspectives.

3. Results

3.1. Routine Bi-Weekly Monitoring in Four Streams

Volume-weighted mean concentrations of measured water quality components in the four forest streams were generally different from those for throughfall and forest floor leachate (Table 1). The stream water pH was usually circumneutral with the two-year mean of 6.7, indicating high acid-neutralizing capacity of the soil given the much lower pH values found in throughfall (5.6) and forest floor leachate (5.8). While the mean concentrations of all the measured parameters fell within the usual range found in streams draining temperate forested watersheds [6,12,13,17,22,29], it is notable that the mean concentrations NO_3^--N in stream water were higher than those for throughfall, indicating leaching losses from the forest soils.

Table 1. Means and standard errors ($n = 4$) of volume-weighted mean pH and electrical conductivity (EC) and concentrations of Cl^-, SO_4^{2-}, NO_3^--N, NH_4^+-N, and total suspended solids (TSS) in throughfall (TF), forest floor leachates (FF), and stream water (SW) for the monitoring period from May 2008 to April 2010. Dissolved oxygen (DO). SE indicates standard error.

		pH	EC ($\mu S \cdot cm^{-1}$)	DO	Cl^-	SO_4^{2-}	NO_3^--N	NH_4^+-N	TSS
						($mg \cdot L^{-1}$)			
TF	mean	5.6	22.4	5.1	1.36	2.76	0.51	0.44	14.0
($n = 4$)	SE	0.1	1.9	0.3	0.14	0.13	0.04	0.03	2.5
FF	mean	5.8	89.7	5.4	3.99	5.17	3.23	0.86	7.2
($n = 4$)	SE	0.1	7.3	0.2	0.37	0.48	0.46	0.17	0.2
SW	mean	6.7	35.2	8.7	1.63	2.65	0.70	0.02	7.3
($n = 4$)	SE	0.1	2.6	0.2	0.13	0.19	0.05	0.00	0.6

Over the 2 year monitoring period, water quality components showed similar seasonal patterns at all four streams, including lower EC, DO, and dissolved anion concentrations and higher TSS concentrations during the summer monsoon period (Figure 2). The seasonal patterns observed for the measured anions were not clear for NH_4^+ concentrations. On some occasions during the monsoon period, pH was lower than the usual circumneutral level, but the decrease did not continue throughout the season. The discharge-weighted mean concentrations of TSS in stream water were relatively low (7.3 $mg \cdot L^{-1}$; Table 1), constrained largely by very low concentrations during dry periods (Figure 2).

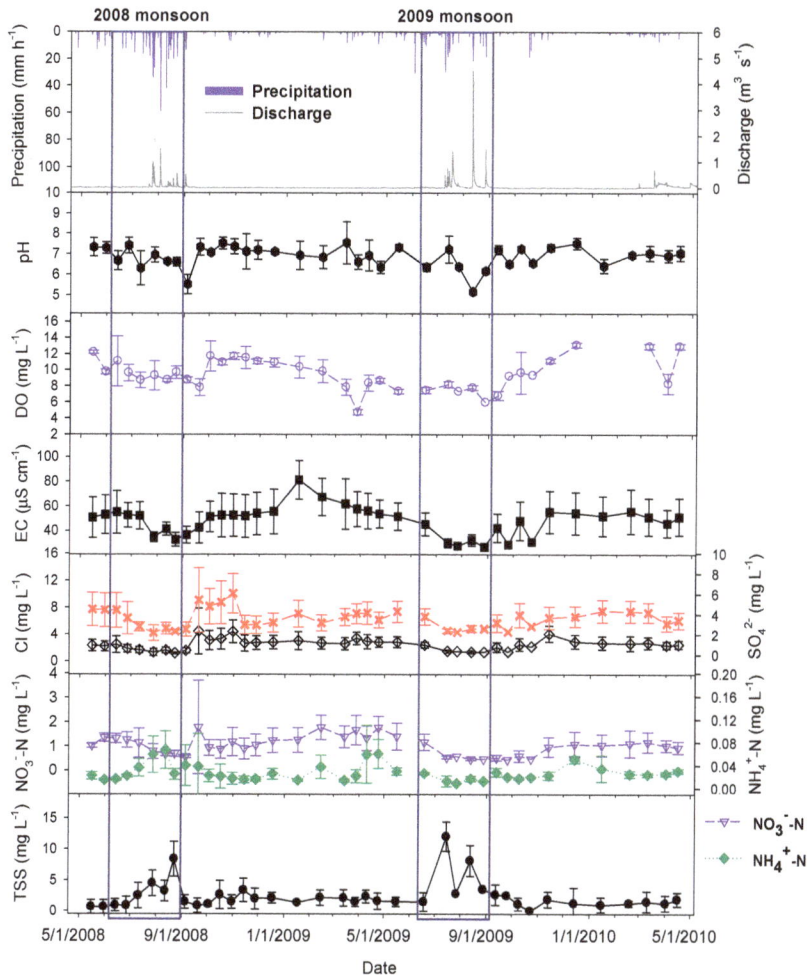

Figure 2. Temporal variations in hourly precipitation (mm), discharge ($m^3 \cdot s^{-1}$), pH, EC ($\mu S \cdot cm^{-1}$), and concentrations ($mg \cdot L^{-1}$) of DO, Cl^-, SO_4^{2-}, $NO_3^- - N$, $NH_4^+ - N$, and TSS in four forest streams from May 2008 to April 2010. Error bars indicate standard deviations. Data collected during the monsoon period from June through August are indicated by the two rectangular boxes.

At least one of the analyzed hydroclimatic variables, including antecedent precipitation, soil moisture, and discharge, had significant negative correlations with pH (at all the sites except B), EC (at all the sites except A), and the concentrations of dissolved anions, but positive correlations with the concentrations of NH_4^+ at site B and TSS at all the sites (Table 2). The DO levels showed significant negative correlations with temperature at all streams except stream A and with soil moisture (SWC) at sites C and D.

Table 2. Correlations between bi-weekly water chemistry data (pH, DO, EC, and concentrations of Cl⁻, SO₄²⁻, NO₃⁻-N, NH₄⁺-N, and TSS) in four streams and climatic variables including air temperature (Temp), 3-day antecedent precipitation (3-day ppt), and 14-day antecedent precipitation (14-ppt), soil water content (SWC) at 30 cm, and discharge. Asterisks indicate statistical significance.

Site	Climatic Variable	Correlation (r) with							
		pH	DO	EC	Cl⁻	SO₄²⁻	NO₃⁻-N	NH₄⁺-N	TSS
A	Temp	−0.26	−0.14	−0.35 *	−0.31	−0.22	0.02	0.01	0.36 *
	3-day ppt	−0.12	−0.23	−0.21	−0.43 **	−0.25	−0.20	0.10	0.75 **
	14-ppt	−0.16	−0.14	−0.20	−0.38 *	−0.34 *	−0.01	0.21	0.71 **
	SWC	−0.37 *	−0.24	−0.11	−0.36 *	−0.40 *	0.39 *	0.24	0.32 *
	Discharge	−0.196	−0.23	−0.20	−0.42 **	−0.16	−0.27	−0.02	0.52 **
B	Temp	−0.29	−0.51 **	−0.77 **	−0.71 **	−0.50 **	−0.61 **	0.23	0.29
	3-day ppt	−0.06	−0.24	−0.62 **	−0.60 **	−0.58 **	−0.49 **	0.08	0.78 **
	14-ppt	−0.25	−0.27	−0.67 **	−0.64 **	−0.62 **	−0.46 **	0.36 *	0.57 **
	SWC	−0.31	−0.34	−0.30	−0.36 *	−0.54 **	−0.07	0.59 **	0.20
	Discharge	−0.15	−0.18	−0.53 **	−0.50 **	−0.45 **	−0.46 **	−0.05	0.64 **
C	Temp	−0.13	−0.35 *	−0.69 **	−0.48 **	−0.25	−0.50 **	0.28	0.47 **
	3-day ppt	−0.24	−0.23	−0.75 **	−0.59 **	−0.58 **	−0.57 **	0.04	0.79 **
	14-ppt	−0.07	−0.22	−0.79 **	−0.64 **	−0.58 **	−0.56 **	0.18	0.64 **
	SWC	−0.28	−0.40 *	−0.66 **	−0.75 **	−0.67 **	−0.50 **	0.13	0.46 **
	Discharge	−0.50 **	−0.24	−0.61 **	−0.48 **	−0.45 **	−0.54 **	−0.01	0.67 **
D	Temp	−0.13	−0.37 *	−0.47 **	−0.12	−0.18	0.02	−0.05	0.37 *
	3-day ppt	−0.35 *	−0.25	−0.64 **	−0.30	−0.49 **	−0.23	−0.18	0.88 **
	14-ppt	−0.20	−0.24	−0.62 **	−0.36 *	−0.56 **	−0.24	−0.15	0.69 **
	SWC	−0.27	−0.35 *	−0.33	−0.28	−0.41 *	0.01	0.15	0.46 **
	Discharge	−0.54 **	−0.23	−0.57 **	−0.23	−0.33	−0.20	−0.10	0.75 **

Notes: *: $p < 0.05$; **: $p < 0.01$; ***: $p < 0.001$.

3.2. Intensive Storm Event Sampling and Continuous Instrumental Measurements

Intensive storm event sampling was carried out at 2 h intervals during eight rainfall events of various duration and intensity (Table 3). The bi-hourly measurements of anion and TSS concentrations showed large changes only during the intense events exceeding a threshold precipitation (total rainfall >100 mm and mean hourly rainfall >5 mm·h^{-1} during Events 3, 4, 6, and 8; Figures 3 and 4). Dissolved anion concentrations tended to decrease upon initiation of the rainfall event during some rainfall events and often exhibited inconsistent patterns depending on ion species. By comparison, TSS concentrations always increased rapidly in response to rising discharge during the peak flow periods of intense rainfall events. POC concentrations were lower than DOC concentrations during baseflow and low-intensity rainfall events, but rapidly increased and exceeded DOC concentrations during the short peak periods of intense rainfall (Event 3, 4 in Figure 3; Event 6, 8 in Figure 4).

Table 3. Summary of hydroclimatic conditions that characterize the eight described rainfall events. Ppt indicates precipitation.

Event	Start Time	Duration (h)	Total ppt (mm)	Maximum Hourly ppt (mm)	Mean Rainfall Intensity (mm·h^{-1})	Antecedent ppt (mm)			
						3-Day	5-Day	7-Day	14-Day
1	03:00 18 June 2008	19	72	12.5	3.8	0	0	0	30
2	13:00 16 July 2008	6	17	11.0	2.8	0	40	40	106
3	02:00 24 July 2008	33	292	33.5	8.9	3	76	86	143
4	04:00 26 July 2008	26	137	26.5	5.3	292	294	378	435
5	06:00 20 June 2009	11	15	2.5	1.4	3	29	55	105
6	05:00 09 July 2009	14	167	23.5	11.9	1	8	14	15
7	00:00 18 July 2009	30	69	7.5	2.3	61	178	285	459
8	14:00 11 August 2009	24	210	19.5	8.7	0	6	6	7

Discharge-weighted event mean concentrations of TSS had a strong positive relationship with the mean rainfall intensity levels ($R^2 = 0.92$, $p = 0.0002$), but only a weak positive relationship with the total rainfall amounts ($R^2 = 0.46$, $p = 0.049$; Figure 5). While the discharge-weighted event mean concentrations of DOC did not have any significant relationship with both rainfall indices, POC concentrations exhibited a strong positive relationship with the mean rainfall intensity ($R^2 = 0.89$, $p = 0.0004$).

Continuous *in situ* measurements of pH, EC, and turbidity at 5 min intervals exhibited very rapid changes in stream water quality during intense rainfall events (Figures 3 and 4). EC and pH changed little during small rainfall events when the total precipitation was less than 100 mm, but showed large decreases during the peak flow periods of larger rainfall events. In contrast, turbidity rapidly increased in response to increasing rainfall intensity (Figures 3 and 4). During three intensive rainfall events (Events 3, 6, and 8), turbidity levels showed much larger fluctuations than those for TSS measurements and exceeded the upper detection limit (1000 NTU)

189

during several short peak flow periods. When all data collected during the eight rainfall events were combined, turbidity measurements exhibited significant positive relationships with the concentrations of TSS ($R^2 = 0.92$, $p < 0.0001$) and POC ($R^2 = 0.77$, $p < 0.0001$) (Figure 6).

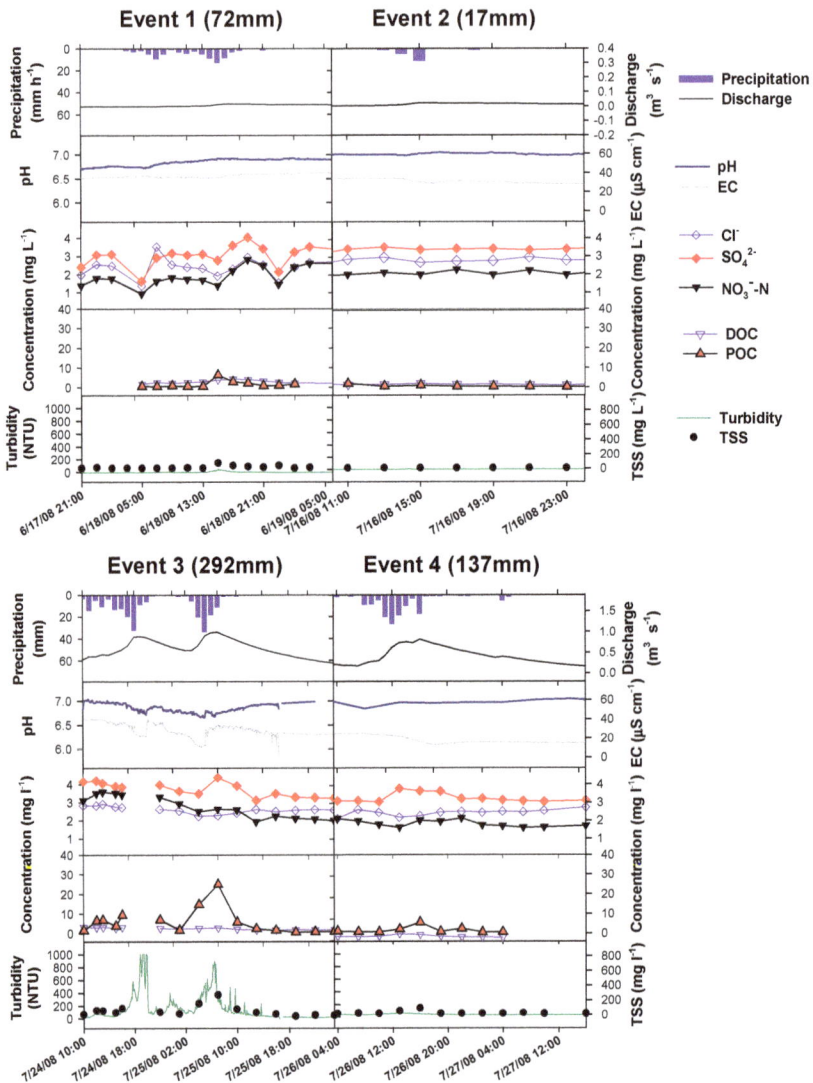

Figure 3. Temporal variations in hourly precipitation (mm), discharge ($m^3 \cdot s^{-1}$), pH, EC ($\mu S \cdot cm^{-1}$), concentrations ($mg \cdot L^{-1}$) of dissolved ions, DOC, and POC, turbidity (NTU), and TSS concentration ($mg \cdot L^{-1}$) during four rainfall events in 2008. DOC and POC data of the four storm events have been published in a previous report [19].

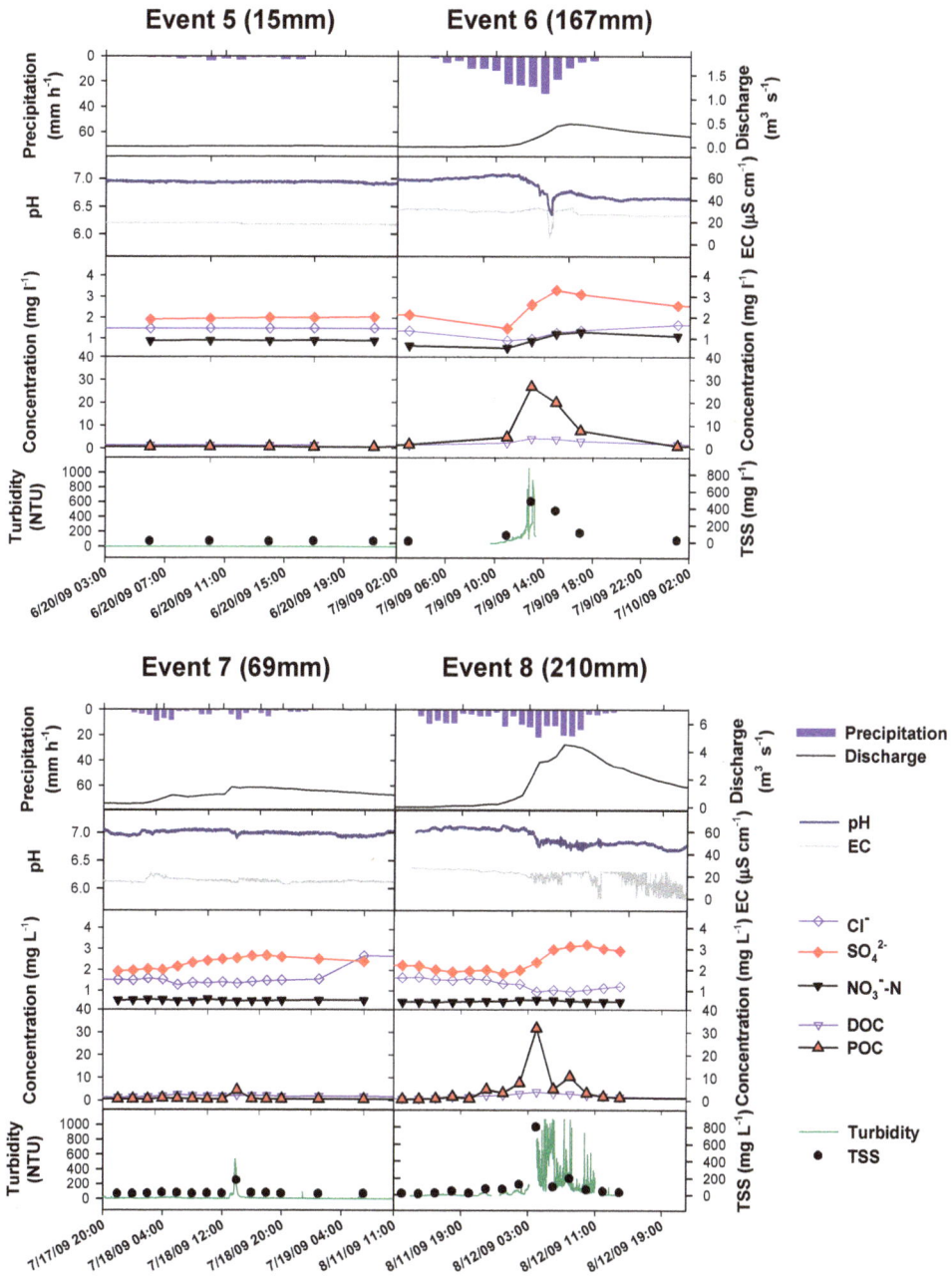

Figure 4. Temporal variations in hourly precipitation (mm), discharge (m^3·s^{-1}), pH, EC (µS·cm^{-1}), concentrations (mg·L^{-1}) of dissolved ions, DOC, and POC, turbidity (NTU) and TSS concentration (mg·L^{-1}) during four rainfall events in 2009. DOC and POC data of Event 8 have been published in a previous report [11].

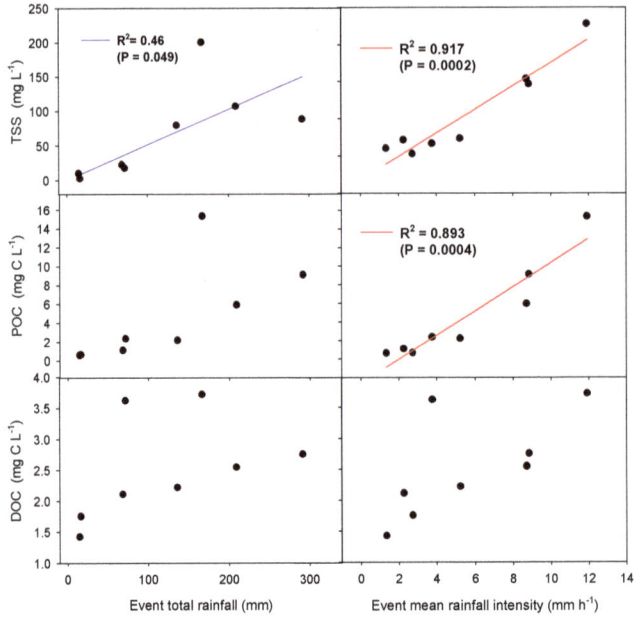

Figure 5. Relationships between event total precipitation (mm) or mean rainfall intensity (mm·h^{-1}) and event discharge-weighted mean concentration (mg·L^{-1}) of TSS, POC, and DOC. Significant relationships at $p < 0.05$ were indicated by drawing a regression line through the plot.

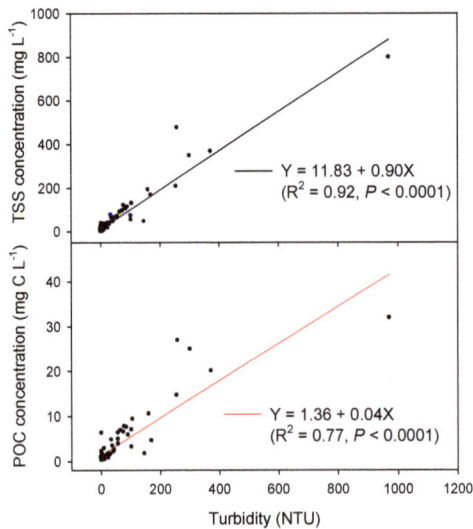

Figure 6. Relationships between turbidity (NTU) and the concentrations (mg·L^{-1}) of TSS ($n = 75$) or POC ($n = 74$) measured during eight rainfall events.

4. Discussion

The comparisons of stream water chemistry data collected on different timescales suggest that conventional routine monitoring at weekly to monthly intervals can adequately describe seasonality and long-lasting or slowly changing patterns of the stream water chemistry, but its low-resolution data are limited in providing accurate estimates of chemical fluxes pulsing during storm events. While bi-weekly monitoring data showed only marginal temporal variations for many parameters, intensive storm event sampling at 2 h intervals and continuous *in situ* measurements at 5 min intervals allowed us to capture rapid storm-induced changes in stream water chemistry at finer temporal resolutions. High-frequency monitoring data are essential in understanding short-term dynamics in both the hydroclimatic conditions and the watershed hydro-biogeochemical responses to rainfall events [10,22,41]. For example, Nagorski *et al.* [41] detected short-duration metal toxicity increases in mining-impacted streams in western Montana during high-flow periods of the spring runoff events by increasing the sampling frequency.

The results from both the routine (Figure 2) and storm event samplings (Figures 3 and 4) indicated potential water quality deterioration during extreme climatic events including intense monsoon rainfalls and extended droughts from the winter to spring. While DO was generally low during the summer, owing to decreased oxygen solubility with increasing temperature as indicated by the significantly negative correlations between temperature and DO (Table 2), an unusually low level of DO (4.6 mg·L^{-1} on 28 March 2009) occurred during the extended spring drought in 2009 (Figure 2), degrading the stream water quality to a highly contaminated level ("Grade 4") according to Korean river water quality standards [42]. This very low level of DO could not be explained solely by the temperature-DO relationship, because the observed DO level on the sampling day with a relatively low water temperature (4.6 °C) accounted for 39.7% of the saturation level determined by the temperature. EC and dissolved ion concentrations tended to be higher during droughts than during wetter periods, but usually within the range of uncontaminated streams (Figure 2). However, eutrophication could significantly degrade drinking water quality in downstream rivers and reservoirs, if increased concentrations of dissolved nutrients should persist over a prolonged period coupled with unusually high temperatures. This condition occurred in December 2011 in Lake Paldang, which receives water from both the North and South Han Rivers and provides drinking water resources to the metropolitan population of Seoul.

Compared to large storm-induced changes in TSS and POC concentrations, the concentrations of dissolved ions showed relatively small changes during rainfall events, often exhibiting ion-specific fluctuations with changing rainfall intensity and discharge (Figures 3 and 4). As with the lowered EC and dissolved ion concentrations during the summer monsoon period (Figure 2), rainfall-induced initial decreases in EC and dissolved ion concentrations can be explained by rainfall-induced dilution [43]. This rainfall-induced dilution might also explain the significantly negative correlations between hydroclimatic variables and dissolved anion concentrations measured biweekly at the four streams (Table 2). Although correlations should not be equated to any causal relationship, the consistent patterns of lower EC and anion concentrations under wetter conditions point to storm-induced dilution overriding flushing of major anions from soil pools of limited size. Compared to consistent decreases in EC with increasing hourly rainfalls, concentrations of different dissolved ion species exhibited different response timing and durations during intense rainfall events (Figures 3 and 4). As previous studies have suggested, rainfall-induced dilution of dissolved ions can be complicated by the flushing of soil-derived chemicals [43] and temporal variations in biogeochemical exports from different sources depending on the rising or falling limbs of the hydrograph [44,45].

Drastic increases in both the TSS concentration and turbidity during monsoon rainfall events (Figures 3 and 4) represent the most outstanding, climate-related water quality issue among all of the monitored water chemistry components. Considering that high levels of suspended sediment cause a suite of water quality deterioration effects, including siltation and sorption of hazardous materials [16], transient but recurrent surges of suspended sediment export during intense rainfall events should be taken into account in preparing climate change adaptation strategies for drinking water facilities that use stream water from mountainous watersheds. In another study conducted at the same watershed, Jo and Park [20] found that during extreme rainfall events, suspended sediment in the forest stream (site A) contained substantial amounts of lead (Pb) that otherwise would have been retained in the forest soil for a long period. At an outlet of a highly turbid agricultural stream downstream of the forest stream, the peak concentrations of both suspended sediment and Pb during storm events exceeded those observed for the forest stream by an order of magnitude [20]. This agricultural stream loaded with suspended sediments contributed to several occasions of serious downstream siltation in a local river during a series of exceptionally intense storm events in 2006, resulting in the construction of a new drinking water facility taking source water from an upstream reach rather than more vulnerable downstream reach.

Although bi-weekly monitoring at the four streams showed slightly higher TSS concentrations during the summer monsoon period than during the other period, as shown by the positive relationships between hydroclimatic variables and TSS (Table 2), only intensive stream sampling and high-frequency turbidity measurements could capture the extraordinary increases in suspended sediment during the intense rainfall events (Events 3, 6, and 8; Figures 3 and 4). Some mismatches between storm magnitude and peak concentrations of TSS and turbidity (e.g., Event 3 in 2008 *vs.* Event 8 in 2009) might reflect differences in the response of discharge and solute export to the varying intensity of the monitored rainfall event and the amount of antecedent rainfalls. For example, more intense rainfall occurring during a very short period following a relatively dry condition might have resulted in more intense responses of both discharge and sediment export during Event 8 compared to Event 3. The positive relationships between the event mean rainfall intensity and discharge-weighted mean concentrations of TSS and POC (Figure 5) suggest that suspended sediments from various soil sources within the watershed can be exported rapidly when a threshold level of rainfall intensity is passed. In addition, these sediments transport large amounts of soil organic carbon to the streams. Large differences in the POC concentrations between low- and high-flow conditions relative to the small range of DOC concentrations correlated with discharge (Figures 3 and 4) were also observed in another study that employed high-frequency, in-stream DOC and POC measurements [11]. Steeper increases in POC export compared to the gradual increase in DOC leaching were attributed to the relatively high threshold energy required to initiate erosion of soil and organic matter particles from potential sediment sources [11]. Storm pulses of POC can represent a transient but dominant pathway of hydrologic export of soil organic carbon, which can increase disproportionately during short peak flow periods in response to more frequent occurrences of extreme rainfall events [19].

During the intense rainfall events when rainfall intensity exceeded a threshold level, the event mean concentrations of TSS exceeded 100 $mg \cdot L^{-1}$, a high contamination level according to the local water quality standard [42]. In accordance with the increasing occurrence of extreme rainfall events observed over a large part of the Northern Hemisphere as a consequence of global climate change [8,34], the occurrence of extreme rainfall events has increased, along with summer precipitation, across South Korea [35]. Using daily precipitation data averaged across 61 weather stations, Choi *et al.* [35] showed that the cumulative amount of precipitation on extremely wet days (those above the 99th percentile) had increased 36.1 mm per decade over the period from 1973 to 2007, with many of the significant very-wet-day precipitation events observed around the mountain regions. If monsoon rainfalls in East Asia continue to intensify in the coming decades, as predicted for tropical precipitation regimes by model simulations combined with satellite observations [34], we can expect more frequent occurrences of intense rainfall events. Intense rainfalls which exceed the threshold rainfall intensity level for initiating the export of erodible soils and associated organic carbon can result in siltation and large organic carbon inputs in receiving waters, until repeated storm-induced flushing processes reduce or deplete soil pools exceeding the capacity of soils to replenish organic matter [11,32].

In a review on forested watersheds in North America, Binkley and Brown [6] found that most forest streams have relatively low annual mean concentrations of suspended sediment (<5 mg·L^{-1}, with stormflow peaks reaching up to 100 mg·L^{-1}), although there were some rare studies reporting unusually high outliers exceeding 1000 mg·L^{-1} even in undisturbed forests. Much wider variations of suspended sediment concentrations ranging from less than 1 mg·L^{-1} to 5000 mg·L^{-1} have been observed in large river systems globally [46]. Large increases in suspended sediment concentrations have also been observed following disturbances caused by forest management activities such as harvesting and road construction [6,47]. The exceptionally high TSS concentrations observed during the peak flow periods of some of the intensive rainfall events we monitored suggest that soil erosion in steep mountainous watersheds can be vulnerable to extreme rainfall events, even without any disturbances over recent years. Similar high rates of sediment export during intense storm events have been observed in steep mountainous watersheds in Japan [12,48,49]. In a Korean natural forest dominated by 80–200-year old trees of similar species to the study site, Kim *et al.* [50] also observed high export rates of suspended sediment (measured as POC) during intense storm events.

Potential sources of suspended sediment in mountainous forested watersheds include the forest floors, trails, streambanks, and streambeds [7,48]. Other studies that have traced sediment sources in the Haean watershed using Pb stable isotopes [20] or C and N stable isotopes [19] suggested streambanks and bare surfaces as major sources of sediments. In a study that traced sediment sources using ^{137}Cs and ^{210}Pb in a Japanese cypress plantation, Mizugaki *et al.* [48] found that suspended sediment was derived from a mixture of sources including interrills on the forest floor, truck trails, and streambanks. Relatively high contributions of suspended sediment from the forest floor during heavy rainfalls were attributed to overland flow and interrill erosion due to low organic matter accumulation and understory vegetation coverage on the forest floor under the dense cover of cypress canopies [48].

High-frequency turbidity measurements on the timescale of minutes had several advantages over the lower temporal resolution of intense hourly storm event sampling, which often did not indicate rapid changes in suspended sediment transport during intense rainfall events (Figures 3 and 4). Care should be taken when turbidity is used as a surrogate of suspended sediment, because turbidity represents the light scattering properties of all the suspended matter unlike TSS that represents the mass of suspended sediment per unit volume [16]. However, there are some advantages of high-resolution time series data of turbidity including the application potential of real-time turbidity measurements as early warning signals of suspended sediment and organic carbon surges in drinking water facilities and their source areas. If we can establish empirical relationships between turbidity levels and the concentrations of TSS or POC, as shown in Figure 6, the relatively cost-effective and robust turbidity measurement system can be employed as a surrogate to estimate the concentrations and fluxes of suspended sediment or organic carbon in streams and rivers. While turbidity time series data have been widely used to estimate suspended sediment loads in polluted waters (e.g., [51]), this study provided a rare empirical relationship between turbidity and the concentrations of POC. Sediment surges in drinking water source areas during extreme rainfall events can have devastating impacts on drinking water facilities and aquatic ecosystems. This occurred in 2006 in the Lake Soyang Watershed where our study site is located. Early warning signals from real-time turbidity measurements in headwater streams can facilitate proactive protection of waterways and water treatment facilities. For example, early warning signals can allow water treatment operators to add treatment modules or shut down the intake valves until "slugs" of suspended sediment pass by. These signals can also help predict potential increases in disinfection byproducts during rainfall events based on the well-established relationship between the amount of organic substances (particularly humic substances) and disinfection byproduct formation potentials [38,39,52].

5. Conclusions

While bi-weekly monitoring data captured long-lasting, substantial decreases in DO and large increases in dissolved nutrients during extended droughts from winter to spring, very rapid, drastic increases in TSS concentrations and turbidity during intense monsoon rainfall events might represent a key climate-related water quality issue in the studied mountainous watersheds. Given the potential of suspended sediment to transport heavy metals and other toxic contaminants, rainfall-induced pulses of suspended sediment can pose further risks to downstream water quality, particularly in mountainous areas under poor forest growth conditions or dotted with abandoned mines.

Comparing stream water chemistry data collected at different timescales allowed us to evaluate the importance of sampling frequency in assessing climate-induced risks to stream water quality arising from either intense monsoon rainfalls or large seasonal variations in rainfall. Based on the comparison of intensive rainfall event sampling at 2 h intervals, *in situ* instrumental measurements at 5 min intervals, and bi-weekly monitoring data, we suggest that routine monitoring based on weekly to monthly sampling should be supplemented with high-frequency sampling or continuous instrumental measurements to provide more accurate estimates of material transport, particularly those related to suspended sediment, during periods when hydrologic conditions vary very rapidly due to frequent occurrence of storm events. In particular, high-frequency, *in situ* measurements of turbidity can provide high-resolution time series data that can be used to estimate the transport of suspended sediment and POC based on the empirical relationships established for several rainfall events of various duration and intensity. These data can also be used as early warning signals for suspended sediment and POC surges during intense rainfall events and to identify potential increases in disinfection byproduct formation by natural organic matter in drinking water source areas based on some empirical relationships between suspended sediment (or POC) and disinfection byproduct formation potentials that were established in a previous study conducted at the same watershed [39]. The positive relationships between the concentrations of suspended sediment and metals (Pb) reported from the same study site [20] might be used to predict the potential range of metal fluxes carried by suspended sediment.

In summary, we propose that high-frequency instrumental monitoring, such as continuous turbidity measurements, should be complemented to the conventional routine monitoring to capture rapid water quality changes in headwater streams draining mountainous watersheds during rainfall events. Within this proposed multiple time-scale monitoring scheme, low-frequency routine monitoring data can be used to establish long-term or seasonal patterns for chemical constituents showing strong seasonal trends, including DO and dissolved ion concentrations.

Acknowledgments: This research was supported by the National Research Foundation of Korea, funded by the Ministry of Education, Science and Technology (2014R1A2A2A01006577; ERC 2009-0083527). We appreciate all the students who were involved in the multi-year projects, including Kyeong-Won Jo, Jong-Jin Jeong, and Byung-Joon Jeong just to name the most important contributors. We also thank two anonymous reviewers for their constructive comments that helped us improve an earlier version of the manuscript.

Author Contributions: Ji-Hyung Park conceived and designed the experiments; Hyun-Ju Lee and Ji-Hyung Park performed the experiments, analyzed the data, and wrote the manuscript; Kun-Woo Chun and Christopher L. Shope contributed additional data and provided comments on the manuscript.

Conflicts of Interest: The authors declare no conflict of interest.

References

1. Gomi, T.; Sidle, R.C.; Richardson, J.S. Understanding processes and downstream linkages of headwater systems. *BioScience* **2002**, *52*, 905–916.
2. Postel, S.L.; Thompson, B.H., Jr. Watershed protection: Capturing the benefits of nature's water supply services. *Nat. Resour. Forum* **2005**, *29*, 98–108.
3. Viviroli, D.; Weingartner, R.; Messerli, B. Assessing the hydrological significance of the world's mountains. *Mt. Res. Dev.* **2003**, *23*, 32–40.
4. Bradshaw, C.J.A.; Sodhi, N.S.; Peh, K.S.H.; Brook, B.W. Global evidence that deforestation amplifies flood risk and severity in the developing world. *Glob. Chang. Biol.* **2007**, *13*, 2379–2395.
5. Eisenbies, M.H.; Aust, W.M.; Burger, J.A.; Adams, M.B. Forest operations, extreme flooding events, and considerations for hydrologic modeling in the Appalachians—A review. *For. Ecol. Manag.* **2007**, *242*, 77–98.
6. Binkley, D.; Brown, T.C. Forest practices as nonpoint sources of pollution in North America. *Water Resour. Bull.* **1993**, *29*, 729–740.

7. Sidle, R.C.; Ziegler, A.D.; Negishi, J.N.; Nik, A.R.; Siew, R.; Turkelboom, F. Erosion processes in steep terrain—Truths, myths, and uncertainties related to forest management in Southeast Asia. *For. Ecol. Manag.* **2006**, *224*, 199–225.

8. Min, S.-K.; Zhang, X.; Zwiers, F.W.; Hegerl, G.C. Human contribution to more-intense precipitation extremes. *Nature* **2011**, *470*, 378–381.

9. Park, J.-H.; Duan, L.; Kim, B.; Mitchell, M.J.; Shibata, H. Potential effects of climate change and variability on watershed biogeochemical processes and water quality in Northeast Asia. *Environ. Int.* **2010**, *36*, 212–225.

10. Kaushal, S.S.; Pace, M.L.; Groffman, P.M.; Band, L.E.; Belt, K.T.; Mayer, P.M.; Welty, C. Land use and climate variability amplify contaminant pulses. *Eos* **2010**, *91*, 221–228.

11. Jeong, J.-J.; Bartsch, S.; Fleckenstein, J.; Matzner, E.; Tenhunen, J.; Lee, S.D.; Park, S.-K.; Park, J.-H. Differential storm responses of dissolved and particulate organic carbon in a mountainous headwater stream, investigated by high-frequency in-situ optical measurements. *J. Geophys. Res.* **2012**, *117*.

12. Fukuyama, T.; Onda, Y.; Gomi, T.; Yamamoto, K.; Kondo, N.; Miyata, S.; Kosugi, K.; Mizugaki, S.; Tsubonuma, N. Quantifying the impact of forest management practice on the runoff of the surface-derived suspended sediment using fallout radionuclides. *Hydrol. Process.* **2010**, *24*, 596–607.

13. Zhang, Z.; Fukushima, T.; Onda, Y.; Gomi, T.; Fukuyama, T.; Sidle, R.; Kosugi, K.; Matsushige, K. Nutrient runoff from forested watersheds in central Japan during typhoon storms: Implications for understanding runoff mechanisms during storm events. *Hydrol. Process.* **2007**, *21*, 1167–1178.

14. Goldsmith, S.T.; Carey, A.E.; Lyons, W.B.; Kao, S.-J.; Lee, T.-Y.; Chen, J. Extreme storm events, landscape denudation, and carbon sequestration: Typhoon Mindulle, Choshui River, Taiwan. *Geology* **2008**, *36*, 483–486.

15. Tsai, C.-J.; Lin, T.-C.; Hwong, J.-L.; Lin, N.-H.; Wang, C.-P.; Hamburg, S. Typhoon impacts on stream water chemistry in a plantation and an adjacent natural forest in central Taiwan. *J. Hydrol.* **2009**, *378*, 290–298.

16. Bilotta, G.S.; Brazier, R.E. Understanding the influence of suspended solids on water quality and aquatic biota. *Water Res.* **2008**, *42*, 2849–2861.

17. Ide, J.; Haga, H.; Chiwa, M.; Otsuki, K. Effects of antecedent rain history on particulate phosphorus loss from a small forested watershed of Japanese cypress (*Chamaecyparis obtuse*). *J. Hydrol.* **2008**, *352*, 322–335.

18. Hilton, R.G.; Galy, A.; Hovius, N.; Chen, M.C.; Horng, M.J.; Chen, H. Tropical-cyclone -driven erosion of the terrestrial biosphere from mountains. *Nat. Geosci.* **2008**, *1*, 759–762.

19. Jung, B.-J.; Lee, H.-J.; Jeong, J.-J.; Owen, J.S.; Kim, B.; Meusburger, K.; Alewell, C.; Gebauer, G.; Shope, C.; Park, J.-H. Storm pulses and varying sources of hydrologic carbon export from a mountainous watershed. *J. Hydrol.* **2012**.

20. Jo, K.-W.; Park, J.-H. Rapid release and changing sources of Pb in a mountainous watershed during extreme rainfall events. *Environ. Sci. Technol.* **2010**, *44*, 9324–9329.

21. Bormann, F.H.; Likens, G.E. Nutrient cycling. *Science* **1967**, *155*, 424–429.

22. Peterson, D.; Smith, R.; Hager, S.; Hicke, J.; Dettinger, M.; Huber, K. River chemistry as a monitor of Yosemite Park mountain hydroclimates. *Eos* **2005**, *86*, 285–292.

23. Likens, G.E.; Driscoll, C.T.; Buso, D.C. Long-term effects of acid rain: Response and recovery of a forest ecosystem. *Science* **1996**, *272*, 244–246.

24. Park, J.-H.; Mitchell, M.J.; McHale, P.J.; Christopher, S.F.; Myers, T.P. Impacts of changing climate and atmospheric deposition on N and S drainage losses from a forested watershed of the Adirondack Mountains, New York State. *Glob. Chang. Biol.* **2003**, *9*, 1602–1619.

25. Murdoch, P.S.; Baron, J.S.; Miller, T.L. Potential effects of climate change on surface-water quality in North America. *J. Am. Water Resour. Assoc.* **2000**, *36*, 347–366.

26. Campbell, J.L.; Rustad, L.E.; Boyer, E.W.; Christopher, S.F.; Driscoll, C.T.; Fernandez, I.J.; Groffman, P.M.; Houle, D.; Kiekbusch, J.; Magill, A.H.; *et al.* Consequences of climate change for biogeochemical cycling in forests of northeastern North America. *Can. J. For. Res.* **2009**, *39*, 264–284.

27. Van Verseveld, W.J.; McDonnell, J.J.; Lajtha, K. The role of hillslope hydrology in controlling nutrient loss. *J. Hydrol.* **2009**, *367*, 177–187.

28. Hornberger, G.M.; Bencala, K.E.; McKnight, D.M. Hydrological controls on dissolved organic carbon during snowmelt in the Snake River near Montezuma, Colorado. *Biogeochemistry* **1994**, *25*, 147–165.

29. Inamdar, S.P.; Christopher, S.F.; Mitchell, M.J. Export mechanisms for dissolved organic carbon and nitrate during summer storm events in a glaciated forested catchment in New York, USA. *Hydrol. Process.* **2004**, *18*, 2651–2661.

30. Inamdar, S.P.; Mitchell, M.J. Hydrologic and topographic controls on storm-event exports of dissolved organic carbon (DOC) and nitrate across catchment scales. *Water Resour. Res.* **2006**, *42*.

31. Butturini, A.; Alvarez, M.; Bernal, S.; Vazquez, E.; Sabater, F. Diversity and temporal sequences of forms of DOC and NO_3^- discharge responses in an intermittent stream: Predictable or random succession? *J. Geophys. Res. Biogeosci.* **2008**, *113*.

32. Dhillon, G.S.; Inamdar, S. Extreme storms and changes in particulate and dissolved organic carbon in runoff: Entering uncharted waters? *Geophys. Res. Lett.* **2013**, *40*, 1322–1327.

33. Choi, G.; Collins, D.; Ren, G.; Trewin, B.; Baldi, M.; Fukuda, Y.; Afzaal, M.; Pianmana, T.; Gomboluudev, P.; Huong, P.T.T.H.; *et al.* Changes in means and extreme events of temperature and precipitation in the Asia-Pacific Network region, 1955–2007. *Int. J. Climatol.* **2009**, *29*, 1906–1925.

34. Allan, R.P.; Soden, B.J. Atmospheric warming and the amplification of precipitation extremes. *Science* **2008**, *321*, 1481–1484.

35. Choi, K.; Kwon, W.-T.; Boo, K.-O.; Cha, Y.-M. Recent spatial and temporal changes in means and extreme events of temperature and precipitation across the Republic of Korea. *J. Korean Geogr. Soc.* **2008**, *43*, 681–700.

36. Baborowski, M.; von Tümpling, W.; Friese, K., Jr. Behavior of suspended particulate matter (SPM) and selected trace metals during the 2002summer flood in the River Elbe (Germany) and Magdeburg monitoring station. *Hydrol. Earth Syst. Sci.* **2004**, *8*, 135–150.

37. Presley, S.M.; Rainwater, T.R.; Austin, G.P.; Platt, S.G.; Zak, J.C.; Cobb, G.P.; Marsland, E.J.; Tian, K.; Zhang, B.; Anderson, T.A.; *et al.* Assessment of pathogens and toxicants in New Orleans, LA following Hurricane Katrina. *Environ. Sci. Technol.* **2006**, *40*, 468–474.

38. Nguyen, H.V.M.; Lee, M.H.; Hur, J.; Schlautman, M.A. Variations in spectroscopic characteristics and disinfection byproduct formation potentials of dissolved organic matter for two contrasting storm events. *J. Hydrol.* **2013**, *481*, 132–142.

39. Jung, B.-J.; Lee, J.-K.; Kim, H.; Park, J.-H. Export, biodegradation, and disinfection byproduct formation of dissolved and particulate organic carbon in a forested headwater stream during extreme rainfall events. *Biogeosciences* **2014**, *11*, 6119–6129.

40. Seibert, J.; Stendahl, J.; Sørensen, R. Topographical influences on soil properties in boreal forests. *Geoderma* **2007**, *141*, 139–148.

41. Nagorski, S.A.; Moore, J.N.; McKinnon, T.E. Geochemical response to variable streamflow conditions in contaminated and uncontaminated streams. *Water Resour. Res.* **2003**, *39*, 1044.

42. Korea Ministry of Environment. White Paper of Environment 2010. Available online: http://library.me.go.kr/search/DetailView.Popup.ax?cid=5256302 (accessed on 1 September 2015).

43. Walling, D.E.; Foster, I.D.L. Variations in the natural chemical concentration of river water during flood flows, and the lag effect: Some further comments. *J. Hydrol.* **1975**, *26*, 237–244.

44. Whitfield, P.H.; Schreier, H. Hysteresis in relationship between discharge and water chemistry in the Fraser River basin, British Columbia. *Limnol. Oceanogr.* **1981**, *26*, 1179–1182.

45. Butturini, A.; Gallart, F.; Latron, J.; Vazquez, E.; Sabater, F. Cross-site comparison of variability of DOC and nitrate c-q hysteresis during the autumn-winter period in three Mediterranean headwater streams: A synthetic approach. *Biogeochemistry* **2006**, *77*, 327–349.

46. Meybeck, M. Carbon, nitrogen, and phosphorus transport by world rivers. *Am. J. Sci.* **1982**, *282*, 401–450.

47. Nisbet, T.R. The role of forest management in controlling diffuse pollution in UK forestry. *For. Ecol. Manag.* **2001**, *143*, 215–226.

48. Mizugaki, S.; Onda, Y.; Fukuyama, T.; Koga, S.; Asai, H.; Hiramatsu, S. Estimation of suspended sediment sources using ^{137}Cs and $^{210}Pb_{ex}$ in unmanaged Japanese cypress plantation watersheds in southern Japan. *Hydrol. Process.* **2008**, *22*, 4519–4531.

49. Ide, J.; Kume, T.; Wakiyama, Y.; Higashi, N.; Chiwa, M.; Otsuki, K. Estimation of annual suspended sediment yield from a Japanese cypress (*Chamaecyparis obtuse*) plantation considering antecedent rainfalls. *For.Ecol. Manag.* **2009**, *257*, 1955–1965.

50. Kim, S.J.; Kim, J.; Kim, K. Organic carbon efflux from a deciduous forest catchment in Korea. *Biogeosciences* **2010**, *7*, 1323–1334.

51. Lacour, C.; Joannis, C.; Chebbo, G. Assessment of annual pollutant loads in combined sewers from continuous turbidity measurements: Sensitivity to calibration data. *Water Res.* **2009**, *43*, 2179–2190.

52. Chow, A.T.; Dahlgren, R.A.; Harrison, J.A. Watershed sources of disinfection byproduct precursors in the Sacramento and San Joaquin Rivers, California. *Environ. Sci. Technol.* **2007**, *41*, 7645–7652.

Effects of the "Run-of-River" Hydro Scheme on Macroinvertebrate Communities and Habitat Conditions in a Mountain River of Northeastern China

Haoran Wang, Yongcan Chen, Zhaowei Liu and Dejun Zhu

Abstract: The main objective of this study was to quantify the impacts of the run of river (ROR) scheme on the instream habitat and macroinvertebrate community. We sampled the macroinvertebrate assemblages and collected the habitat variables above and below an ROR hydropower plant: Aotou plant in the Hailang River, China. The effects of the ROR scheme on habitat conditions were examined using regulation-related variables, most of which, particularly the hydrological variables and substrate composition, presented spatial variations along the downstream direction, contributing to heterogeneous conditions between reaches. The macroinvertebrate richness, the density and the diversity metrics showed significant decreases in the "depleted" reach compared with the upper and lower reaches. Approximately 75% of reach-averaged densities and 50% of taxa richness suffered decreases in the "depleted" reach compared with the upper reach. Furthermore, functional feeding groups also showed distinct site differences along the channel. The relative abundance of both collector-gatherers and the scrapers reduced considerably at the "depleted" sites, particularly at the site immediately downstream of the weir. The total variance in the the functional feeding group (FFG) data explained by Canonical correlation analysis (CCA) was more than 81.4% and the high-loadings factors were depth, flow velocity, DO and substrate composition. We demonstrated that flow diversion at the 75% level and an in-channel barrier, due to the ROR scheme, are likely to lead to poor habitat conditions and decrease both the abundance and the diversity of macroinvertebrates in reaches influenced by water diversion.

Reprinted from *Water*. Cite as: Wang, H.; Chen, Y.; Liu, Z.; Zhu, D. Effects of the "Run-of-River" Hydro Scheme on Macroinvertebrate Communities and Habitat Conditions in a Mountain River of Northeastern China. *Water* **2016**, *8*, 31.

1. Introduction

Hydropower is the most common renewable source in the world and accounts for 16% of the total electricity production [1]. Because hydropower is commonly associated with river regulation, numerous studies have addressed the ecological impacts from flow manipulation and fragmentation [2–4]. Despite the broad recognition of the ecological consequences of hydropower, most studies focused on the impacts of large-scale hydropower on the habitats and behaviors of valuable fishes and relatively few studies paid attention to small-hydro [5–8].

Small-hydro is, in most cases, "run of river" (ROR). ROR schemes use in-stream flow and operate with little or no water storage. Channel obstructions include small dams, weirs and other barriers, which are associated with the secondary channel/tube to divert a proportion of flow to turbines in the powerhouse [1,9]. This small-hydro scheme is regarded as environmentally friendly, because it does not use significant d amming [9], but international studies and other reports are scarce to support this view. The manipulation of flow diversions alters the natural flow regime and will potentially change downstream habitat conditions, and both, in turn, may present threats to ecological processes and river organisms. Although relatively little attention has been given to the ecological impacts of the ROR scheme on river organisms, other relevant studies of water diversions and artificial drought, due to river regulations, revealed some potentially significant ecological impacts. For example, Dewson [10] used whole-channel flow manipulations to imitate real water abstractions and found that significant *Ephemeroptera, Plecoptera,* and *Trichioptera* (EPT) individuals decreased in response to reduced flows. Finn and Boulton [11] compared two Australian streams influenced with or without water extraction, and revealed that artificial drought resulted in declines in macroinvertebrate richness and density but increases in the representation by drought-tolerant groups. These studies provide useful perspectives and references for studying the impacts of an ROR scheme.

China deserves special attention toward the ecological impacts of hydropower. By the end of 2014, China has 27% of the hydropower-installed capacity and has installed a capacity of more than 300 million kW (National Energy Administration, China). Moreover, approximately 40% of small-hydro capacities exist in China, and most of them operate with the ROR scheme. However, the effects of the ROR scheme on river habitats and freshwater species are completely lacking in the rivers of China.

To access the regulation impacts of the ROR scheme on the river ecosystem and to reach a better and effective regulation management, more studies are needed on a case-by-case basis concerning indicator species and meso-habitats within rivers influenced by specific hydropower projects. The macroinvertebrate community is an important component of freshwater ecosystems, and it is widely used in environmental and ecological assessments in freshwater ecosystems [12]. By understanding the consequences of ROR operations on the alterations in the flows and habitat conditions, it may be possible to make inferences on the changes of macroinvertebrate communities associated with habitat variables. The perspective of species–habitat interaction compared with a single biological perspective should be more beneficial to the understanding of the ROR eco-impacts and potential regulation decisions.

The main objectives of this study were to understand the environmental and ecological impacts of ROR operations by comparing macroinvertebrate assemblages and habitat conditions above and below an ROR plant, the Aotou hydropower plant, situated in the Hailang River, northern China, from the middle of June to July 2014. Physico-chemical and biological data were gathered through field investigations and observations at designed sampling sites. The relationship between habitat environments and macroinvertebrate assemblages was also assessed. We hypothesized that flow diversions due to ROR operations could change habitat variables and then impact macroinvertebrate assemblages, which leads to reduced macroinvertebrate biodiversity and poor habitat quality in dewatering reaches. The present study will enrich the knowledge of river ecosystems in northern China.

2. Materials and Methods

2.1. Study Area

The field data were collected from the Hailang River. It is the largest tributary of the Mudan River in northern China, flowing approximately 210 km from the Changbai Mountain to the Mudan River. The Hailang River subcatchment drains 5225 square km of land, and has an annual precipitation of 800 mm. The river freezes from late November until early April. The highest flows in the Hailang River occur when the snow melts during the spring thaw.

As a mountain river, the Hailang River has a mean slope of 2.52‰. The elevation is from 773 m at the waterhead area to 243 m at the mouth. Due to the steep slope and high elevation range from the headwaters to the mouth, the Hailang River has abundant waterpower and a cascade of nine power plants is planned in the near future years. The Aotou Plant is a small ROR hydropower plant situated in a lower gradient reach of the Hailang River with a designed head of 5.5 m and a peak capacity of 1225 KW. The main channel is obstructed with an in-channel weir to regulate water levels, allowing a proportion of flow to be diverted down a "Left Bank" diversion channel to turbines before it is returned to the main channel, 3.7 km further downstream. The Aotou Plant operates without water storage but creates a 3.7 km-long depleted stretch from the main channel weir to return point. In the water-depleted reach, the natural flow regime reduced significantly, and little overflow and seepage are the main types of discharge. This phenomenon is particularly severe in the dry season.

Benthic macroinvertebrate samples and environmental variables were collected along three reaches near the Aotou Plant in the Hailang River: (1) 6.3-km-long reach upstream of the weir; (2) 3.7-km-long "depleted" reach; and (3) 7.6-km-long reach below the flow returning point. Three sampling sites were selected over each sampling reach (Figure 1). Diversion channel was not included in the investigation due to the application of a rectangular reinforced concrete structure, which causes a steep slope and deep water levels, making only a small area available for sampling work.

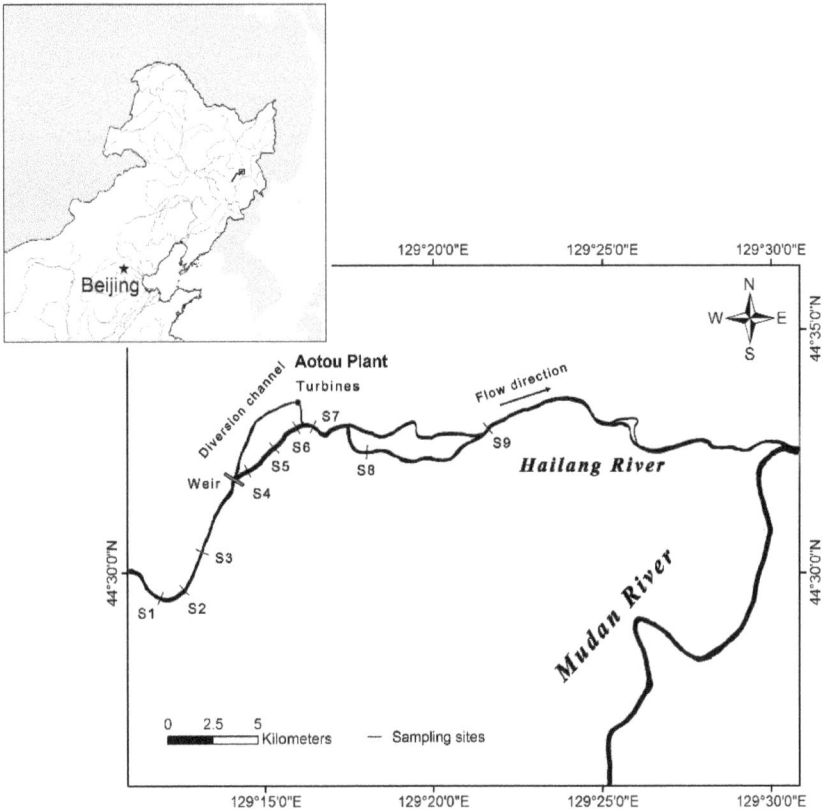

Figure 1. Locations of the study area and sites.

2.2. Sampling and Identification

At each sampling site, three replicates spaced at least 3 m apart were randomly selected from fast-flow habitats, such as riffles and runs. A modified kick-net (mesh size = 0.375 mm, area = 1 m^2), constructed out of a PVC frame and polyethylene net was used to collect benthic macroinvertebrate samples in areas with hard-bottomed substrate where the water depth was less than 0.7 m. Considering the reliability of the research results and sampling conditions, a sampling area of one square meter was chosen at each sampling location [13]. Additionally, the kicking intensity and duration were kept as similar as possible to ensure effectiveness and consistency.

Each mixed sample of macroinvertebrates and debris was obtained from the net following a timed (1.5 min) disturbance of 0.2 m-depth of the substrate upstream from the kick-net. The debris and macroinvertebrates were rinsed through a sieve (mesh size = 0.50 mm) and subsequently moved into the labeled sample containers and preserved in a 5% formaldehyde solution.

All of the faunal samples were counted, sorted and identified in 70% alcohol under a stereoscopic microscope in the laboratory. Macroinvertebrates were identified to the lowest possible taxonomic classification, mostly species-level or genus-level. The sorted taxa were assigned to the functional feeding group (FFG) categories, proposed by Cummins [14,15], to describe feeding structure variations between the study sites. Five following groups were introduced: predators (prd), collector-gatherers (c-g), collector-filterers (c-f), scrapers (scr) and shredders (shr).

2.3. Physical Habitat Assessment

For each replicate, the sampling position was extended into a square cell, with a side length of 1.5 m. Each cell was considered to be a distinct habitat to allow a qualitative comparison of habitat types and quantitative assessments of physical and chemical variables. Hydraulic parameters, including water depth and flow velocity (at 0.6 of the depth by LS300, a portable flow meter), were measured and recorded *in situ* at each cell. The dissolved oxygen (DO) and water temperature (WT) were also detected *in situ* by a dissolved oxygen meter, YSI PRO-ODO. Qualitative records were also made in each cell in the presence or absence of hydrophyte, the coverage of riparian vegetation, the ratio of pool/riffle and the embeddedness of the substrate.

The water samples and substrata samples were collected at each site for further analysis. PH, chemical oxygen demand by the potassium permanganate method (COD_{Mn}), chemical oxygen demand (COD), 5-day biochemical oxygen demand (BOD_5), ammonia nitrogen (NH_3-N), total phosphorus (TP) and total nitrogen (TN) were introduced for water-quality determination. All of these parameters were examined in the water quality analysis laboratory according to State Environmental Protection Administration of China (SEPA) standard methods. Substrata composition were measured and classified following the EPA standard in the laboratory. Four classes of particle sizes were introduced: CB Cobbles (>64 to 250 mm), CG Coarse Gravel (>16 to 64 mm), FG Fine Gravel (>2 to 16 mm) and SA Sand (>0.06 to 2 mm).

2.4. Methods of Analysis

Shannon-Wiener H', Pielou evenness J and Margalef richness d_M were used to evaluate the biodiversity of macroinvertebrate communities between the study sites. A one-way analysis of variance (ANOVA) was introduced to assess the reach differences in both habitat variables and macroinvertebrate data. A *posteriori* Tukey's HSD test was run when the difference was found.

Multivariate methods were used to determine the spatial and temporal patterns underlying abiotic and biotic data. Principal components analysis (PCA) of the physical and chemical habitat variables was used to summarize the total variation in the habitat data and identify major environmental gradients. Prior to the PCA, a Pearson correlation matrix of the environmental variables was introduced to determine the significantly correlated ones. The correlations of COD and COD_{Mn} (correlation coefficient is 0.987, $p < 0.01$), Coarse Gravel and Fine Gravel (correlation coefficient is 0.803, $p < 0.01$) were proved to be strong, so only COD and Fine Gravel were used in the analysis. BOD_5 was also excluded because of its constant value (2.00 mg/L). In total, 12 variables were included in the PCA. Canonical correlation analysis (CCA) was applied to examine the relative importance of environmental conditions in determining the differences in macroinvertebrates' FFG structure between the study sites. A direct gradient analysis using the coefficients for taxa and coefficients for environmental variables of habitats was used to maximize the species–environment correlation [16]. The data matrix of site environmental variables and the data matrix of site macroinvertebrate abundances in terms of FFG were used in the analysis. The significance of all primary CCA axes was determined by the Monte Carlo permutation testing (499 permutations) of the eigenvalues. Prior to the PCA or CCA, all of the data (habitat data in the PCA, habitat data and macroinvertebrate data in the CCA) were logarithmically transformed [$\log_{10} (x + 1)$] to standardize the scales.

3. Results

3.1. Physical and Chemical Variables

Our samplings were conducted from the middle of June to July 2014 (wet season) when the daily average flow ranged from 51.1 to 75.6 m^3/s, which was abundant enough to make the operation continuous. During the sampling period, the proportion of the total flow diverted from main channel to the turbines was approximately more than 75% which largely changed the hydrological regime in the reach below the weir. The average wetted area in the "depleted" reach obviously reduced compared with the upper reach and lower reach, particularly at S4 and S5.

The values of depth, velocity, DO and water temperature were averaged among the three replications at each site, whereas the water chemistry variables and substrata data were recorded once at each site. There were significant effects of the ROR scheme on both the reach-averaged hydraulic parameters water depth (ANOVA: $F = 21.246$, $p < 0.001$) and flow velocity ($F = 10.917$, $p < 0.001$). The water depth was significantly reduced in the "depleted" reach compared with the upper reach (*Tukey's HSD*, $p < 0.001$) and lower reach ($p < 0.05$) (Table 1), although there was no significant difference between the upper reach and lower reach, and the flow velocity was significantly reduced in the "depleted" reach ($p < 0.001$) and lower reach ($p < 0.001$) compared with the upper reach.

Table 1. Summary of habitat variables (SD) for all sites throughout the study period.

Variables	Units	Upper Sites			Depleted Sites			Lower Sites		
		S1	S2	S3	S4	S5	S6	S7	S8	S9
Depth	m	0.34 (0.03)	0.31 (0.06)	0.24 (0.03)	0.09 (0.02)	0.17 (0.04)	0.18 (0.03)	0.28 (0.04)	0.19 (0.04)	0.23 (0.02)
Velocity	m/s	0.56 (0.12)	0.53 (0.11)	0.56 (0.01)	0.21 (0.01)	0.24 (0.06)	0.27 (0.02)	0.34 (0.09)	0.21 (0.01)	0.37 (0.05)
DO	mg/L	10.13 (0.01)	10.38 (0.09)	10.20 (0.08)	9.22 (0.08)	9.29 (0.01)	9.57 (0.04)	9.27 (0.01)	9.19 (0.03)	9.11 (0.02)
Temp	°C	20.23 (0.03)	20.80 (0.31)	19.50 (0.20)	19.97 (0.12)	19.73 (0.03)	20.07 (0.17)	19.43 (0.03)	19.37 (0.03)	19.33 (0.03)
PH	—	7.69	7.36	7.30	7.27	7.23	7.19	7.25	7.27	7.25
COD_{Mn}	mg/L	5.70	4.60	4.70	4.50	4.80	4.80	4.60	4.50	4.40
COD	mg/L	19.50	16.10	16.10	15.40	16.30	16.50	15.90	15.40	14.60
BOD_5	mg/L	2.00	2.00	2.00	2.00	2.00	2.00	2.00	2.00	2.00
NH_3-N	mg/L	0.12	0.17	0.16	0.13	0.14	0.15	0.19	0.17	0.15
TP	mg/L	0.08	0.08	0.07	0.07	0.07	0.11	0.10	0.11	0.12
TN	mg/L	0.24	0.25	0.25	0.25	0.25	0.28	0.30	0.29	0.26
CB Cobble	%	60.00	37.00	42.40	22.89	32.92	28.32	30.73	34.61	27.92
GC Gravel	%	27.01	38.86	35.35	50.66	45.73	48.16	38.90	34.10	39.70
GF Gravel	%	7.80	15.69	13.23	21.94	17.20	19.90	20.90	16.20	20.70
SA Sand	%	4.00	8.10	9.02	4.20	3.70	3.10	9.10	14.50	11.10

The values of PH ($p < 0.001$) and DO ($p < 0.01$) were relatively higher in the upper reach than in either the "depleted" reach or the lower reach. Other water chemistry variables, including COD_{Mn}, COD, BOD_5, NH_3-N, TP, and TN, were not different between the upper reach and the "depleted" reach, however, these variables were significantly different between the upper reach and lower reach, with significantly higher values of COD_{Mn} ($p < 0.01$), COD ($p < 0.01$) in the upper reach and higher values of TP ($p < 0.001$), TN ($p < 0.001$) and NH_3-N ($p < 0.05$) in the lower reach.

The substrata percentage composition showed reach differences. The cobble percentage was significantly higher in the upper reach than the "depleted" reach ($p < 0.001$) and the lower reach ($p < 0.001$). However, the sand/silt percentage presented a significantly high value in the lower reach than the "depleted" reach ($p < 0.001$) and the upper reach ($p < 0.01$).

In addition, qualitative records suggested that the riparian and instream habitat conditions changed distinctly along the Hailang River (Figure 2). The habitat conditions, in terms of wetted area, vegetation coverage, pool/riffle ratio, embeddedness and riverbank stability, degraded sharply from the "optimal" in the upstream to "poor" in the immediate downstream of the weir, and upgraded gradually further downstream with flow returning, based on the criteria of the US EPA [17].

(a)   (b)   (c)

Figure 2. Sequential photographs showing the changing habitat character in a downstream direction in the Hailang River: (a) S2 at the upper reach; (b) S5 at the "depleted" reach; and (c) S7 at the lower reach.

An ordination by the PCA of the physical-chemical habitat variables (the 12 parameters mentioned above) explained 69.8% of the cumulative variance in the data by the first two principal component axes (Figure 3). The variance explained by axis1 was 49.5%. Significant loadings on axis1 showed a positive gradient of increasing water depth, flow velocity, DO, PH, COD, and cobble percentage. Significantly negative loadings exerted an increasing gradient of fine gravel percentage and TN. Axis 2 accounted for 20.3% of the data variance, and three variables, namely NH_3-N, TP and the sand percentage, represented significantly positive loadings on the axis.

The PCA of the physical-chemical variables' metrics indicated considerable site differences. The plot of sample scores demonstrated that the sampling locations of the upper reach, the "depleted" reach and the lower reach clustered into three groups. With respect to the axis1 scores, upper locations typically occupied further right positions compared to locations within the "depleted" reach and lower reach, characterizing the upper habitat conditions by higher velocity, water depth, DO and cobble percentage but lower TN and fine gravel percentage. With respect to the axis 2 scores, locations from the lower reach were located more towards the positive end of the axis than the others, characterizing the lower habitat conditions by higher concentrations of NH_3-N, TP and sand/silt percentage.

210

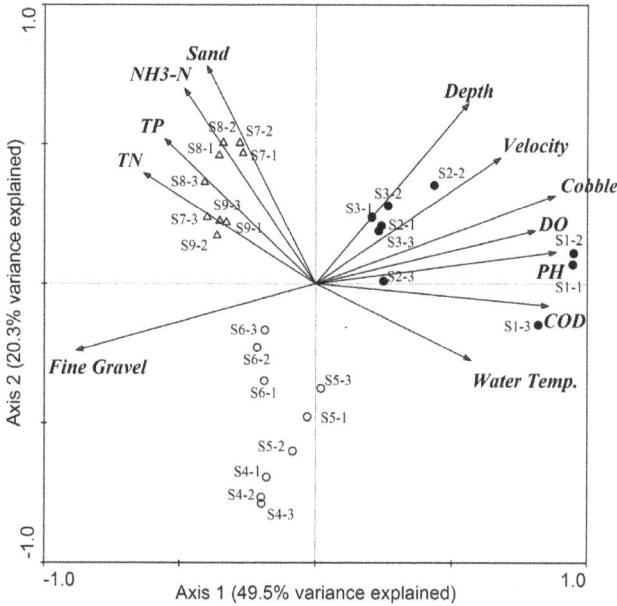

Figure 3. Plot of samples scores from the PCA of physical and chemical habitat variables along principal component axes 1 and 2. Cumulative variance in the data was explained 69.8% by two axes. Each *point* represents a different sampling location (replicate). *Closed circles* represent upper locations, *open circles* represent locations in the "depleted" reach, and *open triangles* represent lower locations.

3.2. Assemblage Composition

In total, 25 taxa were recorded in the study area of the Hailang River, which belonged to 13 phyla, five classes, and 18 families (Table 2). *Insecta* was the dominant taxonomic group and accounted for 87.97% of the total captured individuals. *Ephemerellidae*, *Chironomidae* and *Heptageniidae* were the most abundant representative families, comprising 31.18%, 15.82% and 14.23% of the total fauna, respectively.

Great differences in the taxonomic composition of major communities presented within each reach (Figure 4). The relative abundance of *Ephemeroptera* was consistently high in the upper reach and lower reach, particularly in the upper reach, with a high proportion of 69.57%. *Diptera* had the second highest abundance followed by *Ephemeroptera* in the upper reach and lower reach, presenting 12.21% and 16.67%, respectively. In contrast, *Diptera* was the most numerous group in the "depleted" reach below the regulating weir, with a relatively high abundance of 34.32%. The percentage of *Ephemeroptera* presented a considerable reduction compared with that in other reaches, comprising only 23.78% of the total fauna. The

211

relative abundance of all the other taxonomic groups, which primarily consisted of *Coleoptera, Trichoptera, Oodonata, Oligochaeta*, were consistently low in the reaches.

Table 2. Taxonomic composition (relative abundance) of macroinvertebrates between reaches in the Hailang River.

Class *Genus/Species*	Relative Abundances (%)		
	Upper Reach	**Depleted Reach**	**Lower Reach**
Insecta			
Polypedilum sordens	0.79	6.49	2.19
Cryptochironomus defectus	3.95	14.32	4.50
Chironomus plumosus	5.89	9.19	7.30
Pocladius choreus	1.58	2.43	2.68
Cinygma sp1	4.58	–	–
Cinygma sp2	4.42	–	–
Epeorus uenoi	11.26	7.03	3.53
Drunella sp1	7.53	0.00	5.60
Drunella sp2	14.32	0.00	17.76
Ephemerella sp	14.47	3.24	8.52
Ephemera sp	10.53	7.84	10.22
Baetis sp	0.63	0.00	4.38
Potamanthus huoshanensis	1.84	5.68	2.07
Elmidae	2.16	–	0.73
Dytiscidae	0.11	2.97	–
Ieptoceridae	0.37	–	–
Hydropsychidae	4.32	2.16	9.85
Gomphidae	3.16	7.03	7.54
Muscidae	–	1.89	–
Oligochaeta			
Tubificidae	2.47	8.11	1.82
Clitellata			
Glossiphonia sp	1.42	–	2.43
Whitmania sp	1.95	–	0.61
Gastropoda			
Radix ovata	1.00	5.41	2.92
Oncomelania	1.26	15.41	5.35
Bivalvia			
Corbicula	–	0.81	–

The ROR scheme impacted the distribution of fauna, and the effects could be reflected by density, taxon richness, EPT richness and other common biodiversity indices (Figure 5). For all of the patterns of site density, richness and derived biodiversity indices, the ANOVA indicated significant differences between reaches (density $F = 98.712$, $p < 0.001$; taxon richness $F = 64.012$, $p < 0.001$; EPT richness $F = 78.301$, $p < 0.001$; d_M $F = 23.515$, $p < 0.001$; $H'F = 18.363$, $p < 0.001$; and $JF = 9.284$, $p < 0.001$).

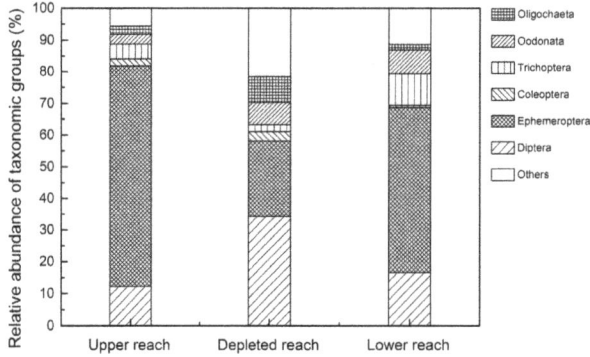

Figure 4. Relative abundance of major taxonomic groups between reaches. Percentage values are for the abundance pooled across the nine samples collected from each reach.

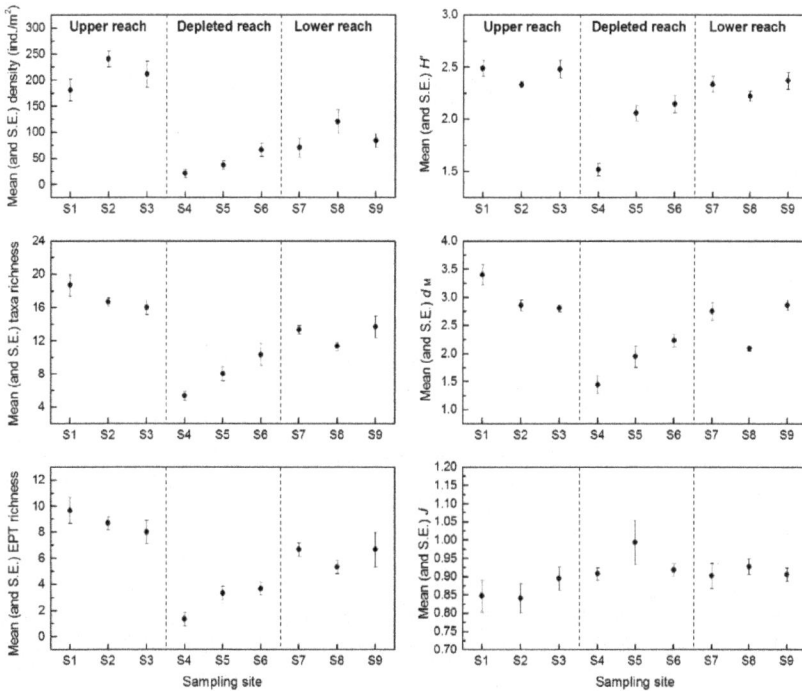

Figure 5. Macroinvertebrate abundance, taxon richness, EPT richness and other indices of diversity from each sampling site in the Hailang River. Mean (± 1 SE) are given for samples collected from nine sites on three reaches.

In terms of density, taxon richness, EPT richness, Shannon-Wiener index H' and Margalef richness d_M, there were significant reductions in the "depleted" reach compared with the upper reach (*Tukey's HSD*, for all indices, $p < 0.001$), particularly at S4, where all of the indices were consistently the lowest compared with other sites. Compared to the upper reach, the indexes reduced significantly in the lower reach (density, taxon richness, EPT richness $p < 0.001$; d_M $p < 0.05$) with the exception of Shannon-Wiener index, which showed no significant difference. From S4 to S9, with flow returning, the levels of these indexes increased gradually in the downstream direction and were significantly higher in the lower reach than the "depleted" reach (for all indices, $p < 0.001$). For Pielou evenness J, significantly higher values were found in the "depleted" reach ($p < 0.001$) and the lower reach ($p < 0.05$) than in the upper reach, although there was no significant difference between the "depleted" reach and lower reach.

3.3. FFG Variations

In total, scrapers, collector-gatherers and predators were the first three predominant functional feeding groups in benthic samples, comprising 41.46%, 35.22% and 15.62% of the macroinvertebrate assemblage, respectively. Collector-filterers and shredders were relatively uncommon and only consisting of 5.63% and 2.07% of the total fauna, respectively. Collector-filterers were primarily at lower sites and appeared to be poor at other sites. Shredders were distributed unevenly and appeared to be present at "depleted" sites and absent at most other sites.

On a reach-scale, the spatial variations of the FFG composition observed at the "depleted" sites (S4–S6) were more distinct than either the upper sites (S1–S3) or the lower sites (S7–S9). The relative abundance of predators decreased from 49.18% at S4 to 22.73% at S6; however, both scrapers and collector-gatherers had a two-fold increase from S4 to S6. The relative abundances of FFG were relatively stable both at the upper sites and the lower sites (Figure 6).

Macroinvertebrates' abundances classified according to the FFG were included in multivariate analyses (Figure 7). A CCA of the physical-chemical habitat data with macroinvertebrate data based on FFG abundance explained 81.4% of the total variance in the first two canonical variables. The first canonical variable captured 57.2% of the variance and the strong positive loadings were from hydraulic variables depth, flow velocity, DO and NH_3-N. The second canonical variable accounted for 24.2% of the total variance. For canonical variable 2, COD_{Mn}, sand proportion and cobble proportion presented the highest positive loadings, whereas water temperature and TP presented the highest negative loadings. The bi-plot of the macroinvertebrate data with the physical-chemical habitat data revealed that scrapers and collector-gatherers were situated to the positive end of variable 1 and appeared to be positively associated with hydraulic variables and DO. In contrast, predators and shredders appeared to be negatively associated with hydraulic variables and DO, especially for shredders, which exerted the highest negative loading. Partial correlations of FFG with each strongly correlated physical-chemical variable revealed a significantly positive correlation between velocity, DO, cobble proportion with scrapers and collector-gatherers. Shredders were negatively correlated with NH_3-N. Collector-filterers were positively correlated with water temperature, NH_3-N and the sand proportion. Predators were also positively correlated with the coarse gravel proportion (Table 3).

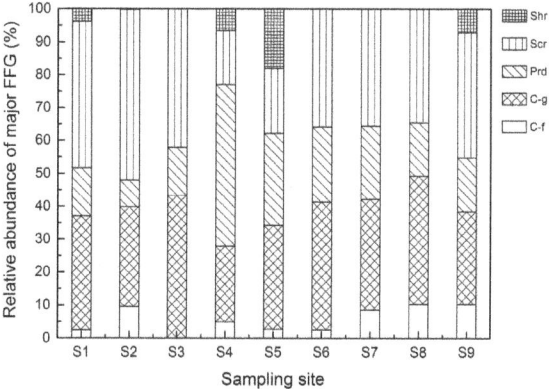

Figure 6. Relative abundance of functional feeding groups between sites. Percentage values are for the abundance pooled across the three samples collected from each site.

Table 3. Partial correlations between FFG and the physical and chemical habitat variables measured in the study from the CCA. Variables with absolute correlation coefficients greater than 0.3 are listed.

Group	Positive Correlation Coefficients	Negative Correlation Coefficients
Scr	Velocity * (0.453) DO ** (0.783) Cobble * (0.408)	TN [†] (−0.321) Fine Gravel [†] (−0.346)
Shr	—	NH_3-N * (−0.373)
Prd	Coarse Gravel * (0.377)	—
C-g	Velocity * (0.452) DO ** (0.646) Cobble * (0.434)	Fine Gravel * (−0.370)
C-f	Water Temp.** (0.770) NH_3-N * (0.138) Sand ** (0.694)	—

[†] Correlation is significant at the 0.1 level (2-tailed); * Correlation is significant at the 0.05 level (2-tailed); and ** Correlation is significant at the 0.01 level (2-tailed).

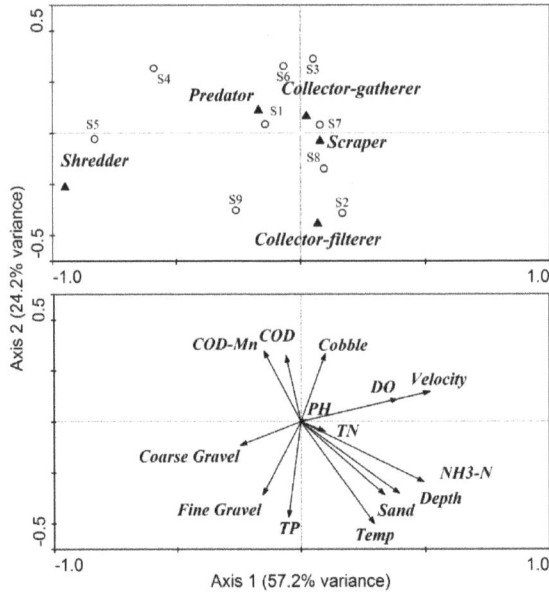

Figure 7. Biplot of FFG data with habitat variables by the CCA along principal component axes 1 and 2. Data from all of the sites were included and the two axes explained 73.9% of the cumulative variance in the data. Test of significance: axis1: $p < 0.005$; all canonical axes: $p < 0.005$.

4. Discussion

4.1. The Effects of ROR Scheme on Spatial Variations in Habitat Conditions

The most critical process affecting downstream habitats within the "depleted" reach of Aotou plant was the flow reduction. In general, natural rivers have relatively stable flow regimes, mostly running at base flow levels [18,19]. However, the operations of the ROR scheme disturb the natural flow regime through in-channel barriers associated with flow diversion by secondary diversion channels. The collected data in the study showed that over the entire sampling period (June to July, 2014), the diverted flow was estimated at 70%–80%, which meant that only 20%–30% of the total flow fed the downstream channel of the weir. Flow regimes are regarded as playing a fundamental role in determining river habitat availability and ecological integrity [20–22]. The direct consequences of flow variations are the alterations of hydrological habitat factors, generally including flow velocity, water depth and wetted widths, which were commonly recognized as key physical habitat factors in river ecosystems. At depleted sites, particularly site S4, the recorded hydrological variables declined dramatically compared with the upper sites due to flow reductions. The mean flow velocities and water depths in riffles located at the "depleted" sites were 3–4 times lower than the upper ones, and the wetted channel widths of these dewatered sites also experienced a reduction of 30%–60%. Decreased flow created a large proportion of lentic or slow-flow habitats in the "depleted" reach, which was largely inconsistent with upper ones that featured riffles and runs. This declining trend is consistent with the results of other case studies [23,24]. With flow return, water depths and flow velocities tended to increase gradually at lower sites, although the magnitudes of increases were heavily dependent on the channel morphology, and the habitat conditions at the lower reach upgraded significantly from the "depleted" reach.

The results from the present study identified spatial variations in DO concentrations, with levels that were significantly higher in the upper reach. The DO concentration in rivers is influenced by many biological, chemical, and physical interactions, and in terms of the physical process, it is heavily controlled by water temperature, water pressure and flow velocity [25,26]. In this case, with little water temperature difference, the lower velocities and induced relatively weak disturbances in downstream reaches are likely to decrease oxygen aeration and, hence, reduce DO levels. Other chemical water chemistry variables (including COD_{Mn}, COD, BOD_5, NH_3-N, TP, and TN) showed no significant differences between the upper reach and the "depleted" reach. These results suggest that the ROR scheme contributed little to level variations of nutrients and organic contaminants. Increasing levels of nutrients (TN, TP and NH_3-N) were found in the lower reach, mainly due to agriculture practices along both the diversion channel and the downstream channel, which contributed to more nutrient loadings to the receiving water.

In addition to changing the flow regime, in-channel barriers are effective sediment traps and are known to have an important effect on the downstream channel [27]. The longitudinal characteristics of substrate compositions were remarkable in the study. Large-diameter sediments, such as cobbles and coarse gravel, were the dominant compositions in the upper reach and were accompanied by great embeddedness. In contrast, cobble compositions significantly decreased in the lower reach, but sand compositions increased. We are unsure as to what extent these changes depended on the ROR plant; therefore, more studies are needed to further explain the shifting mechanisms. Unstable riverbanks with poor riparian vegetation were detected in the "depleted" reach, particularly at sites immediately downstream of the weir. The volume flux of overflow and the resulted sediment composition may be the best explanation for the poor condition. The flow over regulating weir was low in the load of suspended sediment, and such flow can easily lead to erosion and channel incision [27,28]. With respect to the riverbank condition, the duration of periods with no surface flow controls vegetation structure along the "depleted" channel [29]. In the "depleted" reach, the surface flow is intermittent, groundwater levels along the riverbank show strong declines, and these hydraulic conditions are less available to the riparian vegetation. Flow reductions are usually associated with decreases in the riparian water table at higher elevations and losing riparian vegetation due to drought stress [30,31]. Riverbank collapse occurs when the driving forces exceed the resisting forces. Poor vegetation is widely believed to weaken the resisting force, thus decreasing the stability of riverbanks [32]. The interaction of both processes of vegetation loss and erosion acceleration leads to degradation of riparian habitats and exerts impacts on the riverine ecology.

Comparisons of habitat conditions, in terms of physical and chemical variables between reaches, demonstrate that the ROR scheme of the Aotou plant could exert impacts on river habitats in the "depleted" reach and further downstream in terms of three aspects: changing the hydrological regime through flow diversion and return; degrading the riparian condition (poor vegetation and rich bank erosion) in the "depleted" reach; and shifting levels of some water chemistry variables, such as PH and DO. The results of the PCA showed that the key factors, such as flow velocity, cobble composition, DO, and COD, were mainly responsible for the variations obtained in the habitat conditions. The results show that no one factor exerts a supreme influence on habitat conditions, whereas combinations and interactions of these variables appear to fully account for the characteristic differences between reaches.

4.2. Response of Macroinvertebrate Structure and Biodiversity to ROR Scheme

In total, 25 macroinvertebrate taxa were recorded from the nine sampling sites during the survey. The dominance feature of *Insecta* group is similar to the other studies conducted in the Hailang River [33,34]. Tangbin Huo [33] recorded about 51 taxa throughout the whole river during August 2011 and found that 75.44% of them belonged to *Insecta*. Teng Fei [34] record 36 taxa in summer 2010 and *Insecta* accounted for 65.5% of the fauna. The mean density of total macroinvertebrates in the "depleted" reach and lower reach were significantly reduced compared with the upper reach. Total macroinvertebrates density in the "depleted" reach was only 25% of the upper one, despite the increased percentage in the lower reach, which was 43%. This decreasing trend was not consistent with some previous studies concerning the impacts of water diversion and artificially reduced flow regime on stream macroinvertebrates [35,36]. Their cases revealed that the density of macroinvertebrates showed no significant decreases, or even increases under reduced flow. However, the decreasing trend was found in other studies. McIntosh & Benbow [37] found that the mean density of total macroinvertebrates above the diversion was 46% greater than below the diversion, while Cazaubon & Giudicelli [38] found macroinvertebrates in regulated sites had lower densities and diversity compared with natural ones in the same district. Similarly, some previous studies also found taxa richness reductions in low flow conditions. McKay & King [36] compared reaches above and below a diversion and found a low family richness in the 'diverted' treatment reach. In this study, the total taxa richness, EPT richness and d_M were introduced to access the richness from multiple perspectives. These indices also consistently suffered sharp reductions in the "depleted" reach. Chemical variables were considered to make little contribution to the EPT richness differences, because they varied within a limited range and satisfied the same quality standard, given to measure the surface water quality of China. The levels of total taxa richness, EPT richness and d_M increased further in the lower reach, but still remained at relatively lower levels compared the upper reach. Comparisons of density and richness between reaches suggested that both the density and richness (total taxa and EPT taxa) may change in response to the flow variations resulting from the ROR scheme, and reduced flows were prone to decrease both indices.

In general, losses of macroinvertebrate density and richness in the downstream reaches appear to be attributed to two major causes. First, the physical barrier represented by the weir may disturb the river connectivity and may restrict macroinvertebrate drifts from the upstream reach to the downstream reach. This physical isolation and restricted movement may contribute to the insufficient recolonization of macroinvertebrates downstream of the weir and then bring about poor density and richness. Second, intra-species and inter-species competitions are likely to become more intense in the low-flow area due to the limited habitat area and food resources [18,37,39]. Competition may influence the community structure and combine the fauna into fewer species that dominant the confined area.

Additionally, the downstream reaches were always accompanied by low macroinvertebrate diversity, which could be explained by the degradation of habitat diversity. River habitat primarily depends on the channel morphology and hydrological conditions and flow decreases can reduce habitat diversity. In the "depleted" reach, extreme low-flow conditions facilitate the replacement of lotic habitats with lentic ones, which is not suitable for the taxa preferring fast-flow conditions. We assume poor habitat diversity to be a primary factor contributing to the low richness and low diversity.

On the reach-scale, the macroinvertebrate community structure found in the "depleted" reach was distinct from those in other reaches. *Diptera* was the dominant group in the "depleted" reach mainly because of the contributions of *Chironomidae*, which were wildly distributed and abundant. Other taxa with low-flowing preferences (e.g., *Oligochaeta* and *Coleoptera*) also increased in relative abundance in the "depleted" reach. However, the relative abundance of EPT taxa associated with higher flow velocities and heterogeneous instream habitats, such as *Ephemeroptera* and *Trichoptera*, decreased dramatically. The result clearly suggests that macroinvertebrate distribution and community structure are sensitive to flow alterations, and heavily depend on habitat conditions.

In biomonitoring, the environmental quality of a given site is judged from its species assemblages [40]. Further analyses of community composition alterations between sites according to the functional feeding group were made. Functional feeding group classification was helpful for ecological assessments about the river habitat conditions and widely used in previous studies regarding the ecological impacts of water diversion or river regulation [37,41].

A CCA of the physical-chemical habitat data with the macroinvertebrate data based on the FFG abundance explained 81.4% of the total variance in the first two canonical axes. The strong association of macroinvertebrate data with habitat conditions suggests that habitat changes due to the ROR scheme could exert large impacts on the distributions of major functional feeding groups of macroinvertebrates. Partial correlations between habitat variables and functional feeding groups provide a perspective to find the main factors that determine the distribution of the specific group. Different functional feeding groups have different habitat preferences [42]. In upper sites, where flow velocity was the highest and the cobble composition and DO were also higher, the collector-gatherer species and the scraper species were most common. This result is consistent with previous studies [42,43]. For example, Quinn [42] found that filter-feeding species had strong preferences for habitats with high velocity and seston. As for scraper species, Heino [43] found that their composition showed a strong positive relationship with the habitat heterogeneity and water depth. In contrast, the relative abundance of both collector-gatherers and the scrapers reduced at the "depleted" sites, particularly at the site immediately downstream of the weir. Although these reductions were predictable and similar to a previous study [44], predators at these sites were prone to be the dominant group. Due to extremely low density and diversity at these sites, the predators' dominance could not be attributed to either the adaptation ability of some predatory taxa (e.g., *Chironomidae*) or the high competition advantages, due to uncertainty. In the lower reach, the habitat diversity tended to be higher with flow increase, and the compositions of FFG, particularly collector-gatherers and scrapers, tended to be similar to the upper reach. This trend suggests the functional feeding group structure and composition downstream of the diversion reach were resilient to flow alterations. Another feature found in the lower reach was the relatively higher proportion of collector-gatherers, which was positively associated with the sand composition. This may be partially explained by the study of Likens [45], who studied the invertebrate community composition in sand or silt habitats and found that collector-gatherers (e.g., *Chironomids*) were the primary residents in sand habitats. In general, DO, velocity and substrate compositions seem to be the key factors that are positively correlated with FFG groups.

The habitat preferences of macroinvertebrates depend on the balance of various requirements of macroinvertebrates, including the lentic or lotic area preferences, food resources, thermal condition, oxygen acquisition of maintaining position, water quality, substrate and biotic interactions [19,42,46–49]. The ROR scheme changed the natural flow regime and river connectivity through a weir and flow-diversion channel, resulting in distinct habitat conditions between reaches, with particularly low habitat diversity and poor habitat quality in the "depleted" reach.

The changes in the habitat conditions exerted pronounced effects on macroinvertebrate density, richness, diversity and composition structure. The comparisons between reaches can provide insight in order to assess the ecological impacts of the ROR scheme. This study also clearly indicates that macroinvertebrate distribution and community structure are largely affected by habitat variables; thus, they can fulfill a role as indicators for habitat conditions. A series of new ROR plants will be constructed in the near future. The cumulative impacts of hydroelectric development and longitudinal habitat fragmentations on macroinvertebrate communities along the regulated river should be considered in future studies.

Acknowledgments: We acknowledge the financial support of the National Natural Science Foundation of China (No. 51579130 and No. 51279078), the Program for New Century Excellent Talents in University of China (2012490811) and the Program for Changjiang Scholars and Innovative Research Team in the University of Ministry of Education of China (IRI13025).

Author Contributions: Haoran Wang and Zhaowei Liu conducted the field sampling, collected and analyzed the data. Yongcan Chen contributed to the data collection, and supervised the research. Dejun Zhu provided important advice on the structures of the manuscript and writing. The author and co-authors all contributed to the preparation of the manuscript.

Conflicts of Interest: The authors declare no conflict of interest.

References

1. Anderson, D.; Moggridge, H.; Warren, P.; Shucksmith, J. The impacts of "run-of-river" hydropower on the physical and ecological condition of rivers. *Water Environ. J.* **2015**, *29*, 268–276.
2. Humborg, C.; Ittekkot, V.; Cociasu, A.; Bodungen, B. Effect of danube river dam on black sea biogeochemistry and ecosystem structure. *Nature* **1997**, *386*, 385–388.
3. Nilsson, C.; Reidy, C.A.; Dynesius, M.; Revenga, C. Fragmentation and flow regulation of the world's large river systems. *Science* **2005**, *308*, 405–408.
4. Stone, R. The legacy of the three gorges dam. *Science* **2011**, *333*, 817.
5. Parsley, M.J.; Beckman, L.G.; McCabe, G.T., Jr. Spawning and rearing habitat use by white sturgeons in the columbia river downstream from mcnary dam. *Trans. Am. Fish. Soc.* **1993**, *122*, 217–227.

6. Kanehl, P.D.; Lyons, J.; Nelson, J.E. Changes in the habitat and fish community of the milwaukee river, wisconsin, following removal of the woolen mills dam. *N. Am. J. Fish. Manag.* **1997**, *17*, 387–400.

7. Beasley, C.A.; Hightower, J.E. Effects of a low-head dam on the distribution and characteristics of spawning habitat used by striped bass and american shad. *Trans. Am. Fish. Soc.* **2000**, *129*, 1316–1330.

8. Zhang, G.; Wu, L.; Li, H.; Liu, M.; Cheng, F.; Murphy, B.R.; Xie, S. Preliminary evidence of delayed spawning and suppressed larval growth and condition of the major carps in the yangtze river below the three gorges dam. *Environ. Biol. Fish.* **2012**, *93*, 439–447.

9. Paish, O. Micro-hydropower: Status and prospects. *Proc. Inst. Mech. Eng. Part A J. Power Energy* **2002**, *216*, 31–40.

10. Dewson, Z.S.; James, A.B.; Death, R.G. Invertebrate community responses to experimentally reduced discharge in small streams of different water quality. *J. N. Am. Benthol. Soc.* **2007**, *26*, 754–766.

11. Finn, M.A.; Boulton, A.J.; Chessman, B.C. Ecological responses to artificial drought in two australian rivers with differing water extraction. *Fundam. Appl. Limnol.* **2009**, *175*, 231–248.

12. Resh, V.H.; Norris, R.H.; Barbour, M.T. Design and implementation of rapid assessment approaches for water resource monitoring using benthic macroinvertebrates. *Aust. J. Ecol.* **1995**, *20*, 108–121.

13. Duan, X.; Wang, Z.; Xu, M.; Zhang, K. Effect of streambed sediment on benthic ecology. *Int. J. Sedim. Res.* **2009**, *24*, 325–338.

14. Cummins, K.W.; Klug, M.J. Feeding ecology of stream invertebrates. *Ann. Rev. Ecol. Syst.* **1979**, 147–172.

15. Cummins, K.W.; Merritt, R.W.; Andrade, P.C. The use of invertebrate functional groups to characterize ecosystem attributes in selected streams and rivers in south brazil. *Stud. Neotro. Fauna Environ.* **2005**, *40*, 69–89.

16. Ter Braak, C.J.; Verdonschot, P.F. Canonical correspondence analysis and related multivariate methods in aquatic ecology. *Aquat. Sci.* **1995**, *57*, 255–289.

17. Barbour, M.T.; Gerritsen, J.; Snyder, B.; Stribling, J. *Rapid Bioassessment Protocols for Use in Streams and Wadeable Rivers*; USEPA: Washington, DC, USA, 1999.

18. Lake, P. Disturbance, patchiness, and diversity in streams. *J. N. Am. Benthol. Soc.* **2000**, *19*, 573–592.

19. Lytle, D.A.; Poff, N.L. Adaptation to natural flow regimes. *Trends Ecol. Evol.* **2004**, *19*, 94–100.

20. Poff, N.L.; Allan, J.D.; Bain, M.B.; Karr, J.R.; Prestegaard, K.L.; Richter, B.D.; Sparks, R.E.; Stromberg, J.C. The natural flow regime. *BioScience* **1997**, 769–784.

21. Bunn, S.E.; Arthington, A.H. Basic principles and ecological consequences of altered flow regimes for aquatic biodiversity. *Environ. Manag.* **2002**, *30*, 492–507.

22. Arthington, A.H.; Bunn, S.E.; Poff, N.L.; Naiman, R.J. The challenge of providing environmental flow rules to sustain river ecosystems. *Ecol. Appl.* **2006**, *16*, 1311–1318.

23. Anderson, D.; Moggridge, H.; Shucksmith, J.; Warren, P. Quantifying the impact of water abstraction for low head "run of the river" hydropower on localized river channel hydraulics and benthic macroinvertebrates. *River Res. Appl.* **2015**, *2015*.

24. Irvine, R.L.; Oussoren, T.; Baxter, J.S.; Schmidt, D.C. The effects of flow reduction rates on fish stranding in british columbia, canada. *River Res. Appl.* **2009**, *25*, 405–415.

25. Nakamura, Y.; Stefan, H.G. Effect of flow velocity on sediment oxygen demand: Theory. *J. Environ. Eng.* **1994**, *120*, 996–1016.

26. Matthews, K.; Berg, N.H. Rainbow trout responses to water temperature and dissolved oxygen stress in two southern california stream pools. *J. Fish Biol.* **1997**, *50*, 50–67.

27. Williams, G.P.; Wolman, M.G. Downstream Effects of Dams on Alluvial Rivers. Avaliable online: http://relicensing.pcwa.net/documents/Library/PCWA-L-307.pdf (accessed on 11 November 2011).

28. Friedman, J.M.; Lee, V.J. Extreme floods, channel change, and riparian forests along ephemeral streams. *Ecol. Monogr.* **2002**, *72*, 409–425.

29. Stromberg, J.; Beauchamp, V.; Dixon, M.; Lite, S.; Paradzick, C. Importance of low-flow and high-flow characteristics to restoration of riparian vegetation along rivers in arid south-western united states. *Freshw. Biol.* **2007**, *52*, 651–679.

30. Rood, S.B.; Mahoney, J.M.; Reid, D.E.; Zilm, L. Instream flows and the decline of riparian cottonwoods along the st. Mary river, Alberta. *Can. J. Bot.* **1995**, *73*, 1250–1260.

31. Jansson, R.; Nilsson, C.; Dynesius, M.; Andersson, E. Effects of river regulation on river-margin vegetation: A comparison of eight boreal rivers. *Ecol. Appl.* **2000**, *10*, 203–224.

32. Simon, A.; Collison, A.J. Quantifying the mechanical and hydrologic effects of riparian vegetation on streambank stability. *Earth Surf. Proces. Landf.* **2002**, *27*, 527–546.

33. Huo, T. Community structures of macrobenthos and biodiversity in natural aquatic germplasm resource reserve zone in hailang river. *Chin. J. Fish.* **2012**, *3*, 006.

34. Fei, T. Macrobenthos Community Structure and Biological Evaluational Research of Water Quality in Hailang River. Master's Thesis, Northeast Forestry University, Harbin, China, 2012.

35. Wright, J.; Berrie, A. Ecological effects of groundwater pumping and a natural drought on the upper reaches of a chalk stream. *Regul. Rivers Res. Manag.* **1987**, *1*, 145–160.

36. McKay, S.; King, A. Potential ecological effects of water extraction in small, unregulated streams. *River Res. Appl.* **2006**, *22*, 1023–1037.

37. McIntosh, M.D.; Benbow, M.E.; Burky, A.J. Effects of stream diversion on riffle macroinvertebrate communities in a maui, hawaii, stream. *River Res. Appl.* **2002**, *18*, 569–581.

38. Cazaubon, A.; Giudicelli, J. Impact of the residual flow on the physical characteristics and benthic community (algae, invertebrates) of a regulated mediterranean river: The durance, france. *Regul. Rivers Res. Manag.* **1999**, *15*, 441–461.

39. Dewson, Z.S.; James, A.B.; Death, R.G. A review of the consequences of decreased flow for instream habitat and macroinvertebrates. *J. N. Am. Benthol. Soc.* **2007**, *26*, 401–415.

40. Pilière, A.; Schipper, A.M.; Breure, A.M.; Posthuma, L.; de Zwart, D.; Dyer, S.D.; Huijbregts, M.A. Comparing responses of freshwater fish and invertebrate community integrity along multiple environmental gradients. *Ecol. Indic.* **2014**, *43*, 215–226.

41. Miller, S.W.; Wooster, D.; Li, J.L. Does species trait composition influence macroinvertebrate responses to irrigation water withdrawals: Evidence from the intermountain west, USA. *River Res. Appl.* **2010**, *26*, 1261–1280.

42. Quinn, J.M.; Hickey, C.W. Hydraulic parameters and benthic invertebrate distributions in two gravel-bed new zealand rivers. *Freshw. Biol.* **1994**, *32*, 489–500.

43. Heino, J. Lentic macroinvertebrate assemblage structure along gradients in spatial heterogeneity, habitat size and water chemistry. *Hydrobiologia* **2000**, *418*, 229–242.

44. Englund, G.; Malmqvist, B. Effects of flow regulation, habitat area and isolation on the macroinvertebrate fauna of rapids in north swedish rivers. *Regul. Rivers Res. Manag.* **1996**, *12*, 433–446.

45. Likens, G.E. *River Ecosystem Ecology: A global Perspective*; Academic Press: Cambridge, MA, USA, 2010.

46. Morin, A. Intensity and importance of abiotic control and inferred competition on biomass distribution patterns of simuliidae and hydropsychidae in southern quebec streams. *J. N. Am. Benthol. Soc.* **1991**, *10*, 388–403.

47. Parr, L.; Mason, C. Long-term trends in water quality and their impact on macroinvertebrate assemblages in eutrophic lowland rivers. *Water Res.* **2003**, *37*, 2969–2979.

48. González, R.A.; Díaz, F.; Licea, A.; Re, A.D.; Sánchez, L.N.; García-Esquivel, Z. Thermal preference, tolerance and oxygen consumption of adult white shrimp litopenaeus vannamei (boone) exposed to different acclimation temperatures. *J. Therm. Biol.* **2010**, *35*, 218–224.

49. Korte, T. Current and substrate preferences of benthic invertebrates in the rivers of the hindu kush-himalayan region as indicators of hydromorphological degradation. *Hydrobiologia* **2010**, *651*, 77–91.

Effects of Land Use Types on Community Structure Patterns of Benthic Macroinvertebrates in Streams of Urban Areas in the South of the Korea Peninsula

Dong-Hwan Kim, Tae-Soo Chon, Gyu-Suk Kwak, Sang-Bin Lee and Young-Seuk Park

Abstract: Benthic macroinvertebrates were collected from streams located in an urban area from regions featuring different environmental conditions. Physicochemical variables and land use types pertaining to sampling sites were analyzed concurrently. Multivariate analyses (cluster analysis and non-metric multidimensional scaling) and rank-abundance diagrams were used to characterize community patterns to assess ecological integrity in response to environmental conditions. Species composition patterns were mainly influenced by both the gradient of physicochemical variables (e.g., altitude, slope, conductivity) and the proportion of forest area. Community structure patterns were further correlated to the proportion of urbanization and to biological indices (e.g., diversity, number of species). Land use preferences of benthic species were identified based on the indicator values and weighted averaging regression models. Plecoptera species were representative of undisturbed streams in forest areas, whereas Tubificidae species and filtering collector caddis flies were indicator taxa in severely polluted and agricultural areas, respectively. The analyses of community structures and indicator species effectively characterized community properties and ecological integrity following natural and anthropogenic variability in urban stream ecosystems.

Reprinted from *Water*. Cite as: Kim, D.-H.; Chon, T.-S.; Kwak, G.-S.; Lee, S.-B.; Park, Y.-S. Effects of Land Use Types on Community Structure Patterns of Benthic Macroinvertebrates in Streams of Urban Areas in the South of the Korea Peninsula. *Water* **2016**, *8*, 187.

1. Introduction

Due to unprecedented industrial development and human aggregation, ecosystems have been severely damaged and destabilized both locally and globally. Urbanization causes rapid changes in land use via expansion of commercial/residential/industrial areas along with decimation of forests and natural areas [1]. The anthropogenic impact of urbanization upon aquatic ecosystems is critical, considering the importance of resource usage and the diffusion of pollutants to the surrounding environment [2,3]. Ecological degradation of stream drainage was especially affected by urbanization, causing the so-called 'urban stream syndrome' [3]. The symptoms of the urban stream syndrome have been broadly characterized in various fields, including physicochemical and ecological processes such as hydrology [3,4], biodiversity conservation [5–7], and ecosystem processes [8,9].

The conversion of natural areas (e.g., forest) to urban or agricultural areas severely affects both in-stream habitat and macroinvertebrate communities in various ways [10,11]. Both the abundance and the distribution of aquatic organisms are strongly influenced by land use [12,13]. Benthic macroinvertebrates adapt to the environmental changes caused by various anthropogenic impacts, resulting in an increase of tolerant species in the community. Therefore, community composition serves as a good predictor of habitat quality and biotic integrity [14,15]. During the last decade, studies demonstrated that a change in land use was one of the main driving forces behind the loss of regional biodiversity [11,16].

Community data are difficult to analyze because communities consist of numerous species varying in a complex and stochastic manner in response to environmental factors. Computational methods have been developed to analyze community data efficiently; these include predictive models [17], multivariate statistical methods [18], statistical learning models [19,20], and species abundance models [21]. Many studies have performed multivariate analyses on freshwater benthic community data to assess water quality [18,22], spatial and temporal dynamics [23,24], trait patterns [25], and community response to disturbance [18,26]. A rank abundance diagram (RAD), a type of species abundance model, has been applied to community data to elucidate alterations in community structure resulting from environmental change [27]. Several studies using RADs have been applied to various taxa, including freshwater macroinvertebrates [26,28].

In this study, we extended the results of previous studies [21,28] to further investigate the changes in macroinvertebrate community patterns in response to urbanization in the southeast part of the Korean Peninsula. Specifically, we presented the environmental and biological characteristics, and examined the effects of anthropogenic disturbance on benthic macroinvertebrate communities by concurrently measuring hydrological/physical/chemical factors and land use types in various streams located in the Busan metropolitan area.

2. Materials and Methods

2.1. Field Sampling

Benthic macroinvertebrates were collected at 46 sampling sites in the Busan metropolitan area, located in the southeast part of the Korean Peninsula (Figure 1a). The geography of the Busan metropolitan area is covered in mountainous regions (47%), and the southeast part of Korea features a temperate climate with high rainfall

226

in July because of the East Asian Monsoon. Sampling was carried out twice, namely, in late summer (August) and in winter (November–December), in 2013 and 2014. Sampling sites were chosen across various altitudes (1–315 m) and for different land use types. Among the 46 sampling sites, 10 were located on four tributaries of the Nakdong River, which flows through the Busan metropolitan area, and is the longest river in South Korea. Twenty-one sampling sites were located in the Suyeong River of mid-eastern Busan, and 15 sites were located on short streams flowing directly into the East Sea (Figure 1). The overall environmental conditions of each sampling site are shown in Appendix.

We collected macroinvertebrate samples in triplicate from a riffle habitat within 10 m reaches using a Surber sampler (30 × 30 cm^2, 500 μm mesh). Ten environmental variables, including geological, landscape, hydrological, and physicochemical factors, were measured concurrently (Appendix). Water quality parameters, including electrical conductivity (YSI 30, Yellow Springs Instruments, Yellow Springs, OH, USA) and dissolved oxygen (DO) (YSI 550A, Yellow Springs Instruments, Yellow Springs, OH, USA), were measured *in situ*, along with hydrological variables such as velocity (Swoffer 2100 LX, Swoffer Instruments, Seattle, WA, USA), depth, and substrata. Hydrological variables were measured for each replicate. Substrates collected by the Surber sampler were classified in five categories according to Jowett *et al.* [29]; a substrate index was calculated based on the substrate composition [30]. Topological and geological variables, including altitude, slope, and land use types, were extracted from a digital map obtained from the Korea Ministry of Environment using ArcGIS version 9.3 (ESRI). Proportions (%) of land use types (specifically, forest, urban, agriculture, grassland, wetland, and bare land) within a 1 km-long riparian zone (100 m width) were calculated from the digital map. This riparian sub-corridor was selected, because variation in land use is often greater at catchment scales than it is at riparian scales [2,31].

Benthic macroinvertebrate specimens were mostly identified either to the species level or to the lowest possible taxonomic level according to Han *et al.* [32], Won *et al.* [33], and Yoon [34]. In addition, Merritt and Cummins [35], and Brigham *et al.* [36] were used as ancillary literature. Chironomidae and Oligochaeta were identified to the family or class level, owing to difficulty in identification. A list of taxa identified in the study sites are given with their literature in the Supplementary Material (Table S1).

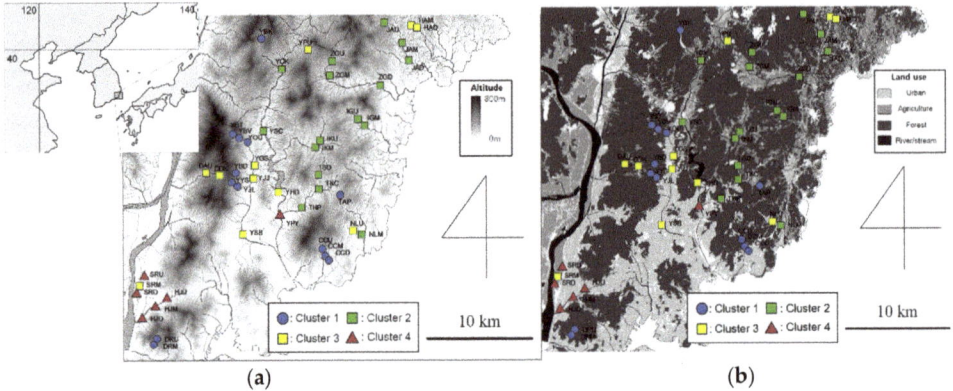

Figure 1. Location of sampling sites for each cluster with altitude (**a**); and land use type (**b**).

2.2. Data Analysis

The community data pooled at each site were used in the analyses. The Shannon diversity index, using a common logarithm [37], the dominance index [38], and Ephemeroptera, Plecoptera, and Trichoptera (EPT) richness, were calculated at each sampling site from the community data. Spearman rank correlation coefficients were obtained to show associations between the biological indices and the environmental variables.

Considering the complexity of the community data due to environmental variability, we applied multivariate analysis and RAD to determine community patterns. Before applying cluster analysis and nonmetric multidimensional scaling (NMS), the community abundance data were log transformed with $\ln(x + 1)$ to reduce the high variance caused by the exceptionally large number of individuals collected at the sampling sites, and to fulfill the assumption of normality for statistical analyses. The sampling sites were classified with a hierarchical cluster analysis, using Ward's linkage method with a Bray-Curtis distance measure, based on the similarity of their species compositions. A Multi-response Permutation Procedure (MRPP) was used to test whether or not there were statistically significant differences among the clusters [39]. NMS was then applied to characterize the overall relationships among communities in association with specific environmental factors. Cluster analyses and NMS were performed using the PC-ORD software [40]. A Monte Carlo test was implemented using both randomized data (one thousand runs) and real data to estimate the significance of the output of the NMS [40]. Kruskal-Wallis test and a multiple comparison test (Dunn's *post hoc* test) were applied to identify differences among clusters using the environmental variables and biological water quality indices ($p < 0.05$).

228

Slope lines of RADs were obtained to show community responsiveness to environmental conditions [26,41,42]. The relative abundance of species (i) in relation to the total abundance for all species in each sampling site was obtained and rearranged according to rank (in order from highest to lowest proportion) to calculate RAD slopes. The slope (S) was calculated with Equation (1):

$$S = \Delta \log (p_i/n) \tag{1}$$

where n is the number of species and p_i is the relative abundance of species i according to the rank. Diversity is maximized when the slope of the rank abundance curve approaches zero (species having a greater chance of being evenly represented), while slopes become increasingly negative when total diversity decreases [26]. For convenience of expression, we used the absolute value to present the slopes. Therefore, larger values represented steeper rank abundance curves, thus showing a tendency toward less diverse assemblages.

The indicator species analysis was conducted to define representative species in different land use types. Indicator values (IndVal) for each species i in cluster k were calculated based on Equation (2) [43,44]:

$$IndVal_{ki} = \left(\frac{Nindividuals_{ki}}{Nindividuals_{+k}} \right) \times \left(\frac{Nsites_{ki}}{Nsites_{k+}} \right) \times 100 \tag{2}$$

where $Nindividuals_{ki}$ is the mean abundance of species i across the sites pertaining to cluster k and $Nindividual_{s+k}$ is the sum of the mean abundances of species i within the various clusters. $Nsites_{kj}$ is the number of sites in cluster k where species i is present and $Nsites_{k+}$ is the total number of sites in that cluster. IndVal ranges from 0 (no indication) to 100 (perfect indication). A Monte Carlo method using 300 permutation runs was also implemented to evaluate the statistical significance of IndVal.

Weighted averaging regression models (WARMs) were applied to quantify the preference of each indicator species for each land use type [44,45]. The optimal preference value of the species was calculated as the mean of the measured land use proportion weighted by the abundance of this species at the study sites, according to Equation (3) [46]:

$$WA_i = \frac{\sum_{j=1}^{n} y_j \times x_{ij}}{\sum_{j=1}^{n} x_{ij}} \tag{3}$$

where WA_i is the weighted average (estimate of optimum preference of species i) of each land use type, y_j is the proportion of each land use type at site j, and x_{ij} is the abundance of species i at site j.

The tolerant condition (TOL_i) of each land use type for species i was calculated as the weighted standard deviation of the species abundance at all sites according to Equation (4) [46]:

$$TOL_i = \sqrt{\frac{\sum_{j=1}^{n}(y_j - WA_i)^2 \times x_{ij}}{\sum_{j=1}^{n} x_{ij}}} \qquad (4)$$

WARMs were performed using the C2 program environment [47]. The coefficients of determination (R^2) were used to estimate the precision of WARMs. Model errors were estimated by bootstrapping with 1000 cycles [48]. If the regression coefficients estimated by the bootstrapping model, and by the original model, were similar (about 80%), then the results of WARMs could be used [45].

3. Results

3.1. Community Composition

A total of 136 taxa were identified to the species or genus level, except for the family Chironomidae and class Oligochaeta, with an average of 22.2 (± 10.5) taxa, and an average density of 1386.7 individuals/m^2 (± 1380.7) at each site. Twenty-six non-insecta species (19%) were collected, including *Physa acuta*, *Asellus* sp., and *Semisulcospira libertina*. Three species (*Baetis fuscatus*, *Hydropsyche kozhantschikovi*, and *Physa acuta*), and the species of family Chironomidae and class Oligochaeta were widely distributed, and showed a high frequency of occurrence (>70%) in the study sites. The Ephemeroptera species, *Baetis fuscatus* and *Baetiella tuberculata*, as well as the Trichoptera species *Hydropsyche kozhantschikovi* were the most abundant. More than half of the species had a low frequency of occurrence (<10%), and ten singletons (7%) were collected, including four Trichoptera species. The values of community indices, including the number of species and diversity, were higher for the sampling sites at mountain streams than in the other areas. Biological indices varied according to disturbance effects (e.g., high values of conductivity and proportion of urban land use) of sampling sites.

3.2. Relationships between Variables

Altitude and slope had strong negative correlations with electrical conductivity ($r = -0.69$ and -0.52 respectively; $p < 0.01$), but positive correlations with SI ($r = 0.61$, $p < 0.01$ and $r = 0.31$, $p < 0.05$, respectively) (Table 1). Low altitude and slope indicated the strong anthropogenic impacts. Conductivity, representing the anthropogenic impact, was negatively correlated with SI ($r = -0.57$; $p < 0.01$) and velocity ($r = -0.50$; $p < 0.01$). The proportion of forest area in the riparian zone was positively correlated with the variables related to the mountain streams (altitude: $r = 0.70$, slope: $r = 0.71$ with $p < 0.01$). The overall proportion of urban area had a negative correlation with natural environmental factors, including altitude ($r = -0.62$; $p < 0.01$) and slope ($r = -0.59$; $p < 0.05$), but it was positively correlated with anthropogenic factors, such as conductivity ($r = 0.58$; $p < 0.01$).

Conductivity, however, was consistently and strongly negatively correlated with most biological indices ($p < 0.01$), including the number of species ($r = -0.65$), diversity ($r = -0.56$), dominance ($r = -0.52$), and EPT% ($r = -0.78$). Depth and conductivity were similarly correlated with the biological indices. SI and velocity, however, were negatively correlated with depth. The proportion of forest area was strongly correlated with EPT% ($r = 0.52$; $p < 0.01$). Negative correlations between urban and biological indices were detected in all cases except for the dominance index. No strong correlation was observed between agricultural land use and the biological variables (Table 1).

3.3. Community Classification

The communities were classified into four clusters based on Bray-Curtis dissimilarities in the dendrograms of hierarchical cluster analysis (MRPP, $A = 0.42$, $p < 0.01$) (Figures 1 and 2). The classification reflected community similarities as well as environmental gradients. Altitude, slope, and DO were considerably higher in cluster 1, indicating that communities in this cluster were mainly collected from mountain streams in forest-dominant areas (Table 2). In contrast, communities in cluster 4 were from lowland areas with urban environments, having low values for altitude, slope, and DO.

231

Table 1. Spearman rank correlations between variables ($n = 46$). Significant correlations have been presented ($p < 0.05$). Bold letters indicate statistically significant differences ($p < 0.01$).

Variables	Physicochemical Variables						Land Use Types			Density (indi./m²)	Biological Indices			
	Slope (%)	DO (mg/L)	Conductivity (μs/cm)	Substrate Index	Depth (cm)	Velocity (cm/s)	Forest (%)	Urban (%)	Agriculture (%)		Taxa Richness	Shannon Diversity	Dominance Index	EPT% *
Altitude	**0.72**	**0.48**	**−0.69**	**0.61**	–	–	**0.70**	**−0.62**	–	–	–	–	–	**0.68**
Slope	–	0.30	**−0.52**	0.38	–	**0.56**	**0.71**	**−0.59**	–	0.30	0.32	–	–	**0.54**
Dissolved oxygen (DO)	–	–	−0.38	0.31	–	–	0.47	−0.34	–	–	–	–	–	–
Conductivity	–	–	–	**−0.57**	0.38	**−0.50**	**−0.63**	**0.58**	–	–	**−0.65**	**−0.56**	**0.52**	**−0.78**
Substrate index	–	–	–	–	−0.44	**0.54**	0.44	−0.37	–	–	0.47	**0.56**	**−0.58**	**0.59**
Depth	–	–	–	–	–	–	–	–	–	−0.34	−0.40	**−0.51**	**0.51**	−0.45
Velocity	–	–	–	–	–	–	0.38	**−0.48**	–	0.36	0.42	0.33	−0.34	**0.48**
Forest	–	–	–	–	–	–	–	**−0.78**	−0.38	–	0.34	0.31	−0.30	**0.56**
Urban	–	–	–	–	–	–	–	–	–	–	−0.42	−0.45	0.41	−0.46
Agriculture	–	–	–	–	–	–	–	–	–	–	–	0.30	−0.31	–

* Proportion (%) of Ephemeroptera, Plecoptera, and Trichoptera (EPT) richness.

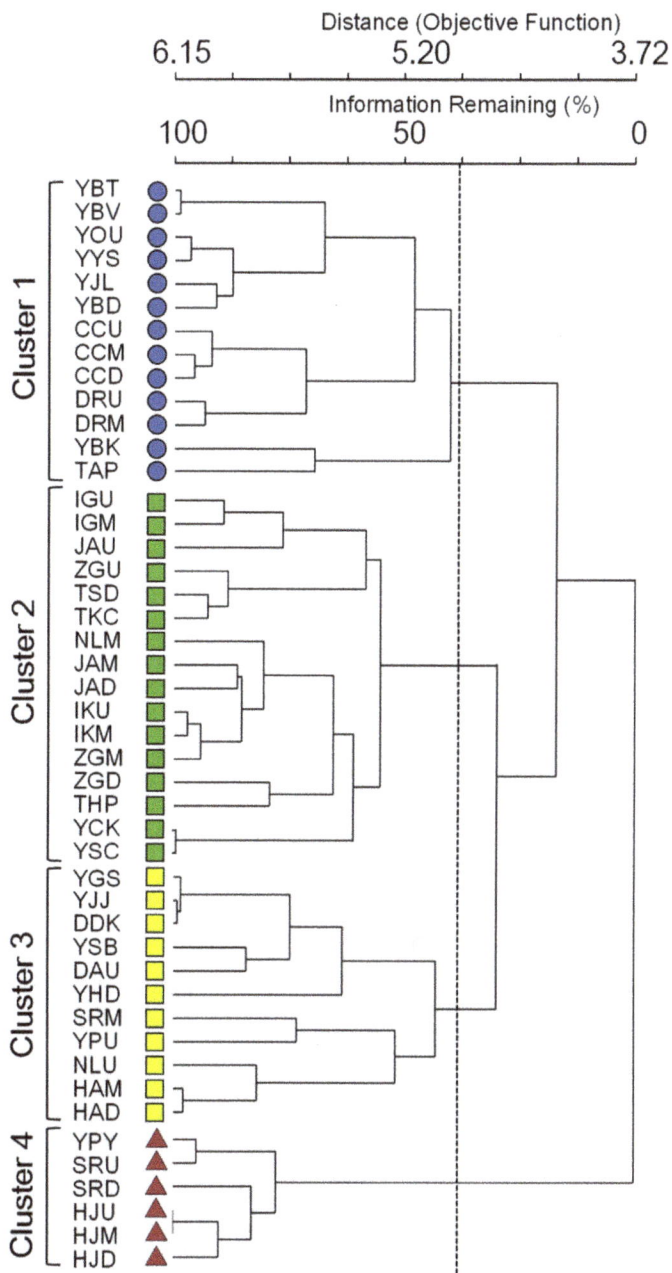

Figure 2. Classification of sampling sites with densities of benthic macroinvertebrates through a hierarchical cluster analysis, using the Ward linkage method with Bray-Curtis distance.

Table 2. Differences in parameters (mean ± standard error) for different clusters defined by clustering analysis.

Variables		Clusters				p *
		1	2	3	4	
Physicochemical variables	Altitude (m)	131.8 (±21.8) [c] **	60.8 (±9.6) [b]	91.0 (±32.2) [b]	7.2 (±2.2) [a]	<0.001
	Slope (%)	8.7 (±2.0) [c]	3.3 (±0.6) [b]	3.2 (±0.9) [b]	0.1 (±0.1) [a]	<0.001
	DO (mg/L)	9.2 (±0.4) [b]	8.6 (±0.3) [ab]	8.6 (±0.2) [ab]	7.8 (±0.3) [a]	0.048
	Conductivity (µs/cm)	105.7 (±12.8) [a]	254.0 (±28.0) [b]	292.1 (±27.5) [bc]	455.8 (±43.0) [c]	<0.001
	Substrate index	5.8 (±0.1) [b]	5.9 (±0.1) [b]	5.5 (±0.2) [b]	4.6 (±0.1) [a]	0.001
	Depth (cm)	16.8 (±1.2) [a]	14.7 (±1.3) [a]	19.6 (±4.2) [a]	30.4 (±2.1) [b]	0.003
	Velocity (cm/s)	22.5 (±2.5) [b]	25.0 (±3.2) [b]	17.7 (±3.7) [b]	2.9 (±0.7) [a]	0.001
Land use types	Forest (%)	85.4 (±4.6) [c]	51.8 (±7.9) [b]	37.7 (±9.6) [ab]	11.6 (±6.4) [a]	<0.001
	Urban (%)	4.4 (±2.6) [a]	3.9 (±1.4) [a]	19.5 (±6.7) [ab]	40.9 (±8.9) [b]	0.001
	Agriculture (%)	2.7 (±1.5) [a]	31.0 (±6.0) [b]	20.1 (±7.4) [ab]	7.0 (±6.0) [a]	0.001
Biological indices	Number of taxa	30.3 (±2.0) [c]	25.6 (±1.9) [c]	17.2 (±1.8) [b]	4.8 (±0.9) [a]	<0.001
	Density (indi./m²)	2155.4 (±504.8) [b]	1455.9 (±317.7) [ab]	964.0 (±211.7) [ab]	311.4 (±85.1) [a]	0.004
	Diversity (H′)	2.0 (±0.1) [bc]	2.3 (±0.1) [c]	1.6 (±0.1) [b]	0.6 (±0.1) [a]	<0.001
	Dominance	59.2 (±3.8) [ab]	46.7 (±3.2) [a]	71.9 (±4.4) [b]	97.1 (±1.1) [c]	<0.001
	EPT%	66.6 (±2.5) [c]	56.0 (±3.0) [bc]	43.5 (±4.7) [b]	14.9 (±6.1) [a]	<0.001

* p values indicate significant differences among clusters based on Kruskal-Wallis test. ** The different alphabets indicate significant difference according to the Dunn's *post hoc* multiple comparison tests.

The values of the environmental variables in clusters 2 and 3 fell in the middle range between clusters 1 and 4. Conductivity was significantly lower in cluster 1 than in cluster 4. SI and the current velocity were also significantly higher in cluster 1 than in the other clusters, while depth was notably high in cluster 4. Clusters 2 and 3 predominated in the agricultural areas in riparian zones (Table 2). It is noteworthy that diversity values in the highest range were observed in cluster 2 rather than cluster 1. Overall, the samples in cluster 1 from the mountainous areas had the highest values for the biological parameters including density, number of species and EPT%, whereas, except for the dominance index, the lowest values for the biological parameters were observed in cluster 4 (Table 2).

3.4. Community Ordination

For the NMS ordination, the sampling sites were distinguished based on similarities in species composition of communities (Figure 3). The final stress value obtained in the NMS ordination was 18.9 (two axes), which is an acceptable stress value for ecological community data, since it falls within the range of 10–20. A two-dimensional ordination explained 66.7% of the variance (R^2 = 0.48 for axis 1, R^2 = 0.19 for axis 2), and the Monte Carlo test implied that the axes explained significantly more variance than by chance alone (p = 0.01). The sampling sites in cluster 1 were aggregated on the upper left part of map, whereas sampling sites in cluster 4 were located on the bottom right part. Clusters 2 and 3 lay on the middle part, between clusters 1 and 4, with cluster 2 to the left and cluster 3 to the right.

The environmental factors had significant correlations with the NMS axes (Figure 3). The main variability in the community data was observed along axis 1 in response to anthropogenic (urban area) and habitat factors. The environmental factors including slope, conductivity, altitude, and the proportion of forest area were related to the bottom and top positions along axis 2. The variability of axis 1 was positively correlated with the proportion of urban area ($r = 0.50$; $p < 0.05$), depth ($r = 0.44$; $p < 0.05$), and conductivity ($r = 0.26$; $p < 0.05$) on the right side with samples in cluster 4, but was negatively correlated with the substrate index ($r = -0.52$; $p < 0.05$) and the current velocity ($r = -0.42$; $p < 0.05$) on the left side of NMS ordination. The variability on axis 2 was lesser than that on axis 1, but it was positively correlated with the slope ($r = 0.69$; $p < 0.05$), followed by the proportion of forest area ($r = 0.61$; $p < 0.05$), altitude ($r = 0.43$, $p < 0.05$), and substrate size ($r = 0.41$; $p < 0.05$) on the upper side with samples in cluster 1, and negatively correlated with conductivity ($r = -0.72$; $p < 0.05$) and the proportion of the urban area ($r = -0.44$; $p < 0.05$) on the lower side (Figure 3). The proportion of agricultural area alone had low level negative correlation on axis 2 ($r = 0.29$; $p < 0.05$). It is notable that the substrate size and the proportion of urban area had strong correlations along two NMS axes.

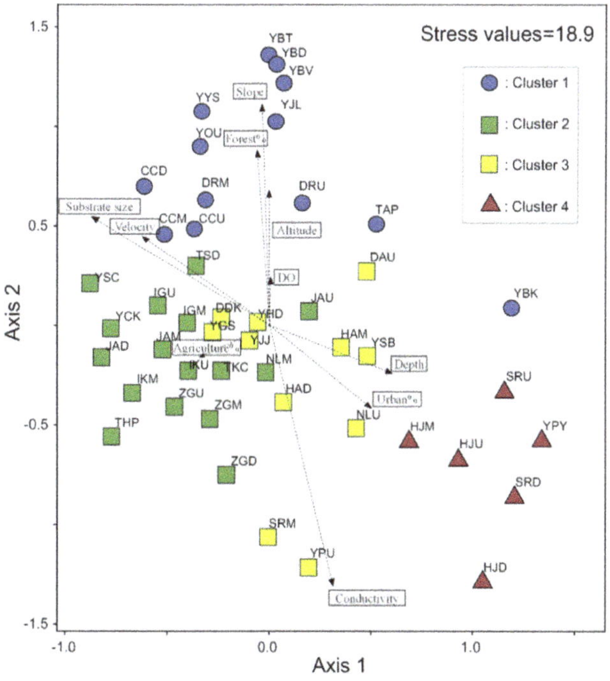

Figure 3. Nonmetric multidimensional scaling (NMS) ordination based on macroinvertebrate communities with fitted vectors of environmental variables ($R^2 > 0.07$). Different colors and symbols represent the four clusters.

3.5. Species Abundance Distribution

RADs were obtained as log abundance scales across different clusters (Figure 4), and reflected community conditions pertaining to specific clusters. RADs have low slope and long length, indicating even, highly diverse communities. All slopes were fitted with linear models using higher values of R^2 (average of 0.90 ± 0.07). The slopes were similarly lower in the sampling sites of clusters 1 and 2, which represented the less polluted streams. RADs at the polluted sites in cluster 3 had relatively steeper slopes, with fewer species compared with the less polluted sites in clusters 1 and 2 (Figures 4 and 5). Communities in cluster 4, representing severely polluted sites, had steeper RAD curves and a minimal number of species. All slope values ranged from -0.12 (at YBV) to -1.90 (at YPY). The absolute values of the slopes in clusters 1 and 2, representing the less polluted sites, fell within the range of 0.12–0.32, whereas the values in cluster 3 were relatively higher (within the range of 0.20–1.00). The slopes in cluster 4 were significantly higher than those of clusters 1 and 2, ranging from 0.64–1.90.

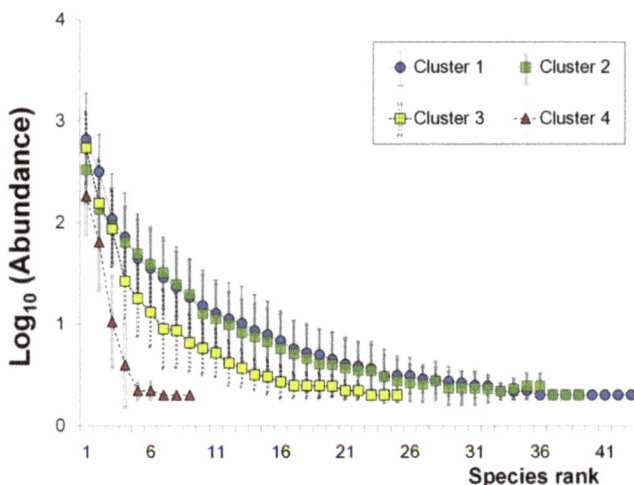

Figure 4. Rank abundance diagram (RAD) curves with mean and standard deviation (error bar) of each ranked species for three different clusters.

3.6. Indicator Species Analysis

Forty species were identified as indicators ($p < 0.05$) for four different clusters, according to the indicator species analysis (Figure 5). Among the indicator species, 22 were chosen in cluster 1 with an IndVal of 23.1–89.7, and 14 in cluster 2 with an IndVal of 34.5–71.5 (Figure 5). Meanwhile, three and one indicator species were identified in cluster 3 and cluster 4 (representing polluted streams), respectively. The

Plecoptera species *Nemoura* KUa and *Neoperla quadrata* showed high indicator values for cluster 1. Ephemeroptera (*Caenis nishinoae*, *Uracanthella rufa*, and *Ecdyonurus levis*) and Tricoptera species (*Hydroptila* KUa, *Cheumatopsyche* KUa, and *Hydropsyche kozhantschikovi*) showed high IndVal (>50) in cluster 2. No aquatic insect was selected as indicator species in clusters 3 and 4. Tubificidae species were considered indicator species for cluster 4 that represented polluted sites (Figure 5).

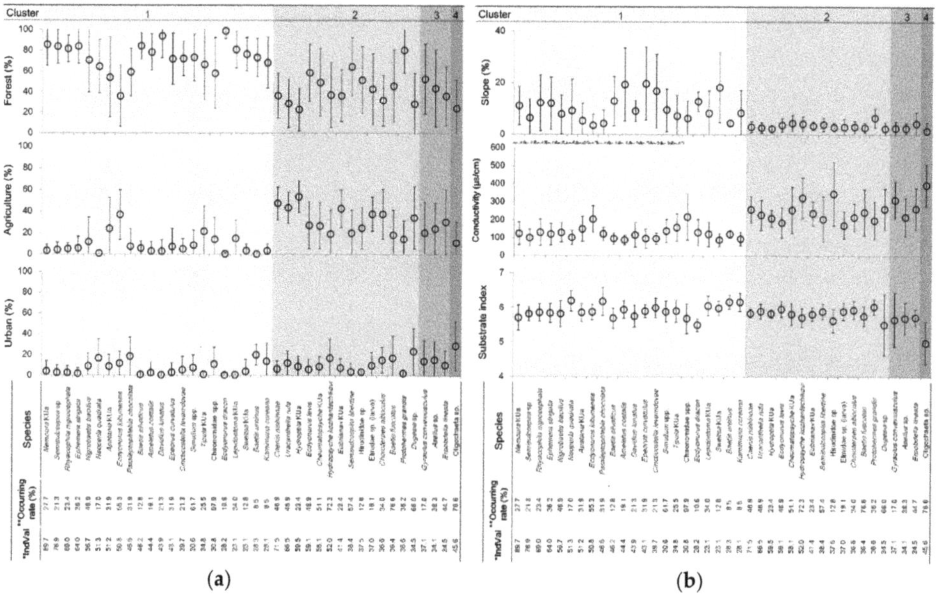

Figure 5. Optimal and tolerance ranges of (**a**) land use types and (**b**) environmental variables for the selected indicator species. Dot indicates the optimal value, and the solid line indicates the tolerance value; * IndVal: indicator value, ** Occurring rate is frequency of occurrence (% of sites).

According to WARMs, the indicator species in each cluster displayed different preferences toward land use types and three environmental variables (slope, conductivity, and substrate index) based on regression coefficients (Figure 5). Regression coefficients between the inferred and the observed proportions of land use types and environmental variables were relatively high, and were similar with and without bootstrapping tests (Table 3). This observation indicated that the optimal variables for each species were estimated reliably by the WARMs. The indicator species in cluster 1 showed a higher optimum preference in mountain forest areas with lower conductivity than in agricultural and urban areas with higher conductivity, ranging between 170–395 µs/cm. The indicator species, including *Ecdyonurus kibunensis*, *Apatania* KUa, and *Tipula* KUa, had high preferences in agricultural lowland areas with relatively high conductivity. Three species, *Neoperla quadrata*, *Paraleptophlebia chocolata*, and *Baetis ursinus*, had relatively high optimum preference in the urban areas of cluster 1. Although clusters 1 and 2 have similar RADs, their habitat preference scopes were substantially different (Figure 5). The proportion of forest area was generally lower in cluster 2 than in cluster 1, whereas that of agricultural area was higher in cluster 2. Species in cluster 1 preferred low conductivity and high slope, whereas the substrate preference was positive in clusters 1 and 2. *Hydroptila* KUa, *Caenis nishinoae*, *Uracanthella rufa*, and *Eubrianax* KUa showed high optimal preference values for the agriculture areas of cluster 2 (>40%). *Hydropsyche kozhantschikovi* and Hirudinidae species were well adapted to disturbed streams with high conductivity. In cluster 3, however, the preference was not much different from that of cluster 2, while the number of indicator species was substantially lower in cluster 3 than it was in cluster 2 (as stated above). In cluster 4, representing severe pollution, Oligochaeta species were exceptionally well adapted to urban areas with the highest conductivity and the lowest substrate size, showing a relatively high IndVal of 45.6 (Figure 5).

Table 3. Predictive power of weighted averaging regression models for land use types and environmental variables.

Variables	R^2 *	R^2_{boot}	RMSE **	$RMSE_{boot}$
Forest (%)	0.56	0.44	31.3	35.6
Urban (%)	0.54	0.35	17.2	21.7
Agriculture (%)	0.46	0.28	24.2	28.0
Slope (%)	0.60	0.36	4.2	5.1
Conductivity (μs/cm)	0.64	0.58	103.5	117.2
Substrate index	0.61	0.36	0.4	0.6

* R^2: regression coefficient between the inferred and measured values. ** RMSE: root mean squared error.

4. Discussion

4.1. Community Patterns and Indicator Species

The applications of multivariate analysis, RADs, and indicator analysis effectively accounted for the environmental impact of human activity on benthic macroinvertebrates. Cluster analysis and NMS were useful to determine the changing patterns of benthic species composition influenced by physicochemical gradients and anthropogenic disturbance. Structural properties of communities described by RAD slopes, however, were feasible for assessing disturbance due to urbanization, and these results provided an additional dimension of biological responses in characterizing environmental impacts for monitoring ecosystems [21,26]. Indicator species analysis using Indval and WARM determined the habitat preference and tolerance range across an environmental gradient [45].

Differentiation patterns on multivariate analyses can be further explained according to the properties of indicator species. Indicator species in each cluster presented higher values of occurrence rates and densities at sites in the same cluster than in other clusters. This indicated that indicator species with higher Indval play a key role in the community differentiation by multivariate analysis. Habitat preference range of indicator species based on WARM also provided additional information for patterning communities responding to different environmental impacts. For instance, the sampling sites belonging to clusters 1 and 2 were located on the left side of NMS axis 1, and had similar RAD patterns. Clusters 1 and 2, however, were clearly differentiated based upon habitat preferences. Species in cluster 2 had a relatively high preference for agricultural areas with low conductivity, whereas species in cluster 1 showed higher preferences for forest areas. It can be noted that community composition and structure are determined by indicator species analysis.

Indicator values provide an additional dimension for addressing ecological integrity in response to different environmental impacts. Community structure patterns based on RADs also show the relations between biological responses and environmental factors linked with NMS. Although numerous studies have reported species abundance distribution, few were concerned with integrating community structure, RADs, and indicative species.

Several species abundance distribution models have been proposed to identify patterns in community structures, including log-normal distribution, log series based on statistical models, and geometric series to show biological processes such as niche subdivision [27,49]. Various parameters of these models (e.g., k values in the geometric series, a and γ values in the log-normal distribution) have been used as indicators for community structure [28,49,50]. In this study, we extended the scope of RADs so that it would offer more information on community structure [21,27]. The application of RAD slopes was supported by other studies that identified community responsiveness to environmental variability [26,41,42]. This approach could also be used for the health assessment of various ecosystems.

4.2. Influence of Land Use Types

Patterns of land use in riparian areas can influence the suitability of habitats for stream communities, and change the distribution and abundance of species [51,52]. In this study, most environmental variables, except depth, showed statistically significant correlations with forest and urban areas in the land use types (Table 1). Differences in land use types influenced biological properties, including community patterns (Figures 2 and 3 Table 2), indicator species (Figure 5) and environmental properties (Table 1). The proportions of urban and forest areas, which had high correlations with the first and second axes on NMS, and relatively strong correlations with RAD slopes, were considered as key factors in the maintenance of biodiversity and community stability overall [26]. Similar observations were reported for numerous streams and rivers [53,54], confirming that the pattern of land use is a critical factor in the conservation of freshwater communities [11,55,56].

The land use type may affect biological trait patterns of benthic communities, such as functional feeding groups. Some shredders, such as Plecoptera and Trichoptera species in mountain streams, would be sensitive to organic pollution [57,58] and could serve as bioindicators in undisturbed forest areas, whereas filter feeders, like caddisflies and Tubificidae species (non-insect gatherers/collectors), may function as indicators for agricultural areas and urban streams contaminated with organic pollution, respectively [59]. Further studies are warranted to learn more about the biological properties (e.g., traits) of benthic macroinvertebrate communities and their adaptation to land use types.

4.3. Geological Features and the Management of Urban Streams

Conserving biodiversity is possible if ecosystem management planning is conducted efficiently, even when the sample sites are located in the vicinity of human settlements. In this study, sampling sites such as YOU, YGS, YJJ, YSB, CCM, and CCD are located in the urban residential areas with apartment complexes and high population densities. These sites, however, displayed high biodiversity and biological indices, and were classified in cluster 1, which represents less polluted streams (Figure 2, Table 2). This phenomenon was the result of an active urban ecosystem management program initiated by the local government, such as the Stream Health Evaluation project, Dong-Cheon Stream Restoration project, or similar plans. These projects concentrated on the restoration and conservation of stream habitats resulting in increased habitat diversity, removing artificial structures, or reviving connectivity between streams and land [60].

However, some study sites were found to be polluted despite the fact that they were located in relatively less polluted areas. The study sites in clusters 2 and 3, which have relatively low biodiversity and biological indices, are in fact surrounded by mountains. This indicates that the areas were not properly ecologically managed. Specifically, these areas were disturbed by organic pollution due to excessive commercial development (DDK and DAU), sedimentation from cemetery construction (YPU), disturbance by local residents (JAU), and organic pollution due to agricultural land use (ZGU and TSD), *etc.* In the Busan region, human settlements occur mostly near the streams since the mountains are dispersed and there is little lowland area. Considering that there is a high risk of biological organisms being exposed to various anthropogenic pollution sources in aquatic ecosystems, these results demonstrate that urban ecosystem management plans should be carefully conducted to efficiently protect biodiversity and ecological integrity [56,61].

5. Conclusions

We investigated the changes on benthic macroinvertebrate communities in streams of urban areas. The changes of land use types in riparian zones due to the urbanization influenced on the species composition as well as community structure. RADs were effective for reflecting the overall degradation of community structure due to urbanization, and indicator species analysis provided additional information for identifying the gradient of habitat preference of species. These results suitably presented the ecological state of streams in urban areas in a comprehensive manner, including particular patterns such as the positive status of stream ecosystems in urban residential areas due to ecosystem management, or negative status in mountainous areas. These approaches would be suitable in assessing community structure properties and species composition, enabling a comprehensive evaluation of the ecological integrity of urban streams.

Supplementary Materials: The following are available online at http://www.mdpi.com/2073-4441/8/5/187/s1, Table S1: A list of taxa identified in the study sites.

Acknowledgments: This study was supported by Busan Metropolitan city (project: Surveys of Natural Environment in the Eastern Zone in the Busan Metropolitan Area) and by the Ministry of Environment and the National Institute of Environmental Research, Korea (project: National Aquatic Ecosystem Health Survey and Assessment).

Author Contributions: Dong-Hwan Kim, Gyu-Suk Kwak, and Sang-Bin Lee conducted data collection, and Dong-Hwan Kim, Tae-Soo Chon, and Young-Seuk Park conducted the data analyses. All authors participated in discussions and approved the final manuscript.

Conflicts of Interest: The authors declare no conflict of interest.

Appendix

Table A1. Average values of physicochemical variables and land use types at all sampling sites. The sampling sites are in alphabetical order.

Sampling Sites	Sampling Dates		Altitude (m)	Physicochemical Variables						Land Use Types (Riparian Sub-Corridors; 1000 m)		
	Summer	Winter		Slope (%)	DO (mg/L)	Conductivity (µs/cm)	Substrate Index	Depth (cm)	Speed (cm/s)	Forest (%)	Urban (%)	Agriculture (%)
CCD	13-August-2014	8-December-2014	46	3.0	6.2	126.9	6.40	15.0 (±1.8)	28.3 (±2.1)	47.2	29.8	0.0
CCM	13-August-2014	8-December-2014	57	4.6	8.9	122.1	6.15	20.0 (±2.9)	26.7 (±5.4)	75.8	19.6	0.0
CCU	13-August-2014	8-December-2014	78	2.8	8.5	65.1	5.83	20.8 (±1.5)	18.3 (±4.9)	92.4	0.0	0.0
DAU	13-August-2014	8-December-2014	263	5.8	9.7	231.5	6.00	33.3 (±4.4)	42.5 (±8.5)	90.3	0.0	8.3
DDK	13-August-2014	8-December-2014	314	4.9	7.9	205.5	5.83	12.5 (±1.1)	30.0 (±3.7)	50.5	0.0	42.0
DRM	13-August-2014	8-December-2014	126	3.1	9.4	152.3	5.48	20.8 (±2.0)	8.5 (±2.3)	88.7	0.3	2.2
DRU	13-August-2014	8-December-2014	170	3.3	10.0	112.6	5.63	20.0 (±2.9)	7.5 (±3.1)	92.7	1.4	3.4
HAD	12-August-2013	9-December-2013	65	0.2	8.1	254.5	5.90	8.8 (±0.9)	13.0 (±1.7)	41.5	7.0	45.8
HAM	12-August-2013	9-December-2013	72	0.7	9.1	115.0	5.80	5.5 (±1.2)	2.0 (±1.0)	23.1	5.0	66.3
HJD	13-August-2014	8-December-2014	9	0.0	7.8	505.5	4.30	23.3 (±3.3)	2.7 (±0.8)	1.7	38.5	0.0
HJM	13-August-2014	8-December-2014	11	0.0	7.5	540.5	4.38	32.5 (±3.1)	3.8 (±1.1)	39.6	15.5	0.0
HJU	13-August-2014	8-December-2014	14	0.4	7.3	325.0	4.43	30.0 (±2.6)	4.0 (±1.4)	8.6	22.1	0.1
IGM	12-August-2013	9-December-2013	31	3.2	9.1	116.3	5.73	19.2 (±1.5)	65.0 (±3.7)	64.1	0.0	31.3
IGU	12-August-2013	9-December-2013	58	3.3	9.6	113.0	6.28	15.0 (±1.8)	30.0 (±3.7)	71.7	0.0	27.8
IKM	12-August-2013	9-December-2013	82	4.0	9.8	304.0	5.80	11.1 (±2.1)	25.0 (±2.6)	43.2	4.0	49.2
IKU	12-August-2013	9-December-2013	83	3.0	9.6	239.0	6.03	4.2 (±1.2)	11.2 (±2.0)	4.2	2.0	78.9
JAD	12-August-2013	9-December-2013	14	2.4	5.7	214.0	5.80	6.7 (±1.2)	31.7 (±1.2)	21.8	5.3	60.7
JAM	12-August-2013	9-December-2013	32	0.2	9.2	211.5	5.93	7.3 (±1.1)	15.8 (±2.0)	49.9	2.0	38.0
JAU	12-August-2013	9-December-2013	119	2.1	9.4	119.0	6.10	7.7 (±1.1)	11.2 (±1.7)	97.9	0.6	1.1
NLM	12-August-2013	9-December-2013	10	0.0	8.1	390.0	5.73	16.2 (±1.7)	21.7 (±3.3)	56.1	0.0	0.0
NLU	12-August-2013	9-December-2013	41	8.4	7.4	390.0	5.58	16.7 (±2.8)	14.2 (±2.7)	69.2	0.0	4.3
SRD	13-August-2014	8-December-2014	1	0.0	8.0	331.0	4.90	25.8 (±5.2)	2.2 (±0.9)	0.0	49.7	36.8
SRM	13-August-2014	8-December-2014	1	0.0	8.6	331.5	4.20	55.0 (±3.4)	0.3 (±0.2)	0.0	52.2	46.2
SRU	13-August-2014	8-December-2014	1	0.0	7.0	462.5	4.45	37.5 (±2.5)	5.0 (±2.6)	0.0	77.0	4.1
TAP	15-August-2013	11-December-2013	91	6.8	12.5	71.0	6.00	16.7 (±1.7)	28.3 (±2.8)	100.0	0.0	0.0
THP	15-August-2013	11-December-2013	24	4.1	8.5	513.5	5.35	13.2 (±1.6)	20.5 (±2.4)	43.4	6.1	17.3

Table 1. *Cont.*

Sampling Sites	Sampling Dates		Altitude (m)	Physicochemical Variables						Land Use Types (Riparian Sub-Corridors; 1000 m)		
	Summer	Winter		Slope (%)	DO (mg/L)	Conductivity (µs/cm)	Substrate Index	Depth (cm)	Speed (cm/s)	Forest (%)	Urban (%)	Agriculture (%)
TKC	15-August-2013	11-December-2013	49	4.1	8.7	417.0	5.83	22.5 (±2.8)	24.3 (±2.9)	59.6	5.0	20.3
TSD	15-August-2013	11-December-2013	135	7.7	9.5	272.0	6.05	12.0 (±2.1)	12.7 (±2.5)	92.4	0.9	6.2
YBD	15-August-2014	20-November-2014	106	15.6	10.2	67.9	5.33	7.5 (±1.1)	25.8 (±2.7)	100.0	0.0	0.0
YBK	15-August-2013	11-December-2013	222	9.5	9.0	44.0	6.05	13.5 (±1.9)	17.8 (±2.9)	98.4	0.0	0.0
YBT	13-August-2014	20-November-2014	315	27.7	8.3	74.6	5.93	20.0 (±2.2)	33.3 (±2.1)	76.7	0.0	0.9
YBV	13-August-2014	20-November-2014	220	8.1	9.6	85.4	5.78	16.7 (±4.8)	28.3 (±4.6)	75.6	2.5	13.2
YCK	15-August-2013	11-December-2013	90	4.6	8.1	275.3	5.50	8.0 (±1.6)	10.0 (±1.3)	29.6	5.7	48.5
YGS	13-August-2014	20-November-2014	37	5.6	8.5	330.5	5.25	12.5 (±2.1)	23.3 (±4.0)	7.5	33.8	0.2
YHD	13-August-2014	20-November-2014	26	6.1	8.0	239.5	5.70	17.5 (±5.0)	24.2 (±2.0)	51.4	8.0	5.0
YJ	13-August-2014	20-November-2014	20	0.0	8.4	384.0	5.65	10.0 (±1.3)	20.0 (±6.2)	8.7	52.4	0.0
YJL	15-August-2014	20-November-2014	70	11.6	9.0	219.5	5.88	16.7 (±3.6)	22.5 (±5.7)	100.0	0.0	0.0
YOU	13-August-2014	20-November-2014	105	4.4	9.5	107.6	6.00	21.7 (±4.2)	35.8 (±3.3)	63.3	4.1	15.7
YPU	15-August-2014	20-November-2014	153	3.3	9.4	319.0	4.90	8.7 (±1.5)	4.8 (±1.2)	72.1	8.3	2.9
YPY	15-August-2014	20-November-2014	7	0.0	9.3	570.0	5.23	33.3 (±3.3)	0.0 (±0.0)	19.6	42.4	0.7
YSB	15-August-2014	20-November-2014	9	0.0	9.9	412.5	5.95	23.3 (±2.5)	15.8 (±2.7)	0.0	48.1	0.0
YSC	15-August-2013	11-December-2013	44	0.8	8.1	197.5	5.50	5.3 (±1.1)	7.5 (±1.9)	10.5	23.9	45.2
YYS	15-August-2014	20-November-2014	107	13.1	8.7	125.5	5.65	11.3 (±3.2)	11.7 (±1.1)	100.0	0.0	0.0
ZGD	12-August-2013	9-December-2013	29	1.8	8.3	292.0	5.65	19.7 (±1.6)	41.7 (±4.8)	86.8	0.0	12.4
ZGM	12-August-2013	9-December-2013	67	2.6	7.3	210.7	5.78	23.8 (±1.8)	20.0 (±1.3)	4.5	3.6	57.1
ZGU	12-August-2013	9-December-2013	105	9.5	9.2	179.0	6.30	16.7 (±2.4)	38.3 (±3.9)	93.3	3.6	1.6

243

References

1. Foley, J.A.; DeFries, R.; Asner, G.P.; Barford, C.; Bonan, G.; Carpenter, S.R.; Chapin, F.S.; Coe, M.T.; Daily, G.C.; Gibbs, H.K. Global consequences of land use. *Science* **2005**, *309*, 570–574.

2. Allan, J.D. Landscapes and riverscapes: The influence of land use on stream ecosystems. *Annu. Rev. Ecol. Syst.* **2004**, *35*, 257–284.

3. Walsh, C.J.; Roy, A.H.; Feminella, J.W.; Cottingham, P.D.; Groffman, P.M.; Morgan, R.P. The urban stream syndrome: Current knowledge and the search for a cure. *J. N. Am. Benthol. Soc.* **2005**, *24*, 706–723.

4. Roy, A.H.; Freeman, M.C.; Freeman, B.J.; Wenger, S.J.; Ensign, W.E.; Meyer, J.L. Investigating hydrologic alteration as a mechanism of fish assemblage shifts in urbanizing streams. *J. N. Am. Benthol. Soc.* **2005**, *24*, 656–678.

5. Chessman, B.C.; Williams, S.A. Biodiversity and conservation of river macroinvertebrates on an expanding urban fringe: Western Sydney, New South Wales, Australia. *Pac. Conserv. Biol.* **1999**, *5*, 36–55.

6. Walsh, C.J. Biological indicators of stream health using macroinvertebrate assemblage composition: A comparison of sensitivity to an urban gradient. *Mar. Freshw. Res.* **2006**, *57*, 37–47.

7. Wang, L.; Lyons, J. *Fish and Benthic Macroinvertebrate Assemblages as Indicators of Stream Degradation in Urbanizing Watersheds*; CRC Press: Boca Raton, FL, USA, 2003; pp. 227–249.

8. Meyer, J.L.; Paul, M.J.; Taulbee, W.K. Stream ecosystem function in urbanizing landscapes. *J. N. Am. Benthol. Soc.* **2005**, *24*, 602–612.

9. Miller, W.; Boulton, A.J. Managing and rehabilitating ecosystem processes in regional urban streams in Australia. *Hydrobiologia* **2005**, *552*, 121–133.

10. Benke, A.C.; Willeke, G.; Parrish, F.; Stites, D. *Effects of Urbanization on Stream Ecosystems*; Georgia Institute of Technology: Atlanta, GA, USA, 1981.

11. Park, Y.-S.; Grenouillet, G.; Esperance, B.; Lek, S. Stream fish assemblages and basin land cover in a river network. *Sci. Total Environ.* **2006**, *365*, 140–153.

12. Petersen, I.; Masters, Z.; Hildrew, A.; Ormerod, S. Dispersal of adult aquatic insects in catchments of differing land use. *J. Appl. Ecol.* **2004**, *41*, 934–950.

13. Wenger, S.J.; Peterson, J.T.; Freeman, M.C.; Freeman, B.J.; Homans, D.D. Stream fish occurrence in response to impervious cover, historic land use, and hydrogeomorphic factors. *Can. J. Fish. Aquat. Sci.* **2008**, *65*, 1250–1264.

14. Rosenberg, D.M.; Resh, V.H. *Freshwater Biomonitoring and benthic Macroinvertebrates*; Chapman & Hall: London, UK, 1993.

15. Allan, J.D.; Castillo, M.M. *Stream Ecology: Structure and Function of Running Waters*; Springer Science & Business Media: Dordrecht, The Netherlands, 2007.

16. Van Diggelen, R.; Sijtsma, F.J.; Strijker, D.; van den Burg, J. Relating land-use intensity and biodiversity at the regional scale. *Basic Appl. Ecol.* **2005**, *6*, 145–159.

17. Wright, J.; Furse, M.; Armitage, P.; Moss, D. New procedures for identifying running-water sites subject to environmental stress and for evaluating sites for conservation, based on the macroinvertebrate fauna. *Archiv für Hydrobiol.* **1993**, *127*, 319–326.

18. Cao, Y.; Bark, A.W.; Williams, W.P. Measuring the responses of macroinvertebrate communities to water pollution: A comparison of multivariate approaches, biotic and diversity indices. *Hydrobiologia* **1996**, *341*, 1–19.

19. Park, Y.-S.; Céréghino, R.; Compin, A.; Lek, S. Applications of artificial neural networks for patterning and predicting aquatic insect species richness in running waters. *Ecol. Model.* **2003**, *160*, 265–280.

20. Chon, T.-S. Self-organizing maps applied to ecological sciences. *Ecol. Inform.* **2011**, *6*, 50–61.

21. Kim, D.-H.; Cho, W.-S.; Chon, T.-S. Self-organizing map and species abundance distribution of stream benthic macroinvertebrates in revealing community patterns in different seasons. *Ecol. Inform.* **2013**, *17*, 14–29.

22. Reynoldson, T.; Norris, R.; Resh, V.; Day, K.; Rosenberg, D. The reference condition: A comparison of multimetric and multivariate approaches to assess water-quality impairment using benthic macroinvertebrates. *J. N. Am. Benthol. Soc.* **1997**, *16*, 833–852.

23. Richards, C.; Minshall, G.W. Spatial and temporal trends in stream macroinvertebrate communities: The influence of catchment disturbance. *Hydrobiologia* **1992**, *241*, 173–194.

24. Marshall, J.C.; Sheldon, F.; Thoms, M.; Choy, S. The macroinvertebrate fauna of an Australian dryland river: Spatial and temporal patterns and environmental relationships. *Mar. Freshw. Res.* **2006**, *57*, 61–74.

25. Beche, L.A.; Mcelravy, E.P.; Resh, V.H. Long-term seasonal variation in the biological traits of benthic-macroinvertebrates in two Mediterranean-climate streams in California, USA. *Freshw. Biol.* **2006**, *51*, 56–75.

26. Sponseller, R.; Benfield, E.; Valett, H. Relationships between land use, spatial scale and stream macroinvertebrate communities. *Freshw. Biol.* **2001**, *46*, 1409–1424.

27. McGill, B.J.; Etienne, R.S.; Gray, J.S.; Alonso, D.; Anderson, M.J.; Benecha, H.K.; Dornelas, M.; Enquist, B.J.; Green, J.L.; He, F. Species abundance distributions: Moving beyond single prediction theories to integration within an ecological framework. *Ecol. Lett.* **2007**, *10*, 995–1015.

28. Qu, X.-D.; Song, M.-Y.; Park, Y.-S.; Oh, Y.; Chon, T.-S. Species abundance patterns of benthic macroinvertebrate communities in polluted streams. *Ann. Limnol. Int. J. Limnol.* **2008**, *44*, 119–133.

29. Jowett, I.G.; Richardson, J.; Biggs, B.J.; Hickey, C.W.; Quinn, J.M. Microhabitat preferences of benthic invertebrates and the development of generalised *Deleatidium* spp. Habitat suitability curves, applied to four New Zealand rivers. *N. Z. J. Mar. Freshw. Res.* **1991**, *25*, 187–199.

30. Suren, A.M. Bryophyte distribution patterns in relation to macro-, meso-, and micro-scale variables in south island, New Zealand streams. *N. Z. J. Mar. Freshw. Res.* **1996**, *30*, 501–523.

31. Stauffer, J.; Goldstein, R.; Newman, R. Relationship of wooded riparian zones and runoff potential to fish community composition in agricultural streams. *Can. J. Fish Aquat. Sci.* **2000**, *57*, 307–316.

32. Han, M.; Na, Y.; Bang, H.; Kim, M.; Kang, K.; Hong, H.; Lee, J.; Ko, B. *Aquatic Invertebrate in Paddy Ecosystem of Korea*; Kwang Monn Dang Press: Suwon, Korea, 2010.

33. Won, D.; Kwon, S.; Jun, Y. *Aquatic Insects of Korea*; Korea Ecosystem Service: Seoul, Korea, 2005.

34. Yoon, I. *Aquatic Insects of Korea*; Jeonghaengsa: Seoul, Korea, 1995.

35. Merritt, R.W.; Cummins, K.W. *An Introduction to the Aquatic Insects of North America*; Kendall Hunt: Dubuque, ID, USA, 1996.

36. Brigham, A.R.; Brigham, W.U.; Gnilka, A. *Aquatic Insects and Oligochaetes of North and South Carolina*; Midwest Aquatic Enterprises: Mahomet, IL, USA, 1982.

37. Shannon, C.E.; Weaver, W. *The Mathematical Theory of Information*; University of Illinois Press: Urbana, IL, USA, 1949.

38. McNaughton, S. Relationships among functional properties of Californian grassland. *Nature* **1967**, *216*, 168–169.

39. Biondini, M.E.; Mielke, P.W., Jr.; Berry, K.J. Data-dependent permutation techniques for the analysis of ecological data. *Vegetatio* **1988**, *75*, 161–168.

40. McCune, B.; Grace, J.B.; Urban, D.L. *Analysis of Ecological Communities*; MjM Software Design: Gleneden Beach, OR, USA, 2002.

41. Hawkins, C.P.; Murphy, M.L.; Anderson, N. Effects of canopy, substrate composition, and gradient on the structure of macroinvertebrate communities in cascade range streams of Oregon. *Ecology* **1982**, 1840–1856.

42. Marsh-Matthews, E.; Matthews, W.J. Spatial variation in relative abundance of a widespread, numerically dominant fish species and its effect on fish assemblage structure. *Oecologia* **2000**, *125*, 283–292.

43. Dufrêne, M.; Legendre, P. Species assemblages and indicator species: The need for a flexible asymmetrical approach. *Ecol. Monogr.* **1997**, *67*, 345–366.

44. Legendre, P. Indicator species: Computation. *Encycl. Biodivers.* **2013**, *4*, 264–268.

45. Li, F.; Chung, N.; BAE, M.J.; KWON, Y.S.; PARK, Y.S. Relationships between stream macroinvertebrates and environmental variables at multiple spatial scales. *Freshw. Biol.* **2012**, *57*, 2107–2124.

46. Ter Braak, C.J.; Juggins, S. Weighted averaging partial least squares regression (wa-pls): An improved method for reconstructing environmental variables from species assemblages. *Hydrobiologia* **1993**, *269*, 485–502.

47. Juggins, S. *C2 Version 1.5: Software for Ecological and Palaeoecological Data Analysis and Visualisation*; University of Newcastle: Newcastle, UK, 2007.

48. Birks, H. Quantitative palaeoenvironmental reconstructions. *Stat. Model. Quat. Sci. Data Tech. Guide* **1995**, *5*, 161–254.

49. Magurran, A.E. Measuring biological diversity. *Afr. J. Aquat. Sci.* **2004**, *29*, 285–286.

50. Tang, H.; Song, M.-Y.; Cho, W.-S.; Park, Y.-S.; Chon, T.-S. Species abundance distribution of benthic Chironomids and other macroinvertebrates across different levels of pollution in streams. *Ann. Limnol. Int. J. Limnol.* **2010**, *46*, 53–66.

51. Hall, M.J.; Closs, G.P.; Riley, R.H. Relationships between land use and stream invertebrate community structure in a south island, New Zealand, coastal stream catchment. *N. Z. J. Mar. Freshw. Res.* **2001**, *35*, 591–603.

52. Miserendino, M.L.; Casaux, R.; Archangelsky, M.; Di Prinzio, C.Y.; Brand, C.; Kutschker, A.M. Assessing land-use effects on water quality, in-stream habitat, riparian ecosystems and biodiversity in Patagonian northwest streams. *Sci. Total Environ.* **2011**, *409*, 612–624.

53. Collier, K.J.; Quinn, J.M. Land-use influences macroinvertebrate community response following a pulse disturbance. *Freshw. Biol.* **2003**, *48*, 1462–1481.

54. Stewart, P.; Butcher, J.; Swinford, T. Land use, habitat, and water quality effects on macroinvertebrate communities in three watersheds of a lake Michigan associated marsh system. *Aquat. Ecosyst. Health Manag.* **2000**, *3*, 179–189.

55. Lenat, D.R.; Crawford, J.K. Effects of land use on water quality and aquatic biota of three North Carolina piedmont streams. *Hydrobiologia* **1994**, *294*, 185–199.

56. Moore, A.A.; Palmer, M.A. Invertebrate biodiversity in agricultural and urban headwater streams: Implications for conservation and management. *Ecol. Appl.* **2005**, *15*, 1169–1177.

57. Fochetti, R.; De Figueroa, J.M.T. Global diversity of stoneflies (plecoptera; insecta) in freshwater. *Hydrobiologia* **2008**, *595*, 365–377.

58. Li, F.; Cai, Q.; Liu, J. Temperature-dependent growth and life cycle of *Nemoura sichuanensis* (plecoptera: Nemouridae) in a Chinese mountain stream. *Int. Rev. Hydrobiol.* **2009**, *94*, 595–608.

59. Martins, R.; Stephan, N.; Alves, R. Tubificidae (annelida: Oligochaeta) as an indicator of water quality in an urban stream in southeast Brazil. *Acta Limnol. Brasil.* **2008**, *20*, 221–226.

60. Paul, M.J.; Meyer, J.L. Streams in the urban landscape. *Annu. Rev. Ecol. Syst.* **2001**, *32*, 333–365.

61. Alvey, A.A. Promoting and preserving biodiversity in the urban forest. *Urban For. Urban Gree.* **2006**, *5*, 195–201.

Examining the Relationships between Watershed Urban Land Use and Stream Water Quality Using Linear and Generalized Additive Models

Sun-Ah Hwang, Soon-Jin Hwang, Se-Rin Park and Sang-Woo Lee

Abstract: Although close relationships between the water quality of streams and the types of land use within their watersheds have been well-documented in previous studies, many aspects of these relationships remain unclear. We examined the relationships between urban land use and water quality using data collected from 527 sample points in five major rivers in Korea—the Han, Geum, Nakdong, Younsan, and Seomjin Rivers. Water quality data were derived from samples collected and analyzed under the guidelines of the Korean National Aquatic Ecological Monitoring Program, and land use was quantified using products provided by the Korean Ministry of the Environment, which were used to create a Geographic Information System. Linear models (LMs) and generalized additive models were developed to describe the relationships between urban land use and stream water quality, including biological oxygen demand (BOD), total nitrogen (TN), and total phosphorous (TP). A comparison between LMs and non-linear models (in terms of R^2 and Akaike's information criterion values) indicated that the general additive models had a better fit and suggested a non-linear relationship between urban land use and water quality. Non-linear models for BOD, TN, and TP showed that each parameter had a similar relationship with urban land use, which had two breakpoints. The non-linear models suggested that the relationships between urban land use and water quality could be categorized into three regions, based on the proportion of urban land use. In moderate urban land use conditions, negative impacts of urban land use on water quality were observed, which confirmed the findings of previous studies. However, the relationships were different in very low urbanization or very high urbanization conditions. Our results could be used to develop strategies for more efficient stream restoration and management, which would enhance water quality based on the degree of urbanization in watersheds. In particular, land use management for enhancing stream water quality might be more effective when urban land use is in the range of 1.1%–31.5% of a watershed. If urban land use exceeds 31.5% in a watershed, a more comprehensive approach would be required because water quality would not respond as rapidly as expected.

Reprinted from *Water*. Cite as: Hwang, S.-A.; Hwang, S.-J.; Park, S.-R.; Lee, S.-W. Examining the Relationships between Watershed Urban Land Use and Stream Water Quality Using Linear and Generalized Additive Models. *Water* **2016**, *8*, 155.

1. Introduction

Land use can have direct impacts on hydrologic systems within a watershed [1–4]. The negative impacts of urban land use in watersheds on adjacent reservoirs, streams, and rivers have been well-documented and are a key concern for restoration and management. In general, previous studies have reported that watersheds with high percentages of developed areas (e.g., urban areas and agricultural areas) tend to have higher concentrations of water pollutants and nutrients [1–10]. Different types of urban land use, including commercial, residential, and industrial development, have significant impacts on water quality. The proportion of land use type in a watershed has been shown to be closely associated with many water quality parameters in various aquatic systems [2,4]. The usefulness of water quality indices as indicators of water pollution has been verified for assessing spatial changes and for classifying water quality. In many countries, chemical parameters, such as dissolved oxygen (DO), pH, biochemical oxygen demand (BOD), chemical oxygen demand (COD), total nitrogen (TN), and total phosphorous (TP), have served as the main criteria for determining the condition of rivers and for managing aquatic ecosystem resources. Environmental Policy Law (EPL) in Korea has also used various chemical criteria (e.g., pH, BOD, COD, TOC, SS, DO, TP, *etc.*) to manage the water quality of rivers and streams.

In previous studies, the most commonly used techniques to determine the relationships between land uses in watersheds and water quality indicators were correlation or regression analyses. These approaches assume linear relationships between land uses in watersheds and water quality indicators, suggesting that the degree of water quality variance is the same regardless of the degree of land use intensity in watersheds. Recently, some studies have reported that the relationships between urban land uses and the chemical, biological, and physical characteristics of streams might not be linear [6,9,11–15]. It has been reported that the average threshold of imperviousness at which water quality degradation first occurs is 10% [11,16]. Similarly, Coles *et al.* [12] reported that significant changes in aquatic health were observed between low and moderate levels (0 to 35) of urbanization intensity (0–100 scale, low to high urbanization intensity) in New England coastal streams. In addition, they found a "threshold effect" in which the water quality indicators no longer changed as the intensity of urbanization increased. More recently, Crim [13] confirmed the presence of a threshold and suggested that it might be much lower than 10% for an impervious surface. In his study, the concentrations of water quality indicators increased considerably as the impervious surface in a watershed increased from 0 to 4% in west-central Georgia, USA. Together with previous studies, his results imply that even a small increase in the impervious surface in a watershed might have significant impacts on the chemical characteristics of water and the biota of streams. This nonlinearity may be derived from the random nature of the hydrodynamic conditions of river systems, meteorological processes, and a shortage of available monitoring data [15,17,18]. It is also noteworthy that there was only one breakpoint (*i.e.*, threshold) in the relationships between land uses and water quality indicators, regardless of the different thresholds reported in previous studies. The presence of a threshold and non-linear relationships between land use and water quality increases the uncertainty and the degree of complexity in water quality management and land use planning for decision makers and policy makers when attempting to enhance water quality and minimize the adverse impacts of various land uses [19].

Despite the possible presence of a non-linear relationship between water quality and land uses, linear correlation and regression analyses have been broadly used to investigate such relationships in various fields of study, including water chemistry, ecology, and hydrology. It is very clear that linear correlation and regression are useful techniques when quantifying the magnitude, direction, and significance of the relationships between land use variables (*i.e.*, impervious areas, developed areas, agricultural areas, *etc.*) and water quality in a number of previous studies [1–4,20]. Despite the various benefits, conventional linear-type approaches may not accurately represent the true nature of the relationships between land uses and water quality [6,9,21]. In addition, this may lead to a misunderstanding by stream managers, land use planners, and decision makers about the impact of different land uses on water quality, particularly during the stages of development with small and large extents of urbanization.

To avoid nonlinearity issues when dealing with the relationships between land uses and water quality, several approaches have been proposed, including stochastic, fuzzy, and interval mathematical programs [19,22–26]. One popular method is to transform the data when making non-linear models linear in the analyses of the relationships between urban land use and river water quality [22], and when determining interval parameters in non-linear optimization models of stream water quality management [15]. However, it is very clear that the presence and shape of the non-linearity in the relationships should be examined prior to applying these methods.

Our first goal was to test for the presence of non-linear relationships between urban land use and water quality indicators in streams in Korea. To test for the presence of non-linearity, this study compared the ability of linear and non-linear models to explain the variance in water quality indicators when responding to the degree of urbanization in watersheds. We hypothesized that a non-linear model would explain the variance in water quality parameters in response to the degree of urbanization better than would a linear model (LM), if the relationships between urban land use and water quality indicators were non-linear. Otherwise, a LM would outperform a non-linear model.

Second, this study also investigated the number of breakpoints in the relationships when non-linear relationships were present. Previous studies have reported that there is either one breakpoint (*i.e.*, threshold) or no breakpoint (*i.e.*, linear regression or correlation) in the relationship. However, if the relationships between land uses in watersheds and water quality indicators are sufficiently complex, more than one breakpoint can exist in the relationships. Thus, the number of breakpoints can represent the complexity of the relationships. If there is only one breakpoint (*i.e.*, threshold), we need to consider only two intervals, including areas with a small and large extent of urbanization in stream management processes. We believe that the presence of non-linearity and the number of breakpoints can provide useful insights into land use planning and stream management. Land managers, planners, stream managers, and policy makers can apply different strategies for different levels of urbanization to enhance the water quality and to minimize the adverse impacts of urban land uses on streams.

We adopted generalized additive models (GAMs) for investigation in this study. GAMs have been shown to be very flexible, providing an excellent fit for non-linear relationships and for datasets with significant noise among the predictor variables [22]. This model is a generalization of multiple regressions, in which the additive nature of the model is maintained, but the simple lines of the linear regression are replaced by nonparametric functional curves with multiple parameters. Compared with an LM, GAMs are data-driven rather than model-driven, and GAMs allow determination of the shape of the response curves from the data instead of fitting an *a priori* parametric model, which is limited in its available shape of response [27]. GAMs have been widely used in various fields, such as species distribution [28–33], plant ecology [34–36], and water quality dynamics [21,37,38]. For example, Murase *et al.* [29] applied GAMs to fishery-survey data to reveal the influences of environmental factors, including surface water temperature, salinity, chlorophyll, near-seabed water temperature, salinity, and depth, on the distribution patterns of Japanese anchovy, sand lance, and krill. The results of their study showed a non-linear response of fishes to environmental factors. Richard *et al.* [21] applied GAMs to explore the functional relationships between four water quality indicators (TN, TP, ammonia, nitrate) and environmental factors, such as catchment inflow, wind speed, and tidal current in the Broadwater Estuary in the Gold Coast region of Australia, using short-term monitoring data. Based on a GAM assessment, they reported that nutrient concentrations within a subtropical estuary were non-linear for various environmental factors and were most dependent on catchment inflow.

2. Materials and Methods

2.1. Study Streams and Sampling Sites

South Korea is located between 127°30′ E and 37°00′ N and occupies an area of about 100,032 km², covering almost the entire southern half of the Korean Peninsula (Figure 1). Approximately two-thirds of the annual precipitation (1388.7 mm) is concentrated in the summer (June through September). Thus, seasonal precipitation and water flow levels fluctuate widely, and stream flow generally diminishes during drought periods, which are characteristic of winter and early spring. The annual average temperature for 2006–2010 was 12.8 °C, with monthly averages ranging from a low of −12.8 °C in January to a high of 29.32 °C in August.

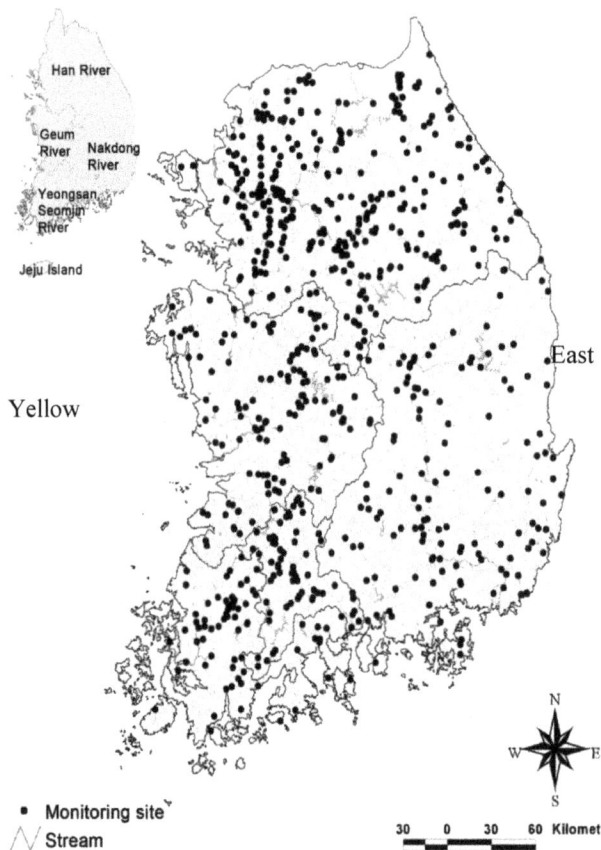

Figure 1. The five major river systems and the locations of National Aquatic Ecological Monitoring Program (NAEMP) monitoring sites. Most streams on the east side can be characterized by a short length, low water temperature, and fast flow rate.

Five major rivers (*i.e.*, the Han, Geum, Nakdong, Youngsan, and Seomjin Rivers) and their independent tributaries and small streams are distributed throughout the country. The Youngsan and Seomjin Rivers are usually treated as one river system (Youngsan–Seomjin River) because their watersheds are located close to one another. Among the five major rivers, the Han River has the largest basin, occupying approximately a quarter of the country. The east side of the country is mountainous, with watersheds that are less disturbed and are covered by dense pine, oak, and mixed forest. In the eastern mountainous areas, most streams are small, flow down steep slopes, and run directly into the East Sea. Most river systems and streams in the western and southwestern areas flow toward the Yellow Sea. Seasonal fluctuations in water levels in the small streams in the eastern areas are particularly extreme because of the steep slopes and low groundwater levels. The headstreams of the five major rivers are located in similar areas in the central part of the eastern mountains.

2.2. Water Quality Variables

As part of the National Aquatic Ecological Monitoring Program (NAEMP), South Korea's Ministry of the Environment (MOE) has monitored numerous aspects of streams and rivers using biochemical, physical, and biological indicators at 720 long-term monitoring sites in tributaries and the main stem of five rivers across the country. The assessment criteria and sampling protocol used by the NAEMP were developed in a preliminary study from 2003 to 2006, and a geographic information system (GIS) database for the locations of all sampling sites was also constructed in the study. The first nationwide monitoring under this protocol started in 2007 at 720 preselected sampling sites. According to the NAEMP protocol, all field survey teams consisted of staff from five universities who had to complete the field survey and water sampling within a month, twice a year. Water samples used for the determination of BOD, TN, TP, and other water quality variables were collected in prewashed 2ℓ bottles. All water samples collected from the five river systems by staff from the five universities were transported in a cooler and analyzed in a commercial laboratory (Chungmyung Environmental Co. Ltd., Seoul, Korea). Laboratory measurements were conducted to determine BOD, TN, and TP following Standard methods [39]. BOD was determined by the difference of dissolved oxygen concentration after a five-day incubation. TP was measured in the unfiltered water by ascorbic acid method after persulfate oxidation. TN was determined by UV spectrophotometric method after potassium sulfate digestion.

Five major river systems that included 527 of the 720 NAEMP sampling sites in 2007 were investigated in this study. We excluded data sampled from sites in islands and estuary areas. We found that some sites only had data during the spring or no data during the fall due to the streams drying up. We excluded these data from our dataset for analysis. We also excluded data sampled from small independent streams running directly into the sea rather than into one of the five major river systems. These streams were mostly located in the eastern mountainous areas and were characterized by a very short length, high flow rate, and low water temperature.

In this study, we focused on the common water quality indicators, including BOD, TN, and TP, monitored in 2007 under the NAEMP. The reason for the use of a sampling dataset collected in 2007 was to match it with the year of the Land Use/Land Cover (LULC) GIS map released by the Ministry of Environment (MOE), Korea. The MOE releases the LULC digital map irregularly, and they released the 2007 LULC digital map in 2009.

BOD is a measure of the amount of dissolved oxygen required by aerobic biological organisms in a body of water to break down organic material, where higher values indicate poorer water conditions and higher pollution levels. TN is a measure of the mixture of organic, ammoniac, nitrite, and nitric nitrogen, which contribute to the eutrophication of a water body. Nitrogen oxidizes into NO_3 when discharged into streams or lakes and consumes dissolved oxygen, acting like organic matter. The rapid and high rate of consummation of dissolved oxygen degrades aquatic habitats. Representing the total quantity of phosphorus compounds, TP is also used as an index of eutrophication in streams and lakes. Phosphorus, together with nitrogen, is known as a nutritive salt that can cause eutrophication and red tides. Phosphorus acts as a limiting factor in algae growth in water systems. Consequently, the internal concentration of TP in a water body is a crucial element in controlling algae growth [40]. TP is discharged as organic phosphorus and $PO_4–P$, with organic phosphorus being a component of agricultural fertilizer that can be toxic in water bodies.

2.3. Measuring the Proportion of Urban Land Use

To calculate the proportions of each type of land use at each sampling site, we used the digital land use and land cover (LULC) map from the Korean MOE. The LULC map is a representation of the land surface based on satellite imagery and data from photographic analyses, and it has been widely used for environmental management. According to the MOE classification, land use types were divided into seven major categories and 23 subcategories. The seven main LULC categories were the following: (1) urban areas; (2) agricultural areas; (3) paddy areas; (4) forested areas; (5) grassland; (6) wetland; and (7) bare soils. In this study, urban land use includes residential areas, commercial areas, roads, and industrial areas.

Water contamination in a stream is highly dependent on storm water runoff in the surrounding drainage areas. It is likely that land uses in drainage areas that are in close proximity to the streams are more likely to have stronger influences on the chemical and biological conditions of streams than are those farther away. Thus, we focused on the land uses in sub-drainage areas adjacent to the sampling site.

Another reason for using these sub-drainage areas was the policy issue of riparian land management. The MOE has prioritized the management of riparian areas as an urgent policy area and managing the entire watershed of a stream is a long-term policy in their stream and watershed management strategies. At the same time, some streams have more than one sampling site. Thus, using a watershed could be problematic, because water quality indicators could vary while land uses are the same within the watershed.

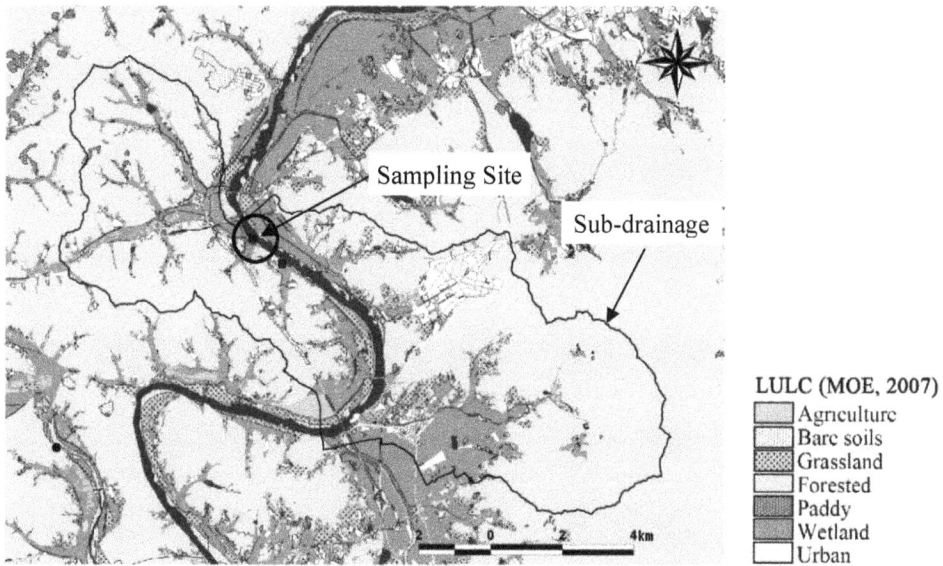

Figure 2. An example of a sub-drainage area used to measure the proportion of urban land use in an area adjacent to a monitoring site. Sub-drainage areas were delineated using a geographical information system (GIS) with a digital elevation map.

To capture the proportion of each type of land use, we used sub-drainage areas from the sampling site rather than the entire watershed area of the stream because small drainage areas can reveal the effects of land use on adjacent streams more clearly. We delineated the drainage areas from the locations of sampling sites using a GIS and digital elevation model (50 m resolution), and these small drainage boundaries were overlaid on the LULC map (Figure 2). The NAEMP monitoring protocol allows field surveyors of each monitoring area to select the best sampling point within a 50 m radius of the sampling site. Figure 2 shows representative locations from all sampling areas (*i.e.*, water quality, biological indicators, and riparian habitats), rather than the exact location of each sampling site. Despite the protocol recommendation that all areas use the same sampling site, there is the possibility that each sampling area can have a different sampling location within a 100m radius from a sampling site. In delineating the sub-drainage area for each sampling site, we tried to draw the boundary slightly upward from the sampling site. Thus, the location of the sampling site within the sub-drainage boundary used in this study is slightly upstream (approximately <200-m) from the outlet point of each sub-drainage boundary.

2.4. Data Distributions and Transformation

Water quality data are often bounded at zero and highly skewed, containing infrequent points at high values. This skewedness in the data is not surprising given that many water quality indices are strongly related to stream flow, which is typically modeled as a lognormal or other highly skewed distribution. Thus, data for groundwater quality are typically log-transformed prior to statistical analysis [41]. A preliminary analysis indicated that the proportion of urban land use and all water quality indicators in the study areas were considerably skewed, and thus all data used in the study were log-transformed before the analysis.

2.5. Linear Models (LMs) and Generalized Additive Models (GAMs)

In this study, the LM and GAM were analyzed using the statistical software R (R Core Development Team). To compare models, we used the coefficient of determination (R^2) and Akaike's information criterion (AICc). The AICc was derived from information theory, which differs from statistical hypothesis testing. The AICc method can be used to determine the relative likelihood that two (or more) models can explain the data. This method can show whether one model fits the data significantly better than another, allowing the user to reject unlikely models. The larger the AICc value for a model, the less probable it is.

3. Results

3.1. Descriptive Statistics

The mean area of small zones was 130.08, and the mean proportion of urban land use was 8.74%, with a maximum of 75.42% within the sub-drainage areas. Urbanization values were relatively normally distributed around the mean value, but the proportion of urban land use was high in some particular sites. The mean values for BOD, TN, and TP were 2.09 mg/L, 2.52 mg/L, and 0.13 mg/L, respectively, in the study areas (Table 1). According to the criteria of the Korean MOE, these values were categorized as level II (moderately good), VI (extremely poor), and III (normal) for BOD, TN, and TP, respectively. These results show that water quality was good based on BOD, but poor based on TN and TP levels. The descriptive statistics and box plots for the water quality indicators suggest that the physiochemical properties of the five rivers varied greatly from site to site.

Table 1. Descriptive statistics for water quality indices and the proportions of urban land use. A large variation in the variables was observed in the study areas.

Variables		Min.	Max.	Mean	Std. D.
Watershed	Small Zone (ha)	7.46	573.79	130.08	71.41
Land Use	Urban Land Use (%)	0.38	75.42	8.74	0.097
Water quality	BOD $(mg \cdot L^{-1})$	0.0	18.7	2.09	1.64
	TN $(mg \cdot L^{-1})$	0.3	29.0	2.52	2.31
	TP $(mg \cdot L^{-1})$	0.1	1.75	0.13	0.19

Note: $n = 527$.

3.2. Relationships between Urban Land Use and Water Quality Indices

A correlation analysis was conducted to examine the relationships among water quality variables and the proportion of urban land use (Table 2). The results indicated that the proportion of urban land use values were significantly correlated with all water quality variables, including BOD ($r = 0.419$), TN ($r = 0.445$), and TP ($r = 0.438$). The results also suggested a strong correlation among water quality indicators. For example, BOD was correlated with TN ($r = 0.592$) and TP ($r = 0.459$). TN was also strongly correlated with TP ($r = 0.613$). These results indicated that higher proportions of urban land use were associated with higher concentrations of BOD, TN, and TP in streams. Overall, the correlation analysis revealed close relationships between urban land use types with extensive human activities and poor water quality.

Table 2. Pearson-correlations between the proportion of urban land use and water quality indicators. All variables used in the study displayed strong correlations with each other.

Variables	TN	TP	Urban
BOD	0.59 **	0.46 **	0.42 **
TN	1	0.48 **	0.45 **
TP		1	0.44 **
Urban land use			1

Notes: $n = 527$. * $p < 0.05$, ** $p < 0.01$.

3.3. The Linear and Non-Linear Models

3.2.1. Linear Models (LMs)

The proportion of urban areas within a sub-drainage area was regressed against BOD, TN, and TP (Table 3). In the LMs, the variables for urban land use significantly explained the variance in all water quality indices(Table 3). Based on the linear models, the relationship with the percentage of urban land use explained 18% of the variation in BOD and 20% of the variation in both TN and TP ($p < 0.01$). In these models, urban land use negatively affected all water quality indices, including BOD (b = 0.33, β = 0.42), TN (b = 0.29, β = 0.45), and TP (b = 0.45, β = 0.44).To compare the goodness-of-fit between LMs and GAMs, the AICc values of linear models were calculated for BOD (288.32), TN (69.88), and TP (561.72).

Table 3. Outputs from linear models of the relationships between the proportion of urban land use and water quality indicators. All water quality variables were strongly influenced by the proportion of urban land use.

Variable and Criteria	BOD		TN		TP	
	b	β	b	β	b	β
Urban land use	0.33	0.42	0.29	0.45	0.45	0.44
F	111.59 **		129.47 **		124.82 **	
R^2	0.18		0.2		0.2	
AICc	288.32		69.88		561.72	

Notes: $n = 527$. * $p < 0.05$, ** $p < 0.01$.

3.2.2. Generalized Additive Models (GAMs)

Statistical link function selection for assessing non-linear models was conducted according to the lower AICc and deviance values (goodness of fit) in Table 4. For BOD, there was no difference in the deviances of the identity function (50.31) and log function (50.31), but the identity function had a lower AICc value (273.69). For TN, the values of the AICc of the identity function (43.85) and inverse function (43.95) were almost identical, but the identity function had a lower deviance value (32.64). Similarly, the deviance values of the identity function and log function were the same (82.81), but the identity function had a lower AICc value (537.82). The comparison of AICc and deviance values among link functions indicated that the use of identity functions was appropriate to fit the models.

Table 4. AICc and deviance values for BOD, TN, and TP used to select the statistical function for assessing GAMs. The identity function had the lowest AICc and deviance values.

Link Function	BOD		TN		TP	
	AICc	Deviance	AICc	Deviance	AICc	Deviance
Identity	273.69	50.31	43.85	32.54	537.82	82.81
Inverse	274.10	51.2	43.93	35.17	545.12	87.54
Log	274.63	50.31	47.52	34.69	538.18	82.81

From Table 5, it can be seen that urban land use had a negative impact on water quality. The mean effects of urban land use on BOD, TN, and TP in the GAMs were 6.02, 5.94, and 6.76, respectively. All models including BOD ($F = 19.02$, $p < 0.01$), TP ($F = 23.87$, $p < 0.01$), and TP ($F = 20.73$, $p < 0.01$) were statistically significant, and the model explained 21.3%, 25.1%, and 24.5% of the variance of the BOD, TN, and TP in streams, respectively. The AICc values of the GAM models for BOD, TN, and TP were 273.86, 44.02, and 537.99, respectively.

Table 5. The results of the generalized additive models (GAMs) for the relationships between the proportion of urban land use and water quality indicators. The non-linear model for TN had the highest R^2 value.

Variable and Criteria	BOD	TN	TP
F	19.02 **	23.87 **	20.73 **
R^2	0.21	0.251	0.245
AICc	273.86	44.02	537.99

Notes: $n = 527$. * $p < 0.05$, ** $p < 0.01$.

3.2.3. Comparison between LMs and GAMs

In the two types of regression models (linear and non-linear models), the urban land use had a negative impact on water quality in terms of the BOD, TN, and TP in streams. The LM and GAM explained 18% and 21.3% of the variance in the BOD level in streams, respectively. Compared with the R^2 of the LM (20%), the higher R^2 of the GAM (25%) indicated that the non-linear model had a higher explanatory power than that of the LM for the variance in TN. Similarly, the higher R^2 of the GAM ($R^2 = 0.24$) better explained the variance in the TP concentration in streams than did the LM ($R^2 = 0.2$). Thus, it was clear that, compared with the LM, the non-linear model (*i.e.*, GAM) better explained the variances in the BOD, TN, and TP in streams according to the proportion of urban land use in sub-drainage areas. The greatest improvement in explanatory power was observed between the LM and non-linear model (*i.e.*, GAM) for the variance in the TN concentration in streams.

All AICc values of the assessed GAMs (Table 4) appeared to be lower than those of the LMs (Table 2). Specifically, the non-linear model of BOD (AICc = 273.86) had lower AICc values than those of the LM (288.32). A considerable decrease in AICc values was also observed between linear and non-linear models for TN and TP. Specifically, the AICc value of the non-linear model (44.02) for TN was lower than that of the LM (69.88). Similarly, the AICc value of the non-linear model for TP (537.99) was significantly lower than that of the LM (561.72).

In all cases, the non-linear models (*i.e.*, GAMs) had higher coefficients of determination (R^2) and lower AICc values than those of the LMs. These results were indicative of the presence of non-linear relationships between the proportion of urban land use and water quality indicators in the study areas.

4. Discussion

The GAMs better explained the relationships between the proportion of urban land use and water quality and suggested that these relationships were non-linear. Interestingly, the shapes of all the relationships between urban land use and water quality variables in the assessed GAMs were similar. In addition, the shapes of the relationships in GAMs suggested that there was more than one breakpoint that divided the relationships of urban land use and water quality variables into several regions. However, we divided the scatter plots of GAMs for BOD, TN, and TP into three regions using two breakpoints (0 and 1.5 of the log transformed percentage urban land use) to characterize the intervals, despite the possibility of another breakpoint around 0.7 of the log-transformed percentage of urban land use (Figure 3).

In Region 1 (0% ⩽ urban land use ⩽1%), each water quality variable almost invariably responded non-linearly to a gradient of the proportion of urban land use, or even indicated a positive impact of urban land use on water quality variables. However, only a few cases fell into Region 1, and the range (±95% confidence intervals) of the cases was relatively large. The relationships in Region 1 were quite different from our expectation, and the findings of many previous studies that have reported a negative influence of urban land use on water quality indicators (e.g., [1–4,15,20]). Such a relationship was not observed in Region 1. Areas where the proportion of urban land use in the watershed was less than 1% might be undeveloped natural areas, which are very rare in Korea. The sampling sites falling into the Region 1 category were in headstreams located in mountainous areas. The influence of urban land use on stream water quality in Region 1 cases was likely to be modest at best, and stream water quality would therefore be more affected by other environmental and anthropogenic variables, such as agricultural land use [42], geology [43], soil type [44], plant litter [45], and waste water released from scattered rural houses.

In Region 2 (1.1% ⩽ urban land use ⩽31.5%), the relationships between urban land use and water quality displayed similar patterns to those reported in many previous studies. An increase in the amount of urban land use in the watershed had a significant negative impact on the BOD, TN, and TP in streams. Most cases in this study fell into Region 2, and the range (±95% confidence intervals) of the cases was relatively small. The BOD, TN, and TP rapidly increased as urban land use increased. However, the slopes of the relationships were slightly different. Specifically, the slope of the relationship between urban land use and BOD was relatively gentle, while the slope of the relationships between urban land use and the concentration of TP was steep. In Figure 3, Region 2 of the BOD and TN model are divided into sub-regions at approximately 0.5 of the log-transformed urban land use (approximately 3.2% of the actual percentage urban land use). Slopes were relatively gentle until the percentage urban land use reached 3.2%, beyond which they became steeper.

In Region 3 (31.6% ⩽ urban land use), the relationships between urban land use and water quality variables dramatically changed direction at the 1.5 breakpoint in the log-transformed percentage of urban land use. Like Region 1, water quality variations at very high levels of urbanization in Region 3 were somewhat different from those reported in previous studies, which used mostly LMs. In the GAMs, the variations in BOD, TN, and TP were independent of the variation in urban land use, or even decreased as the proportion of urban land use increased. Compared with Region 1, few cases fell in this region, and the range (±95% confidence intervals) of the cases was relatively large. In Region 3, reducing urbanized areas might not be effective for enhancing stream water quality. There should be additional considerations, such as the placement of riparian vegetation buffers.

The most important parameter in determining the abstraction of urban land use is frequently the area of the impervious surface connected directly to the drainage system. This is because impervious surfaces connected to the drainage system allow for a runoff volume that closely approximates the amount of incident precipitation [6]. In contrast, precipitation that falls on pervious surfaces or on areas not directly connected to the drainage system will infiltrate the ground surface and will not contribute to the immediate runoff. Previous studies have shown that the effects of impervious surface areas on stream water quality differ depending on the watershed, based on random effect solutions and random coefficient model simulations [46].

As discussed earlier, previous studies indicated that a 10% cover of impervious areas in a watershed is the average threshold at which water quality degradation first occurs [11,16]. Coles *et al.* [12] reported that significant changes in aquatic health could occur at low and moderate levels (0 to 35%) of urban land cover. Crim [13] suggested the threshold might be much lower than 10% cover of impervious surfaces.

In his study, the concentrations of water quality indicators increased considerably as the amount of impervious surface in a watershed increased from 0 to 4% in west-central Georgia, USA. Similarly, Nagy *et al.* [6] reported that an alteration in stream conditions can occur at low levels of development. It is difficult to compare our results directly with previous studies due to the different measurements (e.g., proportion of impervious areas, degree of urbanization, and proportion of urban land use) and different spatial scales (e.g., entire watershed, buffer zones, sub-drainage areas in riparian areas) used in the analyses. Our results suggest that water quality degradation could occur at extremely low levels of urban development (around 1% urban land cover), particularly in sub-drainage areas near streams or riparian zones.

It was slightly surprising to observe the pattern of the relationships between urban land use and water quality indicators, which were downward slopes in Regions 1 and 3. In this study, we were unable to identify the cause of the patterns displayed in Regions 1 and 3. One possible explanation could be the type of land cover across the entire watershed, for example, a high proportion of urban land in riparian areas and a high proportion of forested area throughout the entire watershed. Other variables could be the presence of sewage treatment facilities, drainage systems, pollution control systems for non-point source pollutants established by local authorities, and a high vegetation density in riparian areas. Thus further studies considering these factors are needed to explain the patterns in Regions 1 and 3. The results of this study also suggested that different strategies should be used corresponding to different degrees of urbanization for enhancing stream water quality. Decreasing urban land use in a watershed could be an effective way to improve the water quality in moderately urbanized areas. However, decreasing urban land use in a watershed might not be effective in highly urbanized areas, because water quality might not be improved as much as expected.

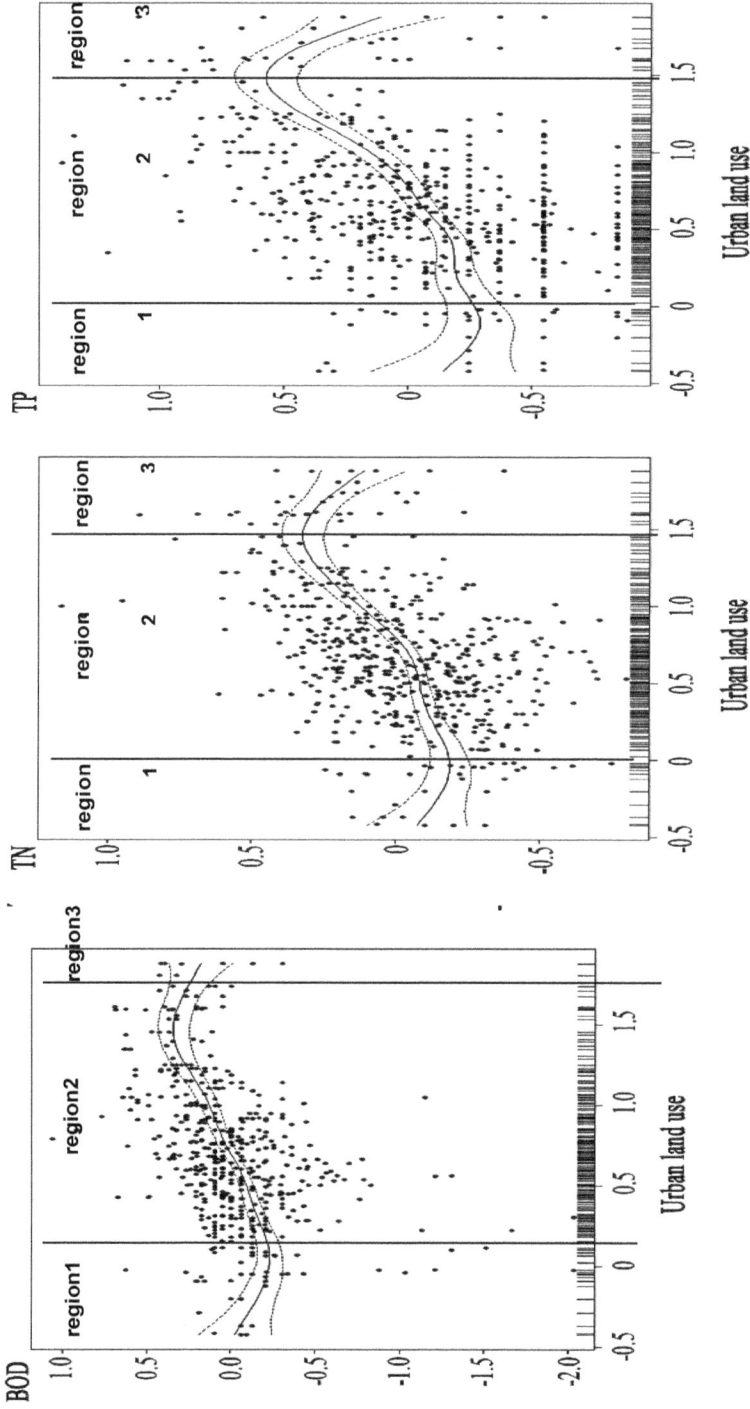

Figure 3. Regionalized zones in the non-linear relationships between urban land use and biological oxygen demand (BOD), total nitrogen (TN), and total phosphorous (TP) in streams. All variables were log-transformed.

5. Conclusions

In general, streams in urbanized areas are likely to have higher levels of oxygen demand, nutrients, suspended solids, ammonium, hydrocarbons, and metals. The negative impacts of urban land use on adjacent reservoirs, streams, and rivers have been well-documented and are a key concern for stream restoration, stream management, land planners, and land managers [2,5–10,47–49]. To establish effective water quality management policies, it is essential to understand the true nature of the relationship between water quality and urban land use.

In this study, we assessed LMs and non-linear models (GAMs) for the associations of BOD, TN, and TP with urban land use in the sub-drainage areas of five major river systems in Korea. Regardless of the type of model used, a higher proportion of urban land use had a significant impact on the degradation of stream water quality. Comparisons between LMs and non-linear models, based on R^2 and AICc values, indicated that the non-linear models (GAMs) could describe the relationships between urban land use and water quality more accurately. The GAMs demonstrated non-linear relationships between urban land use and water quality indicators (*i.e.*, BOD, TN, and TP) in streams and also revealed several breakpoints in the relationships. Based on two breakpoints, the relationships could be categorized into three regions. Only Region 2 showed similar relationships between land use and water quality to those reported in many previous studies using linear models. Regions with extremely low or extremely high levels of urban land use had a somewhat different relationship with the findings of previous studies. Stream restoration, stream management, and watershed land use policies should differ among these different regions. Water quality might not be improved as much as expected by reducing the extent of the urban area in areas with extremely low or high levels of urban land use. In particular, a comprehensive approach, including the installation of sewage treatment facilities or establishing riparian vegetation for filtering non-point source pollutants should be used.

In this study, we were not able to identify the cause of the unexpected pattern seen among the relationships between urban land use and water quality in areas with extremely low or high levels of urban land use. Further studies are needed, with a consideration of sewage treatment facilities, drainage systems, and the land cover across the entire watershed. It is also noteworthy that previous studies indicated that 3%–4% of impervious area cover in a watershed could cause degradation of water quality in streams. Interestingly enough, our GAMs suggested that this value might be even lower than 3%–4%. To understand the threshold value of urban areas, GAMs may need to be assessed at other spatial scales.

The results of this study are useful for stream restoration and management, because they highlight the negative impacts of urban land use and the non-linear relationships between urban land use and water quality. Water quality variance might differ with the degree of urbanization. Thus, improved water quality could be attainable by crafting management plans according to a region's specific urbanization characteristics.

Acknowledgments: This study was conducted under the project, "National Aquatic Ecosystem Health Survey and Assessment" in Korea, and was supported by the Ministry of the Environment and the National Institute of Environmental Research, Korea.

Author Contributions: All of the authors contributed extensively to the work. Sun-Ah Hwang and Sang-Woo Lee designed the research, performed the data analysis, and wrote the manuscript. Soon-Jin Hwang and Se-Rin Park contributed to the data interpretation, discussion, and editing of the manuscript.

Conflicts of Interest: The authors declare no conflict of interest.

References

1. Bolstad, P.V.; Swank, W.T. Cumulative impacts of land use on water quality in a southern Appalachian watershed. *J. Am. Water Resour. Assoc.* **1997**, *33*, 519–533.

2. Lenat, D.R.; Crawford, J.K. Effects of land use on water quality and aquatic biota of three North Carolina Piedmont streams. *Hydrobiologia* **1994**, *294*, 185–199.

3. Paul, M.J.; Meyer, J.L. Streams in the urban landscape. *Annu. Rev. Ecol. Evol. Syst.* **2001**, *32*, 333–365.

4. Tong, S.T.Y.; Chen, W. Modeling the relationship between land use and surface water quality. *J. Environ. Manag.* **2002**, *66*, 377–393.

5. Meierdiercks, K.L.; Smith, J.A.; Baeck, M.L.; Miller, A.J. Analyses of urban drainage network structure and its impact on hydrologic response. *J. Am. Water Resour. Assoc.* **2010**, *46*, 932–943.

6. Nagy, R.C.; Lockaby, B.G.; Kalin, L.; Anderson, C. Effects of urbanization on stream hydrology and water quality: The Florida Gulf Coast. *Hydrol. Process.* **2012**, *26*, 2019–2030.

7. Roberts, A.D.; Prince, S.D. Effects of urban and non-urban land cover on nitrogen and phosphorus runoff to Chesapeake Bay. *Ecol. Indic.* **2010**, *10*, 459–474.

8. Sun, R.H.; Chen, L.D.; Chen, W.L.; Ji, Y.H. Effect of land-use patterns on total nitrogen concentration in the upstream regions of the Haihe river basin, China. *Environ. Manag.* **2013**, *51*, 45–58.

9. Wu, J.; Thompson, J.; Kolka, R.; Franz, K.; Stewart, T. Using the Storm Water Management Model to predict urban headwater stream hydrological response to climate and land cover change. *Hydrol. Earth Syst. Sci.* **2013**, *17*, 4743–4758.

10. Wu, J.; Stewart, T.; Thompson, J.; Kolka, R.; Franz, K. Watershed features and stream water quality: Gaining insight through path analysis in a Midwest urban landscape, U.S.A. *Landsc. Urban Plan.* **2015**, *143*, 219–229.

11. Bledsoe, B.P.; Watson, C.C. Effects of urbanization on channel instability. *J. Am. Water Resour. Assoc.* **2001**, *37*, 255–270.

12. Coles, J.F.; Cuffney, T.F.; McMahon, G.; Beaulieu, K.M. *The Effects of Urbanization on the Biological, Physical, and Chemical Characteristics of Coastal New England Streams*; U.S. Geological Survey Professional Paper 1695; USGS: Reston, VA, USA, 2004.

13. Crim, J.F. Water Quality Changes across an Urban-Rural Land Use Gradient in Streams of the West Georgia Piedmont. Master's Thesis, Auburn University, Auburn, AL, USA, 17 December 2007.

14. Walsh, C.J.; Fletcher, T.D.; Ladson, A.R. Stream restoration in urban catchments through redesigning storm water systems: Looking to the catchment to save the stream. *J. North Am. Benthol. Soc.* **2005**, *24*, 690–705.

15. Qin, X.; Huang, G.; Chen, B.; Zhang, B. An interval-parameter Waste-Load-Allocation model for river water quality management under uncertainty. *Environ. Manag.* **2009**, *43*, 999–1012.

16. Arnold, C.L.; Gibbons, C.J. Impervious surface coverage. *J. Am. Plan. Assoc.* **1996**, *62*, 243–258.

17. Revelli, R.; Ridolfi, L. Stochastic dynamics of BOD in a stream with random inputs. *Adv. Water Resour.* **2004**, *27*, 943–952.

18. Karmakar, S.; Mujumdar, P.P. Grey fuzzy optimization model for water quality management of a river system. *Adv. Water Resour.* **2006**, *29*, 1088–1105.

19. Loucks, D.P. Managing America's rivers: Who's doing it? *Int. J. River Basin Manag.* **2003**, *1*, 21–31.

20. Ahearn, D.S.; Sheibley, R.W.; Dahlgren, R.A.; Anderson, M.; Johnson, J.; Tate, K.W. Land use and land cover influence on water quality in the last free-flowing river draining the western Sierra Nevada, California. *J. Hydrol.* **2005**, *313*, 234–247.

21. Richards, R.; Chaloupka, M.; Strauss, D.; Tomlinson, R. Using Generalized Additive Modelling to understand the drivers of long-term nutrient dynamics in the Broadwater Estuary (a subtropical estuary), Gold Coast, Australia. *J. Coast. Res.* **2014**, *30*, 1321–1329.

22. Hill, T.; Lewicki, P. *STATISTICS Methods and Applications*; StatSoft: Washington, DC, USA, 2007.

23. Huang, G.H. A hybrid inexact-stochastic water management model. *Eur. J. Oper. Res.* **1998**, *107*, 137–158.

24. Huang, G.H.; Loucks, D.P. An inexact two-stage stochastic programming model for water resources management under uncertainty. *Civ. Eng. Environ. Syst.* **2000**, *17*, 95–118.

25. Li, Y.P.; Huang, G.H.; Nie, S.L. Mixed interval-fuzzy two-stage integer programming and its application to flood-diversion planning. *Eng. Optim.* **2007**, *39*, 163–183.

26. Li, Y.P.; Huang, G.H.; Nie, S.L.; Liu, L. Inexact multistage stochastic integer programming for water resources management under uncertainty. *J. Environ. Manag.* **2008**, *88*, 93–107.

27. Lehmann, A. GIS modeling of submerged macrophyte distribution using Generalized Additive Models. *Plant Ecol.* **1998**, *139*, 113–124.

28. Guisan, A.; Edwards, T.C.; Hastie, T. Generalized linear and generalized additive models in studies of species distributions: Setting the scene. *Ecol. Model.* **2002**, *157*, 89–100.

29. Murase, H.; Nagashima, H.; Yonezaki, S.; Matsukura, R.; Kitakado, T. Application of a generalized additive model (GAM) to reveal relationships between environmental factors and distributions of pelagic fish and krill: A case study in Sendai Bay, Japan. *ICES J. Mar. Sci.* **2009**, *66*, 1417–1424.

30. Swartzman, G. Analysis of the summer distribution of fish schools in the Pacific Eastern Boundary Current. *ICES J. Mar. Sci.* **1997**, *54*, 105–116.

31. Swartzman, G.; Brodeur, R.; Napp, J.; Hunt, G.; Demer, D.; Hewitt, R. Spatial proximity of age-0 walleye Pollock (Theragra Chalcogramma) to zooplankton near the Pribilof Islands, Bering Sea, Alaska. *ICES J. Mar. Sci.* **1999**, *56*, 545–560.

32. Taylor, J.C.; Rand, P.S. Spatial overlap and distribution of anchovies (Anchoa spp.) and copepods in a shallow stratified estuary. *Aquat. Living Resour.* **2003**, *16*, 191–196.

33. Winter, A.; Coyle, K.O.; Swartzman, G. Variations in age-0 pollock distribution among eastern Bering Sea nursery areas: A comparative study through acoustic indices. *Deep. Sea Res. II* **2007**, *54*, 2869–2884.

34. Yee, T.W.; Mitchell, N.D. Generalized additive models in plant ecology. *J. Veg. Sci.* **1991**, *2*, 587–602.

35. Hastie, T.J.; Tibshirani, R.J. *Generalized Additive Models*; Chapman & Hall: New York, NY, USA, 1990.

36. Chambers, J.M.; Hastie, T.J. *Statistical Models in S*; Chapman & Hall: New York, NY, USA, 1993.

37. Richard, E.; Hughes, L.; Gee, D.; Tomlinson, R. Using Generalized Additive Models for water quality assessment: A case study example from Australia. In Proceedings of the 12th International Coastal Symposium (ICS2013), Plymouth, UK, 8–12 April 2013.

38. Morton, R.; Henderson, B. Estimation of nonlinear trends in water quality: An improved approach using generalized additive models. *Water Resour. Res.* **2009**, *44*, w07420.

39. APHA. *Standard Methods for the Examination of Water and Wastewater*, 21st ed.; American Public Health Association: Washington, DC, USA, 2001.

40. Horne, A.J.; Goldman, C.R. *Limnology*; McGraw-Hill, Inc.: New York, NY, USA, 1994.

41. Gaugush, R.F. *Statistical Methods for Reservoir Water Quality Investigations*; Instruction Report E-86-2; U.S. Army Engineer Waterways Experiment Station: Vicksburg, MS, USA, 1986.

42. Lassaletta, L.; Garcı́a-Go´mez, H.; Gimeno, B.S.; Rovira, J.B. Headwater streams: Neglected ecosystems in the EU Water Framework Directive. Implications for nitrogen pollution control. *Environ. Sci. Policy* **2010**, *13*, 423–433.

43. Liu, Z.J.; Weller, D.E.; Correll, D.L.; Jordan, T.E. Effects of land cover and geology on stream chemistry in watersheds of Chesapeake Bay. *J. Am. Water Resour. Assoc.* **2000**, *36*, 1349–1365.

44. Mayer, P.M.; Reynolds, S.K.; McCutchen, M.D.; Canfield, T.J. Meta-analysis of nitrogen removal in riparian buffers. *J. Environ. Qual.* **2007**, *36*, 1172–1180.

45. Huang, W.; McDowell, W.H.; Zou, X.; Ruan, H.; Wang, J.; Ma, Z. Qualitative differences in headwater stream dissolved organic matter and riparian water-extractable soil organic matter under four different vegetation types along an altitudinal gradient in the Wuyi Mountains of China. *Appl. Geochem.* **2015**, *52*, 67–75.

46. Schueler, T.R. The importance of imperviousness. *Watershed Prot. Tech.* **1994**, *1*, 100–111.

47. United States Geological Survey (USGS). *The Quality of Our Nation's Waters–Nutrients and Pesticides*; USGS Circular 1225; USGS: Reston, VA, USA, 1999.

48. Porcella, D.B.; Sorensen, D.L. *Characteristics of Non-Point Source Urban Runoff and Its Effects on Stream Ecosystems*; EPA-600/3-80-032; EPA: Washington, DC, USA, 1980.

49. Latimer, J.S.; Quinn, J.G. Aliphatic petroleum and biogenic hydrocarbons entering Narragansett Bay from tributaries under dry weather conditions. *Estuaries* **1998**, *21*, 91–107.

Roles of N:P Ratios on Trophic Structures and Ecological Stream Health in Lotic Ecosystems

Young-Jin Yun and Kwang-Guk An

Abstract: Little is known about the functions of N:P ratios in determining trophic structures and ecological health in lotic ecosystems, even though N:P ratios have been frequently used as a stoichiometric determinant in ambient water for trophic allocation of low-level organisms such as phytoplankton or zooplankton. In this study, nutrients (N, P) and sestonic chlorophyll (CHL) from 40 different streams in the Geum-River watershed were measured from 2008 to 2011. Fish compositions and stream health were also assessed, based on the multi-metric modeling of an index of biological integrity. Land use patterns in these watersheds were a key factor regulating nutrient contents and N:P ratios in ambient water, and also influenced empirical relationships between N:P ratios (or nutrients) and sestonic CHL. Land use patterns in forested, urban and wastewater treatment plant regions were associated with significant differences in stream N:P ratios, and the ratios were mainly determined by phosphorus. Sestonic CHL was significantly correlated with nutrient level (N, P); the ratios had a positive linear relationship with the proportion of omnivores, and a negative relationship with the proportion of insectivores. A similar trend in the N:P ratios was observed in indicator fishes such as *N. koreanus* and *Z. platypus*. Overall, the N:P ratio may be a good surrogate variable of ambient concentrations of N or P in assessing trophic linkage and diagnosing the ecological stream health in aquatic ecosystems.

Reprinted from *Water*. Cite as: Yun, Y.-J.; An, K.-G. Roles of N:P Ratios on Trophic Structures and Ecological Stream Health in Lotic Ecosystems. *Water* **2016**, *8*, 22.

1. Introduction

During the past two decades, nutrient regime of nitrogen (N) or phosphorus (P) in freshwater ecosystems is one of the most important factors regulating the ecosystem production and biological diversity [1,2]. Thus, the concept of nutrient loading as a factor controlling trophic state has been a key theory in numerous studies of aquatic ecosystems [3]. Furthermore, the importance of 16N:1P molar ratios as well as ambient nutrient concentrations in aquatic ecosystems [4], and suggested their key roles in primary production, nutrient cycling, resource competition, and animal growth in the systems, in spite of partial limitation on biological unavailability of some forms of nutrients. The various roles of N:P ratios, therefore, have been frequently tested in various trophic linkages as criteria of nutrient limitation on phytoplankton growth (*i.e.*, bluegreens; Smith [5]), invertebrate compositions [6], consumer-resource [7], and fish trophic guilds [8,9], and the ecological stream health [8]. In other words, the N:P ratios determined the production of primary producers [10], and this influenced the compositions and guilds in the higher trophic consumers. The ratios of N:P are associated with differences in biotic components between ecosystems and are closely linked with land use activities [11]. Therefore, these ratios have been used as a stoichiometric determinant in ambient water for trophic allocation of low-level organisms [5,12] through to higher trophic level organisms [6,7,9]. Previous studies on lotic ecosystems have demonstrated that the complexity of the in-stream environment is largely influenced by land use patterns within the watershed. Wastewater disposal plants and urban runoff or cropland, as significant non-point or point sources, increase N and P enrichment in stream and river environments [13,14], resulting in an alteration of mass N:P ratios in ambient water [15].

Rapid industrialization and dense industrial complexes have caused chemical pollution and habitat disturbances in urban regions, along with intense agricultural activities. Wastewater treatment plants (WTPs) are needed to reduce nutrient (N, P) and organic matter discharge from urban polluted water [16]. Haggard *et al.* [17] and Ekka *et al.* [18] demonstrated that the most significant sources of N and P are point-sources of WTPs, even though nutrient contents vary largely depending on the treatment methods of effluents [16,19]; furthermore, the effluents may directly or indirectly influence chemical pollution and biological disturbances downstream in the watershed [20]. Under these conditions, nitrogen and phosphorus levels are generally high [18,21] and the N:P ratios are relatively low in stream and river ecosystems. For this reason, a low N:P ratio may be a good surrogate variable of ambient concentrations of N or P in diagnosing and assessing anthropogenic nutrient pollution and eutrophication in lentic [22] and lotic ecosystems [15].

N:P ratios have also been used as ecological indicators to identify how aquatic organisms are regulated by N:P stoichiometry, through thresholds and spectrums. Redfield [23] found that the stoichiometric threshold for phytoplankton growth was identified as 16N:1P molar ratio, which is frequently considered to be a trophic interaction from an ecological perspective. Numerous studies on lentic ecosystems [24–27] have demonstrated that N:P ratios are a key index of nitrogen or phosphorus limitation in algal populations. However, these stoichiometric thresholds vary according to N-limitation and P-limitation criteria in aquatic ecosystems [10,28], as frequently shown by actual field data. Despite regional and seasonal variations in stoichiometric indices of the N:P ratio, it has been used to determine the abundance of specific taxa and trophic levels in the food chain [12,29,30]. Typical examples of N:P ratios are shown by bluegreen algae dominance when the ratios drop below 30 [5], and by empirical models of bluegreen algae [31]. Similarly, specific taxa of lotic periphyton [12] are regulated directly by N and P stoichiometry, and aquatic biota is associated with specific optimal N:P ratios [32]. Within high trophic organisms, their growth and abundance may be affected by stoichiometric N:P ratios as well as by absolute ambient nutrients; furthermore, N:P ratios may determine the food quality of aquatic insects [7] and fishes [8,9]. These studies have shown that trophic interactions, in functional taxa at the species level, are related to specific N:P ratios in aquatic environments. The trophic dynamic concept [33] has been used to demonstrate bioavailable energy and nutrient transfers to higher trophic organisms, representing both specific trophic interactions and the effects of community on aquatic ecosystems; however, the role of N:P ratios is unknown in the context of trophic interactions. Despite the importance of N:P ratios in terms of trophic linkages within the food chain/web of aquatic ecosystems, little is known about the effects of trophic compositions and fish tolerance on the range of N:P ratio values in ambient stream water [8].

Aquatic environmental stressors such as nutrients and N:P ratios may affect aquatic biota of low trophic levels and/or higher trophic level organisms and their stoichiometry, and are regarded as being among the most important regulating factors of stream ecosystems [8]. The nutrient regime regulates various fish compositions and their abundance in aquatic ecosystems [34], and these parameters may be closely associated with the eutrophication processes of N or P, and with N:P ratios. Primary productivity, regulated by the N or P contents and N:P ratios, increased as fish abundance increased [35] or decreased [36], depending on the regional scale and fish species. Tolerance, or trophic levels of fish related to nutrient regimes, has been used to assess stream conditions ranging from pristine to polluted [35]. Conventional criteria of fish tolerance (divided into three categories: sensitive, intermediate, and tolerant species [37,38]) are closely associated with eutrophication in water bodies, which are judged according to N or P contents or N:P ratios [8,39]. Noble et al. [40] demonstrated that the trophic compositions of fish were affected by their available food items and feeding habitats, and that changes in nutrient regimes or N:P ratios may modify the proportions of insectivorous and omnivorous fish due to changes in their feeding resources (in accordance with water chemistry [8,40]). These findings suggest that nutrients and stoichiometric N:P ratios may alter species compositions, tolerance level, and trophic compositions in aquatic ecosystems.

The objectives of this study were to assess the influence of land use patterns on nutrient contents and N:P ratios in stream ecosystems, and to determine the empirical relationships between N:P ratios and nutrients (total phosphorous; TP) and sestonic algal biomass (chlorophyll-a; CHL). Furthermore, the influence of the chemical regime on stoichiometric N:P ratios was elucidated by analyzing trophic composition and fish tolerance, and the fish bio-indicators of pollution, in lotic ecosystems.

2. Materials and Methods

2.1. Study Area and Selection of Sampling Site

This study was conducted in the Geum-River watershed, South Korea (Figure 1). The Geum-River watershed (36°–37° N; 127°–128° E) is located in the mid-western part of South Korea, and consists of a main stem length of 414 km and catchment area of 9886 km². Concerning the sampling site, both the site and the land use patterns were considered to be possible factors affecting the aquatic environment. This research was an ideal case study because of the morphology with longitudinal gradients and the diversity of land uses.

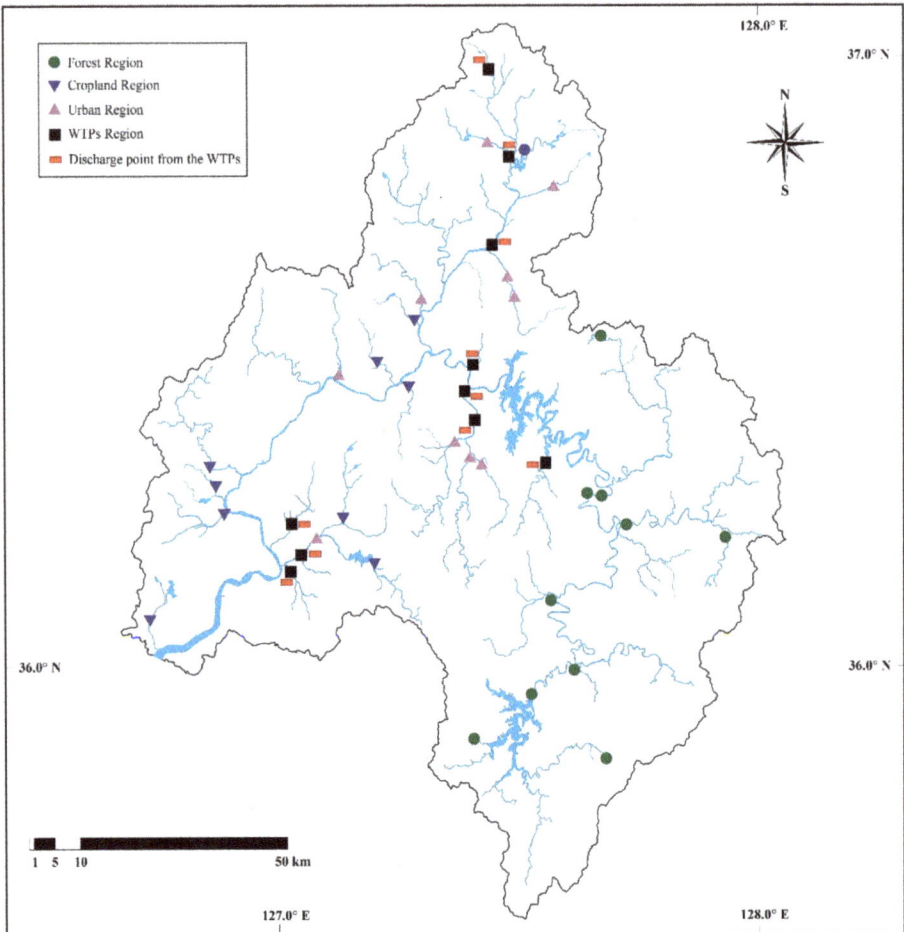

Figure 1. Catchment-scale map showing the location of the four different land types, among the 40 sampling sites of the Geum-River watershed.

Land use patterns were analyzed by calculating the proportion of forest, cropland, and urban land (a 500-m buffer) around the stream boundaries [11]. Land use for the sampling site was categorized according to the dominant land cover type: forest region (proportion of forest cover >50%), cropland region (proportion of cropland cover >50%), or urban region (proportion of urban cover >50%). Municipal WTPs were regarded only as the point source for defining the WTP region. The regions were divided equally into 10 sampling sites.

2.2. Analysis of Water Quality

Physicochemical and biological water quality data for the sampling sites were obtained from the National Institute of Environmental Research (NIER). Stream water sampling (surface water at 0.5 m) was conducted monthly from 2008 to 2010 in 40 different streams, and electrical conductivity (EC) was simultaneously measured using a portable multiparameter analyzer (YSI Sonde Model 6600: Yellow Springs, OH, USA). TP was determined using the ascorbic acid method after persulfate oxidation [41], and total nitrogen (TN) was determined using the second derived procedure after persulfate digestion [41]. Biological oxygen demands at 5 days (BOD$_5$) were measured per the method of Eaton and Franson [41]. Sestonic CHL concentrations were determined using a spectrophotometer (DU-530; Beckman Coulter Inc., Brea, CA, USA) after sampled water was processed through a GF/C filter and ethanol extraction in hot water [42]. Nutrient (N, P) and sestonic CHL analyses were performed in triplicate; EC and BOD$_5$ were applied in duplicate.

2.3. Analysis of Physical Habitat Conditions

Physical habitat conditions at the sampling sites were assessed using a qualitative health evaluation index (QHEI). The QHEI assessment was conducted in 2009; this study used a 6-metric model of QHEI for application at a regional level [43–45], modified from the original 10-metric model [46,47]. To assess the physical habitat condition, primary, secondary, and tertiary attributes were included in the model and their metrics were composed of substrate structure and vegetation coverage, channel and bank characteristics, and bank structure. All metric characteristics have been described previously [43,48]. The metrics of QHEI comprised M_1–M_6, evaluating epifaunal substrate cover, pool substrate, channel flow status, channel alteration, and sediment deposition. An additional metric was also included to account for the effects of dam construction. The health conditions of the habitat were evaluated by summing the scores obtained from the six metric scores (M_1–M_6) and then categorizing the system as "excellent" (A; score 120–96), "good" (B; 80–66), "fair" (C; 60–36), or "poor" (D; 30–6) based on the recommendations of MOE/NIER [44]. The final scores were transformed to a 0–1 scale for more comparable analysis.

2.4. Fish Collection and Sampling Method

Fish assemblages were collected twice at each site from 2008 to 2011 during the pre- and post-monsoon seasons; these seasons produce a hydrologically-stable aquatic environment. The sampling approach followed the modified protocols of the Ohio Environmental Protection Agency [49]. Casting nets (mesh size, 5×5 mm) and kick nets (mesh size, 4×4 mm) were used for sample collection, following the standard method for ecological fish health assessment proposed by the MOE/NIER [43]. Casting nets were used in various types of deep (2–3 m) and shallow (<0.5 m) habitat, and kick nets were used at locations with fast current velocities or in stream vegetation zones [43]. The sampling and handling techniques were based on catch per unit effort methods [50], with a sampling period of 50–60 min at each location. Fish were collected from all types of habitat, including riffles, runs, and pools, using the wading method [49]. All fish specimens were preserved in neutral-buffered 10% formalin and returned to the laboratory for identification [51,52]. Currently used scientific names, such as genus *Nipponocypris*, *Tanakia* spp., were employed [53]. The external characteristics of individual fish were examined in the laboratory for deformities, erosions (skin, barbels), lesions (open sores, ulcerations) and tumors [54].

2.5. Analysis of Trophic Composition and Tolerance Level

For the classification of fish trophic compositions and tolerance levels, the approach of the US EPA [49] and Karr [55] was used. Trophic composition was classified into four categories: insectivores, omnivores, piscivores, and herbivores, all of which were determined in accordance with the primary feeding resource. Tolerance levels were classified into sensitive, intermediate and tolerant species; this approach was based on the principle that an increase in the number of species and individuals in the first two categories indicates better ecosystem health, whereas an increase in omnivores indicates a degradation of ecosystem health [49]. Information on the classification of guild compositions for freshwater fishes in both trophic categories and tolerance levels are available [52].

2.6. Multi-Metric Fish Index of Biological Integrity (IBI) Model

The biological health of the lotic ecosystem was evaluated using the multi-metric fish IBI model. Ten metric models of IBI [55,56] were developed on the basis of regional application [38,48,56]. The metrics consisted of the three major ecological characteristics: species richness with magnitude of stream order (M_1, M_2 and M_7), trophic/tolerance guild compositions (M_3–M_6), and fish abundance according to health conditions (M_8). The following metrics were used: M_1, total number of native species; M_2, number of riffle-benthic dwelling species; M_3, number of sensitive species; M_4, proportion of tolerant species; M_5, proportion of omnivorous species; M_6, proportion of native insectivorous species; M_7, total number of native individuals; and M_8, proportion of abnormal individuals. Each metric was assigned a score of 1, 3, or 5, and five classification criteria—"excellent" (A; score = 40–36), "good" (B; 34–28), "fair" (C; 26–20), "poor" (18–14), and "very poor" (13 or below)—were used. Detailed descriptions of specific metric characteristics and scoring criteria for the model are available [38].

2.7. Statistical Analysis

A one-way analysis of variance (ANOVA), with *Scheffe*'s post-hoc test applied, was used to assess differences in water quality, habitat conditions, and biological components between the different regions (forest, cropland, urban, and WTP regions), using the SPSS in the Windows software package (ver. 22.0; IBM Corp., Armonk, NY, USA). Simple linear regression analysis and Spearman's correlation analysis were also conducted.

3. Results and Discussion

3.1. Influence of Land Use on Water Chemistry, Habitat Conditions, and Biological Components

The effects of land use pattern on water chemistry, habitat conditions, and biological components are shown in Table 1. TP, BOD_5 and EC were lowest in the forest region, and were significantly higher (*Scheffe*'s test, $p < 0.001$) in the WTP region compared to all forest, cropland, and urban regions. The mean mass ratio of N:P was highest in the forest region, and lowest in the WTP region (*Scheffe*'s test, $p < 0.001$); these results indicate that water chemistry was directly influenced by the type of land in the watershed. Physical habitat health, based on QHEI scores, was superior for the forest area than for any other region, but the differences were not significant (Table 1). Biological components of sestonic CHL (as a primary producer) and the IBI (based on fish assemblages) reflected the water chemistry, and were directly influenced by the type of land. Thus, the mean value of IBI was significantly greater, and the sestonic CHL significantly lower, in the forest region than in any other region (both $p < 0.05$; Table 1).

Table 1. Water chemistry, physical habitat condition, and biotic components according to the type of land in the Geum River watershed. Data are provided as means $\pm$ SE (range: 5%–95%) in each region. The superscript letters indicate significant post-hoc differences according to land use type.

Variables	Forest Region Mean $\pm$ SE Range	Cropland Region Mean $\pm$ SE Range	Urban Region Mean $\pm$ SE Range	WTPs Region Mean $\pm$ SE Range
Total nitrogen ($\mu g \cdot L^{-1}$)	1821 $\pm$ 33 [a] (1130–2836)	1740 $\pm$ 52 [a] (545–3454)	2605 $\pm$ 65 [b] (1002–5024)	6792 $\pm$ 225 [c] (1713–14782)
Total phosphorus ($\mu g \cdot L^{-1}$)	26 $\pm$ 2 [a] (4–76)	67 $\pm$ 4 [a, b] (11–164)	85 $\pm$ 3 [b] (21–212)	462 $\pm$ 27 [c] (66–1698)
N:P ratios in ambient water	143 $\pm$ 8 [c] (29–388)	53 $\pm$ 4 [b] (5–175)	46 $\pm$ 3 [b] (10–110)	24 $\pm$ 1 [a] (5–58)
Electrical Conductivity ($\mu s \cdot cm^{-1}$)	136 $\pm$ 2 [a] (74–212)	210 $\pm$ 4 [b] (133–342)	263 $\pm$ 4 [c] (148–383)	482 $\pm$ 13 [d] (237–935)
BOD$_5$ ($mg \cdot L^{-1}$)	0.9 $\pm$ 0.02 [a] (0.4–1.4)	1.9 $\pm$ 0.08 [b] (0.5–4.4)	2.5 $\pm$ 0.10 [c] (0.9–5.5)	4.1 $\pm$ 0.16 [d] (1.1–9.6)
QHEI	90 $\pm$ 5 [b] (70–108)	67 $\pm$ 5 [a] (45–87)	61 $\pm$ 5 [a] (40–82)	72 $\pm$ 6 [a,b] (47–92)
Sestonic CHL ($\mu g \cdot L^{-1}$)	2.4 $\pm$ 0.2 [a] (0.2–7.9)	3.4 $\pm$ 0.3 [a] (0.1–10.5)	7.0 $\pm$ 0.7 [a] (0.2–30.7)	18.9 $\pm$ 2.4 [b] (0.1–182.4)
Sestonic CHL:TP	0.15 $\pm$ 0.010 [b] (0.00877–0.44200)	0.08 $\pm$ 0.007 [a] (0.00095–0.27789)	0.10 $\pm$ 0.008 [a] (0.00313–0.42713)	0.08 $\pm$ 0.009 [a] (0.00035–0.44106)
Sestonic CHL:TN	0.0015 $\pm$ 0.0001 [a] (0.00006–0.00485)	0.0025 $\pm$ 0.0003 [a,b] (0.00004–0.01016)	0.0039 $\pm$ 0.0005 [b] (0.00006–0.01919)	0.0039 $\pm$ 0.0004 [b] (0.00002–0.01967)
Index of biotic integrity	32 $\pm$ 0.7 [c] (24–38)	23 $\pm$ 0.6 [b] (16–30)	19 $\pm$ 0.5 [a] (12–24)	20 $\pm$ 0.6 [a] (14–26)

Notes: WTPs, wastewater treatment plants; QHEI, Qualitative Habitat Evaluation Index; BOD$_5$, biological oxygen demand at 5 days; [a.] statistical result (post hoc test) of the smallest value on the land use, [b] (or [c,d.])statistical result (post hoc test) of the largest value on the land use

3.2. Effect of Mass Ratios of N:P on Nutrient Regimes and Their Influence on Land Use

The average concentration of TP in the WTPs region was 462 $\pm$ 27 $\mu g \cdot L^{-1}$, which was 10 times greater than the TP in the forest region (Table 1). TN concentrations were much higher than TP concentrations among all of the different types of land, and the regional mean N:P ratios were greater than 24–143, indicating a nitrogen-rich system. In fact, concentrations of TN had a weak relation with TP in the watershed, and the relation of TN *vs.* TP showed that the N:P ratios were mainly distributed in the range of 100–400 in the forest area but in the range of 10–20 in the region of WTPs. (Figure 2). Regression analysis of the association between log-transformed N:P ratios and nutrients (Figure 2) indicated that the N:P ratios were directly affected by TP, but not by TN. The N:P ratios, which are widely used as an index of nutrient limitations, were negatively related to TP ($p < 0.001$, $F = 1009.1$, $R^2 = 0.64$), but were not related to TN ($n = 569$, $p > 0.05$). Thus, the association between N:P ratios and TP was strongest in the forest regions (the green circles in Figure 2b) and lowest in the

WTP region (dark squares in Figure 2b), indicating that the variation in N:P ratios is explained by the land use patterns in this watershed. Strong negative regression coefficients ($R^2 = 0.64$, $p < 0.001$) for the N:P ratios on TP were found in numerous lentic ecosystems [22,57]. These studies indicated that the N:P ratio is a key regulator of the nutrient regime, and of primary production, in ambient water [11,12,58], and is also a major indicator of ecological river health in multi-metric fish models [8]. Our results indicate that the N:P ratios in forest region (high N:P) were clearly segregated from the WTPs region (low N:P), but partial overlaps in the N:P ratios were also shown in the cropland and urban regions.

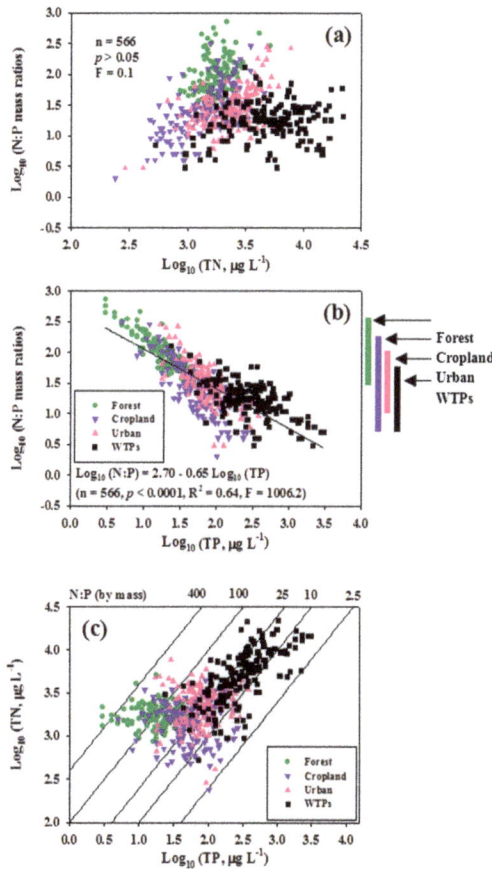

Figure 2. Relationships between log_{10}-transformed N:P ratios and total nutrient concentrations (TN = total nitrogen, TP = total phosphorus), and between log_{10}-transformed TN and TP in the Geum River. (**a**) log_{10} (TN)–log_{10} (N:P mass ratios); (**b**) log_{10} (TP)–log_{10} (N:P mass ratios); and (**c**) log_{10} (TP)–log_{10} (TN). Vertical bar (right side of panel b) indicates N:P mass ratios on the relations of TN *vs.* TP among the different land uses.

3.3. Effects of Nutrients and N:P Ratios on Sestonic CHL

Sestonic CHL concentrations were more affected by nutrients in ambient water, and the variations in sestonic CHL were better explained by variations in TP ($p < 0.001$, $F = 31.0$, $R^2 = 0.46$) than TN ($p < 0.001$, $F = 17.5$, $R^2 = 0.32$; Figure 3a,b). When the log-transformed TP values were high, as seen in WTP regions, sestonic CHL values were also high (>46 µg· L^{-1}). Similar patterns were observed with TN, except for differences in R^2 values (Figure 3b). In contrast, sestonic CHL values were low (<2.2 µg· L^{-1}) in the forest regions when N:P ratios were high (>60; Figure 3c). The regression coefficients (R^2) for the relationship between sestonic CHL and TP, TN, and the N:P ratios, were 0.46, 0.32, and 0.32, respectively (Figure 3a–c). The nutrients N and P were also important limiting factors influencing algal productivity, although current velocity and the light regime in a lotic environment are primary physical factors regulating sestonic phytoplankton and algal biomass. These results suggest that sestonic CHL in the watershed was increased by high P or N and low N:P [59].

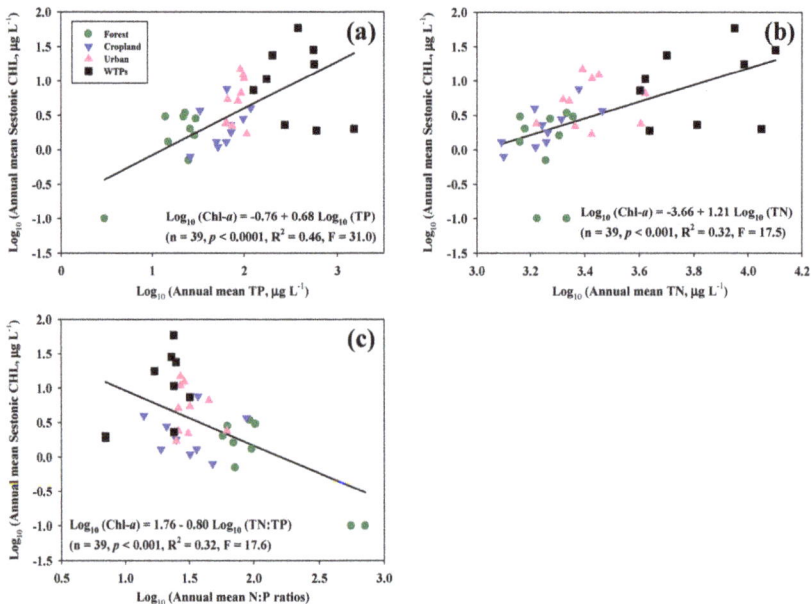

Figure 3. Simple linear regression models of log$_{10}$-transformed annual mean (**a**) TP-sestonic CHL; (**b**) TN-sestonic CHL; and (**c**) N:P mass ratios-sestonic CHL.

3.4. Relationships between Fish Communities and Compositions and N:P Ratios

Fish communities and community structures in the watershed were reflected in the type of land (Table 2). The forest region was designated as a *Nipponocypris-Zacco* community, whose dominant species was *Nipponocypris koreanus* (34.5% of the total), a sensitive species that is known to dwell in pristine chemical conditions. The average N:P ratio in ambient waters of the forest region was 143, which was higher than in any other type of land. In contrast, the dominant species in the WTPs region were *Zacco platypus*, (51.1%) and *Hemiculter eigenmanni* (6.2%), which are tolerant, omnivorous species with high abundance in polluted aquatic environments. The N:P ratio was lowest (24) in the *Zacco-Hemiculter* community due to the high degree of phosphorus enrichment from the effluents of WTPs. The forest region contained the largest proportion of insectivorous species (56.2%) and sensitive species (47.0%) (Figure 4a), whereas the WTP region showed a predominance of omnivorous species (76.8%) and tolerant species (78.6%) (Figure 4d). These preliminary results indicate that the type of land directly affects both trophic compositions and fish tolerance guilds [38,47]. In the meantime, other physical factors as well as the land use pattern might also have influenced the fish community and species compositions, even if we did not put the data of physical variables such as stream discharge, water temperature, dissolved oxygen(DO), mean depth, and canopy cover, *etc*.

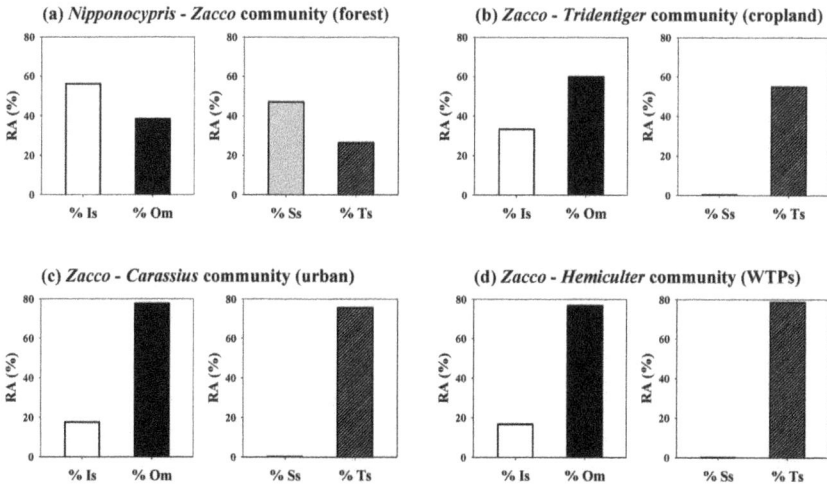

Figure 4. Trophic composition (**left** panel) and tolerance level (**right** panel) of fish communities in the Geum River watershed. % I_s, proportion of insectivores; % O_m, proportion of omnivores; % S_s, proportion of sensitive species; % T_s, proportion of tolerance species. (**a**) *Nipponocypris-Zacco* community (forest); (**b**) *Zacco-Tridentiger* community (Cropland); (**c**) *Zacco-Carassius* community (urban); and (**d**) *Zacco-Hemiculter* community (WTPs).

3.5. Multi-Metric IBI Model and Its Chemical Effects

The multi-metric river health model, based on the IBI, showed that IBI values were determined by land use patterns (Table 3). The mean value of the IBI model was

32.3 ± 4.6 in the forest region, which was judged to be in a "fair-excellent condition" with respect to river health. In contrast, the mean IBI value ranged from 19.2 to 22.6 in the cropland, urban, and WTP regions; the mean IBI values in all three of these regions was significantly (*Scheffe's* test, $p < 0.05$) lower than that of the forest region. However, there were no significant differences (one-way ANOVA, $p < 0.001$) in the mean IBI values of the cropland, urban and WTP regions (Table 3). Statistical analysis of each model parameter showed that there were significant differences (one-way ANOVA, $p < 0.001$) for M_1-M_6, and M_8, but not for M_7 (Table 3). In the forest region, the highest IBI values occurred at M_3 and M_6 (*Scheffe's* test, $p < 0.01$), and the lowest values occurred at M_4 and M_5 (*Scheffe's* test, $p < 0.01$).

Table 2. N:P ratios in the ambient water (means ± SE), fish community, and fish compositions (including tolerance level and trophic composition) according to type of land in the Geum River watershed.

Type of Land	N:P Ratios in the Ambient Water	Community Group	Fish Species	Tolerance Level	Trophic Compositions	RA (%)
Forest Region	143 ± 8	Nipponocypris-Zacco community	*Nipponocypris koreanus*	S_s	I_s	34.5
			Zacco platypus	T_s	O_m	24.1
			Tanakia koreensis	I_n	O_m	6.0
			Coreoleuciscus splendidus	S_s	I_s	4.8
			Pungtungia herzi	I_n	I_s	4.2
			Other species (43)	–	–	26.4
Cropland region	53 ± 4	Zacco-Tridentiger community	*Zacco platypus*	T_s	O_m	38.0
			Tridentiger brevispinis	I_n	I_s	13.8
			Pseudogobio esocinus	I_n	I_s	6.1
			Tanakia lanceolatus	I_{nt}	O_m	5.1
			Opsarichthys uncirostris amurensis	T_s	C_a	3.3
			Other species (51)	–	–	33.8
Urban Region	46 ± 3	Zacco-Carassius community	*Zacco platypus*	T_s	O_m	52.5
			Carassius auratus	T_s	O_m	9.5
			Pseudogobio esocinus	I_n	I_s	6.8
			Tanakia lanceolatus	I_n	O_m	5.7
			Opsarichthys uncirostris amurensis	T_s	C_a	2.9
			Other species (41)	–	–	22.5
WTPs region	24 ± 1	Zacco-Hemiculter community	*Zacco platypus*	T_s	O_m	51.1
			Hemiculter eigenmanni	T_s	O_m	6.2
			Pseudogobio esocinus	I_n	I_s	6.1
			Carassius auratus	T_s	O_m	5.7
			Hemibarbus labeo	T_s	I_s	5.0
			Other species (46)	–	–	25.9

Notes: S_s, sensitive species; I_n, intermediate species; T_s, tolerant species; I_s, insectivore; O_m, omnivore; C_a, carnivore; RA, relative abundance; WTPs, wastewater treatment plants.

Multi-metric IBI values, as an indicator of river health, had negative relationships with TP and BOD_5, and a positive relationship with the N:P ratio (Figure 5). In other words, the river health on IBI was directly affected by nutrient level (P) and organic matter (BOD_5) in the lotic ecosystem (Figure 5). Regression analysis of IBI on TP and BOD_5 showed that the variation in river health (IBI) was accounted for by variations in log_{10}-transformed TP (33%; $n = 216$, $F = 106.0$, $p < 0.001$), and log_{10}-transformed BOD_5 values (33%; $n = 216$, $F = 92.4$, $p < 0.001$; Figure 5). For the linear models of IBI, the mean IBI value was $37.9 - 7.8\ log_{10}$ (TP), and the mean IBI value was $26.4 - 11.8\ log_{10}$ (BOD_5). Mass ratios of N:P in the ambient water had a positive linear relationship with IBI values (mean IBI $= 10.6 + 7.8\ log10$ (N:P), $p < 0.001$), indicating that N:P ratios are an important determinant of river health. Mean IBI had no significant relationship with TN, due to the systems being nitrogen-rich in this watershed regardless of sampling location or season.

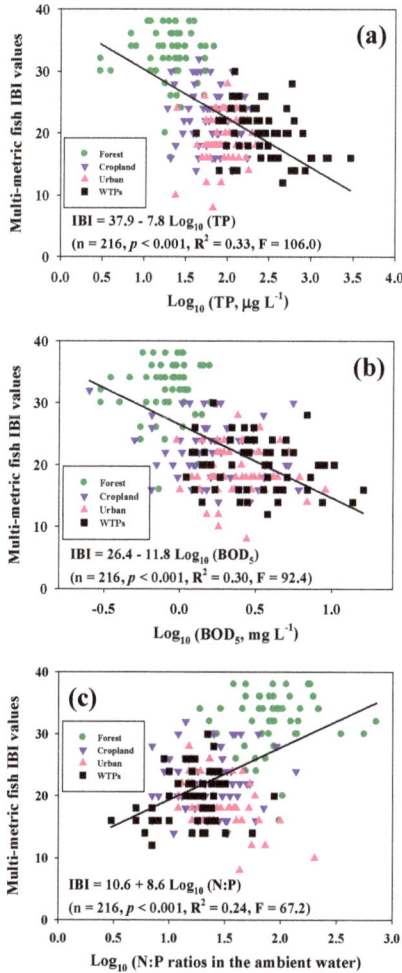

Figure 5. Relationships between multi-metric fish index of biological integrity (IBI) values and log_{10}-transformed (**a**) TP; (**b**) BOD_5; (**c**) N:P ratios in the ambient water.

Table 3. Model metrics and scoring criteria of the multi-metric fish index of biological integrity values according to type of land. The superscript letters indicate significant differences among the four regions on *Scheffe's* post-hoc test.

Category	Model Metric	Scoring Criteria			Forest	Cropland	Urban	WTPs	F-Value	p-Value
		5	3	1						
Species richness and composition	M_1: total number of native fish species	Expectations of M_1 vary with stream order			12.2 ± 4.4 [b]	10.1 ± 3.2 [a]	9.7 ± 3.2 [a]	9.8 ± 4.6 [a]	8.5	***
					(3.8 ± 1.1)	(4.0 ± 1.2)	(3.4 ± 1.2)	(3.3 ± 1.5)		
	M_2: number of riffle benthic dwelling species	Expectations of M_2 vary with stream order			2.7 ± 1.5 [b]	1.2 ± 1.1 [a]	0.8 ± 0.8 [a]	0.9 ± 1.1 [a]	38.3	***
					(3.6 ± 1.6)	(1.7 ± 1.3)	(1.2 ± 0.6)	(1.6 ± 1.3)		
	M_3: number of sensitive species	Expectations of M_3 vary with stream order			3.9 ± 1.6 [b]	0.1 ± 0.3 [a]	0.1 ± 0.3 [a]	0.1 ± 0.4 [a]	308.3	***
					(3.7 ± 1.2)	(1.0 ± 0.0)	(1.0 ± 0.0)	(1.0 ± 0.0)		
	M_4: proportion of individuals as tolerant species	<5	5–20	>20	24.2 ± 21.4 [a]	56.3 ± 21.5 [b]	72.3 ± 19.6 [c]	74.6 ± 18.0 [c]	62.7	***
					(4.2 ± 1.1)	(2.6 ± 1.3)	(1.9 ± 1.1)	(1.8 ± 1.1)		
Trophic composition	M_5: proportion of individuals as omnivore species	<20	20–45	>45	35.7 ± 24.3 [a]	63.7 ± 19.6 [b]	70.4 ± 22.8 [b]	72.5 ± 20.1 [b]	26.9	***
					(3.6 ± 1.3)	(2.3 ± 1.2)	(2.0 ± 1.3)	(1.8 ± 1.1)		
	M_6: proportion of individuals as native insectivore species	>45	20–45	<20	59.1 ± 25.2 [b]	27.2 ± 18.3 [a]	23.2 ± 18.9 [a]	19.5 ± 17.0 [a]	36.4	***
					(4.2 ± 1.2)	(2.5 ± 1.4)	(2.3 ± 1.5)	(1.9 ± 1.4)		
Fish abundance and condition	M_7: total number of native individuals	Expectations of M_7 vary with stream order			191.1 ± 86.3	165.2 ± 130.3	170.7 ± 131.4	194.3 ± 146.5	0.9	NS
					(4.3 ± 1.0)	(4.1 ± 1.4)	(3.7 ± 1.6)	(4 ± 1.4)		
	M_8: proportion of individuals as abnormalities	0	0–1	>1	0.0 ± 0.0 [a]	0.3 ± 0.8 [a]	1.2 ± 2.6 [b]	0.5 ± 1.8 [a,b]	5.1	**
					(5.0 ± 0.0)	(4.4 ± 1.4)	(3.7 ± 1.7)	(4.3 ± 1.2)		
Scores (criteria of multi-metric fish IBI model)					32.3 ± 4.6 [c]	22.6 ± 4.7 [b]	19.2 ± 3.9 [a]	19.7 ± 4.1 [a]	99.2	***
					Fair-excellent	Poor-fair	Poor-fair	Poor-fair		

Notes: The values in parentheses indicate the mean ± SD of the M_n metric (n ranges from 1 to 8); F- and p-values are from one-way ANOVA; ** significant at the 0.01 probability level; *** significant at the 0.001 probability level and below; NS, non-significant ($p > 0.05$); WTPs, wastewater treatment plants; IBI, index of biological integrity; [a], statistical result (post hoc test) of the smallest M_n value on land use, [b] or [c] statistical result (post hoc test) of the largest M_n value on land use.

3.6. Influence of Sestonic CHL on Trophic Compositions and River Health

Sestonic CHL, as a measure of primary production, affected the food chain at higher trophic levels, which in turn influenced the trophic compositions (Figure 6). Concentrations of sestonic CHL had inverse linear relationships with % I_s (R^2 = 0.28, F = 14.7, p < 0.001) and the IBI model values for stream health (R^2 = 0.21, F = 9.8, p < 0.01). The results of simple linear models of insectivores and IBI values were as follows: % I_s = 44.9 − 22.5 $\log_{10}$ (CHL) and IBI = 25.9 − 5.0 $\log_{10}$ (CHL). In contrast, sestonic CHL had a positive linear relationship (R^2 = 0.25, p < 0.01) with the proportion of omnivore species (% O_m); the result of the simple linear model was as follows: % O_m = 49.0 + 20.6 $\log_{10}$ (CHL). Even if the regression coefficients of both dependent variables were <0.30, increases in sestonic CHL generally decreased the insectivore fish in the oligo- or meso-trophic state [60], and increased the degree of impairment of stream health as shown by the IBI model values [61]. Our results are supported by the studies of Robertson *et al.* [62], who showed that increases in organic solids (such as phytoplankton CHL) reduce the abundance of insectivores, and increase the abundance of omnivores in stream ecosystems. These results indicate that organic solids that originate from the sestonic CHL increase the contents of organic matters; furthermore, their accumulation on the stream bottom favors omnivorous species and affected insectivores, resulting in rapid impairment of stream health.

3.7. Relationships among Lotic N:P Ratios, Trophic Composition and Tolerance Level

Fish trophic compositions closely associated with the food chain were directly affected by the N:P ratios [9], which is directly determined by P rather than N (Figure 7). When N:P ratios in ambient water were less than 30, % I_s was also under 30%; when N:P ratios were greater than 200, the proportion of insectivores exceeded 60%. The variation in % I_s was largely (93%) related to variation in mean N:P ratios (R^2 = 0.93, p < 0.01). The linear regression results were as follows: % I_s = 13.0 + 10.5 $\log_{10}$ (N:P). This effect of N:P ratios was modified by % O_m; the variation in % omnivores was largely related (90%) to variation in the mean N:P ratios (R^2 = 0.90, p < 0.01) and the linear regression result was as follows: % O_m = 77.4–9.1 $\log_{10}$ (N:P). Thus, mean N:P ratios had a positive relationship with % O_m in these streams (Figure 7). The trophic compositions of insectivores and omnivores were matched with fish tolerance. The proportion of sensitive species (% S_s) had a positive relationship with N:P ratios (R^2 = 0.95, p < 0.01; % S_s = -16.4 + 15.2 $\log_{10}$ (N:P)). In contrast, the proportions of tolerant species (% T_s) had a negative relationship with N:P ratios [R^2 = 0.91, p < 0.05; % T_s = 75.4 - 10.4 $\log_{10}$ (N:P)]. These outcomes suggest that N:P ratios determined the degree of fish tolerance (S_s, T_s) as well as trophic compositions (I_s, O_m).

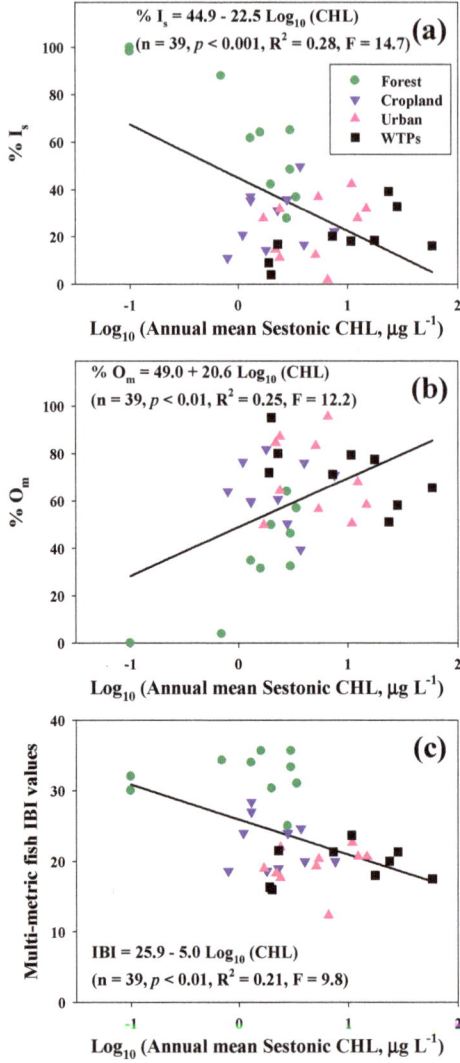

Figure 6. Simple linear regression models of (**a**) $\log_{10}$-transformed annual mean sestonic CHL–% I_s; (**b**) $\log_{10}$-transformed annual mean sestonic CHL–% O_m; (**c**) $\log_{10}$-transformed annual mean sestonic CHL–multi-metric fish IBI values, respectively.

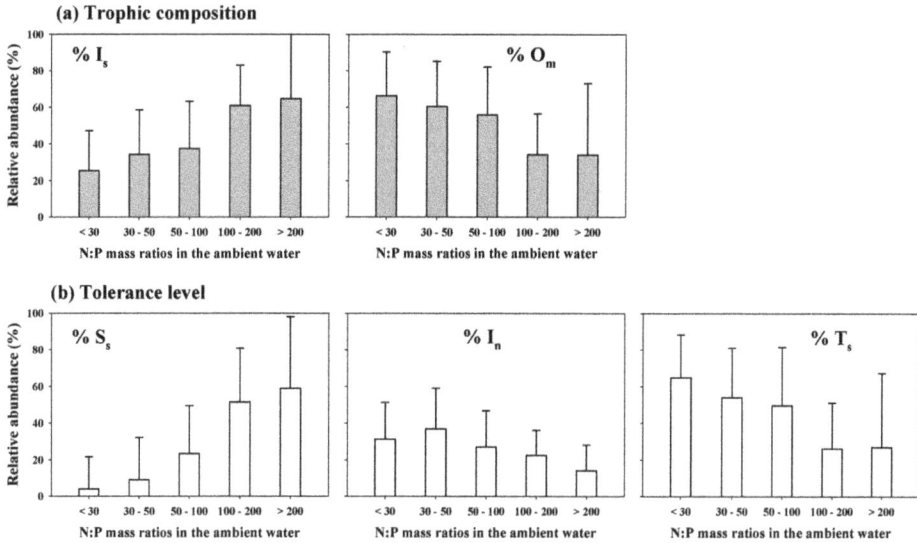

(a) Trophic composition

(b) Tolerance level

Figure 7. Mass ratios of the N:P interval in ambient water, in relation to means ± SE of (**a**) trophic composition (I_s = *Insectivore* sp., O_m = *Omnivore* sp.); (**b**) tolerance level (S_s = *Sensitive* sp., I_n = *Intermediate* sp., T_s, = *Tolerant* sp.) in the Geum River watershed. % I_s, proportion of insectivores; % O_m, proportion of omnivores; % S_s, proportion of sensitive species; % I_n, proportion of intermediate species; % T_s, proportion of tolerance species.

3.8. Influence of Land Use Pattern on N:P Ratios and Linkage with Trophic Compositions and Fish Tolerance

Land use patterns influenced the N:P ratios in ambient water, and determined the trophic compositions and fish tolerance in this study (Figure 8). The N:P ratios in the forest region averaged 143 and ranged from 29 to 388. In contrast, the N:P ratios in the WTP region averaged 24 and ranged from 5 to 58. This indicates that N:P ratios are high in the pristine forest region, and are low in the nutrient-rich WTP regions. Thus, when the N:P ratios were above 130, % I_s was high (79.1%), and the proportion of sensitive species (% S_s) was also high (73.6%). Conversely, when the N:P ratios were below 9, the proportions of both the insectivores (5.8%) and sensitive species (0.3%) were low (Figure 8). Conversely, in the WTPs regions, the % O_m and tolerant species (% T_s) in the WTPs regions were 84.4% and 90.2%, respectively, indicating that omnivore and/or tolerant fish species dominated the fish community in the point-source region with low N:P ratios.

Regression analysis of the association between trophic compositions and fish tolerance and N:P ratios indicated that log_{10}-transformed N:P ratios were significantly ($p < 0.001$) associated with these variables (Figure 8). In the analysis of trophic compositions, the variation in insectivore proportions was positively associated ($R^2 = 0.38$, $p < 0.001$) with the N:P ratios, as follows: % I_s = -29.3 + 39.0 Log_{10} (N:P). However, the proportion of omnivores was negatively associated ($R^2 = 0.31$, $p < 0.001$) with the N:P ratios, as follows: % O_m = 117.1 $-$ 35.9 log_{10} (N:P). In the analysis of tolerance guilds, the variation in sensitive species was positively associated ($R^2 = 0.45$, $p < 0.001$) with the N:P ratios, as follows: % S_s = -60.4 + 46.3 Log_{10} (N:P). The variation in tolerant species was negatively associated ($R^2 = 0.32$, $p < 0.001$) with the N:P ratios, as follows: % T_s = 121.4 $-$ 41.0 Log_{10} (N:P).

3.9. Influence of N:P Ratios on Fish Species Indicators

The N:P ratios in ambient water determined the indicator fish species in each fish community, as shown in Figure 9. The dark chub, *Nipponocypris koreanus*, which is known to be dominant in clean water environments [45], preferred high N:P ratios (>200); under these conditions, the relative abundance (RA) of the dark chub was 49.8%. In contrast, when the N:P ratios were < 30, the mean value of RA was only 0.4%. Thus, the abundance of *N. koreanus* was significantly increased ($p < 0.05$, $r = 0.36$) with high N:P ratios (Figure 9). The pale chub, *Zacco platypus*, which is known to be a dominant species in polluted environments [63], preferred low N:P ratios of <30; under these conditions, the relative abundance (RA) of the pale chub was 44.9%. In contrast, when the N:P ratios were >200, the mean value of RA was only 18.4%. Thus, the abundance of *Z. platypus* was significantly increased ($p < 0.05$, $r = -0.18$) with low N:P ratios. These results suggest that the N:P ratio should be considered as an important factor in determining the indicator fish (sensitive or tolerant species) in fish communities.

(a) Trophic compositions

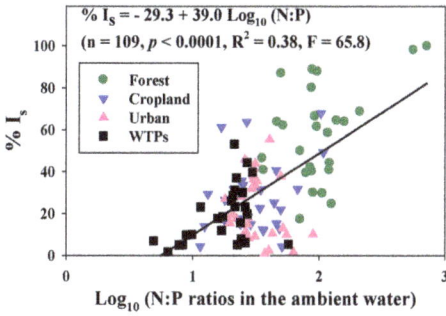

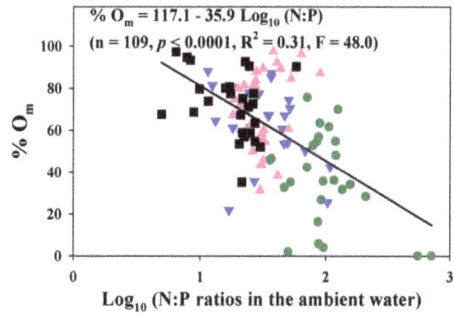

$\% I_s = -29.3 + 39.0 \, Log_{10} \, (N:P)$
$(n = 109, p < 0.0001, R^2 = 0.38, F = 65.8)$

Forest
Cropland
Urban
WTPs

$\% O_m = 117.1 - 35.9 \, Log_{10} \, (N:P)$
$(n = 109, p \leq 0.0001, R^2 = 0.31, F = 48.0)$

(b) Tolerance level

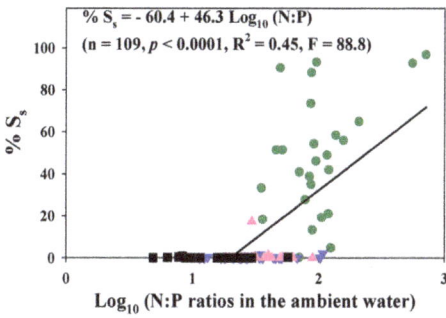

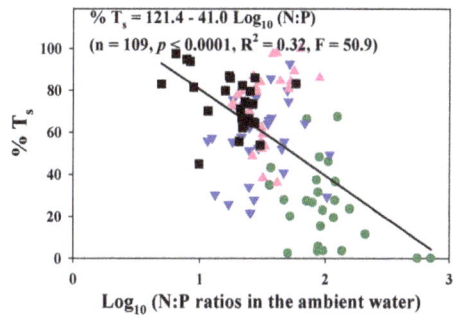

$\% S_s = -60.4 + 46.3 \, Log_{10} \, (N:P)$
$(n = 109, p < 0.0001, R^2 = 0.45, F = 88.8)$

$\% T_s = 121.4 - 41.0 \, Log_{10} \, (N:P)$
$(n = 109, p \leq 0.0001, R^2 = 0.32, F = 50.9)$

Figure 8. Scatter and simple linear regression models of (**a**) log_{10}-transformed N:P ratios-trophic composition (% I_s, **left**; % O_m, **right**); and (**b**) log_{10}-transformed N:P ratios–tolerance level (% S_s, **left**; % T_s, **right**) in the Geum River watershed. Forest = • (closed circle); cropland = ▼ (closed triangle down); urban = ▲ (closed triangle up); wastewater treatment plants = ■ (closed square).

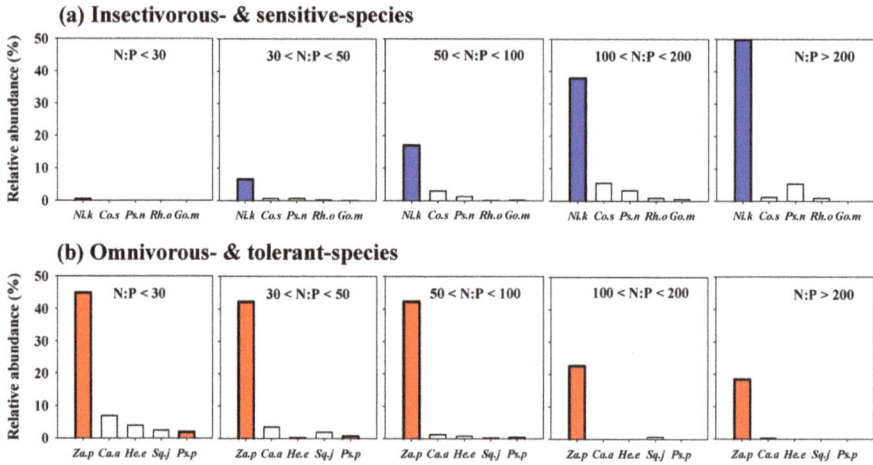

Figure 9. Mass ratios in the stream water of the N:P interval, plus the trophic composition and tolerance level of relative abundance (RA) (%). Upper panels indicate a combination of (**a**) insectivorous and sensitive species of RA (%); (**b**) omnivorous and tolerance species of RA (%). Pearson's correlations of N:P ratios with RA (%) indicate a significant increase ($p < 0.05$, $r > 0$; closed vertical blue bar: (■), a significant decrease ($p < 0.05$, $r < 0$; closed vertical red bar: (■), and no significant change ($p > 0.05$; opened vertical bar: (□). Abbreviations: *Za.P, Zacco platypus; Ca.a, Carassius auratus; He.e, Hemiculter eigenmanni; Sq.j, Squalidus japonicas coreanus; Ps.p, Pseudorasbora parva; Ni.k, Nipponocypris koreanus; Co.s, Coreoleuciscus splendidus; Ps.n, Pseudopuntungia nigra; Rh.o, Rhynchocypris oxycephalus; Go.m, Gobiobotia macrocephala.*

4. Conclusions

This study examined the N:P ratios and biological components from low trophic level of phytoplankton to high trophic level of fish, which are influenced by land use patterns and point/nonpoint sources in the watershed of Geum River. In this watershed, nutrient regime (N, P) influenced by land use patterns (① [14]) and flow regimes (② [64]), as well as point/nonpoint sources (①, ③ [14,17]), altered the ratios of N:P (④ [22]; Figure 10) and higher trophic linkage. The N:P ratios were directly or indirectly associated with the trophic level of phytoplankton production (sestonic CHL; ⑤, ⑥ [32,65]) and were determined by P rather than N, suggesting that differences in TP and N:P ratios were related to land use patterns and the location of WTPs. Furthermore, fish trophic compositions (⑭ [8]), tolerance guilds (⑭ [8]), and fish community (⑮ [9]) were determined by the availability of food resources (⑬ [66]), which are directly influenced by N:P ratios or nutrient regimes (N, P). In other words, the N:P ratios determined the sestonic CHL, which was associated

288

with the food chain at higher trophic levels. Fish trophic compositions and tolerance guilds (⑭ [8]) were closely associated with the food chain and were directly affected by N:P ratios in ambient water, resulting in a modification of stream ecosystem health (based on the IBI multi-metric model, ⑭ [8]). Overall, the N:P ratio may be a good surrogate variable of ambient concentrations of N or P in assessing trophic linkage and diagnosing the ecological stream health in aquatic ecosystem.

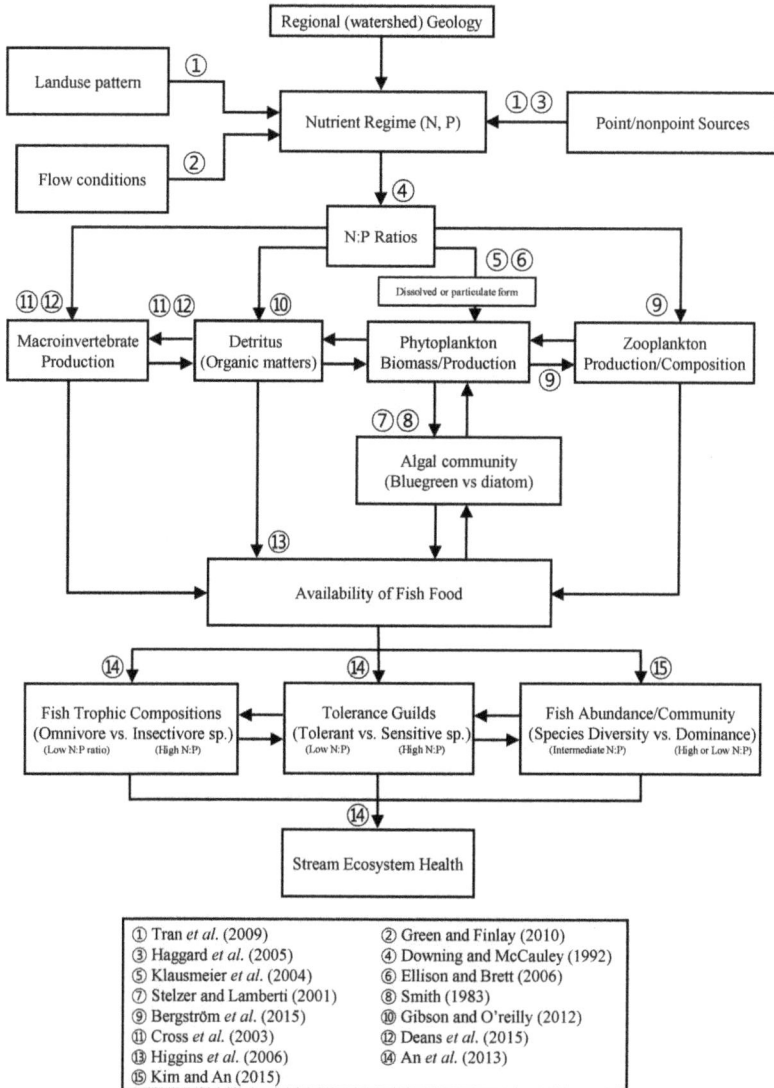

Figure 10. Schematic diagram on the control of N:P ratios on their environmental and ecological variables in aquatic ecosystems.

Acknowledgments: This research was supported by the Basic Science Research Program through the National Research Foundation of Korea (NRF) funded by the Ministry of Education (No. 2013R1A1A4A01012939).

Author Contributions: Young-Jin Yun performed the data analysis and prepared illustrations. Kwang-guk An provided constructive suggestions in the preparation of this manuscript.

Conflicts of Interest: The authors declare no conflict of interest.

References

1. Hecky, R.E.; Kilham, P. Nutrient limitation of phytoplankton in freshwater and marine environments: A review of recent evidence on the effects of enrichment. *Limnol. Oceanogr.* **1988**, *33*, 796–822.
2. Elser, J.J.; Marzolf, E.R.; Goldman, C.R. Phosphorus and nitrogen limitation of phytoplankton growth in the freshwaters of North America: A review and critique of experimental enrichments. *Can. J. Fish. Aquat. Sci.* **1990**, *47*, 1468–1477.
3. Vollenweider, R.A. Advances in defining critical loading levels for phosphorus in lake eutrophication. *Mem. 1st. Ital. Idrobiol.* **1976**, *33*, 53–83.
4. Redfield, A.C. On the proportions of organic derivations in sea water and their relation to the composition of plankton. In *James Johnstone Memorial Volume*, 1st ed.; Daniel, R.J., Ed.; University Press of Liverpool: Liverpool, UK, 1934; pp. 177–192.
5. Smith, V.H. Low nitrogen to phosphorus ratios favor dominance by blue-green algae in lake phytoplankton. *J. Am. Water Resour. Assoc.* **1983**, *221*, 669–671.
6. Frost, P.C.; Tank, S.E.; Turner, M.A.; Elser, J.J. Elemental composition of littoral invertebrates from oligotrophic and eutrophic Canadian lakes. *J. N. Am. Benthol. Soc.* **2003**, *22*, 51–62.
7. Cross, W.F.; Benstead, J.P.; Rosemond, A.D.; Bruce Wallace, J. Consumer-resource stoichiometry in detritus-based streams. *Ecol. Lett.* **2003**, *6*, 721–732.
8. An, K.-G.; Choi, J.-W.; Lee, Y.-J. Modifications of ecological trophic structures on chemical gradients in lotic ecosystems and their relations to stream ecosystem health. *Anim. Cells Syst.* **2013**, *17*, 53–62.
9. Kim, S-.Y.; An, K-.G. Nutrient regime, N:P ratios and suspended solids as key factors influencing fish tolerance, trophic compositions, and stream ecosystem health. *J. Ecol. Environ.* **2015**, *38*, 505–515.
10. Guildford, S.J.; Hecky, R.E. Total nitrogen, total phosphorus, and nutrient limitation in lakes and oceans: Is there a common relationship? *Limnol. Oceanogr.* **2000**, *45*, 1213–1223.
11. Choi, J.-W.; Han, J.-H.; Park, C.-S.; Ko, D.-G.; Kang, H.-I.; Kim, J.Y.; Yun, Y.-J.; Kwon, H.-H.; An, K.-G. Nutrients and sestonic chlorophyll dynamics in Asian lotic ecosystems and ecological stream health in relation to land-use patterns and water chemistry. *Ecol. Eng.* **2015**, *79*, 15–31.
12. Stelzer, R.S.; Lamberti, G.A. Effects of N: P ratio and total nutrient concentration on stream periphyton community structure, biomass, and elemental composition. *Limnol. Oceanogr.* **2001**, *46*, 356–367.

13. Perkins, B.D.; Lohman, K.; van Nieuwenhuyse, E.; Jones, J.R. An examination of land cover and stream water quality among physiographic provinces of Missouri, U.S.A. *Verh. Int. Verein. Limnol.* **1998**, *26*, 940–947.

14. Tran, C.P.; Bode, R.W.; Smith, A.J.; Kleppel, G.S. Land-use proximity as a basis for assessing stream water quality in New York State (USA). *Ecol. Indic.* **2010**, *10*, 727–733.

15. Loehr, R.C. Characteristics and comparative magnitude of non-point sources. *J. Water Pollut. Control Fed.* **1974**, *46*, 1849–1872.

16. Carey, R.O.; Migliaccio, K.W. Contribution of wastewater treatment plant effluents to nutrient dynamics in aquatic systems: A review. *Environ. Manag.* **2009**, *44*, 205–217.

17. Haggard, B.E.; Stanley, E.H.; Storm, D.E. Nutrient retention in a point-source-enriched stream. *J. N. Am. Benthol. Soc.* **2005**, *24*, 29–47.

18. Ekka, S.A.; Haggard, B.E.; Matlock, M.D.; Chaubey, I. Dissolved phosphorus concentrations and sediment interactions in effluent-dominated Ozark streams. *Ecol. Eng.* **2006**, *26*, 375–391.

19. Tchobanoglous, G.; Burton, F.L.; Stensil, H.D. *Wastewater Engineering: Treatment and Reuse,* 4th ed.; McGraw-Hill: New York, NY, USA, 2003.

20. Ra, J.S.; Kim, S.D.; Chang, N.I.; An, K.-G. Ecological health assessments based on whole effluent toxicity tests and the index of biological integrity in temperate streams influenced by wastewater treatment plant effluents. *Environ. Toxicol. Chem.* **2007**, *26*, 2010–2018.

21. Andersen, C.B.; Lewis, G.P.; Sargent, K.A. Influence of wastewater-treatment effluent on concentrations and fluxes of solutes in the Bush River, South Carolina, during extreme drought conditions. *Environ. Geosci.* **2004**, *11*, 28–41.

22. Downing, J.A.; McCauley, E. The nitrogen: Phosphorus relationship in lakes. *Limnol. Oceanogr.* **1992**, *37*, 936–945.

23. Redfield, A.C. The biological control of chemical factors in the environment. *Am. Sci.* **1958**, *46*, 205–221.

24. Sakamoto, M. Primary production by phytoplankton community in some Japanese lakes and its dependence on lake depth. *Arch. Hydrobiol.* **1966**, *62*, 1–28.

25. Forsberg, C.; Ryding, S.O. Eutrophication parameters and trophic state indices in 30 Swedish waste receiving lakes. *Arch. Hydrobiol.* **1980**, *89*, 189–207.

26. Huber, W.C.; Brezonik, P.L.; Heaney, J.P.; Dickinson, R.E.; Preston, S.D.; Dwornik, D.S.; DeMaio, M.A. *A Classification of Florida Lakes,* 1st ed.; Water Resources Research Center, University of Florida: Tallahassee, FL, USA, 1982; pp. 1–547.

27. Canfield, D.E. Prediction of chlorophyll a concentrations in Florida lakes: The importance of phosphorus and nitrogen. *J. Am. Water Resour. Assoc.* **1983**, *19*, 255–262.

28. Biggs, B.J. Eutrophication of streams and rivers: Dissolved nutrient-chlorophyll relationships for benthic algae. *J. N. Am. Benthol. Soc.* **2000**, *19*, 17–31.

29. Volk, C.; Kiffney, P. Comparison of fatty acids and elemental nutrients in periphyton, invertebrates, and cutthroat trout (*oncorhynchus clarki*) in conifer and alder streams of western washington state. *Aquat. Ecol.* **2012**, *46*, 85–99.

30. Deans, C.A.; Behmer, S.T.; Kay, A.; Voelz, N. The importance of dissolved N:P ratios on mayfly (*baetis* spp.) growth in high-nutrient detritus-based streams. *Hydrobiologia* **2014**, *742*, 15–26.

31. Smith, V.H. Predictive models for the biomass of blue-green algae in lakes. *J. Am. Water Resour. Assoc.* **1985**, *21*, 433–439.

32. Klausmeier, C.A.; Litchman, E.; Daufresne, T.; Levin, S.A. Optimal nitrogen-phosphorus stoichiometry of phytoplankton. *Nature* **2004**, *429*, 171–174.

33. Lindeman, R.L. The trophic-dynamic aspect of ecology. *Ecology* **1942**, *23*, 399–417.

34. Jeppesen, E.; Peder Jensen, J.; SØndergaard, M.; Lauridsen, T.; Landkildehus, F. Trophic structure, species richness and biodiversity in Danish lakes: Changes along a phosphorus gradient. *Freshw. Biol.* **2000**, *45*, 201–218.

35. Bachmann, R.W.; Jones, B.L.; Fox, D.D.; Hoyer, M.; Bull, L.A.; Canfield, D.E., Jr. Relations between trophic state indicators and fish in Florida (USA) lakes. *Can. J. Fish. Aquat. Sci.* **1996**, *53*, 842–855.

36. Rohlich, G.A. Fish as indices of eutrophication. In *Eutrophication: Causes, Consequences, Correctives, Proceedings of a Symposium*, 1st ed.; Larkin, P.A., Northcote, T.G., Eds.; National Academy of Sciences: Washington, DC, USA, 1969; Volume 1, pp. 256–273.

37. Oberdorff, T.; Pont, D.; Hugueny, B.; Porcher, J.-P. Development and validation of a fish-based index for the assessment of "river health" in France. *Freshw. Biol.* **2002**, *47*, 1720–1734.

38. Choi, J.-W.; Kumar, H.K.; Han, J.-H.; An, K.-G. The development of a regional multimetric fish model based on biological integrity in lotic ecosystems and some factors influencing the stream health. *Water Air Soil Pollut.* **2011**, *217*, 3–24.

39. Frey, J.W.; Bell, A.H.; Hambrook Berkman, J.A.; Lorenz, D.L. *Assessment of Nutrient Enrichment by Use of Algal-, Invertebrate-, and Fish Community Attributes in Wadeable Streams in Ecoregions Surrounding the Great Lakes*; U.S. Geological Survey Scientific Investigations Report 2011–5009. U.S. Geological Survey: Reston, WV, USA, 2011; pp. 1–49.

40. Noble, R.A.A.; Cowx, I.G.; Goffaux, D.; Kestemont, P. Assessing the health of European rivers using functional ecological guilds of fish communities: Standardising species classification and approaches to metric selection. *Fish. Manag. Ecol.* **2007**, *14*, 381–392.

41. Eaton, A.D., Franson, M.A.H., Eds.; *Standard Methods for the Examination of Water and Wastewater*; American Public Health Association: Washington, DC, USA, 2005.

42. Marker, A.F.H.; Crowther, C.A.; Gunn, R.J.M. Methanol and acetone as solvents for estimating chlorophyll a and phaeopigments by spectrophotometry. *Arch. Hydrobiol. Beih. Ergebn. Limnol.* **1980**, *14*, 52–69.

43. The Ministry of Environment (MOE)/National Institute of Environmental Research (NIER). *Researches for Integrative Assessment Methodology of Aquatic Environments (III): Development of Aquatic Ecosystem Health Assessment and Evaluation System*, 1st ed.; MOE/NIER: Incheon, Korea, 2006.

44. The Ministry of Environment (MOE)/National Institute of Environmental Research (NIER). *The Survey and Evaluation of Aquatic Ecosystem Health in Korea*, 1st ed.; MOE/NIER: Incheon, Korea, 2008.

45. Lee, J.H.; Han, J.H.; Kumar, H.K.; Choi, J.K.; Byeon, H.K.; Choi, J.; Kim, J.K.; Jang, M.H.; Park, H.K.; An, K.-G. National-level integrative ecological health assessments based on the index of biological integrity, water quality, and qualitative habitat evaluation index, in Korea rivers. *Ann. Limnol. Int. J. Lim.* **2011**, *47*, S73–S89.

46. Plafkin, J.L.; Barbour, M.T.; Porter, K.D.; Gross, S.K.; Hughes, R.M. *Rapid Bioassessment Protocols for Use in Streams and Rivers: Benthic Macroinvertebrate and Fish*; EPA/444/4-89-001. Office of Water Regulations and Standards; US EPA: Washington, DC, USA, 1989; pp. 1–34.

47. Barbour, M.T.; Gerritsen, J.; Snyder, B.D.; Stribling, J.B. *Rapid Bioassessment Protocols for Use in Streams and Wadeable Rivers: Periphyton, Benthic Macroinvertebrates and Fish*, 2nd ed.; EPA 841-B-99-002. U.S. Environmental Protection Agency, Office of Water: Washington, DC, USA, 1999; pp. 1–35.

48. An, K.-G.; Park, S.S.; Shin, J.-Y. An evaluation of a river health using the index of biological integrity along with relations to chemical and habitat conditions. *Environ. Int.* **2002**, *28*, 411–420.

49. Ohio, E.P.A. *Biological Criteria for the Protection of Aquatic Life: Volume III. Standardized Biological Field Sampling and Laboratory Method for Assessing Fish and Macroinvertebrate Communities*, 2nd ed.; Ohio, E.P.A., Ed.; Columbus, OH, USA, 2015; pp. 1–64.

50. U.S. EPA. *Fish Field and Laboratory Methods for Evaluating the Biological Integrity of Surface Waters*; EPA 600-R-92-111. Environmental Monitoring Systems Laboratory-Cincinnati office of Modeling, Monitoring Systems, and Quality Assurance Office of Research Development, U.S. EPA: Cincinnati, OH, USA, 1993; pp. 1–348.

51. Kim, I.S.; Choi, Y.; Lee, C.L.; Lee, Y.J.; Kim, B.J.; Kim, J.H. *Illustrated Book of Korean Fishes*, 1st ed.; Kyohak: Seoul, Korea, 2005.

52. Han, J.-H.; Park, C.-S.; An, J.-W.; An, K.-G.; Baek, W.-K. *A Guide Book of Freshwater Fishes*, 1st ed.; National Science Museum: Daejeon, Korea, 2015.

53. FishBase. World Wide Web Electronic Publication. Available online: http://www.fishbase.org (accessed on 18 May 2015).

54. Sanders, R.E.; Miltner, R.J.; Yoder, C.O.; Rankin, E.T. The use of external deformities, erosion, lesions, tumors (DELT anomalies) in fish assemblages for characterizing aquatic resources: A case study of seven Ohio streams. In *Assessing the Sustainability and Biological Integrity of Water Resources Using Fish Communities*, 1st ed.; Simon, T.P., Ed.; CRC: Boca Raton, FL, USA, 1999; pp. 225–245.

55. Karr, J.R. Assessment of biotic integrity using fish communities. *Fisheries* **1981**, *6*, 21–27.

56. Karr, J.R.; Fausch, K.D.; Angermeier, P.L.; Yant, P.R.; Schlosser, I.J. *Assessing Biological Integrity in Running Water: A Method and Its Rationale*, 1st ed.; Illinois National History Survey Special Publication 5: Champaign, IL, USA, 1986; pp. 1–28.

57. An, K.-G.; Park, S.S. Indirect influence of the summer monsoon on chlorophyll-total phosphorus models in reservoirs: A case study. *Ecol. Models* **2002**, *152*, 191–203.

58. An, K.-G. Long-term seasonal and interannual variability of epilimnetic nutrients (N, P), chlorophyll-a, and suspended solids at the Dam site of Yongdam Reservoir and empirical models. *Korean J. Limnol.* **2011**, *44*, 214–225.

59. Van Nieuwenhuyse, E.E.; Jones, J.R. Phosphorus-chlorophyll relationship in temperate streams and its variation with stream catchment area. *Can. J. Fish. Aquat. Sci.* **1996**, *53*, 99–105.

60. Miltner, R.J.; Rankin, E.T. Primary nutrients and the biotic integrity of rivers and streams. *Freshw. Biol.* **1998**, *40*, 145–158.

61. Lee, J.H.; An, K.-G. Integrative restoration assessment of an urban stream using multiple modeling approaches with physical, chemical, and biological integrity indicators. *Ecol. Eng.* **2014**, *62*, 153–167.

62. Robertson, D.M.; Graczyk, D.J.; Garrison, P.J.; Wang, L.; LaLiberte, G.; Bannerman, R. *Nutrient Concentrations and Their Relations to the Biotic Integrity of Wadeable Streams in Wisconsin*, 1st ed.; Professional Paper 1722; U.S. Geological Survey: Reston, VA, USA, 2006; pp. 1–139.

63. Yeom, D.-H.; Lee, S.-A.; Kang, G.S.; Seo, J.; Lee, S.-K. Stressor identification and health assessment of fish exposed to wastewater effluents in Miho Stream, South Korea. *Chemosphere* **2007**, *67*, 2282–2292.

64. Green, M.B.; Finlay, J.C. Patterns of hydrologic control over stream water total nitrogen to total phosphorus ratios. *Biogeochemistry* **2010**, *99*, 15–30.

65. Ellison, M.E.; Brett, M.T. Particulate phosphorus bioavailability as a function of stream flow and land cover. *Water Res.* **2006**, *40*, 1258–1268.

66. Higgins, K.A.; Vanni, M.J.; González, M. Detritivory and the stoichiometry of nutrient cycling by a dominant fish species in lakes of varying productivity. *Oikos* **2006**, *114*, 419–430.

Exploring the Non-Stationary Effects of Forests and Developed Land within Watersheds on Biological Indicators of Streams Using Geographically-Weighted Regression

Kyoung-Jin An, Sang-Woo Lee, Soon-Jin Hwang, Se-Rin Park and Sun-Ah Hwang

Abstract: This study examined the non-stationary relationship between the ecological condition of streams and the proportions of forest and developed land in watersheds using geographically-weighted regression (GWR). Most previous studies have adopted the ordinary least squares (OLS) method, which assumes stationarity of the relationship between land use and biological indicators. However, these conventional OLS models cannot provide any insight into local variations in the land use effects within watersheds. Here, we compared the performance of the OLS and GWR statistical models applied to benthic diatom, macroinvertebrate, and fish communities in sub-watershed management areas. We extracted land use datasets from the Ministry of Environment LULC map and data on biological indicators in Nakdong river systems from the National Aquatic Ecological Monitoring Program in Korea. We found that the GWR model had superior performance compared with the OLS model, as assessed based on R^2, Akaike's Information Criterion, and Moran's I values. Furthermore, GWR models revealed specific localized effects of land use on biological indicators, which we investigated further. The results of this study can be used to inform more effective policies on watershed management and to enhance ecological integrity by prioritizing sub-watershed management areas

Reprinted from *Water*. Cite as: An, K.-J.; Lee, S.-W.; Hwang, S.-J.; Park, S.-R.; Hwang, S.-A. Exploring the Non-Stationary Effects of Forests and Developed Land within Watersheds on Biological Indicators of Streams Using Geographically-Weighted Regression. *Water* **2016**, *8*, 120.

1. Introduction

Land use in watersheds has both direct and indirect impacts on the water quality [1–4] and biological community integrity [5–12] of adjacent streams. The ways in which land is used in watersheds determine the type and quantity of pollutants loaded into streams, and can lead to degradation of water quality and ecological integrity. Previous studies have demonstrated that high levels of urbanization within watersheds are strongly linked to poor water quality and biological conditions. In contrast, a high level of forest coverage within watersheds is closely related to lower concentrations of pollutants and more favorable ecological conditions in streams. Therefore, managing the water quality and ecological integrity of streams requires intelligent watershed management that takes into account the type and extent of land use. Such policies must be rooted in a deep understanding of the relationship between land use and stream conditions.

Most previous studies have adopted conventional statistical tools, including correlation or ordinary least squares (OLS) regression analysis, in assessing the effects of land use on ecological integrity as measured by certain biological indicators. One important assumption of these conventional statistical methods is that the effects of land use on ecosystems are constant (*i.e.*, stationary) within the entire study area; thus, local variations in these effects are ignored. However, in practice, the effects of land use on ecological variables may be spatially heterogeneous; observations from one region might not hold true for other regions, owing to differences in watershed characteristics and pollution sources [13]. Indeed, the relationship between land use and biological variables can vary across space, based on factors such as hydrological systems, watershed characteristics, land use patterns, riparian characteristics, stream types, and precipitation. However, conventional statistical approaches (e.g., Spearman's rank correlations or OLS regression) are unable to capture this spatial variation, because they analyze average values for an entire area. It is, therefore, difficult to develop area-specific watershed management practices for policy-makers, government agencies, and land managers, owing to the discrepancy between area-specific requirements and the conclusions drawn from data averaged across entire regions.

A few statistical techniques have been proposed to address local variations in land use effects, including the expansion method [14,15], spatial adaptive filtering [16,17], and multi-level modeling [18]. One of the most simple and powerful tools for dealing with the spatial heterogeneity of effects is geographically-weighted regression (GWR) [19,20]. GWR estimates parameters for all sample points in a dataset while taking into account non-stationary relationships. GWR can directly and effectively explore the non-stationarity of a regression for spatial data by locally calibrating a spatially-varying coefficient regression model. Due to its simplicity and efficiency, GWR has been successfully applied in fields such as forestry [21], economics [22,23], remote sensing [24], urban studies [25], and water quality assessment [4,13].

In this study, we compared the performances of the OLS and GWR models in explaining variation in biological indicators by the type of land use in the watershed, water quality indicators, and the topographic variables of sampling sites. Using the results of GWR models, we also investigated the contrasting effects of forests and developed areas in watersheds on biological indicators in streams. Forests and land development have long been recognized as competing land uses in Korea. Typically, land development in Korea involves removing forests and developing the land for purposes such as residential housing, roads and industrial land. Therefore, the watershed management practices of local governments must regulate the manner in which forestland is developed to minimize adverse effects and ensure the water quality and ecological integrity of streams. Understanding the area-specific effects of forests and developed land on the ecological integrity of streams may be critical for local governments to effectively regulate land transformation and manage watersheds. One recent study found significant local variation in the effects of land use on water quality in the Boston area [13]. In this study, relationships were not consistent among different water quality parameters and land use indicators, but rather depended on the level of urbanization within watersheds. Therefore, we hypothesize that a similar non-stationarity may exist in the relationship between land use (*i.e.*, forested areas and developed areas) and biological indicators in streams.

In summary, the aims of this study were:

(a) To test the non-stationarity of land use effects on biological indicators through a comparison of the OLS (global model) and GWR (local model) regression models, for the three biological indicators of the trophic diatom index (TDI) (benthic diatom), Korean saprobic index (KSI) (macroinvertebrate), and index of biotic integrity (IBI) (fish) using the three criteria of R^2, the AICc value [20], and Moran's I [26] of the OLS model and GWR model.

(b) To investigate the spatial distribution of land use effects on biological indicators, including the TDI, KSI, and IBI.

2. Materials and Methods

2.1. Study Streams and Sampling Sites

Since 2007, the Ministry of Environment in Korea has monitored various stream characteristics across the entire nation twice a year (in spring and fall) under the National Aquatic Ecological Monitoring Program (NAEMP). The variables measured include aquatic organisms in streams (benthic diatoms, macroinvertebrates, and fish), habitat quality, and various physicochemical parameters. The data generated by this program have been stored in a database in both spatial and non-spatial formats. For monitoring purposes, the NAEMP has identified and hierarchically-structured watersheds across the entire country, including the national watershed management regions (NWMRs), base watershed management regions (BWMRs), and sub-watershed management areas (SWMAs). The total number and average area of the NWMRs are 21 and 5191.74 km^2, respectively; the respective values for BWMRs are 117 and 931.85 km^2, and those for SWMAs are 840 and

129.79 km^2 [27]. Typically, NWMRs and BWMAs fall under the purview of the national government, whereas local governments are concerned with SWMAs [11].

For the current study, we focused on a Nakdong national watershed management region (NWMR) containing 22 BWMRs and 191 SWMAs (Figure 1). We decided to use SWMAs as the study unit, because they are the basic unit of land use management for local governments and the MOE. Furthermore, we expected that the relationships between land use and biological indicators would be clearer at the SWMA level than at the NWMR or BWMR level.

The NAEMP has set up monitoring networks consisting of 1200 target sampling sites for the entire nation and 347 sites for the Nakdong river systems, including reference sites. Although not always the case, the target number of sampling sites under the monitoring program often corresponds approximately to the number of SWMAs. Owing to a limited budget, only 149 sites among the 347 target sites in the Nakdong river systems were monitored in 2011. As described earlier, we selected only 191 SWMAs belonging to the Nakdong river systems for our analysis. Some sampling sites in the study areas were not monitored in 2011, resulting in 148 sampling sites for the analysis (Figure 1).

The Nakdong River is the largest river within the Nakdong NWMR and the longest river system in Korea (525 km), covering most of the southeastern part of the Korean Peninsula. The majority of the streams in this watershed range from second to eighth order. According to the Korean Meteorological Administration, there has been no great change in precipitation over the last 30 years. The average annual precipitation for the last 30 years in the study areas was 1064 mm. However, the annual temperature has been gradually increasing from 12.9 °C in 1981 to 14.4 °C in 2010. The long-term annual average temperature for 1981–2010 was 14.1 °C, with the lowest average monthly temperature in January (−12.8 °C) and the highest in August (29.32 °C). Over the same period, average moisture and average wind speed were 61.6 (%), and 2.7 (m/s), respectively. The annual mean evapotranspiration was 1305.7 (mm). It is also noteworthy that approximately two-thirds of the total annual precipitation occurred during summer (June-September). Significant fluctuations in seasonal precipitation and water-flow levels are common in this area, and droughts often occur during winter and spring.

2.2. Biological Indicators and Water Quality

Using chemical parameters alone in stream management has been criticized as not fully capturing ecosystem dynamics [28,29]. For example, these parameters do not capture any information about the biological communities in streams [27]. Aquatic biota reflects the long-term cumulative effects of various anthropogenic disturbances [30] and are, therefore, crucial indicators of the ecosystem health of streams [28,31,32]. Various individual and aggregated indices for algae, macrophytes, macroinvertebrates, and fish have been proposed for the purposes of evaluating the ecological condition of streams and rivers [33–38]. Accordingly, the NAEMP has adopted and modified biological indicators for benthic diatoms, macroinvertebrates, and fish, as well as metrics for habitat quality based on indicators developed in other geographical areas (mainly North America and Europe). After reviewing a broad range of indices, the NAEMP adopted the TDI of Kelly and Whitton [39] for diatom communities, because it was developed based on the sensitivity and occurrence of a limiting nutrient (PO_4-P) in freshwater systems. The sensitivity of the original TDI was evaluated using diatoms present in Korean rivers and streams, and the values for major taxa were amended accordingly. To evaluate macroinvertebrate communities, the NAEMP also adopted the KSI, which was constructed based on the method of Zelinka and Marvan [40]. Later, the KSI was modified and improved following the German standard method [41]. Since the introduction of the IBI by Karr [42], an IBI-type model using fish assemblages has been adopted by many countries. The 12 metrics originally proposed by Karr [42] were reduced to eight metrics after analysis of their properties according to the ecological characteristics of Korean fish assemblages in the NAEMP (for more detailed information on the biological indicators used by the NAEMP, see Lee *et al.* [27]). In the present study, we analyzed monitoring results from the NAEMP using TDI, KSI, and IBI in 2011, based on the mean values of spring and fall, ranging from 0 (very poor condition) to 100 (excellent condition) (Table 1). We adopted water quality parameters (biological oxygen demand (BOD), total nitrogen (T-N), total phosphorous (T-P), and chlorophyll-a (Chl-a)) as independent variables for biological indicators. We also measured mean elevation and slopes within a 1 km buffer of sampling sites.

Table 1. Aquatic organisms and their biological indicators used in the NAEMP, and their underlying indicators.

Aquatic Organism	Parameters	Descriptions	References
Benthic diatoms	TDI (Trophic Diatom Index)	Diatom sensitivity and occurrence in relation to a limiting nutrient (PO_4-P) in freshwater systems	Kelly and Whitton [39]
Macroinvertebrates	KSI (Korean Saprobic Index)	Index with representative taxonomic groups and their occurrences of macroinvertebrates	Zelinka and Marvan [40]
Fish	IBI (Index of biotic Integrity)	Index using fish assemblages to assess the effect of human disturbance on streams and watersheds with 8 metrics	Karr [42]

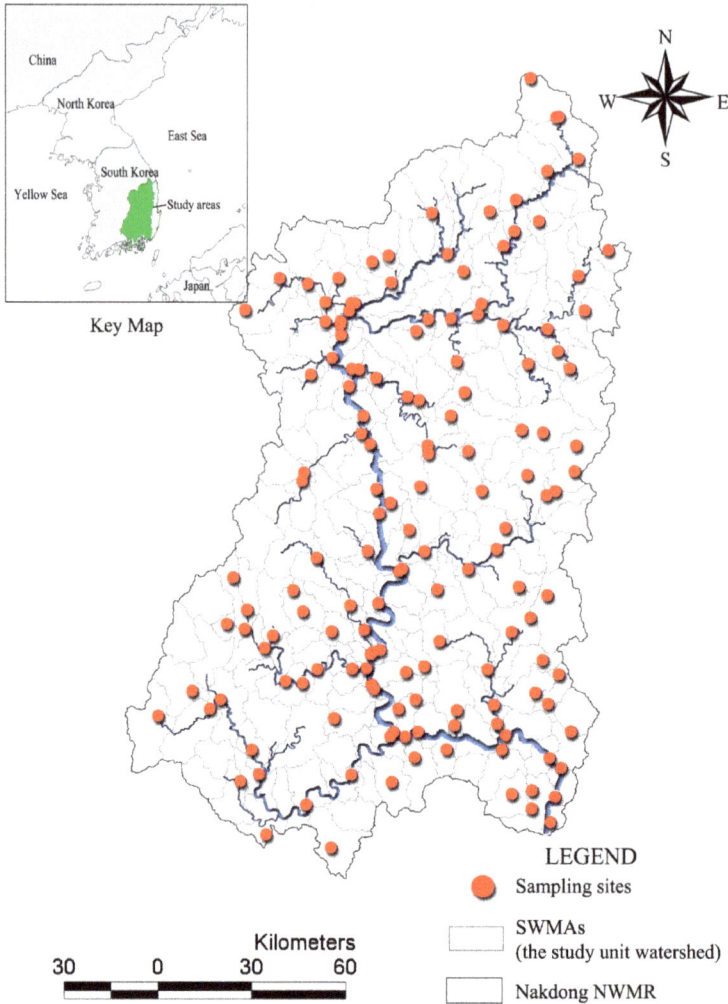

Figure 1. The study area, the Nakdong national watershed management region. The area consists of 191 sub-watershed management areas (SWMAs) and sampling sites in the National Aquatic Ecological Monitoring Program in Korea. The number and locations of sampling sites roughly correspond to the SWMAs, which are the base watershed management unit of the local government.

2.3. Land Uses and Topographic Variables

To calculate the proportions of land use in watersheds at the SWMA level, we integrated the Land Use/Land Cover map released by the Ministry of Environment. The Land Use/Land Cover map of the Ministry of Environment was generated using the Landsat Thematic Mapper (30 m resolution) and Indian Remote Sensing-1C panchromatic (5.8 m resolution) images in 2007 and updated in 2011. The original Land Use/Land Cover map of the Ministry of Environment classified land use and land cover into 23 subcategories. For the current study, we reclassified the 23 subcategories into six categories: (a) developed areas including industrial, residential, roads, commercial, and bare soils; (b) agricultural areas; (c) paddy areas; (d) forested areas, including deciduous forest, coniferous forest, and mixed forest; (e) grassy areas including grassland and golf courses; and (f) wetlands. The area of each reclassified Land Use/Land Cover map category within each SWMA was computed in ArcMap and converted into proportional data for analysis. However, we included only two land use types for the study: forest and developed land. To compute topographic variables, including elevation and slope, we created a 1 km buffer from sampling sites, and computed the mean slope and elevation.

2.4. Geographically Weighted Regression (GWR) Model

OLS regression is the most common statistical tool used to explore the effects of independent variables (e.g., proportions of land use types) on a dependent variable (e.g., biological indicators). It is a global estimation technique that assumes spatial stationarity of the regression relationship and generates a single regression equation that best fits the variables for the entire study area. However, as discussed earlier, this global estimation technique does not capture local variations in the relationship between the proportions of different land uses and biological indicators. A typical OLS model for the relationship between land use and biological indicators might be:

$$y = \beta_0 + \sum_{i=1}^{n} \beta_i x_i + \varepsilon \tag{1}$$

where y is the dependent variable (i.e., variance of biological indicators), β_0 is the intercept, β_1 is the coefficient of variable x_i (i.e., proportion of land use type i), n is the number of independent variables (i.e., number of land use types), and ε is the error term. To estimate the local variation in proportions of land use with the above model, the locations of sampling sites need to be integrated into the equation as well:

$$y_j = \beta_0 \left(u_j + v_j \right) + \sum_{i=1}^{n} \beta_i \left(u_j + v_j \right) x_i + \varepsilon \tag{2}$$

where u_j and v_j are the coordinates for each sampling location j, $\beta_0(u_j+v_j)$ is the intercept for sampling location j, and $\beta_1(u_j+v_j)$ is the local coefficient of proportion land use type i at location j [19,43–48]. From this perspective, the OLS model is a special case of the GWR model in which the parameter surface is assumed to be constant over space [20,44].

Equation (2) is a base GWR model for the biological indicators in this study. Rather than a single equation, it comprises an array of equations corresponding to different sampling sites. In GWR, an observation is weighted according to its proximity to sampling location j, with the result that the weighting of an observation is no longer constant in the calibration but varies with j. GWR is calibrated by weighting all observations around a sampling site using a distance decay function, which assumes that the observed values closer to the sampling location have higher impact on the local parameter estimates for the location [20,49]. The weighting function can be expressed in the exponential distance decay form as follows:

$$w_{jk} = exp\left(-d_{jk}^2 / h^2\right) \tag{3}$$

where w_{jk} is the weight of observation at sampling location k for the sampling location j, d_{jk} is the distance (meters, in this study) between sampling locations j and k, and h is referred to as the kernel bandwidth.

In GWR, there are two types of bandwidth options, fixed and adaptive. Fixed kernel bandwidth uses a constant bandwidth over study areas. In contrast, adaptive kernel bandwidth uses varying bandwidths based on data density: bandwidths are larger in locations where data are sparse and smaller where data are dense. In the current study, we used adaptive kernel bandwidth because of the inconsistent sampling densities over the study areas. The GWR model generated estimates the proportions for each land use type for each sampling site (*i.e.*, local coefficient), the values of *t*-tests on the local parameter estimates, the local R^2, and the local residuals. Non-stationary effects of land uses on biological indicators can, therefore, be visualized by mapping coefficients, t-statistics, and R^2 values over the study areas.

2.5. Model Comparisons

The relative performance of the OLS and GWR models can be assessed by comparing R^2 values, Akaike's Information Criterion (AICc) [20], and spatial autocorrelation of residuals (Moran's *I*). Greater R^2 values indicate that variance in watershed land use explains a larger proportion of the variance in biological indicators in streams. Lower AICc values indicate a closer approximation of the model to the actual nature of the relationships between land uses and biological indicators [50].

Similar to conventional correlation coefficients, Moran's *I*, a measure of spatial autocorrelation, ranges from −1 to 1. When the value of Moran's *I* for the residuals of estimated models is close to zero, it suggests that the residuals are spatially independent. When the value is close to −1 or 1, it indicates that residuals are strongly spatially dependent [51]. We used the embedded software ArcToolbox in ArcMap to estimate GWR-derived and OLS-derived values for the TDI, KSI, and IBI indicators of benthic diatoms, macroinvertebrates, and fish, respectively, in the study areas. ArcMap GIS was also used to compute Moran's I value and to visualize the non-stationarity of the effects of land uses in watersheds.

3. Results

3.1. Descriptive Statistics and Spatial Distributions

Ecological conditions measured by biological indicators, including TDI, KSI, and IBI, varied greatly among sampling sites (Table 2). Very low biological indicator values (near zero: very poor ecological condition) were observed at some sites, while others had very high values (near 100: very good ecological condition). The mean values of the TDI, KSI, and IBI within the study area were 43.27, 63.19, and 56.89, respectively. Overall, the condition of macroinvertebrate assemblies in the study areas was slightly better than the condition of benthic diatom or fish assemblies. The standard deviation of KSI values suggested greater variance in KSI values than in TDI or IBI values. Interestingly, all indicators reflected poor ecological conditions along main streams and in downstream areas in the southeastern part of the study region (Figure 2).

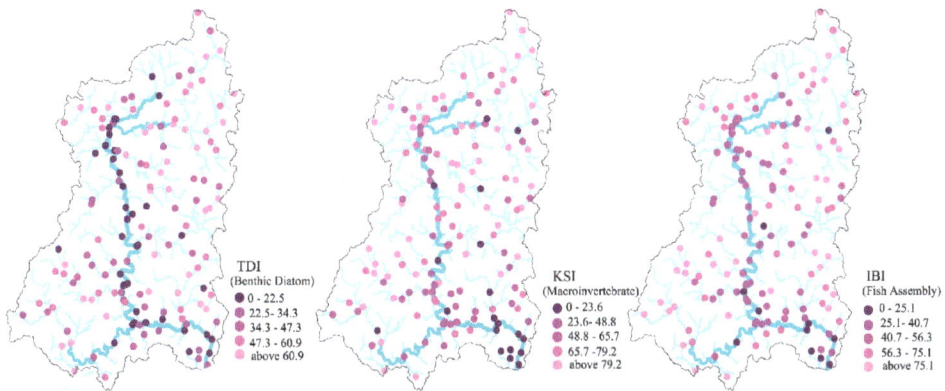

Figure 2. Distribution of biological indicator (Trophic Diatom Index (TDI), Korean Saprobic Index (KSI), and Index of Biotic Integrity (IBI)) values. Lower values (poor ecological conditions) were observed primarily along the main river, while high values (good ecological conditions) were observed in feeding streams. Notably, downstream (southeast) areas tended to have lower values for all indicators.

Table 2. Descriptive statistics of the variables used in the study. TDI, KSI, and IBI are biological indicators. Proportions of land use types were computed at the Sub-Watershed Management Area (SWMA) scale. Study areas varied greatly in the proportion of land use types, biological indicator values, and topographic variables.

	Variables	*min*	*max*	*mean*	*SD*
Biological indicators	TDI (Benthic diatoms)	0.00	76.10	43.41	20.90
	KSI (Macroinvertebrates)	5.80	97.30	62.73	24.14
	IBI (Fish assembly)	6.30	96.90	56.79	22.22
Water quality parameters	BOD (mg·L^{-1})	0.2	7.9	1.00	0.80
	T-N (mg·L^{-1})	0.74	4.95	2.24	0.87
	T-P (mg·L^{-1})	0.01	0.44	0.04	0.05
	Chl-a (mg/m^3)	2	8.2	1.05	0.88
Topographic variables	Elevation (m)	1.57	904.11	154.86	154.05
	Slope (%)	0.33	24.23	8.84	5.37
Proportions of land use types	Forests (%)	26.67	95.54	63.84	15.29
	Developed (%)	0.65	38.78	6.53	5.27

$n = 138$. Abbreviations: TDI, Trophic Diatom Index; KSI, Korean Saprobic Index; IBI, Index of Biotic Integrity; BOD, biological oxygen demand; T-N, total nitrogen; T-P, total phosphorous; Chl-a, chlorophyll-a.

The mean values of the BOD, T-N, and T-P were 1.0, 2.24, and 0.04, respectively, indicating relatively good water quality in the areas investigated. However, the large standard deviation of elevation and slope indicate complex topographic characteristics in the study areas.

The relative proportions of each type of land use in watersheds also varied greatly across the study areas. The dominant land use type in the study area was forest (mean: 63.64%). Developed areas were relatively small, and concentrated at several sites along the main river, particularly near the central and downstream regions (Figure 3). As shown in Figure 3, several cities of various sizes were located near/along the main stream. Daegu and Busan were the two largest cities, with populations of 2.5 million and 3.5 million, respectively.

Before we undertook a detailed analysis, we considered the changes of the biological indicators over the five years from 2008 through to 2012 to understand the nature of the dataset (Figure 4). In 2011, there were changes in the biological indicators. KSI and IBI had slightly higher values, while TDI had slightly lower values, than the other years. Despite this fluctuation in the biological indicators in 2011, we used 2011 monitoring data to match with the most up-to-date land use/land cover data released by the Korean Ministry of Environment. Thus, there is a possibility that models estimated using datasets for other years might be slightly different from the model estimated using the 2011 dataset.

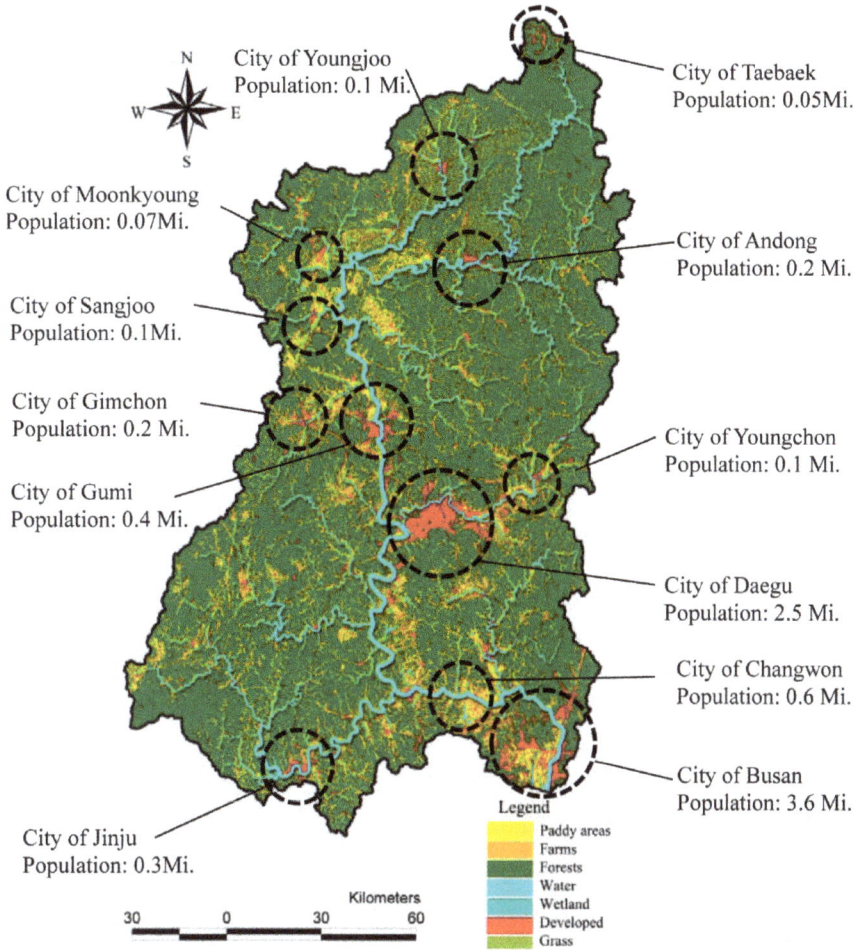

Figure 3. Land Use/Land Cover in the study areas and major cities. The dominant Land Use/Land Cover in the study areas was forest (green), and lumpy developed areas (red) were located in several places along the main stream, particularly in the middle and downstream areas.

3.2. Selecting the Best Predictors for Biological Indicators

Before estimating the OLS and GWR models for each biological indicator, we conducted a preliminary regression analysis using the water quality parameters (*i.e.*, BOD, T-N, T-P), topographic variables (*i.e.*, elevation and slope), and land use parameters (*i.e.*, developed areas, forested areas, agricultural areas, grass, wetland, and bare soils) to select the best predictive variables. To select the best-fit model for each biological indicator, we used the stepwise option in the SPSS for Windows software (SPSS Inc., Chicago, IL, USA) giving the R^2, F-statistics, and t values ($p < 0.05$) of each variable.

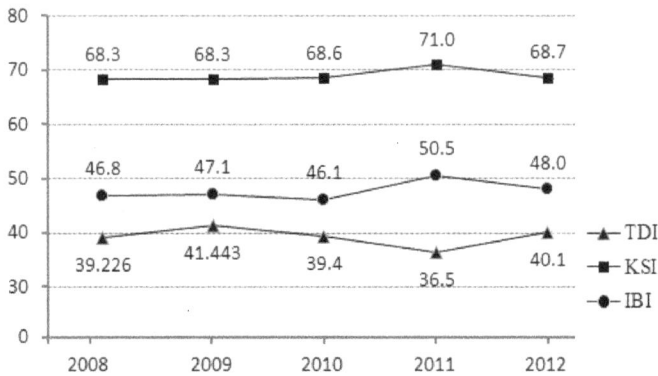

Figure 4. Changes in the averaged biological indicators over five years (2008–2012) in the study areas.

In Table 3, a model with %forests, T-N, T-P, %bare soil, %wetland, and elevation had the highest adjusted-R^2 (0.41) value for the TDI. For the KSI, a model with %developed areas and concentration of T-P had the highest R^2 value (0.32, $F = 22.50$, $p < 0.01$). The percentage of forests, concentration of T-P, BOD, and elevation were the most significant variables for explaining the variance of IBI in the study region (adjusted-R^2 = 0.42, $F = 26.01$, $p < 0.01$). From a land use perspective, the proportions of forests, developed areas, wetland, and grass in the watershed seemed to be significant variables for explaining the variances of the TDI, KSI, and IBI. In terms of water quality parameters, the concentration of T-P was the most significant determinant for all indicators, while the concentration of T-N and BOD was the most significant variable only for TDI and IBI, respectively.

Table 3. Preliminary regression estimations used to select effective variables for model estimations.

Biological Indicator	R^2	Effective Predictors	F-Values
TDI (Benthic diatoms)	0.41	%forests, T-N, T-P, %bare soil, %wetland, elevation	16.71 **
KSI (Macroinvertebrates)	0.32	%developed areas, T-P, %grass	22.50 **
IBI (Fish)	0.42	%forests, T-P, BOD, elevation	26.01 **

$n = 138$, * $p < 0.05$, ** $p < 0.01$. Abbreviations: TDI, Trophic Diatom Index; KSI, Korean Saprobic Index; IBI, Index of Biotic Integrity; BOD, biological oxygen demand; T-N, total nitrogen; T-P, total phosphorous; Chl-a, chlorophyll-a.

306

In most previous studies dealing with the relationships between land use and water quality, the proportion of forest in the watersheds had a strong positive relationship with water quality parameters, while the percentage of developed areas had a strong negative relationship with water quality [1–4] and biological indicators [5–12]. However, it was rare for both variables to be significant in a regression model due to a negative mutual relationship. In our preliminary analysis, there was a strong negative correlation between %forest and %developed areas ($r = 0.61$) in the study areas. Despite there being other variables affecting the biological indicators, we focused on only two contrasting land use types (*i.e.*, forest and developed areas) in our study. In particular, we investigated the spatial pattern of the coefficients of these two variables in GWR models for the TDI, KSI, and IBI when holding the other significant determinants constant.

3.3. Comparison between OLS and GWR Models

We compared the performance of the general OLS (global) and GWR (local) models for the TDI indicators (Table 4). In the global model, forest land had a positive effect (b = 0.25, β = 0.18) on TDI indicators, while the concentration of T-P and T-N had negative effects (b = −111.38, β = −0.32, b = −4.98, β = −0.21). Thus, a higher proportion of forest in watersheds may enhance the benthic diatom communities of streams. Conversely, a higher concentration of T-P and T-N had adverse effects on benthic diatom communities. The percentage of bare soil in the watershed appeared to have a negative impact on the TDI (b = −3.44, β = −0.24), while the percentage of wetland had a positive influence on the TDI (b = 5.27, β = 0.19). Elevation also had a positive effect on the TDI of streams (b = 0.02, β = 0.15). In the global model (OLS model), the intercept, land uses in watersheds (e.g., proportion of forest, bare soils, and wetland), water quality parameters (e.g., concentrations of T-N and T-P), and topographic characteristics (e.g., elevation) were significant at $p < 0.01$, and the global model was also significant overall ($F = 16.71$, $p < 0.01$).

The adjusted R^2 of the global model was 0.41, indicating that 41% of the variance in the TDI across streams in the study area could be explained by three land use variables, two water quality parameters, and one topographic variable, while the remaining 59% was not explainable with these six variables. The R^2 value of the GWR model was 0.44, which was slightly higher than the R^2 value of the OLS model, suggesting that the GWR model performed better than the OLS model in explaining the variance of the TDI in the study areas. Similarly, the AICc values of the global and local models were 1165.43 and 1159.81, respectively. The lower AICc values of the GWR model also suggested a closer approximation of the model to the actual nature of the relationships between the dependent variables and TDI indicators. The Moran's I value of the residuals in the local model was −0.10, which was slightly higher than in the global model (−0 09). However, the difference in the Moran's I values for the two models was very small (0.01).

Table 4. Results of the OLS (global) and GWR (local) models for the TDI. The higher R^2 values and lower AICc values of the GWR model indicates that it explains more of the variance in the TDI.

Variables	OLS Model (Global Model)			GWR model [1] (Local Model)
	b	β	t-value	
Intercept	41.55	-	4.22 **	-
% Forests	0.25	0.18	2.22 **	-
T-P	−111.88	−0.32	−4.34 **	-
T-N	−4.98	−0.21	−3.04 **	
%Bare soil	−3.44	−0.24	−2.95 **	
%Wet land	5.27	0.19	2.38 **	
Elevation	0.02	0.15	2.09 *	
F-value		16.71 **		
Adjusted R^2		0.41		0.44
AICc		1165.43		1159.81
Moran's I [2]		−0.09		−0.10

Abbreviations: T-P, total phosphorous; T-N, total nitrogen; AICc, Akaike's Information Criterion. [1] Coefficients of the intercept and other variables in the GWR model vary from observation to observation; [2] Spatial autocorrelation index of residuals. $n = 138$, * $p < 0.05$, ** $p < 0.01$.

As discussed previously, the relative performance of the OLS and GWR models can be assessed based on the R^2, AICc, and Moran's I values of model residuals. Comparisons of these criteria suggested that the local model (GWR) performed better in explaining the variance of the TDI in the study areas and the presence of non-stationarity in the relationships between dependent variables, including the proportion of forest and TDI over space. The presence of non-stationarity suggested that the influence of forest on the TDI might vary over the study areas.

For the KSI, the proportion of the variation explained by the OLS model was modest ($R^2 = 0.33$) (Table 5). About 32% of the KSI variance could be explained by the percentage of developed areas in the watershed (b = −2.87, β = −0.33), the concentration of T-P (b = −100.96, β = −0.25), and the percentage of grass areas (b = −4.77, β = −0.19), while the remaining 68% could not be explained by these variables. The high F- statistic (23.5, $p < 0.01$) for the OLS model suggests that this model was significant for the KSI. The results of this model further suggested an inverse relationship between the proportion of developed land in watersheds and the KSI values of streams. From a land use perspective, the proportion of developed land had the highest β value (−0.33) among the effective independent variables, including the concentration of T-P ($\beta = −0.25$) and proportion of grass areas ($\beta = −0.19$) in the OLS model.

308

Table 5. Results of the OLS (global) and GWR (local) models for the KSI. The similar values of R^2, AICc and Moran's I of the two models suggest a similar performance in explaining the variance of KSI in the study areas.

Variables	OLS Model (Global Model)			GWR Model [1] (Local Model)
	b	β	t-value	
Intercept	85.31	-	25.56 **	-
% Developed	−2.87	−0.33	−4.15 **	-
T-P	−100.96	−0.25	−3.36 **	-
%Grass	−4.77	−0.19	−2.47 **	
F-value			22.50 **	
Adjusted R^2			0.32	0.32
AICc			1215.70	1217.88
Moran's I [2]			−0.04	−0.06

Abbreviations: T-P, total phosphorous; AICc, Akaike's Information Criterion. [1] Coefficients of the intercept and other variables in GWR model vary from observation to observation; [2] Spatial autocorrelation index of residuals. $n = 138$, * $p < 0.05$, ** $p < 0.01$.

The GWR model had the same R^2 value (0.32) as the OLS model (0.32), suggesting that the GWR model explains almost the same amount of variance as the KSI. Furthermore, the similar AICc values of the OLS (1215.70) and GWR (1217.88) models revealed that both described the relationship between the independent variables and KSI to a similar degree of accuracy. The spatial autocorrelation indexes, measured by Moran's I values, of the OLS (−0.04) and GWR (−0.06) models were very similar, suggesting that there was no significant spatial dependency of the residuals in the two models. The comparison between the two models of the KSI also indicated that non-stationarity effects were not present in the relationships between the KSI and independent variables, including the proportion of developed areas, the concentration of T-P, and the percentage of grass areas in watersheds.

For the IBI, the results of the global (OLS) model indicated that IBI values increased significantly with the proportion of forest within watersheds (b = 0.47, β = 0.32, $p < 0.01$) and elevated land (b = 0.03, β = 0.18, $p < 0.01$). Conversely, the IBI values were inversely related to the concentration of T-P (b = −81.91, β = −0.23, $p < 0.01$) and BOD (b = −5.99, β = −0.22, $p < 0.01$). The adjusted R^2 of the global model was 0.42, indicating that ~42% of the variance in the IBI among streams can be explained by the four variables of forests, T-P, BOD, and elevation. The F-value (26.57) of the OLS model was significant ($p < 0.01$) (Table 6).

The R^2 value of the GWR model (0.49) was considerably higher than that of the OLS, and suggested that ~49% of the variance in the IBI among study sites could be explained by the proportions of forests, T-P, BOD, and elevation. The GWR model also had a lower AIC value (1165.13) than that of the OLS model

(1171.27). Both the higher R^2 value and the lower AIC value strongly indicate that the GWR model performed better in terms of explaining the IBI variance and approximating reality. Furthermore, the lower Moran's I value (−0.01) of the GWR model compared with the OLS model (0.07) indicates that the residuals in the former model exhibited less spatial dependency. The higher R^2, lower AICc and lower Moran's I of the GWR model strongly suggest the presence of non-stationarity between the independent variables and IBI in the study areas. The presence of non-stationarity in the relationships suggests that the influence of the proportion of forest, along with other independent variables, might vary stream by stream in the study areas.

Table 6. OLS (global) and GWR (local) model results for the IBI. The higher R^2 values and lower AICc values indicate that the GWR model explains more of the variance in the IBI. The considerably lower Moran's I value of the GWR model indicates a superior performance compared to the OLS model.

Variables	OLS Model (Global Model)			GWR Model [1] (Local Model)
	b	β	t-value	
Intercept	33.23	-	4.25 **	-
% Forests	0.47	0.32	4.19 **	-
T-P	−81.91	−0.23	−3.00 **	-
BOD	−5.99	−0.22	−2.93 **	-
Elevation	0.03	0.18	2.48 **	-
F-value		26.01 **		
Adjusted R^2		0.42		0.49
AICc		1171.27		1165.13
Moran's I [2]		0.07		−0.01

Abbreviations: T-P, total phosphorous; BOD, biological oxygen demand; AICc, Akaike's Information Criterion. [1] Coefficients of the intercept and other variables in GWR model vary from observation to observation; [2] Spatial autocorrelation index of residuals. $n = 138$, * $p < 0.05$, ** $p < 0.01$.

Overall, the results of the OLS models (Tables 4–6) indicated that the selected variables for each biological indicator could explain ~41%, 32%, and 42% of the variance in the TDI, KSI, and IBI, respectively. No considerable differences between the OLS and GWR models were observed for the KSI indicators. The estimated OLS and GWR models of the KSI had almost the same R^2, AICc, and Moran's I values suggesting that there might be no non-stationary effects of land use, water quality, and topographic variables for the KSI models. Compared to the OLS model, the considerably higher R^2 value and $β$-value of the GWR model for the TDI and IBI indicated that, according to this model, the TDI and IBI were more sensitive to the heterogeneity of forest coverage than the KSI. Therefore, a higher percentage of forest land in watersheds may substantially enhance the ecological conditions as measured by the RDI and IBI, and the relationships between forests and two indicators (*i.e.*, TDI and IBI) might vary over space (*i.e.*, non-stationary effects). Interestingly, the negative impacts of developed areas were found only in the OLS model for the KSI implying that the higher proportion of developed areas in watersheds can adversely affect the KSI to a greater extent than the other biological indicators.

Positive influences of forests were found in the TDI and IBI models, suggesting that a higher proportion of forests in watersheds may enhance the TDI and IBI in streams. Interestingly, land use (developed areas or forests) in watersheds appeared to be a more significant variable than water quality parameters or topographic variables in the KSI and IBI models. In the TDI model, a water quality parameter (*i.e.*, T-P) was more significant than land use or topographic variables.

In contrast to the OLS model, the GWR model assumes non-stationarity in the relationship between a dependent variable (*i.e.*, a biological indicator) and independent variable (*i.e.*, the proportions of forest or developed land in watersheds). In the comparison of the performance of the OLS and GWR models, the GWR models of the KSI were not able to better explain the variances of the KSI in the study areas. However, the GWR model of the TDI and IBI clearly performed better than the OLS model in terms of the R^2, AICc, and spatial autocorrelation index values (*i.e.*, Moran's I). The GWR model is based on non-stationary effects (Equation (2)), while the OLS model is based on stationary effects (Equation (1)). Therefore, the superior performance of the GWR models strongly suggests non-stationarity in the effects of land use (*i.e.*, forests) on biological indicators (*i.e.*, the TDI and IBI). The OLS model might be an effective tool for understanding regionally averaged effects of land use on ecological conditions, but this global model cannot capture local variations in such effects. For some watersheds and indicator types, the OLS model might overestimate or underestimate the effects of land use.

3.4. Description of Local Estimated Land Use Effects in GWR models

In the GWR models, descriptive statistics for local R^2 and land use coefficients for the TDI, KSI, and IBI vary greatly in each GWR model (Table 7). For example, the proportion of forest with other variables in the local (GWR) TDI model could explain about 38% (minimum) of the variance in the TDI among streams in some watersheds, while it could explain 48% (maximum) of the variance in the TDI among other streams. Similarly, the coefficients of the proportions of forest varied to a considerable degree among watersheds, ranging from 0.08 to 0.31. Despite this variation, TDI values always increased with the proportion of forest land. The mean R^2 and coefficient values for forests in the local TDI model were 0.44 and 0.20, respectively.

Table 7. Descriptive statistics for R^2 and coefficients in GWR models. The R^2 and local coefficients of the proportion of forests in the Trophic Diatom Index (TDI) and Index of Biotic Integrity (IBI) models, and developed land in the Korean Saprobic Index (KSI) model vary greatly over space.

Estimated GWR Models		*min*	*max*	*mean*	*SD*
TDI model	Local R^2	0.38	0.48	0.44	0.03
	Coefficient of %Forests	0.08	0.31	0.20	0.08
KSI model	Local R^2	0.15	0.41	0.32	0.09
	Coefficient of %Developed	−3.12	−2.31	−2.79	0.25
IBI model	Local R^2	0.27	0.54	0.49	0.06
	Coefficient of %Forests	−0.03	0.98	0.39	0.30

$n = 138$.

Similarly, we also found a large degree of variance in R^2 values and the coefficients of developed land in the local KSI model. In particular, the changes of R^2 values in the KSI models surprisingly varied watershed by watershed. In some watersheds, the proportion of developed areas and other variables explained a small proportion of the KSI variance (15%), while in other areas they explained up to 41% of the variance. The proportion of developed areas in watersheds had a negative relationship with the KSI, ranging from −3.12 to −2.31.

The R^2 of the GWR model for the IBI also varied considerably among the study areas, ranging from 0.27 to 0.54. This variance indicates that the proportion of forests, T-P, BOD, and elevation were not consistently able to explain IBI values over space. Although the effect varied significantly (−0.03 –0.98), higher proportions of forest in watersheds were associated with increased IBI values in streams. The mean value for the coefficient of the proportion of forest areas was 0.39, the minimum value was −0.03, and the maximum value was 0.98. It is interesting to note that, while in most watersheds the proportion of forests was associated with increased IBI values for streams, in some watersheds forests had a negligible (or even negative) relationship with IBI values (Table 7). The standard deviation of %developed ($SD = 0.25$) land in the KSI local model and %forest ($SD = 0.30$) in the IBI local model were relatively high.

Scatterplots between observed and predicted values of the TDI local model indicate that most sites fell within the 95% confidence range of the estimated GWR model. It seemed that the observed values in the low range of TDI values were underestimated in the GWR model (Figure 5a). The KSI-GWR model showed a clear relationship in the middle–high range of KSI values. The GWR model overestimated in the range of observed KSI values from 20 to 40, while it underestimated in the range from 0 to 20. Seven watersheds were outside the 95% confidence interval (Figure 5b). The IBI-GWR model produced an even more complex estimation pattern. In the high range, three watersheds were overestimated, and one watershed was underestimated. In the middle range, one was overestimated and one was underestimated. In the low range of IBI values, two watersheds were underestimated in the GWR model (Figure 5c).

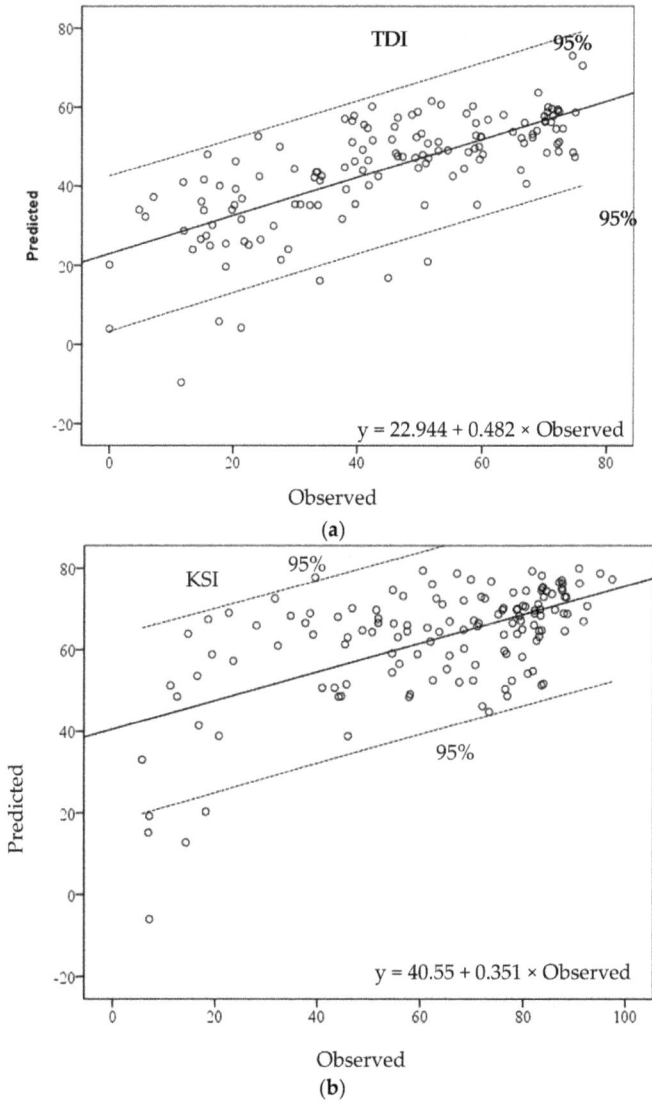

Figure 5. Cont.

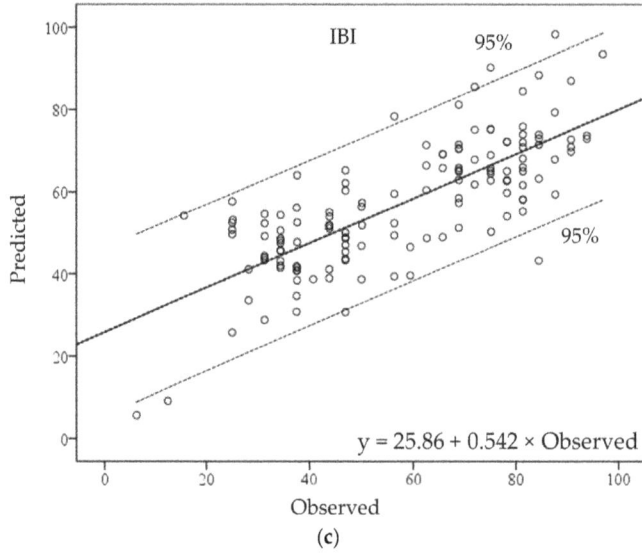

Figure 5. The relationships between observed and predicted values of (**a**) TDI, (**b**) KSI, and (**c**) IBI in estimated GWR models.

3.5. Spatial Distribution of the Estimated Parameters of GWR models

GIS mapping technology is an effective means of visualizing the variability of local R^2 and land use coefficients in local GWR models for the TDI, KSI, and IBI indicators (Figure 6). Within the study region, higher R^2 values (dark red dots) of the GWR model for the TDI indicator (Figure 6a) were observed mostly in the upstream areas, while lower R^2 values (cream color) were located mainly in the middle stream areas. In these regions, the forests and other variables explained a relatively small proportion of the TDI variability. Higher forest coefficient values (dark red) were concentrated in downstream areas, suggesting a relatively higher influence of forests on the TDI in these areas. In contrast, lower coefficients for forest in the estimated GWR model were observed mostly in middle-stream areas, while the upstream areas produced mid-range coefficients for forests (red dots). Despite the spatial variation, TDI values were always increased by forests (Figure 6a).

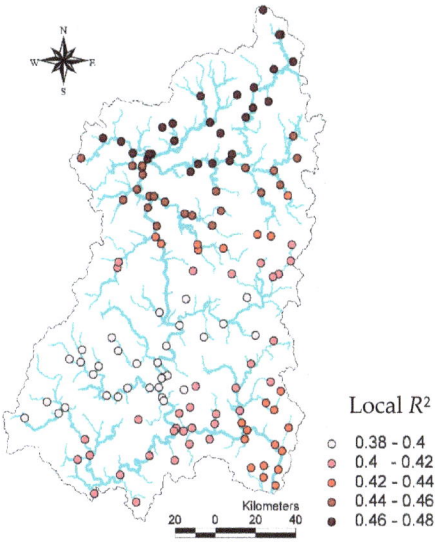
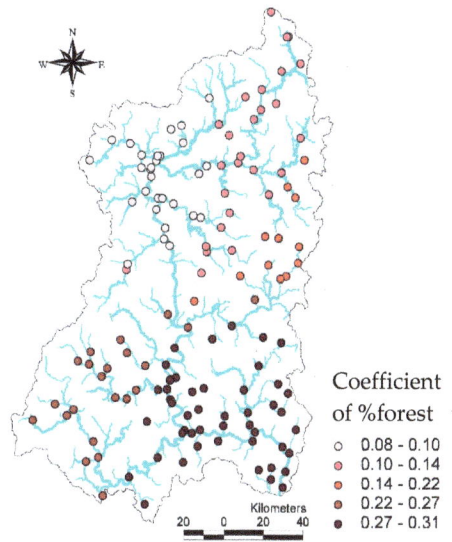

Local R^2

○	0.38 - 0.4
○	0.4 - 0.42
●	0.42 - 0.44
●	0.44 - 0.46
●	0.46 - 0.48

Coefficient
of %forest

○	0.08 - 0.10
●	0.10 - 0.14
●	0.14 - 0.22
●	0.22 - 0.27
●	0.27 - 0.31

(a)

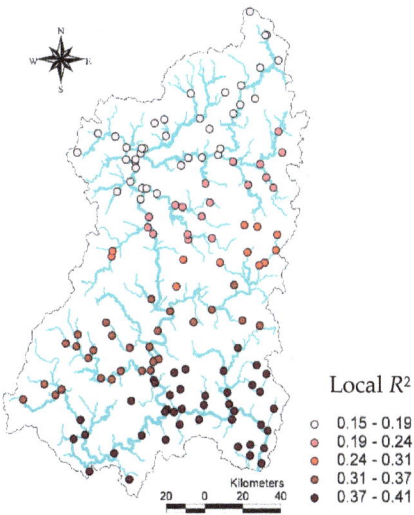
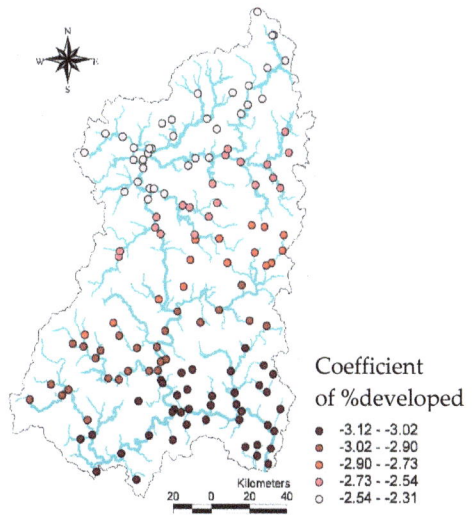

Local R^2

○	0.15 - 0.19
●	0.19 - 0.24
●	0.24 - 0.31
●	0.31 - 0.37
●	0.37 - 0.41

Coefficient
of %developed

●	-3.12 - -3.02
●	-3.02 - -2.90
●	-2.90 - -2.73
●	-2.73 - -2.54
○	-2.54 - -2.31

(b)

Figure 6. *Cont.*

315

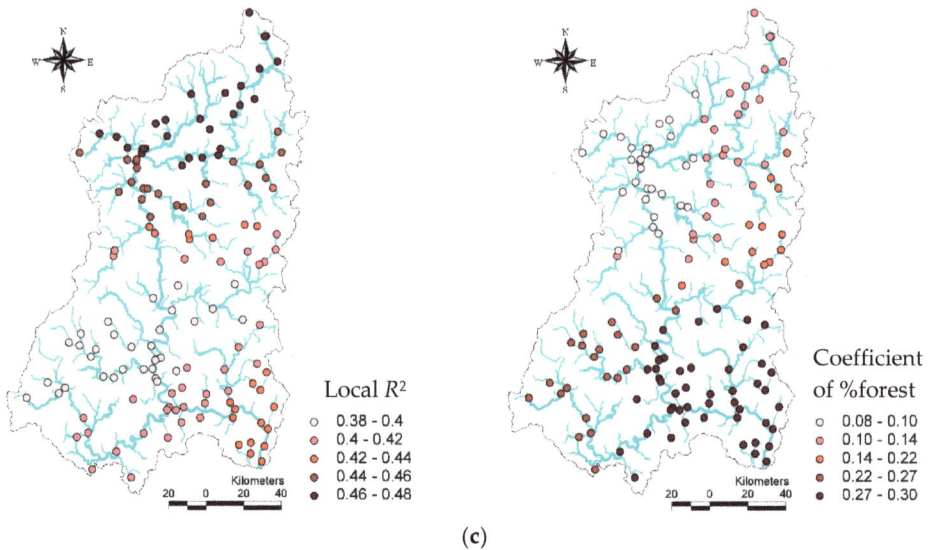

Figure 6. Spatial distribution of R^2 and land use coefficients in local models for the TDI (**a**), KSI (**b**), and IBI (**c**) indicators. Unlike the OLS model, the GWR model showed great spatial variance of the R^2 and land use coefficients for each biological indicator.

Higher R^2 values in estimated GWR models for TDI and IBI were observed in upstream areas, where forests were relatively well preserved and land development was less extensive. Conversely, the higher R^2 values in the estimated GWR model for the KSI were concentrated in downstream areas. Interestingly, both the proportion of forest and developed areas had relatively higher coefficient values in downstream areas in the estimated GWR models for the TDI, KSI, and IBI.

4. Discussion

Forests and developed land have contrasting effects on stream biological communities (e.g., diatoms, macroinvertebrates, and fish) in watersheds. Previous studies have shown that greater proportions of forest coverage are consistently associated with a more favorable ecological status for streams [5–7,52–57]. In contrast, greater proportions of developed land in watersheds are consistently associated with significantly poorer ecological conditions, as measured by various biological indicators [6,7,52,53,56,58–61]. Adverse effects of agricultural land use in watersheds on biological indicators are also well documented [7,11,62,63].

Our OLS models for the TDI, KSI, and IBI indicators confirmed previous research reporting negative effects of developed land (e.g., urbanized or impervious areas) and positive effects of forest on biological indicators in watersheds. Forests had positive impacts on the TDI and IBI, while developed areas had a negative influence on the KSI.

Most previous studies focusing on the relationship between land use in watersheds and the ecological conditions of streams have adopted conventional correlation or regression methods, which are unable to capture spatial variability among study sites. However, land development in watersheds alters not only the watershed itself, but also various environmental conditions in adjacent streams (e.g., water temperature, stream channel morphology, and the dynamics of sediments and water). Land development also affects hydrologic regimes and increases pollutant loads. These factors may have independent or interacting impacts on aquatic organisms [64]. Furthermore, different regions vary greatly in their topography (e.g., elevation, slope, soil type, *etc.*), watershed characteristics (e.g., size, composition, and distance from main cities), and stream environments (e.g., stream order, type and width; channel type; sediments; riparian areas). Therefore, it is unreasonable to assume that all landscapes, watersheds, and streams are affected in the same way and to the same degree by the presence of developed land or forests.

Our GWR models for the TDI and IBI strongly suggested a non-stationarity in the influence of forest in the watershed. Significant differences between the OLS and GWR models for the TDI and IBI in terms of R^2, AIC, and Moran's I values (Tables 4 and 6) suggested a better performance of the GWR models for explaining the variance of TDI and IBI in streams. This provides strong evidence that the effects of forests on the TDI and IBI indicators were non-stationary. Magner *et al.* (2008) [62] reported similar results, indicating that land use effects on biological parameters in streams could vary by location due to the localized geology. In their study on grazed riparian management and stream channel responses in Southeastern Minnesota streams, they found that land use metrics had no significant relationship with ordination axes, and they explained that the localized geology associated with producing stream bed cobbles has a stronger influence on the site than the land uses in the watershed.

However, as pointed out by Brunsdon *et al.* [43,44], it is difficult to measure the underlying mechanisms of spatial variability in the effects of land. Potential contributors to this spatial variability include differences in watershed characteristics, pollution sources, and degrees of urbanization [4,13]. In addition, variables such as topography, stream environments, and precipitation levels may also play a role in spatial variability. We tried to include available water quality variables (*i.e.*, BOD, T-N, and T-P) and topographic variables (elevation and slope) when estimating the OLS and GWR models.

However, the reason for using these models was not to investigate the underlying mechanisms of the variables used in estimating the models. The OLS model was used to determine the overall influence of land uses in water environments (*i.e.*, water quality parameters and topographic variables) on biological indicators, regardless of the location of sampling sites, while the GWR was used to investigate the spatial variation of land uses on biological indicators. To understand the mechanism of land use effects on biological indicators, the use of a structural equation modeling (SEM) [65] and path analysis [66] should be considered. Nevertheless, our GWR models clarify our understanding of the localized responses of stream organisms to land use. Given the complexity of this system, it is beyond the scope of the present study to measure all of the factors affecting the spatial variability of the effects of land use and the underlying mechanism of the pathways that influence them.

5. Conclusions

Many previous studies investigating the relationship between land use and the ecological response of stream biota have used correlation and regression analyses (OLS), which assume stationarity of the effects. However, in the present study, we found that the relationship between land use and stream ecology was spatially variable (*i.e.*, non-stationary). We used OLS models (global) and GWR models (local) to analyze the effects of contrasting land uses (forest and developed land) on the TDI, KSI, and IBI indicators, representing the ecological status of benthic diatom, macroinvertebrate, and fish assemblages, respectively. The performances of these two types of models were then compared based on R^2, AIC, and Moran's I values. Compared with the OLS models of the TDI and IBI, the GWR models of the TDI and IBI were better able to reveal details of the effects of particular land use types in specific locations; however, OLS models may be more useful in other applications. For example, OLS models may provide an effective means of assessing general trends within larger regions. Furthermore, OLS models may be more practical for creating environmental policies, while GWR may be more useful for effectively applying such policies to specific target streams or watersheds.

The effects of land use in watersheds can also vary among stream organisms. R^2 and land use coefficients varied considerably among GWR models for different biological indicators, suggesting that different biological assemblages respond to different land uses in different ways. Differences in the sensitivities of biological assemblages to land use may help explain the observed variations in R^2 and land use coefficients.

Although the present study demonstrates the utility of GWR models for understanding location-specific relationships between land use and stream communities, some critical questions remain to be answered. In particular, it will be important to investigate the underlying mechanisms of the impact of forest and developed land on stream biota, as well as for the spatial variability in these effects. We discuss several potential explanatory factors above; however, these will be challenging to study if they vary significantly among study sites. Further studies are, therefore, needed to elucidate more fully the effects of land use on stream biological communities, the reasons for their spatial variation, and the associated principles that can be derived, to manage the ecological integrity of streams more intelligently and effectively. It may also be possible to use regression tree analysis [67], SEM [65], path analysis [66], and non-metric multidimensional scaling [68,69] to explore the underlying mechanism of land use effects on biological indicators in streams, avoiding spatial autocorrelation issues, and to support decision-makers in watershed-specific land use management in minimizing the adverse impacts of land use on ecological communities of streams.

Acknowledgments: This study was conducted under the project "National Aquatic Ecosystem Health Survey and Assessment" in Korea, and was supported by the Ministry of the Environment and the National Institute of Environmental Research, Korea.

Author Contributions: All of the authors contributed extensively to the work. Kyoung-Jin An and Sang-Woo Lee designed the research, performed the data analysis, and wrote the manuscript. Soon-Jin Hwang, Se-Rin Park, and Sun-Ah Hwang contributed to the data interpretation, discussion, and editing of the manuscript.

Conflicts of Interest: The authors declare no conflict of interest.

References

1. Sliva, L.; Williams, D.D. Buffer zone *versus* whole catchment approaches to studying land use impact on river water quality. *Water Res.* **2001**, *35*, 3462–3472.
2. Tong, S.T.Y.; Chen, W. Modeling the relationship between land use and surface water quality. *J. Environ. Manag.* **2002**, *66*, 377–393.
3. Mehaffey, M.H.; Nash, M.S.; Wade, T.G.; Ebert, D.W.; Jones, K.B.; Rager, A. Linking land cover and water quality in New York City's water supply watersheds. *Environ. Monit. Assess.* **2005**, *107*, 29–44.
4. Tu, J.; Xia, Z.G. Examining spatially varying relationships between land use and water quality using geographically weighted regression I: Model design and evaluation. *Sci. Total Environ.* **2008**, *407*, 358–378.
5. Weaver, L.A.; Garman, G.C. Urbanization of a watershed and historical changes in a stream fish assemblage. *Trans. Am. Fish. Soc.* **1994**, *123*, 162–172.
6. Wang, L.; Lyons, J.; Kanehl, P.; Gatti, R. Influences of watershed land use on habitat quality and biotic integrity in Wisconsin streams. *Fisheries* **1997**, *22*, 6–12.
7. Moore, A.A.; Palmer, M.A. Invertebrate biodiversity in agricultural and urban headwater streams: Implications for conservation and management. *Ecol. Appl.* **2005**, *15*, 1169–1177.
8. Steffy, L.Y.; Kilham, S.S. Effects of urbanization and land use on fish communities in Valley Creek watershed, Chester County, Pennsylvania. *Urban Ecosyst.* **2006**, *9*, 119–133.

9.   Meador, M.R.; Coles, J.F.; Zappia, H. Fish assemblage responses to urban intensity gradients in contrasting metropolitan areas: Birmingham, Alabama, and Boston, Massachusetts. In *Effects of Urbanization on Stream Ecosystems*; Brown, L.R., Gray, R.H., Hughes, R.M., Meador, M.R., Eds.; American Fisheries Society: Bethesda, MD, USA, 2005; pp. 409–422.

10.   Lee, S.W.; Hwang, S.J.; Lee, S.B.; Hwang, H.S.; Sung, H.C. Landscape ecological approach to the relationships of land use patterns in watersheds to water quality characteristics. *Landsc. Urban Plan.* **2009**, *92*, 80–89.

11.   Park, S.R.; Lee, H.J.; Lee, S.W.; Hwang, S.J.; Byeon, M.S.; Joo, G.J.; Jeong, K.S.; Kong, D.S.; Kim, M.C. Relationships between land use and multi-dimensional characteristics of streams and rivers at two different scales. *Ann. Limnol. Int. J. Limnol.* **2011**, *47*, S107–S116.

12.   Utz, R.M.; Hilderbrand, R.H.; Raesly, R.L. Regional differences in patterns of fish species loss with changing land use. *Biol. Conserv.* **2010**, *143*, 688–699.

13.   Tu, J. Spatially varying relationships between land use and water quality across an urbanization gradient explored by geographically weighted regression. *Appl. Geogr.* **2011**, *31*, 376–392.

14.   Casetti, E. Generating models by the expansion method: Applications to geographical research. *Geogr. Anal.* **1972**, *4*, 81–91.

15.   Jones, J.; Casetti, E. *Applications of the Expansion Method*; Routledge: London, UK, 1992.

16.   Foster, S.A.; Gorr, W.L. An adaptive filter for estimating spatially-varying parameters: Application to modeling police hours spent in response to calls for service. *Manag. Sci.* **1986**, *32*, 878–889.

17.   Gorr, W.L.; Olligschlaeger, A.M. Weighted spatial adaptive filtering: Monte Carlo studies and application to illicit drug market modeling. *Geogr. Anal.* **1994**, *26*, 67–87.

18.   Goldstein, H. *Multilevel Models in Educational and Social Research*; Oxford University Press: London, UK, 1987.

19.   Brunsdon, C.; Fotheringham, A.S.; Charlton, M.E. Geographically Weighted Regression: A Method for Exploring Spatial Nonstationarity. *Geogr. Anal.* **1996**, *28*, 281–298.

20.   Fotheringham, A.S.; Brunsdon, C.; Charlton, M.E. *Geographically Weighted Regression: The Analysis of Spatially Varying Relationships*; Wiley: Chichester, UK, 2002.

21.   Propastin, P. Modifying geographically weighted regression for estimating aboveground biomass in tropical rainforests by multispectral remote sensing data. *Int. J. Appl. Earth Obs. Geoinf.* **2012**, *18*, 82–90.

22.   Lu, B.; Charlton, M.; Fortheringham, A.S. Geographically weighted regression using a non-euclidean distance metric with a study on London house price data. *Procedia Environ. Sci.* **2011**, *7*, 92–97.

23.   Saphores, J.D.; Li, W. Estimating the value of urban green areas: A hedonic pricing analysis of the single family housing market in Los Angeles, CA. *Landsc. Urban Plan.* **2012**, *104*, 373–387.

24.   Kamarianakis, Y.; Feidas, H.; Kokolatos, G.; Chrysoulakis, N.; Karatzias, V. Evaluating remotely sensed rainfall estimates using nonlinear mixed models and geographically weighted regression. *Environ. Model. Softw.* **2008**, *23*, 1438–1447.

25. Szymanowski, M.; Kryza, M. Application of geographically weighted regression for modelling the spatial structure of urban heat island in the city of Wroclaw (SW Poland). *Procedia Environ. Sci.* **2011**, *3*, 87–92.

26. Moran, P.A.P. Notes on Continuous Stochastic Phenomena. *Biometrika* **1950**, *37*, 17–23.

27. Lee, S.W.; Hwang, S.J.; Lee, J.K.; Jung, D.I.; Park, Y.J.; Kim, J.T. Overview and application of the National Aquatic Ecological Monitoring Program (NAEMP) in Korea. *Ann. Limnol. Int. J. Limnol.* **2011**, *47*, S3–S14.

28. Davis, W.S.; Simon, T.P. *Biological Assessment and Criteria: Tools for Water Resource Planning and Decision Making*; CRC Press: New York, NY, USA, 1995.

29. McCarron, E.; Frydenborg, R. The Florida bioassessment program: An agent of change. *Hum. Ecol. Risk Assess. Int. J.* **1997**, *3*, 967–977.

30. Karr, J.R.; Chu, E.W. *Restoring Life in Running Waters: Better Biological Monitoring*; Island Press: Washington, DC, USA, 1999.

31. US EPA. *Summary of Biological Assessment Programs and Biocriteria Development for States, Tribes, Territories, and Interstate Commissions: Streams and Wadeable Rivers (EPA-822-R-02-048)*; U.S. Environmental Protection Agency, Office of Water Regulations and Standards: Washington, DC, USA, 2002.

32. Simon, T.P. *Assessing the Sustainability and Biological Integrity of Water Resources Using Fish Communities*; CRC Press: New York, NY, USA, 1999; p. 652.

33. Osborne, P.E.; Suárez-Seoane, S. Should data be partitioned spatially before building large-scale distribution models? *Ecol. Model.* **2002**, *157*, 249–259.

34. Hering, D.; Johnson, R.K.; Kramm, S.; Schmutz, S.; Szoszkiewicz, K.; Verdonschot, P.F.M. Assessment of European streams with diatoms, macrophytes, macroinvertebrates and fish: A comparative metric-based analysis of organism response to stress. *Freshw. Biol.* **2006**, *51*, 1757–1785.

35. Pont, D.; Hugueny, B.; Beier, U.; Goffaux, D.; Melcher, A.; Noble, R.; Rogers, C.; Roset, N.; Schmutz, S. Assessing river biotic condition at a continental scale: A European approach using functional metrics and fish assemblages. *J. Appl. Ecol.* **2006**, *43*, 70–80.

36. Johnson, R.K.; Furse, M.T.; Hering, D.; Sandin, L. Ecological relationships between stream communities and spatial scale: Implications for designing catchment-level monitoring programmes. *Freshw. Biol.* **2007**, *52*, 939–958.

37. Ode, P.; Hawkins, C.; Mazor, R. Comparability of biological assessments derived from predictive models and multimetric indices of increasing geographic scope. *J. North Am. Benthol. Soc.* **2008**, *27*, 967–985.

38. Stoddard, J.L.; Herlihy, A.T.; Peck, D.V.; Hughes, R.M.; Whittier, T.R.; Tarquinio, E. A process for creating multimetric indices for large-scale aquatic surveys. *J. North Am. Benthol. Soc.* **2008**, *27*, 878–891.

39. Kelly, M.G.; Whitton, B.A. The trophic diatom index: A new index for monitoring eutrophication in rivers. *J. Appl. Phycol.* **1995**, *7*, 433–444.

40. Zelinka, M.; Marvan, P. Zur präzisierung der biologischen klassifikation der reinheit fließender gewässer. *Arch. Hydrobiol.* **1961**, *57*, 389–407.

41. *DIN, Biological-Ecological Analysis of Water (Group M): Determination of the Saprobic Index (M2). German Standard Methods for the Examination for Water, Wastewater and Sludge*; DIN 38410, Part 2; Beuth Verlag GmbH: Berlin, Germany, 1990; p. 10.

42. Karr, J.R. Assessment of biological integrity using fish communities. *Fisheries* **1981**, *6*, 21–27.

43. Brunsdon, C.; Fotheringham, S.; Charlton, M. Geographically Weighted Regression-modelling spatial non-stationarity. *J. R. Stat. Soc. Ser. D Stat.* **1998**, *47*, 431–443.

44. Brunsdon, C.; Fotheringham, A.S.; Charlton, M. Spatial nonstationarity and autoregressive models. *Environ. Plan. A* **1998**, *30*, 957–973.

45. Fotheringham, A.S.; Charlton, M.; Brunsdon, C. The geography of parameter space: An investigation of spatial non-stationarity. *Int. J. Geogr. Inf. Syst.* **1996**, *10*, 605–627.

46. Fotheringham, A.S. Trends in quantitative methods I: Stressing the local. *Prog. Hum. Geogr.* **1997**, *21*, 88–96.

47. Fotheringham, A.S.; Charlton, M.E.; Brunsdon, C. Two techniques for exploring non-stationarity in geographical data. *Geogr. Syst.* **1997**, *4*, 59–82.

48. Fotheringham, A.S.; Charlton, M.E.; Brunsdon, C. Geographically Weighted Regression: A Natural Evolution of the Expansion Method for Spatial Data Analysis. *Environ. Plan. A* **1998**, *30*, 1905–1927.

49. Fotheringham, A.S. Quantification, evidence and positivism. In *Approaches to Human Geography*; Valentine, G., Aitken, S., Eds.; Sage: London, UK, 2006.

50. Wang, Q.; Ni, J.; Tenhunen, J. Application of a geographically-weighted regression analysis to estimate net primary production of Chinese forest ecosystems. *Glob. Ecol. Biogeogr.* **2005**, *14*, 379–393.

51. Fortin, M.J.; Drapeau, P.; Legendre, P. Spatial autocorrelation and sampling design in plant ecology. *Vegetatio* **1989**, *83*, 209–222.

52. Lenat, D.R.; Crawford, J.K. Effects of land use on water quality and aquatic biota of three North Carolina Piedmont streams. *Hydrobiologia* **1994**, *294*, 185–199.

53. Wallace, J.B.; Eggert, S.L.; Meyer, J.L.; Webster, J.R. Multiple trophic levels of a forest stream linked to terrestrial litter inputs. *Science* **1997**, *277*, 102–104.

54. Welsh, H.H.; Lind, A.J. Multiscale habitat relationships of stream amphibians in the Klamath-Siskiyou Region of California and Oregon. *J. Wildl. Manag.* **2002**, *66*, 581–602.

55. Wipfli, M.S.; Gregovich, D.P. Export of invertebrates and detritus from fishless headwater streams in southeastern Alaska: Implications for downstream salmonid production. *Freshw. Biol.* **2002**, *47*, 957–969.

56. Allan, J.D. Landscapes and riverscapes: The influence of land use on stream ecosystems. *Annu. Rev. Ecol. Evol. Syst.* **2004**, *35*, 257–284.

57. Olson, D.H.; Weaver, G. Vertebrate assemblages associated with headwater hydrology in western Oregon managed forests. *For. Sci.* **2007**, *53*, 343–355.

58. Yoder, C.O.; Miltner, R.; White, D. Using biological criteria to assess and classify urban streams and develop improved landscape indicators. *National Conference on Tools for Urban Water Resource Management and Protection*; U.S. Environmental Protection Agency: Washington, DC, USA, 2000.

59. Roy, A.H.; Rosemond, A.D.; Paul, M.J.; Leigh, D.S.; Wallace, J.B. Stream macroinvertebrate response to catchment urbanisation (Georgia, USA.). *Freshw. Biol.* **2003**, *48*, 329–346.

60. Morse, C.C.; Huryn, A.D.; Cronan, C. Impervious surface area as a predictor of the effects of urbanization on stream insect communities in Maine, USA. *Environ. Monit. Assess.* **2003**, *89*, 95–127.

61. Miltner, R.J.; White, D.; Yoder, C. The biotic integrity of streams in urban and suburbanizing landscapes. *Landsc. Urban Plan.* **2004**, *69*, 87–100.

62. Magner, J.A.; Vondracek, B.; Brooks, K.N. Grazed riparian management and stream channel response in Southeastern Minnesota (USA) streams. *Environ. Manag.* **2008**, *42*, 377–390.

63. Omernik, J.M.; Abernathy, A.R.; Male, L.M. Stream nutrient levels and proximity of agricultural and forest land to streams: Some relationships. *J. Soil Water Conserv.* **1981**, *36*, 227–231.

64. Walsh, C.J.; Roy, A.H.; Feminella, J.W.; Cottingham, P.D.; Groffman, P.M.; Morgan, R.P. The urban stream syndrome: Current knowledge and the search for a cure. *J. North Am. Benthol. Soc.* **2005**, *24*, 706–723.

65. Rabe-Hesketh, S.; Skrondal, A.; Pickles, A. Generalized multilevel structural equation modeling. *Psychometrika* **2004**, *69*, 167–190.

66. Streiner, D.L. Finding Our Way: An Introduction to Path Analysis. *Can. J. Psychiatry* **2005**, *50*, 115–122.

67. Breiman, L.; Friedman, J.H.; Olshen, R.A.; Stone, C.J. *Classification and Regression Trees*, 2nd ed.; Pacific Grove: Wadsworth, CA, USA, 1984.

68. Kruskal, J.B. Multidimensional scaling by optimizing goodness of fit to a nonmetric hypothesis. *Psychometrika* **1964**, *29*, 1–27.

69. Mather, P.M. *Computational Methods of Multivariate Analysis in Physical Geography*; John Wiley and Sons, Inc.: London, UK, 1976.

Integrated Ecological River Health Assessments, Based on Water Chemistry, Physical Habitat Quality and Biological Integrity

Ji Yoon Kim and Kwang-Guk An

Abstract: This study evaluated integrative river ecosystem health using stressor-based models of physical habitat health, chemical water health, and biological health of fish and identified multiple-stressor indicators influencing the ecosystem health. Integrated health responses (IHRs), based on star-plot approach, were calculated from qualitative habitat evaluation index (QHEI), nutrient pollution index (NPI), and index of biological integrity (IBI) in four different longitudinal regions (Groups I–IV). For the calculations of IHRs values, multi-metric QHEI, NPI, and IBI models were developed and their criteria for the diagnosis of the health were determined. The longitudinal patterns of the river were analyzed by a self-organizing map (SOM) model and the key major stressors in the river were identified by principal component analysis (PCA). Our model scores of integrated health responses (IHRs) suggested that mid-stream and downstream regions were impaired, and the key stressors were closely associated with nutrient enrichment (N and P) and organic matter pollutions from domestic wastewater disposal plants and urban sewage. This modeling approach of IHRs may be used as an effective tool for evaluations of integrative ecological river health.

Reprinted from *Water*. Cite as: Kim, J.Y.; An, K.-G. Integrated Ecological River Health Assessments, Based on Water Chemistry, Physical Habitat Quality and Biological Integrity. *Water* **2015**, *7*, 6378–6403.

1. Introduction

Recent studies of river ecosystems [1–3] pointed out that integrated ecological health assessment is one of the key issues for efficient river management and is frequently used as a tool for the identification of major factors in impaired ecosystems. The degradation of river ecosystem health is largely associated with chemical pollution and physical habitat alterations due to rapid industrialization and urbanization [4–7]. Especially, stream ecosystems are rapidly disturbed by heavy sources of pollution such as industrial effluents [8], municipal wastewater discharges [9] and intense agricultural activities [10]. These sources of pollution may modify longitudinal patterns in nutrients (N and P) and physical habitat from headwaters to downstream near estuaries, and these directly or indirectly influence ecological functions of trophic compositions and tolerance species in aquatic biota [11–13]. Thus, comprehensive indicator analysis of each component in river ecosystems are necessary for assessing and diagnosing the river health, but still little is known about the integrated approach in river health assessments.

Earlier studies on stream/river health have traditionally focused on chemical monitoring due to analytical easiness of chemical condition [14]. Recent paradigms of stream health assessments, however, pointed out that chemical monitoring alone may not be enough for assessing the status of integrative ecological health and thus further biological and ecological health assessments of aquatic systems are necessary for effective management [15–17]. Complex outcomes on habitat modifications arising from channelization, barriers, and altered flow regimes [18] demonstrated partially some reasons why ecological health is modified in the assessments. An integrative ecological health approach is required to identify key factors influencing chemical water quality, physical habitat and biological conditions [15,19,20]. Despite these facts, stream monitoring and assessments for broad goals and management objectives were largely demonstrated by each chemical, physical, and biological criteria, respectively [21].

The assessments of stream and river health were conducted by multi-metic models based on different trophic-level taxa of aquatic organisms along with physical habitat models of Habitat Quality Index (HQIs; [22]). Early studies of Winget and Mangum [23] and Platts *et al.* [24] used Biotic Condition Index (BCI; [23]) for the health assessments, and later biological integrity concepts have been widely applied for evaluating the ecological health of river ecosystems. The concept of "Rapid Bioassessment Protocol" (RBP), developed by US EPA [25], was largely applied to many other countries. This concept, based on the index of biological integrity (IBI) using fish assemblage, was originally developed by Karr [15], and the concept was used with qualitative habitat evaluation index (QHEI; [26]) as an important factor of numerous physical parameters in the health assessments.

The key biota used most frequently in the assessments of river ecosystem health are periphyton or aquatic plants, as an indicator of primary production [25,27,28], macroinvertebrates as an indicator of primary consumers [25,26], and fish as an indicator of primary and top consumers [29,30]. Fish indicator among the biota was most widely used in other Asian [31] and European countries [32,33] as well as in North America (USA [34] and Canada [35]). The biological integrity models using fish assemblages have been regionally developed [36,37] and adapted by many countries in North America [38,39], Europe [40], South America [41,42] and Africa [43–45]. These studies suggested that fish is one of the best indicators for health assessments of aquatic ecosystems due to following characteristics of easiness to collect and identify in the field, longevity in the water during their entire life, and sensitive response to change of water chemistry and physical habitat.

Fish taxa are effectively used in assessing long-term damage with environmental modification, the population growth, obesity and fish health conditions [46]. For these reasons, fish was used in various research approaches from micro-level biomarkers of DNA [47], cellular [48], physiological [25], histopathological assays [49] to macro-level bioindicators of organism, population and community [50,51], and these studies inferred the river/stream health using the different organization of the fish. Low-level health response could identify the potential effects on DNA, cellular and physiological levels of organisms, thus diagnosed the impaired health influenced on chemical pollutants and disturbance [52,53]. Major problem of these studies, however, were short-term response and ecological relevance is low [54]. Thus, Adams and Greeley [48] pointed out that integrative multi-metric modeling, based on population or community-levels is required for actual assessments of ecological health assessments [54].

The objective of this study was to evaluate the ecological health of Nakdong River in Korea using an integrated health responses (IHRs) model based on chemical water quality, physical habitat and biological parameters. We developed an original national model of index of biological integrity (IBI) using fish assemblages in 2006 and the model was applied to more than 1000 wadable streams and rivers in Korea. However, it was not enough to diagnose the stream health using only fish variables. Thus, the government required a new methodology for "integrative stream health assessments" and this research was part of that. Our hypothesis was that the single IBI model might not assess the overall ecological health in Korean stream ecosystems because the model did not cover the physical habitat conditions of the stream and also did not include chemical pollution (nutrient pollution). The integrative health assessments, based on the overall parameters of physical, chemical and biological variables, were required in the national health assessments. Our integrative stream health assessments provide key identification of key factors in a problem in the stream health degradations to the Ministry of Environment, Korea, so our research suggests which factor (physical, chemical or biological components) should be restored in the Korean stream ecosystem. Under the hypothesis, we developed an integrative health assessment methods to evaluate (1) the overall ecological health condition of a specific watershed (Nakdong River) of Korean stream ecosystems using biological assessments; (2) nutrient pollution (N, P, biochemical oxygen demand (BOD), and chemical oxygen demand (COD)); and (3) physical habitat health (QHEI model), which is one of the key stressors damaging the Korean stream ecosystems. The outcomes of this research were intended to serve as a starting point for Korean government to eventually establish overall assessment approaches and ecological health criteria, and specifically a new integrative modeled for Korean stream ecosystem. The comprehensive approach has been used to demonstrate the river ecosystems health. Chemical health was evaluated using the nutrient pollution index (NPI) developed in this study. Physical habitat health was determined using the qualitative habitat evaluation index (QHEI) and biological health was determined using the index of biological integrity (IBI). Based on these models, integrative ecological health was compared using a star-plot approach. This IHRs approach can be used as a key tool for the integrative ecological health assessments of river ecosystems. In addition, these approaches provide valuable results for effective management and restoration of river ecosystems in the future.

2. Materials and Methods

2.1. Study Sites

This study was conducted in the Nakdong River watershed with the length of 525 km and basin area of 23,860 km^2, South Korea (Figure 1), which is located in the southeast of the Korean Peninsula (35°–36° N, 128° E). Nakdong River watershed is influenced by various point/non point sources of pollution such as wastewater disposal plants and urban sewage from several tributaries of Yeong River, Geumho River, Hwang River, Nam River, and Miryang River. Most largest point sources are located on the Geumho River, with a large industrial complex and wastewater disposal plants, so the water quality downstream is rapidly degraded, as suggested in numerous previous studies [8–10]. Intensive agricultural activities, as non-point sources, are concentrated in the zone of Sites 5–8. The total number of sampling sites, consisting of sixth order streams [55], was 21, with 5 reference sites and 16 sites in the main stream. The selection of reference site followed the approach of Hughes *et al.* [56] and U.S. EPA [25].

The reference streams in the watershed were originally designated in 2006 by the Ministry of the Environment, Korea for efficient watershed management. The reference site was defined as a least-disturbed stream with low impact from human activities such as farming, urban development, and forest management. The selection of reference streams in this region was based on overall ecological conditions of chemical water quality (N, P, or organic pollutants), physical habitat conditions, and biota (periphyton, macroinvertebrate, and fish taxa) in Korea.

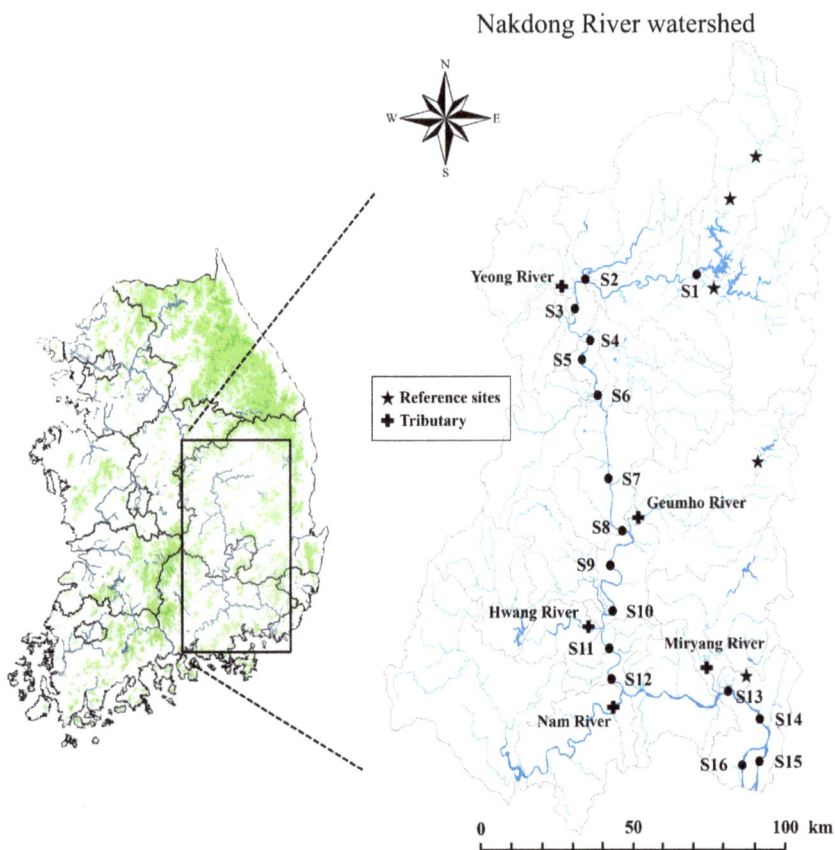

Figure 1. Map showing sampling sites in Nakdong River watershed.

2.2. Fish Sampling

Field sampling for fish and water chemistry was conducted twice in the premonsoon (May) and postmonsoon (September) seasons during 2008–2009. Stream flows were relatively stable in both seasons. Fish samplings were conducted by a modified wading method [57,58] to evaluate the Korean aquatic ecosystem health based on the Ohio EPA method [59]. For the fish sampling, we considered all habitat types, such as riffle, run, and pool, in the same site and directed in an upstream to downstream reach for at least 200 m distance during 50 min for the catch per unit efforts (CPUE). Casting-net (7 × 7 mm, CN) and kick-net (4 × 4 mm, KN), the most popular fish sampling gears in Korea, were applied to sample. All fishes were identified *in situ* and released immediately. All specimens were identified according to the key characteristics of Kim and Park [60] and the classification system of Nelson [61]. However, some ambiguous specimens hard to identify were preserved in 10% formalin solution and then brought to the laboratory for further research. All sampled fishes were examined for anomalous external characteristics such as deformities (D), erosion (E), lesion (L), and tumors (T) (DELT) based on the concept of Sanders *et al.* [62]. Tolerance and trophic species analysis were based on the previous regional studies [63].

2.3. Analysis of Water Quality Parameters

Sampling for water quality was conducted twice at the same time as for fish sampling per watershed in 2008–2009. Ten water chemistry parameters analyzed in this study are as follows: biological oxygen demand (BOD), chemical oxygen demand (COD), total nitrogen (TN), ammonia nitrogen (NH_4–N), nitrate nitrogen (NO_3–N), total phosphorus (TP), ortho phosphorus (PO_4–P), total suspended solids (TSS), electrical conductivity (EC) and chlorophyll-a (Chl-*a*). TSS, EC and Chl-*a* were measured at the time of sample collection with the YSI sonde 6600. TN, total dissolved nitrogen (TDN) and total particle nitrogen (TPN) were measured by second derivative method after a persulfate digestion [64]. TP was determined using the ascorbic acid method after persulfate oxidation [65]. TSS, BOD and COD were measured by the standard methods [66]. Nutrient analyses were performed thrice, and BOD, COD and SS were measured twice [66].

2.4. Nutrient Pollution Index (NPI) for Chemical Health Analysis

To develop chemical health assessment model, multi-metric model of nutrient pollution index (NPI), followed methods used by Dodds *et al.* [67] and Lee and An [68]. The metrics were composed as following; M_1: total nitrogen (TN, mg·L^{-1}), M_2: total phosphorus (TP, μg·L^{-1}), M_3: TN:TP ratio, M_4: BOD (mg·L^{-1}), M_5: total suspended solids (TSS, mg·L^{-1}), M_6: electrical conductivity (μS·cm^{-1}), and M_7: chlorophyll-*a* (μg·L^{-1}). We established the criteria for boundaries and boundary was defined by the third of the observed distribution of the values. Each metric was scored 5, 3 or 1 point, respectively. The health conditions of the chemistry were evaluated by summing the scores obtained from the seven parameters and then categorizing the system as excellent (Ex; 31–35), good (G; 25–29), fair (F; 19–23), poor (P; 13–17), and very poor (VP; 7–11).

2.5. Qualitative Habitat Evaluation Index (QHEI) for Physical Habitat Health Analysis

Physical habitat health, based on the multi-metric model of the qualitative habitat evaluation index (QHEI), was evaluated at the sampling sites. The original 11-metric QHEI model [25,26] was modified as a six metric model for regional application [58,69]. The metric attributes were as follows: M_1: epifaunal substrate/available cover, M_2: pool substrate characterization, M_3: channel flow status, M_4: existence of small-scale dams, M_5: channel alteration, and M_6: sediment deposition. Habitat health conditions were evaluated by summing the scores obtained from the six parameters and divided into 4 categories of excellent (Ex; score 96–120), good (G; 66–80), fair (F; 36–60) or poor (P; 6–30) conditions [58].

2.6. Index of Biological Integrity (IBI) for Biological Health Analysis

Multi-metric fish model was developed for the diagnosis of the ecosystem health in the Nakdong River. Our model, IBI, which was based on the IBI concept [15,70], was modified from the original U.S. EPA [71] model and the regional model of An *et al.* [72]. The metrics (M) were consisted in three major groups as ecological characteristics by species richness and composition, trophic composition, and fish abundance with health condition. The individual metrics were: M_1: total number of native species, M_2: number of riffle-benthic species, M_3: number of sensitive species, M_4: proportion of individuals as tolerant species, M_5: proportion of individuals as omnivore species, M_6: proportion of individuals as insectivore species, M_7: total number of native individuals, and M_8: percent individuals with anomalies. Four of the eight metrics (M_1, M_2, M_3, and M_7) were evaluated by the maximum species richness line (MSRL, [73]) with the stream orders. Each metric was scored 5, 3 or 1 and community-level health conditions were judged using the criteria of Barbour *et al.* [25]. The IBI scores were judged as five categories, excellent (Ex; 36–40), good (G; 28–34), fair (F; 20–26), poor (P; 14–18) and very poor (VP; 8–13). Detailed descriptions of specific metric characteristics and scoring criteria for the model are available in An *et al.* [72].

2.7. Integrated Health Responses (IHRs) Model Using Star-Plot Analysis

The integrated health responses (IHRs) model was developed in this research to enable the multi-metric assessment of ecological health. The IHRs model was composed of multiple functional metrics and was based on the integration of all parameters derived from biological, chemical and physical health parameters. Data-processing step was to generate the assessment scores (*i.e.*, standardized data) followed methods used by Yeom and Adams [74], Lee *et al.* [75] and Lee and An [3]. The area score enclosed by each star-plot was used to compare the assessment results for the difference among sampling sites relative to their ecological health response to environmental conditions at each site. The area score of star-plot was calculated according to methods described by Beliaeff and Burgeot [76] and Kim *et al.* [77]. The integrated health, IHRs model values, was judged as five ranks of excellent (Ex; >90% of reference), good (G; 75%–90% of reference), fair (F; 55%–75% of reference), poor (P; 35%–55% of reference), and very poor conditions (VP; <35% of reference).

Self-organizing map (SOM) was used to analyze the longitudinal patterns of fish composition and water chemistry parameters at the 16 sampling sites. The SOM approach is based on a learning algorithm in an artificial neural network and approximates the probability density function of the input data [78,79]. It has a wide range of engineering applications for handling complex ecological data (e.g., non-linear modeling or optimization) and is typically used for classification, clustering, prediction, modeling, and data mining [80]. The learning process of the SOM was applied using the SOM Toolbox package developed by the Laboratory of Information and Computer Science in the Helsinki University of Technology for Matlab ver. 6.1, and we adopted the initialization and training methods suggested by the authors of the SOM Toolbox that allow the algorithm to be optimized [81]. In addition, the PC-Ord statistical package (Ver. 4.25 for Windows; [82]) was used for principal components analysis (PCA) to identify the major environmental factors influencing ecological parameters on clustered by SOM.

3. Results and Discussion

3.1. Cluster Analysis of Sampling Regions Using a Model of Self-Organizing Maps (SOMs)

The relations between the chemical water quality and biological variables of fish (tolerance and trophic compositions) were analyzed using the modeling approach of Self-Organizing Maps (SOMs). As shown in Figure 2, each variable of BOD, TN, TP, TSS, electrical conductivity (EC) and N:P ratios were patterned according to the similarity of community compositions through training with the SOM. To evaluate the relations between chemical and biological parameters, the classified variables were visualized by the color on the map (Figure 2). Red color regions in the map indicated high values, whereas blue color regions indicated low values. Through the learning process of the SOMs, the clusters were divided into four groups, I–IV along the longitudinal gradients of ecological factors. Undisturbed sites were grouped in Group I of the SOM map, while polluted sites appeared Group IV. Group I, which was located in the headwater region, occurred in the pristine regions with low organic matter (BOD: 0.96 ± 0.15 mg·L^{-1}) and nutrient levels (TN: 1.79 ± 0.12 mg·L^{-1}, TP: 19 ± 0 µg·L^{-1}). In contrast, Group IV, which was located in the downstream region, occurred in the polluted regions with high organic matter (BOD: 2.9 ± 1.33 mg· L^{-1}) and nutrient levels (TN: 2.66 ± 0.48 mg·L^{-1}, TP: 195 ± 80 µg·L^{-1}). Chemical parameters were clearly differentiated between Group I and Group IV. This pattern of downstream degradations was similar to other parameters of BOD, TN and TP. Likewise, the proportion of sensitive species (SS) was higher in the Group I and the proportion of tolerant species (TS) was relatively higher toward the downstream region (Group IV). In the case of omnivores (O), they were widely distributed in every site of the stream of Nakdong River, but the proportion of omnivores was relatively lower in downstream than upstream due to high dominance of carnivorous species (55%). Our results of SOMs model suggest that the clustering of the trained SOMs units reflected the regional differences of water chemistry (chemical parameters) and biological compositions (biological parameters) from the upstream to downstream.

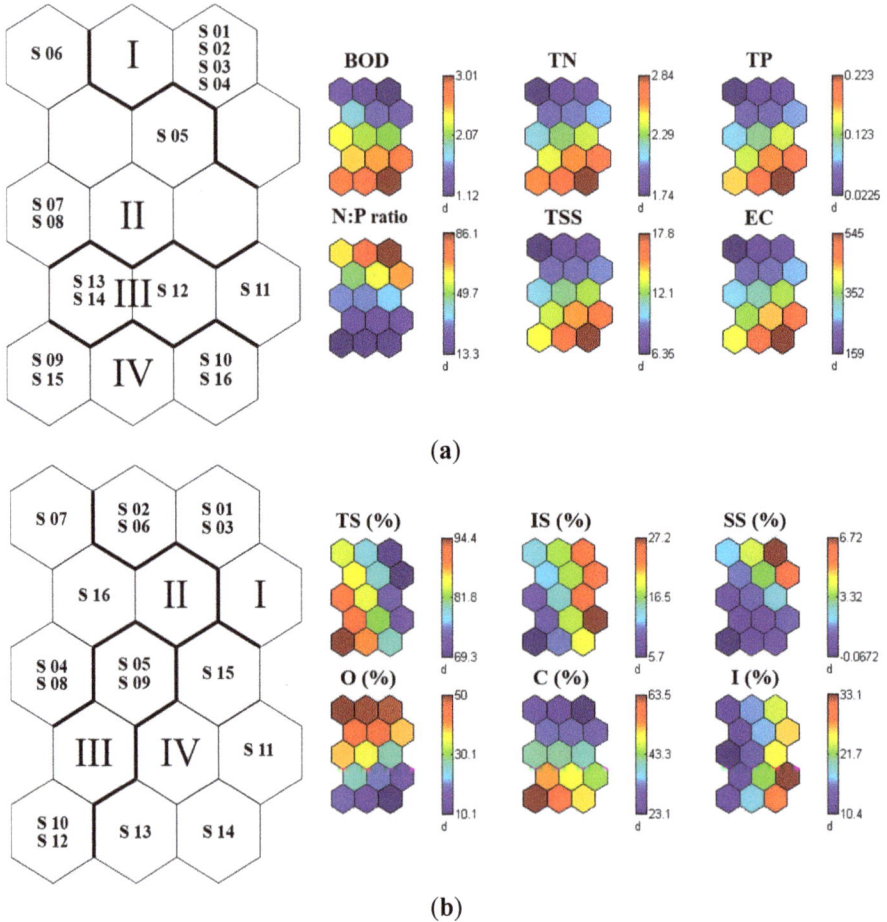

(a)

(b)

Figure 2. Clustering of the trained self-organizing maps (SOMs) units for chemical parameters and biological parameters. The four groups (I–IV) indicate different clusters of ecological characteristics, and the code in each unit of the map refers to the sampling site. The mean value of each variable was calculated from each output neuron of the trained SOM. The red and blue colors indicate a high and low value, respectively, for each environmental parameter: (a) Chemical parameters and (b) biological components.

3.2. Ecological Factor Identification Using a Principal Component Analysis (PCA)

Principal component analysis (PCA) was used to analyze key factors influencing biological components and chemical parameters (Figure 3). Results indicated that the groups of river regions could be divided into Group I, Group II, Group III and Group IV by eigenvalues of >1.0. Results of PCA indicated that three axes explained 80.8% of the variation in our data matrix (eigenvalues of >1.0). The axis-1 on BOD, TN, and TP could explain about 47.4% of the total and the axis-2 on % omnivores, % insectivores, and IBI model values explained 21.6% of the total. Also, the axis-3 on the proportions of tolerant species and sensitive species explained 11.9% of the total (Table 1). The PCA analysis indicated that axis-1 was mainly influenced by organic matter (BOD; -0.3955) and nutrient levels (TN; -0.4420, TP; -0.4131), which had negative responses. The biological responses of % omnivores, % insectivores and IBI model values were useful indicators in the axis-2 analysis. The eigenvalue of omnivore species was -0.5207, which indicated a negative response, while % insectivores and IBI model values of >0.5 were showed positive responses. In the meantime, axis-3 was weakly influenced by tolerant species and sensitive species. Thus, Group I, which is located in the upstream, was directly influenced with mass ratios of N:P and NPI values, and Group IV, which is located in downstream, was directly influenced by organic matter (BOD), high nutrient levels (TN and TP) and suspended solids (TSS). Overall, the results of PCA suggested that greater impacts of chemical pollution were evident in the downstream regions.

Table 1. Principal component analysis (PCA) based on biological and chemical variables. Bold values indicate statistically significant in the level of <0.05.

Principal Component Analysis/Eigenvalue >1.0			
Structure Metrics	Axis 1	Axis 2	Axis 3
% Tolerant species	−0.2508	−0.0625	**0.7282**
% Sensitive species	0.2932	0.0293	**−0.4934**
% Omnivores	0.2854	**−0.5207**	−0.0887
% Insectivores	0.2392	**0.5115**	−0.0667
BOD	**−0.3955**	−0.1632	−0.1394
TN	**−0.4420**	−0.0142	−0.2790
TP	**−0.4131**	−0.0689	−0.3019
IBI score	−0.1209	**0.6550**	0.0112
Eigenvalue:	4.264	1.941	1.070
Proportion of variance	47.380	21.562	11.890

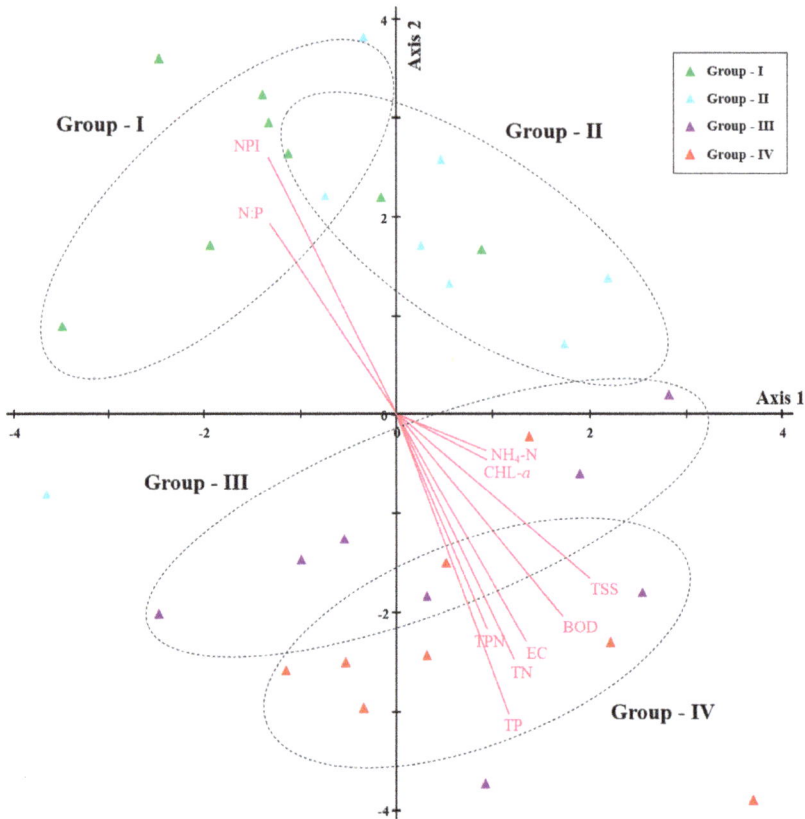

Figure 3. Principal component analysis (PCA) based on biological components (% tolerant species, % sensitive species, % omnivores, % insectivores, IBI value) and chemical factors (BOD = biological oxygen demand, COD = chemical oxygen demand, TN = total nitrogen, NH_4–N = ammonium nitrogen, NO_3–N = nitric nitrogen, TDN= total dissolved nitrogen, TPN = total particle nitrogen, TP = total phosphorus, PO_4–P = ortho phosphorus, N:P = N:P ratio, TSS = total suspended solids, EC = electrical conductivity, CHL-*a* = chlorophyll-*a*, NPI = nutrient pollution index score).

3.3. Chemical Model of Water Quality Index and Its Evaluation

Multi-metric model of nutrient pollution index (NPI) was developed and applied to the model Nakdong River watershed (Figure 4). The metrics of NPI model was composed of seven (M_1–M_7) and were categorized as four groups of nutrient compositions (N and P), organic matter (BOD), inorganic contents/solids, and primary production indicators (Table 2). For variables of the NPI model, we selected total nitrogen (TN) and total phosphorus (TP) along with N:P mass ratios, which are known as key determinants regulating the river water quality and eutrophication [83–86].

Chemical criteria, based on ambient nutrient metrics of TN, were categorized as oligotrophic (<1.5 mg·L^{-1}), mesotrophic (1.5–3 mg·L^{-1}) and eutrophic (>3 mg· L^{-1}), respectively, and these criteria differed from the previous criteria in North America [87–89] and Europe [90,91]. Mean value of TN in Group III and Group IV regions were significantly higher ($p < 0.05$) than those at the regions of Group I and Group II as well as the reference sites (1.82 ± 1.3 mg·L^{-1}). The values of TN, however, were categorized as mesotrophic ("3" in the metric score) in the analysis, indicating no large spatial variations in the model score. In the meantime, TP had large longitudinal variations along the main axis of the headwaters to the downstream, by the criteria of TP (<30, 30–100, >100 µg·L^{-1}); TP was oligotrophic (mean: 12 ± 5 µg·L^{-1}) in the region of Group I, and this condition was similar to the reference sites. In contrast, the mean TP of Group III and Group IV was >10 fold than the reference and Group I streams, indicating severe phosphorus enrichment in the downstream regions. The metric indicator of N:P applied in this study also showed the similar pattern with TP rather than TN (Table 2). Based on the mass ratio metric of TN:TP, reference and Group I streams had high ratios of >100, whereas Group III and Group IV streams had low ratios of <20. Previous studies pointed out that N:P ratio in the ambient stream and river waters is an indirect indicator of limiting nutrient for algal biomass or primary production/growth [86,92,93] and is lower in polluted streams or eutrophic waterbodies [67]. Our outcome of N:P ratios in this watersheds was supported by previous research. In fact, the contents of chlorophyll-a (CHL), as good indicators of primary productivity, were directly determined by mass nutrient ratios (N:P) and TP.

Mean values of CHL in the Group III and Group IV regions were >30 µg·L^{-1}, at high TP and low N:P ratios, and these values were significantly higher ($p < 0.05$) than those at the regions of Group I (7.1 ± 3.4 µg·L^{-1}) and reference (2.6 ± 1.2 µg· L^{-1}) with low TP and high N:P ratios. These results indicate that the high CHL values in our watershed were closely associated with high P and low N:P ratios. The ratio of N:P, thus, is a key chemical health parameter controlling cyanobacterial blooms in the aquatic environment. In the meantime, the metric of biological oxygen demand (BOD), as an indicator for organic matter pollution, showed the distinct differences between the headwaters (Group I) or reference and the downstream regions (Groups III/IV); the metric values were 4–5 in the reference and headwater streams but were 1.5 in the downstream regions (Groups III/IV). Thus, metric values of BOD showed similar spatial patterns with total suspended solids (TSS) as well as the parameter metrics of N:P ratios and TP.

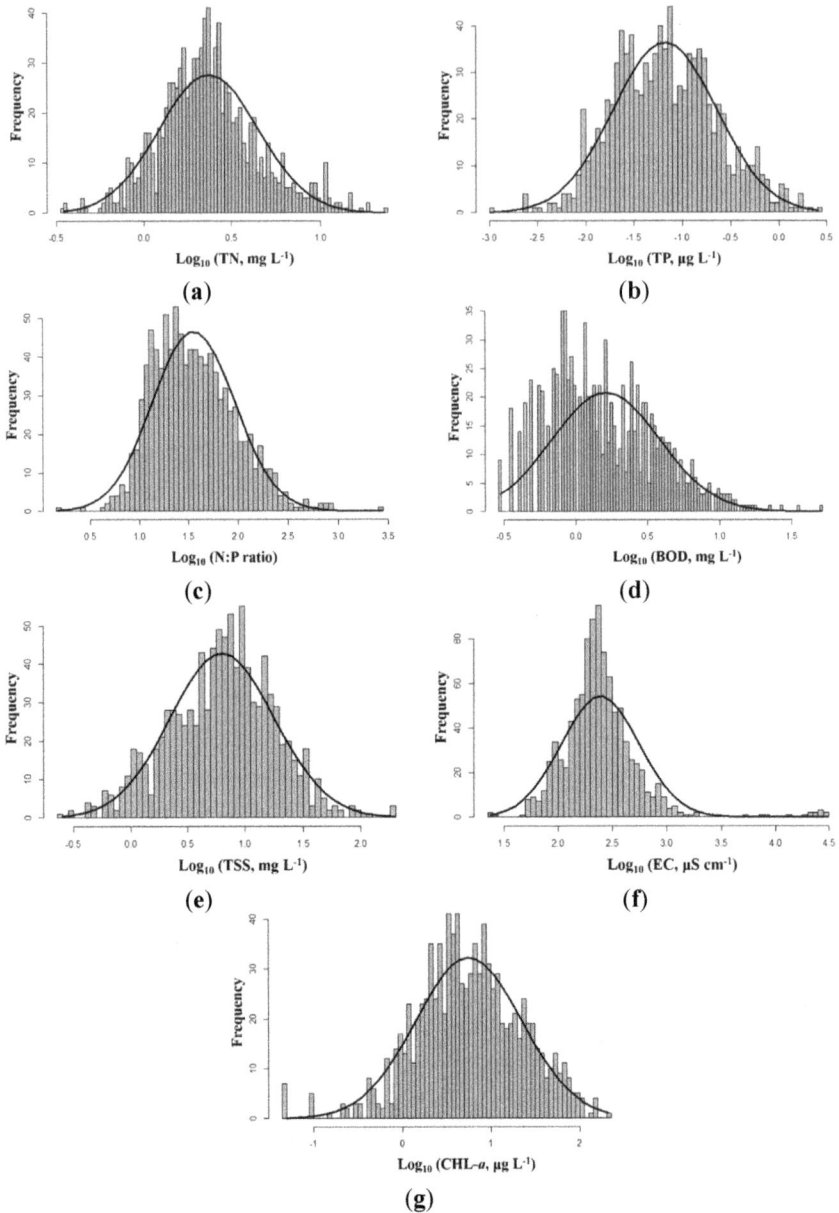

Figure 4. Observed frequency diagram (total nitrogen: $n = 974$, total phosphorus: $n = 973$, TN:TP ratio: $n = 973$, BOD: $n = 974$, total suspended solids: $n = 974$, electrical conductivity: $n = 974$, chlorophyll-a: $n = 969$). (**a**) M_1: Total Nitrogen (mg·L^{-1}); (**b**) M_2: Total Phosphorus (μg·L^{-1}); (**c**) M_3: TN: TP ratio; (**d**) M_4: BOD (mg·L^{-1}); (**e**) M_5: Total Suspended Solids (mg·L^{-1}); (**f**) M_6: Electrical Conductivity (μS·cm^{-1}); and (**g**) M_7: Chlorophyll-a (μg·L^{-1}).

336

Table 2. Chemical health assessment model, based on the Nutrient Pollution Index (NPI), suggested scoring criteria and evaluated score in the watershed of Nakdong River.

Category	Metric	Scoring Criteria			Mean ± SD (Score)				
		5	3	1	R_f	Group I	Group II	Group III	Group IV
Nutrient regime	M_1: Total Nitrogen (mg·L^{-1})	<1.5	1.5–3	>3	1.82 ± 1.33 (3)	1.80 ± 0.1 (3)	1.85 ± 0.33 (3)	2.82 ± 0.42 (2.5)	2.66 ± 0.49 (2)
	M_2: Total Phosphorus (µg·L^{-1})	<30	30–100	>100	12 ± 5 (5)	19 ± 6 (5)	46 ± 18 (3.5)	197 ± 37 (1)	195 ± 95 (1)
	M_3: TN: TP ratio	>50	20–50	<20	188 ± 189 (5)	100 ± 33 (5)	46 ± 21 (3)	14 ± 2 (1)	14 ± 3 (1)
Organic matter	M_4: BOD (mg·L^{-1})	<1	1–2.5	>2.5	0.72 ± 0.22 (5)	0.96 ± 0.20 (4)	1.87 ± 0.80 (3)	2.37 ± 0.56 (2)	2.9 ± 1.55 (2)
Ionic contents and solids	M_5: Total Suspended Solid (mg·L^{-1})	<4	4–10	>10	1.87 ± 1.31 (5)	6.1 ± 5.5 (4)	9.0 ± 2.3 (2)	14.3 ± 2.9 (1.5)	15.8 ± 7.2 (1.5)
	M_6: Electrical conductivity (µS·cm^{-1})	<180	180–300	>300	179 ± 94 (5)	170 ± 28 (4)	231 ± 67 (3)	434 ± 110 (1)	460 ± 245 (2)
Primary production indicator	M_7: Chlorophyll-a (µg·L^{-1})	<3	3–10	>10	2.6 ± 1.2 (5)	7.1 ± 3.4 (3)	36.6 ± 18.7 (1)	37.7 ± 25.0 (1)	32.4 ± 12.3 (1)
Scores (model criteria of NPI)					33 (Ex)	28 (G)	19.5 (F)	10 (VP)	10.5 (VP)

Chemical health, based on seven multi-metric model of Nutrient Pollution Index (NP index), showed distinct spatial differences between the regions of the watershed. Index model values of NP in the region of Group I was 28 and this value was similar to the reference sites (33 score). This indicates that the chemical health was judged as "good condition (G)" in the headwater region (Group I) by the health criteria of five classes. In contrast, the model values of NP in the downstream regions (Groups III and IV) ranged between 10.0 and 10.5, which were judged as "very poor (VP) condition" (most impaired level) among the five classes. The impaired chemical health in the downstream regions was mainly due to effluents from the massive point/non point sources of municipal wastewater disposal plants and urban runoff, which are come from tributary of Geumho River. The degradation of chemical health in the downstream is supported by previous research on chemical water quality [39,94,95].

3.4. Responses of Biological Indicators on Water Chemistry

Responses of biological indicators, as fish tolerance and trophic species, on water chemistry are shown in Figure 5. The proportions of tolerant species (TS), sensitive species (SS) and insectivore species (I) in the watershed were directly determined by chemical water quality parameter. When chemical concentrations of TP, BOD, and electrical conductivity (EC) are low, these three environmental factors showed a wide variation in biological responses between the maximum and minimum values (Figure 5). The responses in the proportion of tolerant species, sensitive species, and insectivore species, however, had direct functional relations with chemical conditions. When values of TP were >200 $\mu g \cdot L^{-1}$, the proportions of tolerant species and insectivore species had positive functional responses to increased TP, but the proportions of sensitive species had negative functional responses to TP (Figure 5). Similar functional responses in the proportions of tolerant species, sensitive species, and insectivore species were shown in BOD, as an indicator for organic matter pollution, and the EC as an indicator of ionic pollution, when BOD and EC values were >2.0 $mg \cdot L^{-1}$ and 270 $\mu S \cdot cm^{-1}$, respectively (Figure 5). Such nutrients of phosphorus directly determined concentrations of sestonic chlorophyll-a (CHL), and the CHL values, in turn, influenced the fish compositions of tolerant species, sensitive species and insectivore species. These responses are supported by findings of US EPA [39] that the proportions of sensitive and/or insectivore fish species decrease with nutrient enrichment and organic matter pollution, and vise verse in tolerant species. In the meantime, the responses on the levels of TN were not shown in this study due to high concentrations of N regardless of location and season. The high nitrogen was more attributed to stream geology rather than degree of nutrient pollution, thus nitrogen contents were high in the pristine regions with 100% forest stream.

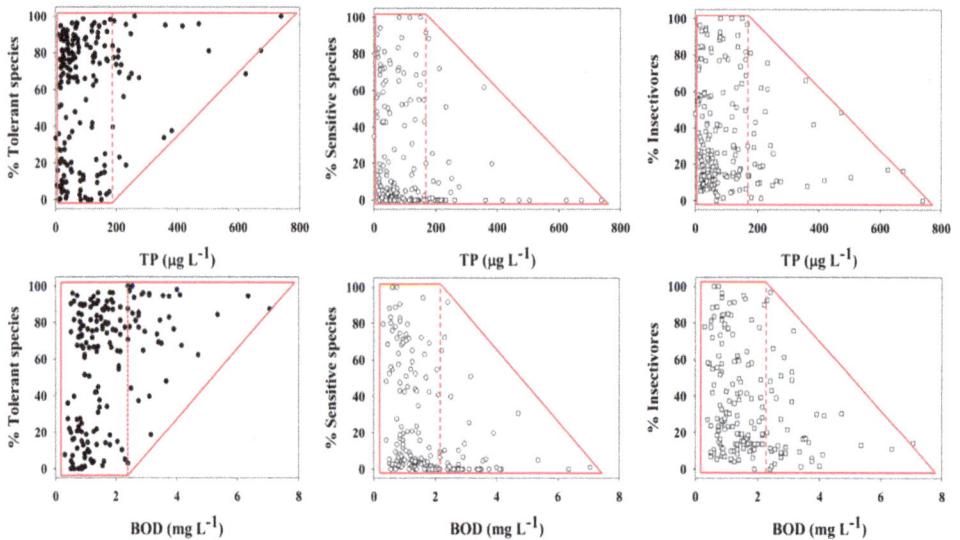

Figure 5. *Cont.*

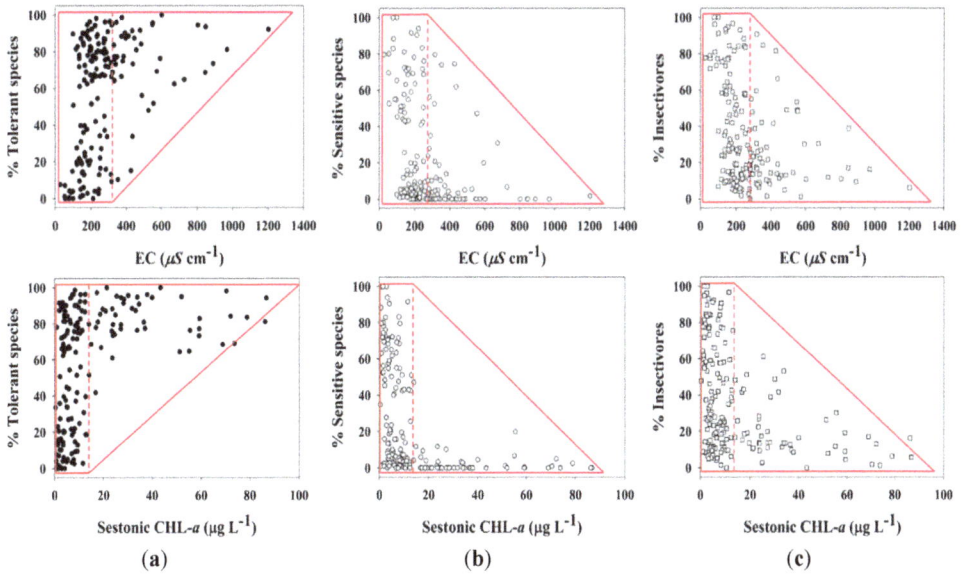

Figure 5. The relation of fish indicators (tolerance and trophic species) to water quality parameters (total phosphorus (TP), biochemical oxygen demand (BOD), electrical conductivity (EC), sestonic chlorophyll-a (Chl-a)): (**a**) % Tolerant Sp.; (**b**) % Sensitive Sp.; and (**c**) % Insectivores Sp.

3.5. Physical Habitat Health Using a Multi-metric Model

Qualitative Habitat Evaluation Index (QHEI), based on a six metric model, was used for the evaluation of physical habitat health in the watershed of Nakdong River. Values of QHEI averaged 67 in the watershed regions of Groups I–IV and ranged between 60 and 75. Thus, physical habitat health in all regions was judged as a "good condition" (G) by the criteria of An *et al.* [96] (Table 3). As shown in Table 3, spatial variations from the headwaters to the downstream were not high, unlike other watersheds in Korea [97,98]. Physical habitat health of Groups I and II showed 30% more degradation, compared to the reference regions, thus some sites were not suitable for fish habitats and this was mainly influenced by human disturbances. Habitat health of Groups III and IV in downstream regions, was better than the regions of Groups I and II. The health impairments in the upstream were mainly due to poor epifaunal substrate/available cover (M_1) and poor pool substrate conditions (M_2) throughout the habitat simplification by sand accumulations. In addition, partial channel alterations and sediment depositions were found in the impaired habitats.

Biological river health assessments, based on multi-metric Index of Biological Integrity (IBI) are shown in Table 4. The river health in the regions of Groups I–IV was compared with the regions of reference sites. The values of IBI model averaged 30 in the reference sites with ranges of each metric value from 3 to 5 (Table 4). The river health, thus, was judged as a "good condition (G)" by the criteria of An *et al.* [72]. Such IBI values in these reference sites were not so high as shown in reference regions of other countries [15,39]. In the meantime, the model values of IBI in all regions of Groups I–IV ranged from 12 to 18, which is corresponding to poor (P) to very poor (VP), respectively. The IBI values averaged 12 in regions of Groups III and IV, and this value was lower than the IBI values of Groups I and II (mean IBI = 16) as well as reference regions (IBI = 30; Table 4). Biological river health in this watershed was more impaired downstream than upstream, and the impairment was mainly attributed to reduced metric values of riffle-benthic species, sensitive species, insectivore species and native species and anomalies. The low values in the metrics were due to chemical degradations of the downstream and the degradations in the main river downstream was closely associated with nutrient-rich effluents of wastewater disposal plants from tributary streams. Such impairments of the river health in the downstream are similar in previous studies [39,97,98], which are directly influenced by large point-source pollutions of wastewater treatment plants and industrial complex.

Table 3. Physical habitat health assessment, based on the Qualitative Habitat Evaluation Index (QHEI), in the watershed of Nakdong River.

Metric	Study Areas				
	R_f	Group I	Group II	Group III	Group IV
M_1: Epifaunal substrate/Available cover	15.3	4.8	6.6	9.4	13.1
M_2: Pool substrate characterization	14.2	7.5	8.8	8.6	10.8
M_3: Channel flow status	8.3	12.8	11.1	13.4	14.8
M_4: Existence of small-scale dams	13.4	12.3	13.0	15.4	13.6
M_5: Channel alteration	11.9	11.5	10.6	13.9	11.1
M_6: Sediment deposition	12.5	13.1	10.3	11.9	11.1
Scores (model criteria of QHEI)	75.6 (G)	61.9 (G-F)	60.4 (G-F)	72.5 (G)	74.5 (G)

Notes: R_f: Reference sites; Group I: 4 site (S1, S2, S3, S4), Group II: 4 site (S5, S6, S7, S8), Group III: 4 site (S9, S10, S11, S12), Group IV: 4 site (S13, S14, S15, S16), Ex: excellent, G: good, F: fair.

In addition, the river health was closely associated with community structures, based on fish compositions of tolerance species and trophic species. In this study, total 45 species and 4610 individuals were collected from the watershed of Nakdong River. The dominant fishes with greater than 5% in relative abundance are shown in

Table 5 in Nakdong River. The highest dominant species was *Opsarichthys uncirostris*, which composed about 30% of the total, and then followed by *Zacco platypus* (28%), *Micropterus salmoides* (5%), and *Squalidus chankaensis tsuchigae* (5%). The fish fauna suggest that the dominant species are composed of more tolerant species on the water chemistry or physical habitat. Meanwhile, key dominant species in the reference streams were *Zacco koreanus* and *Coreoleuciscus splendidus*, which made up 58% of the total and are sensitive species and insectivore species (Table 5). Thus, the reference region was designated as *Zacco-Coreoleuciscus* community, and differed largely from the regions of Groups I and IV, indicating a distinct difference in species composition in the structural aspects of the community (Figure 6). The community of the redions of Group I, with *Zacco-Opsarichthys* domination, showed tolerant species at >70% of the total, while the regions of Group IV, with a *Opsarichthys-Micropterus* community, were composed of a community of only tolerant species.

Table 4. Biological river health assessment, based on the multi-metric fish model of Index of Biological Integrity (IBI), in the watershed of Nakdong River.

Category	Metric	Scoring Criteria			R_f	Mean ± SD (Score)			
		5	3	1		Group I	Group II	Group III	Group IV
Species richness & compositions	M_1: Total number of native species	Expectations of M_1 vary with stream order			11.4 ± 3.2 (3)	11 ± 2.2 (3)	11.8 ± 4.3 (3)	10.5 ± 1.9 (3)	9.8 ± 2.4 (1)
	M_2: Number of riffle-benthic species	Expectations of M_2 vary with stream order			3.2 ± 1.3 (3)	1.5 ± 0.6 (1)	1.5 ± 1.7 (1)	1.3 ± 0.5 (1)	0.8 ± 0.5 (1)
	M_3: Number of sensitive species	Expectations of M_3 vary with stream order			6.4 ± 1.1 (3)	2.8 ± 1.0 (1)	1.5 ± 1 (1)	0.8 ± 1.0 (1)	0.3 ± 0.5 (1)
	M_4: Proportion of individuals as tolerant species	<5	5–20	>20	11 ± 7 (3)	74 ± 9 (1)	77 ± 10 (1)	83 ± 5 (1)	86 ± 7 (1)
Trophic compositions	M_5: Proportion of individuals as omnivore species	<20	20–45	>45	19 ± 12 (5)	54 ± 4 (1)	47 ± 10 (1)	27 ± 18 (3)	18 ± 9 (5)
	M_6: Proportion of individuals as insectivore species	>45	45–20	<20	73 ± 12 (5)	27 ± 5 (3)	17 ± 5 (1)	12 ± 8 (1)	14 ± 5 (1)
Fish abundance & conditions	M_7: Total number of native individuals	Expectations of M_7 vary with stream order			226 ±82 (3)	213.5 ± 59.6 (3)	327.5 ± 118.4 (3)	196 ± 23 (1)	81.5 ± 36.7 (1)
	M_8: Percent individuals with anomalies	0	0–1	>1	0 (5)	0 ± 0 (5)	0.2 ± 0.2 (3)	1.3 ± 2.5 (1)	1.8 ± 3.6 (1)
Scores (model criteria of IBI)					30 (G)	18 (P)	14 (P)	12 (VP)	12 (VP)

Table 5. Fish fauna and dominant species for collected fish population in Nakdong River.

Sample	Fish Community	Dominant Species	To.	Tr.	RA (%)
Reference Sites	Zacco-Coreoleuciscus Community	Zacco koreanus [†]	SS	I	47.8
		Coreoleuciscus splendidus [†]	SS	I	10.6
		Zacco platypus	TS	O	10.4
		Pungtungia herzi	IS	I	9.2
		Niwaella multifasciata [†]	SS	O	4.4
Group I	Zacco-Opsarichthys Community	Zacco platypus	TS	O	46.7
		Opsarichthys uncirostris amurensis	TS	C	15.6
		Pseudogobio esocinus	IS	I	7.8
		Hemibarbus labeo	TS	I	6.3
		Pungtungia herzi	IS	I	5.7
Group II	Zacco-Opsarichthys Community	Zacco platypus	TS	O	38.5
		Opsarichthys uncirostris amurensis	TS	C	31.3
		Rhinogobius brunneus	IS	I	6.1
		Pseudogobio esocinus	IS	I	5.8
		Squalidus chankaensis tsuchigae [†]	IS	O	5.8
Group III	Opsarichthys-Zacco Community	Opsarichthys uncirostris amurensis	TS	C	47.7
		Zacco platypus	TS	O	13.7
		Micropterus salmoides [‡]	TS	C	8.9
		Squalidus chankaensis tsuchigae [†]	IS	O	7.2
		Carassius auratus	TS	O	3.5
Group IV	Opsarichthys-Micropterus Community	Opsarichthys uncirostris amurensis	TS	C	29.8
		Micropterus salmoides [‡]	TS	C	21.0
		Lepomis macrochirus [‡]	TS	I	9.2
		Tridentiger obscurus	TS	I	6.2
		Mugil cephalus	TS	H	5.8

Notes: To.: tolerance species (TS: tolerant species, IS: intermediate species, SS: sensitive species), Tr.: trophic species (C: carnivores, O: omnivores, I: insectivores, H: herbivores), RA: relative abundance, [†]: Korean endemic species, [‡]: exotic species.

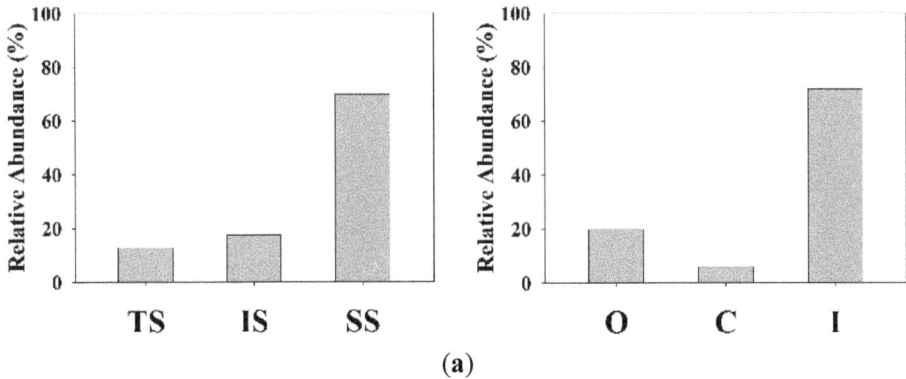

(a)

Figure 6. *Cont.*

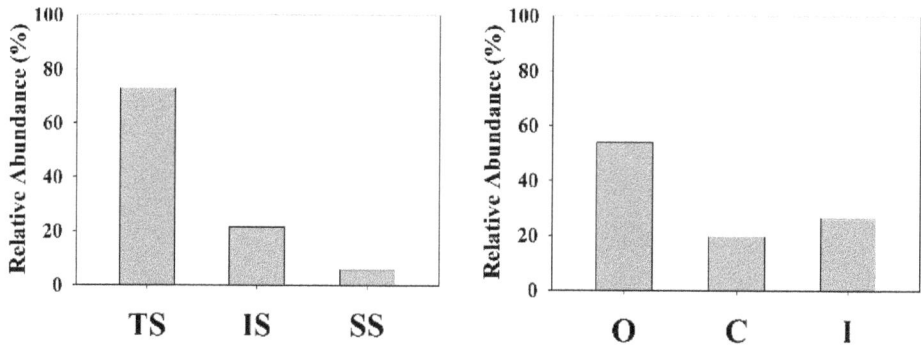

(b)

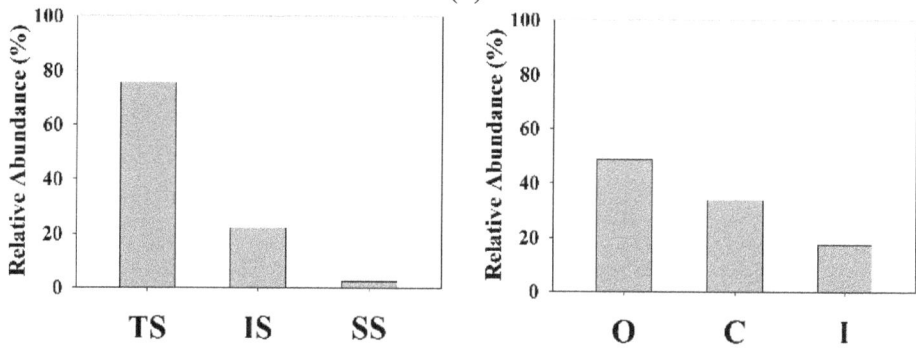

(c)

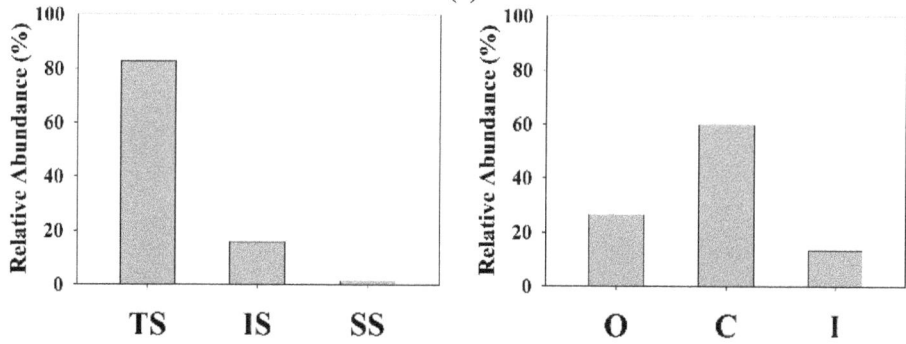

(d)

Figure 6. *Cont.*

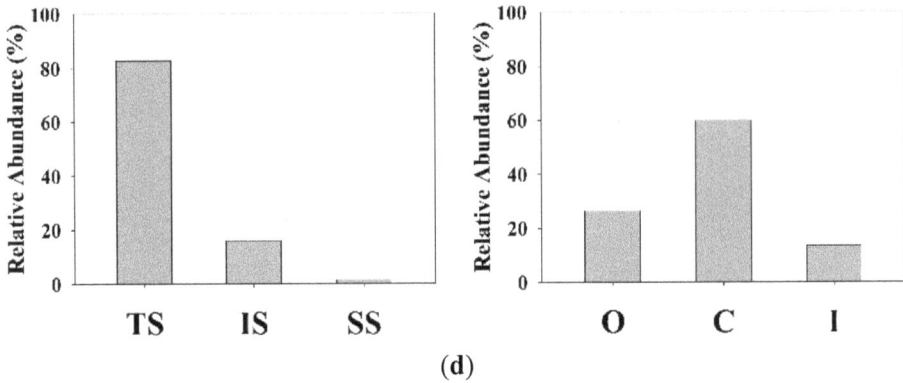

Figure 6. Comparison of tolerance and trophic species analysis in Nakdong River (TS: tolerant species, IS: intermediate species, SS: sensitive species, O: omnivores, C: carnivores, I: insectivores): (**a**) Reference: *Zacco-Coreoleuciscus* Community; (**b**) Group I: *Zacco-Opsarichthys* Community; (**c**) Group II: *Zacco-Opsarichthys* Community; (**d**) Group III: *Opsarichthys-Zacco* Community; and (**e**) Group IV: *Opsarichthys-Micropterus* Community.

3.7. Integrated Health Responses (IHRs), Based on Physical, Chemical, and Biological Parameters

The model of Integrated Health Responses (IHRs), based on the star-plot approach of Beliaeff and Burgeot [76], was used for a diagnosis of overall ecological river health. Mean values of IHRs model for the upstream to downstream were derived by integrating the physical habitat health (Qualitative Habitat Evaluation Index, QHEI), chemical health (nutrient pollution index, NPI), and biological health parameters (Index of Biological Integrity, IBI; Figure 7). Area score of Group I in upstream regions was 0.37, which was less than twice of the reference regions (0.62). In the Group I, the axis values of biological health (0.45) and physical health (0.52) were lower than the value (0.8) of chemical health, and the area of Group I was 62% of the reference condition, indicating a "fair" condition. The axis values of Group II were 0.35, 0.56 and 0.5, respectively, for the three variables of NPI, QHEI, and IBI. The area score of Group II was 2.6 times lower, compared to the values of the reference condition, indicating a "poor" condition, and also was lower than the regions of Group I (0.37). The lowest area score (0.17) in the star-plot were found in the Group III and this area value was similar to the regions of Group IV (area score = 0.18). The integrated health in Group III was judged as a "very poor" condition, and was same as Group IV. The star-plot analysis indicated that integrated river health, based on Integrated Health Responses (IHRs), was more impaired in the downstream regions (Group III and IV) than in the upstream (Group I and II) and reference regions. The impaired river health was due to greater impacts in biological health and chemical health than the physical health. Physical habitat health did not largely differ among the four regions, indicating not so significant in the health gradients of regions. In contrast, chemical health was most pronounced in the downstream of Group III and IV (axis value of Group III = 0.29, Group IV = 0.3) due to nutrient enrichment and organic matter pollutions of tributary river (*i.e.*, Geumho River), which is directly influenced by domestic wastewater disposal plants and the urban sewage.

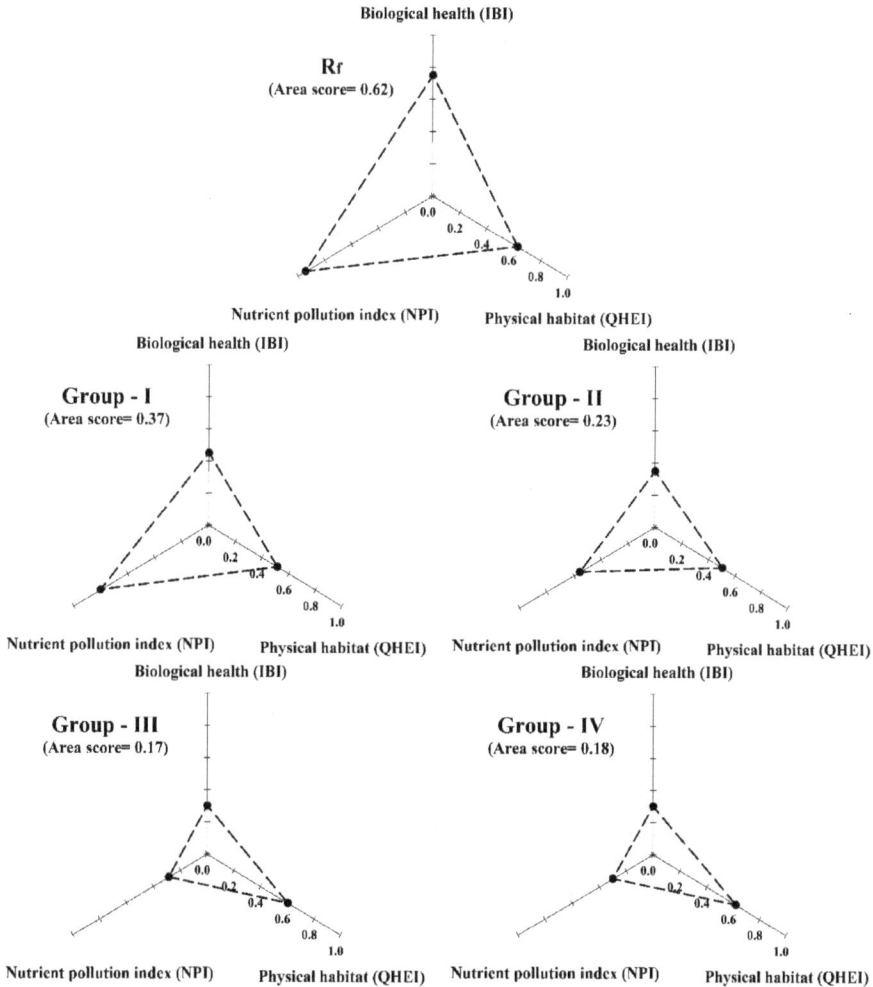

Figure 7. Integrated health responses (IHRs) using the star-plot area analysis in Nakdong River.

4. Conclusions

Integrated Health Responses (IHRs) in this study were determined by the integration of three multi-metric models of chemical water health (NPI), physical habitat health (QHEI) and biological health (IBI). Each metric model was developed separately for the application of IHRs model using a star-plot approach, and then the health conditions were determined by the comparison of the five reference sites. In the data analysis, the integrated ecological health, based on the mean of IHRs, was more impaired downstream than upstream, and this was mainly attributed to influences of point-sources and urban developments downstream. Thus, longitudinal gradients in the health from the upstream to downstream were evident in the three model, NPI, QHEI, and IBI. The model of Self-Organizing Map (SOM) at 16 sampling streams was matched to the longitudinal patterns of chemical and biological parameters from headwater to downstream. Statistical tests of principle component analysis (PCA) indicated that Group I was located in the region of the upstream and was closely associated with high N:P ratios in the ambient water. In contrast, Group IV was located in the downstream with nutrient enrichment and organic matter pollution. These results of PCA were also supported by spatial pattern analysis using the SOM model. Overall, this approach of the IHRs may be used as a key tool for the quantification of integrated ecological river health in the river ecosystems.

Acknowledgments: This research was supported by the projects of "Survey and Evaluation of Aquatic Ecosystem Health in Korea" funded by the Ministry of Environment and National Institute of Environmental Research (Korea) and "Basic Science Research Program (2013R1A1A4A01012939)" through the National Research Foundation of Korea (NRF) funded by the Ministry of Education (Korea).

Author Contributions: Ji Yoon Kim collected the field data, and performed the data analysis and manuscript writing under the supervise of Kwang-Guk An who is Ji Yoon Kim's M.S. thesis advisor.

Conflicts of Interest: The authors declare no conflict of interest.

References

1. Miserendino, M.L.; Brand, C.; di Prinzio, C.Y. Assessing urban impacts on water quality, benthic communities and fish in streams of the Andes Mountains, Patagonia (Argentina). *Water Air Soil Pollut.* **2008**, *194*, 91–110.
2. Lee, J.H.; Han, J.H.; Kumar, H.K.; Choi, J.K.; Byeon, J.K.; Choi, J.S.; Kim, J.K.; Jang, M.H.; Park, J.K.; An, K.G. National-level integrative ecological health assessments based on the index of biological integrity, water quality, and qualitative habitat evaluation index, in Korean rivers. *Ann. Limnol. Int. J. Limnol.* **2011**, *47*, S73–S89.

3. Lee, J.H.; An, K.G. Integrative restoration assessment of an urban stream using multiple modeling approaches with physical, chemical, and biological integrity indicators. *Ecol. Eng.* **2014**, *62*, 153–167.

4. Deacon, J.R.; Soule, S.A.; Smith, T.E. *Effects of Urbanization on Stream Quality at Selected Sites in the Seacoast Region in New Hampshire, 2001–2003*; USGS Scientific Investigations Report 2005-5103; U.S. Geological Survey: Reston, VA, USA, 2005; pp. 1–18.

5. Cooper, M.J.; Uzarski, D.G.; Burton, T.M.; Rediske, R.R. Macroinvertebrate community composition, chemical/physical variables, land use and cover, and vegetation types within a Lake Michigan drowned river mouth wetland. *Aquat. Ecosyst. Health Manag. Soc.* **2006**, *9*, 463–479.

6. Burcher, C.L.; Valett, H.M.; Benfield, E.F. The land-cover cascade: Relationships coupling land and water. *Ecology* **2007**, *88*, 229–242.

7. Lee, J.H.; An, K.G. Analysis of various ecological parameters from molecular to community level for ecological health assessments. *Korean J. Limnol.* **2010**, *43*, 24–34.

8. Yeom, D.H.; Lee, S.A.; Kang, G.S.; Seo, J.; Lee, S.K. Stressor identification and health assessment of fish exposed to wastewater effluents in Miho Stream, South Korea. *Chemosphere* **2007**, *67*, 2282–2292.

9. Kim, J.H.; Yeom, D.H. Population response of Pale Chub (*Zacco platypus*) exposed to wastewater effluents in Gap Stream. *Toxicol. Environ. Health Sci.* **2009**, *1*, 169–175.

10. Lee, J.H.; An, K.G. Seasonal dynamics of fish fauna and compositions in the Gap Stream along with conventional water quality. *Korean J. Limnol.* **2007**, *40*, 503–510.

11. Schlossser, I.J. Fish community structure and function along two habitat gradients in a headwater stream. *Ecol. Monogr.* **1982**, *52*, 395–414.

12. Gelwick, F.P. Longitudinal and temporal comparisons of riffle and pool fish assemblages in a Northeastern Oklahoma Ozark Stream. *Copeia* **1990**, *1990*, 1072–1082.

13. Edds, D.R. Fish assemblage structure and environmental correlations in Nepal's Gandaki River. *Copeia* **1993**, *1993*, 48–60.

14. Yoder, C.O. *The Integrated Biosurvey as a Tool for 852 Evaluation of Aquatic Life use Attainment and Impairment 853 in Ohio Surface Waters*; EPA 440-5-91-005; US EPA: Washington, DC, USA, 1991; pp. 110–122.

15. Karr, J.R. Assessment of biotic integrity using fish communities. *Fisheries* **1981**, *6*, 21–27.

16. Barbour, M.T.; Gerritsen, J.; Griffity, G.E.; Frydenborg, R.; McCarron, E.; White, J.S.; Bastian, M.L. A framework for biological criteria for Florida streams using benthic macroinvertebrates. *J. North Am. Benthol. Soc.* **1996**, *15*, 185–211.

17. Yoder, C.O.; Rankin, E.T. The role of biological indicators in a state water quality management process. *Environ. Monit. Assess.* **1998**, *51*, 61–88.

18. Fore, L.S.; Karr, J.R.; Conquest, L.L. Statistical properties of an index of biological integrity used to evaluate water resources. *Can. J. Fish. Aquat. Sci.* **1993**, *51*, 1077–1087.

19. Gore, J.A. *The Restoration of Rivers and Streams*; Butterworth: Boston, MA, USA, 1985; p. 280.

20. Brookes, A.; Shields, F.D., Jr. *River Channel Restoration: Guiding Principles for Sustainable Projects*; Wiley: Chichester, UK, 1996; p. 433.

21. Suter, G.W., II. A critique of ecosystem health concepts and indices. *Environ. Toxicol. Chem.* **1993**, *12*, 1533–1539.

22. Binns, N.A.; Eiserman, F.M. Quantification of fluvial trout habitat in Wyoming. *Trans. Am. Fish. Soc.* **1979**, *108*, 215–228.

23. Winget, R.N.; Mangum, F.A. *Biotic Condition Index: Integrated Biological, Physical, and Chemical Stream Parameters for Management*; U.S. Department of Agriculture, Forest Service, Intermountain Region: Ogden, UT, USA, 1979; p. 51.

24. Platts, W.S.; Megahan, W.F.; Minshall, G.W. *Methods for Evaluating Stream, Riparian, and Biotic Conditions*; U.S. Department of Agriculture. Forest Service. Intermountain Forest and Range Experiment Station: Ogden, UT, USA, 1983; p. 70.

25. Barbour, M.T.; Gerritsen, J.; Snyder, B.D.; Stribling, J.B. *Rapid Bioassessment Protocols for Use in Streams and Wadeable Rivers: Periphyton, Benthic Macroinvertebrates and Fish*, 2nd ed.EPA 841-B-99-002; U.S. Environmental Protection Agency, Office of Water: Washington, DC, USA, 1999; p. 235.

26. Plafkin, J.L.; Barbour, M.T.; Porter, K.D.; Gross, S.K.; Hughes, R.M. *Rapid Bioassessment Protocols for Use in Streams and Rivers: Benthic Macroinvertebrate and Fish*; EPA/444/4-89-001; Office of Water Regulations and Standards; US EPA: Washington, DC, USA, 1989; pp. 1–34.

27. Kelly, M.G.; Whitton, B.A. The trophic diatom index: A new index for monitoring eutrophication in rivers. *J. Appl. Phycol.* **1995**, *7*, 433–444.

28. Kelly, M.G.; Cazaubon, A.; Coring, E.; Dell'Uomo, A.; Ector, L.; Goldsmith, B.; Guasch, H.; Hurlimann, J.; Jarlman, A.; Kawecka, B.; *et al.* Recommendations for the routine sampling of diatoms for water quality assessments in Europe. *J. Phycol.* **1998**, *10*, 215–224.

29. Lang, C.; l'Eplattenier, G.; Reymond, O. Water quality in rivers of Western Switzerland: Application of an adaptable index based on benthic invertebrates. *Aquat. Sci.* **1989**, *51*, 224–234.

30. Lang, C.; Reymond, O. An improved index of environmental quality for Swiss rivers based on benthic invertebrates. *Aquat. Sci.* **1995**, *57*, 172–180.

31. Munkittrick, K.R.; McMaster, M.E.; Kraak, G.V.E.; Portt, C.; Gibbons, W.N.; Farwell, A.; Gray, M. *Development of Methods for Effects-Driven Cumulative Effects Assessment Using Fish Populations: Moose River Project*; A Technical Publication of SETAC: Pensacola, FL, USA, 2000; p. 236.

32. Larsson, D.G.J.; Hällman, H.; Förlin, L. More male fish embryos near a pulp mill. *Environ. Toxicol. Chem.* **2000**, *19*, 2911–2917.

33. Robinet, T.T.; Feunteun, E.E. Sublethal effects of exposure to chemical compounds: A cause for decline in Atlantic eels. *Ecotoxicology* **2002**, *11*, 265–277.

34. Adams, S.M.; Ham, K.D.; Greeley, M.S.; LeHew, R.F.; Hinton, D.E.; Saylor, C.F. Downstream gradients in bioindicator response: Point source contaminant effects on fish health. *Can. J. Fish Aquat. Sci.* **1996**, *53*, 2177–2187.

35. Environment Canada. *Metal Mining Guidance Document for Aquatic Environmental Effects Monitoring*; Environment Canada: Gatineau, QC, Canada, 2004.

36. Simon, T.P. *Biological Response Signatures: Indicator Patterns Using Aquatic Communities*; CRC Press: Boca Raton, FL, USA, 2003; p. 600.

37. Pyron, M.; Lauer, T.E.; le Blanc, E.; Weitzel, D.; Gammon, J.R. Temporal and spatial variation in an index of biological integrity for the middle Wabash River, Indiana. *Hydrobiologia* **2008**, *600*, 205–214.

38. Karr, J.R.; Dionne, M. Designing surveys to assess biological integrity in lakes and reservoirs. Biological Criteria Research and Regulation, *Proceedings of a Symposium*; EP-440/5-91-005; Arlington, VA, USA, 12–13 December 1990, U.S. EPA: Office of Waters, Washington, DC, USA, 1991; pp. 62–72.

39. U.S. EPA. *Fish Field and Laboratory Methods for Evaluating the Biological Integrity of Surface Waters*; EPA 600-R-92-111; Environmental Monitoring systems Laboratory-Cincinnati office of Modeling, Monitoring systems, and quality assurance Office of Research Development, U.S. EPA: Cincinnati, OH, USA, 1993; p. 348.

40. Oberdorff, T.; Hughes, R.M. Modification of an index of biotic integrity based on fish assemblages to characterize rivers of the Seine Basin, France. *Hydrobiologia* **1992**, *228*, 117–130.

41. Lyons, J.; Navarro-Perez, S.; Cochran, P.A.; Santana, E.; Guzman-Arroyo, M. Index of biotic integrity based on fish assemblages for the conservation of streams and rivers in west-central Mexico. *Conserv. Biol.* **1995**, *9*, 569–584.

42. Soto-Galera, E.; Diaz-Pardo, E.; López-López, E.; Lyons, J. Fish indicator of environmental quality in the Rio Lerna Basin, México. *Aquat. Ecosyst. Health Manag.* **1998**, *1*, 267–276.

43. Hugueny, B.S.; Camara, B.; Samoura, B.; Magassouba, M. Applying an index of biotic integrity based on communities in a West African river. *Hydrobiologia* **1996**, *331*, 71–78.

44. Kamdem-Toham, A.; Teugels, G.G. First data on an index of biotic integrity (IBI) based on fish assemblages for the assessment of the impact of deforestation in a tropical West African system. *Hydrobiologia* **1999**, *397*, 29–38.

45. Kleynhans, C.J. The development of a fish index to assess the biological integrity of South African rivers. *Water SA* **1999**, *25*, 265–278.

46. Munkittrick, K.R.; Dixon, D.G. A holistic approach to ecosystem health assessment using fish population characteristics. *Hydrobiologia* **1989**, *188-189*, 122–135.

47. Singh, N.P.; McCoy, M.T.; Tice, R.R.; Schneider, E.L. A simple technic for quantitation of low levels of DNA damage in individual cells. *J. Exp. Cell Res.* **1988**, *175*, 184–191.

48. Adams, S.M.; Greeley, M.S. Ecotoxicological indicators of water quality: Using multi-reponse indicators to assess the health of aquatic ecosystems. *Water Air Soil Pollut* **2000**, *123*, 103–115.

49. Adams, S.M. *Biological Indicators of Aquatic Ecosystem Stress*; American Fisheries Society: Bethesda, MD, USA, 2002; pp. 1–11.

50. Anderson, R.O.; Gutreuter, S.J. Length, weight, and associated structural indices. In *Fisheries Techniques*; Nielsen, L.A., John, D.L., Eds.; American Fisheries Society: Bethesda, MD, USA, 1983; pp. 283–300.

51. Goede, R.W.; Barton, B.A. Organismic indices and autopsy-based assessment as indicators of health and condition in fish. In *Biological Indicator of Stress in Fish*; Adams, S.M., Ed.; American Fisheries Society: Bethesda, MD, USA, 1990; pp. 93–108.

52. U.S. Geological Survey (USGS). *Biomonitoring of Environmental Status and Trends (BEST) Program: Selected Methods for Monitoring Chemical Contaminants and Their Effects in Aquatic Ecosystems*; UITR-2000-0005; Information and Technology Report: Columbia, OR, USA, 2000; p. 81.

53. Triebskorn, R.; Böhmer, J.; Braunbech, T.; Honnen, W.; Köhler, H.R.; Lehmann, R.; Oberemm, A.; Schwaiger, J.; Segner, H.; Schüürmann, G.; *et al*. The project VALIMAR (VALIdation of bioMARkers for the assessment of small stream pollution): Objectives, experimental design, summary of results, and recommendations for the application of biomarkers in risk assessment. *J. Aquat. Ecosyst. Stress Recovery* **2001**, *8*, 161–178.

54. Adams, S.M. Biomarker/bioindicator response profiles of organisms can help differentiate between sources of anthropogenic stressors in aquatic ecosystems. *Biomarkers* **2001**, *6*, 33–44.

55. Strahler, A.N. Quantitative analysis of watershed geomorphology. *Am. Geophys. Union Trans.* **1957**, *38*, 913–920.

56. Hughes, R.M.; Heiskary, S.A.; Mathews, W.J.; Yoder, C.O. Use of ecoregions in biological monitoring. In *Biological Monitoring of Aquatic Systems*; Loeb, S.L., Spacie, A., Eds.; CRC Press: Boca Raton, FL, USA, 1994; pp. 125–151.

57. An, K.G.; Yeom, D.H.; Lee, S.K. Rapid Bioassessments of Kap Stream Using the Index of Biological Integrity. *Korean J. Environ. Biol.* **2001**, *19*, 261–269.

58. MOE/NIER. *The Survey and Evaluation of Aquatic Ecosystem Health in Korea*; The Ministry of Environment/National Institute of Environmental Research (NIER): Incheon, Korea, 2008.

59. Ohio EPA. *Biological Criteria for the Protection of Aquatic life: Volume III. Standardized Biological Field Sampling and Laboratory Method for Assessing Fish and Macroinvertebrate Communities*; Division of Surface Water, Ecological Assessent Section: Ohio, OH, USA, 1989; p. 66.

60. Kim, I.S.; Park, J.Y. *Freshwater Fish of Korea*; Kyohak Publishing: Seoul, Korea, 2002; p. 465.

61. Nelson, J.S. *Fishes of the World*, 4th ed.; John Wiley & Sons: New York, NY, USA, 2006; p. 624.

62. Sanders, R.E.; Milter, R.J.; Yondr, C.O.; Rankin, E.T. The use of external deformities, erosion, lesions, and tumors (DELT anormalies) in fish assemblages for characterizing aquatic resources: A case study of seven Ohio streams. In *Assessing the Sustainability and Biological Integrity of Water Resources Using Fish Communities*; Simon, T.P., Ed.; CRC Press: Boca Raton, FL, USA, 1999; pp. 225–245.

63. An, K.G.; Kim, D.S.; Kong, D.S.; Kim, S.D. Integrative assessments of a temperate stream based on a multimetric determination of biological integrity, physical habitat evaluations, and toxicity tests. *Bull. Environ. Contam. Toxicol.* **2004**, *73*, 471–478.

64. Crumpton, W.G.; Isenhart, T.M.; Mitchell, P.D. Nitrate and organic N analyses with second-derivative spectroscopy. *Limnol. Oceanogr.* **1992**, *37*, 907–913.

65. Prepas, E.E.; Rigler, F.A. Improvements in qualifying the phosphorus concentration in lake water. *Can. J. Fish. Aquat. Sci.* **1982**, *39*, 822–829.

66. APHA. *Standard Methods for the Examination of Water and Wastewater*, 21st ed.; American Public Health Association: New York, NY, USA, 2005.

67. Dodds, W.K.; Jones, J.R.; Welch, E.B. Suggested classification of stream trophic state: Distributions of temperate stream types by chlorophyll, total nitrogen and phosphorus. *Water Res.* **1998**, *32*, 1455–1462.

68. Lee, H.J.; AN, K.G. The Development and Application of Multi-metric Water Quality Assessment Model for Reservoir Managements in Korea. *Korean J. Limnol.* **2009**, *42*, 242–252.

69. MOE/NIER. *Researches for Integrative Assessment Methodology of Aquatic Environments (III): Development of Aquatic Ecosystem Health Assessment and Evaluation System*; The Ministry of Environment/National Institute of Environmental Research (NIER): Incheon, Korea, 2006.

70. Karr, J.R.; Fausch, K.D.; Angermeier, P.L.; Yant, P.R.; Schlosser, I.J. Assessing biological integrity in running waters: A method and its rationale. *Ill. Nat. Hist. Surv. Spec. Publ.* **1986**, *5*, 28.

71. U.S. EPA. *Environmental Monitoring and Assessment Program: Integrated Quality Assurance Project Plan for the Surface Waters Resource Group*; 1994 Activities, Rev.2.00. EPA 600/X-91/080; U.S. EPA: Las Vegas, NV, USA, 1994.

72. An, K.G.; Lee, J.Y.; Bae, D.Y.; Kim, J.H.; Hwang, S.J.; Won, D.H.; Lee, J.K.; Kim, C.S. Ecological assessments of aquatic environment using multi-metric model in major nationwide stream watersheds. *J. Korean Soc. Water Qual.* **2006**, *22*, 796–804.

73. Rankin, E.T.; Yoder, C.O. Adjustments to the index of biotic integrity: A summary of Ohio experiences and some suggested modifications. In *Assessing the Sustainability and Biological Integrity of Water Resources Using Fish Communities*; Simon, T.P., Ed.; CRC Press: Boca Raton, FL, USA, 1999; p. 672.

74. Yeom, D.H.; Adams, S.M. Assessing effects of stress across levels of biological organization using an aquatic ecosystem health index. *Ecotoxicol. Environ. Saf.* **2006**, *67*, 286–295.

75. Lee, J.H.; Kim, J.H.; Oh, H.M.; An, K.G. Multi-level stressor analysis from the DNA/biochemical level to community levels in an urban stream and integrative health response (IHR) assessments. *J. Environ. Sci. Health Part A* **2013**, *48*, 211–222.

76. Beliaeff, B.; Burgeot, T. Integrated biomarker response: A useful tool for ecological risk assessment. *Environ. Toxicol. Chem.* **2002**, *21*, 1316–1322.

77. Kim, J.H.; Yeom, D.H.; An, K.G. Diagnosis of Sapgyo Stream watershed using the approach of integrative star-plot area. *Korean J. Limnol.* **2010**, *43*, 356–368.

78. Kohonen, T. Self-Organized Formation of Topologically Correct Feature Maps. *Biol. Cybern.* **1982**, *43*, 59–63.

79. Kohonen, T. *Self-Organizing Maps*; Springer: Berlin, Germany, 2001; p. 501.

80. Vesanto, J. Data Exploration Process Based on the Self-Organizing Map. Ph.D. Thesis, Helsinki University of Technology, Espoo, Finland, 2002. p. 86.

81. Vesanto, J.; Alhoniemi, E.; Himbrg, J.; Kiviluoto, K.; Parviainen, J. Self-organizing map for data mining in MATLAB: The SOM toolbox. *Simul. News Eur.* **1999**, *99*, 16–17.

82. McCune, B.; Mefford, M.J. *PC-ORD. Multivariate Analysis of Ecological Data. Version 4.0*; MjM Software, Gleneden Beach: Oregon, UT, USA, 1999.

83. An, K.G.; Jones, J.R. Temporal and spatial patterns in ionic salinity and suspended solids in a reservoir influenced by the Asian monsoon. *Hydrobiologia* **2000**, *436*, 179–189.

84. Forsberg, G.; Ryding, S.O. Eutrophication parameter and trophic state indices in 30 Swedish waste receiving lakes. *Arch. Hydrobiol.* **1980**, *89*, 189–207.

85. Sakamoto, M. Primary production by phytoplankton community in some Japanese lakes and its dependence on lake depth. *Arch. Hydrobiol.* **1966**, *62*, 1–28.

86. Smith, V.H. Low nitrogen to phosphorus ratios favor dominance by blue-green algae in lake phytoplankton. *Science* **1983**, *221*, 669–671.

87. Cude, C. Oregon water quality index: A tool for evaluating water quality management effectiveness. *J. Am. Water Res. Assoc.* **2001**, *37*, 125–137.

88. Dodds, W.K.; Welch, E.B. Establishing nutrient criteria in stream. *J. N. Am. Bethol. Soc.* **2000**, *19*, 186–196.

89. Dodds, W.K. Trophic state, eutrophication and nutrient criteria in streams. *Trends Ecol. Evol.* **2007**, *22*, 669–675.

90. Nives, S.G. Water quality evaluation by index in Dalamatia. *Water Res.* **1999**, *33*, 3423–3440.

91. Walker, D.; Jakovljević, D.; Savić, D.; Radovanović, M. Multi-criterion Water Quality Analysis of the Danube River in Serbia: A Visualisation Approach. *Water Res.* **2015**, *79*, 158–172.

92. Fujimoto, N.; Sudo, R. Nutrient-limited growth of Microcystis aerugimosa and phormidium tenue and competition under narious N:P supply ratios and temperatures. *Limnol. Oceonogr.* **1997**, *42*, 250–256.

93. Seppala, J.; Tamminen, T.; Kaitala, S. Experimental evaluation of nutrient limitation of phytoplankton communities in the Gulf of Riga. *J. Mar. Syst.* **1999**, *23*, 107–126.

94. Han, J.H.; An, K.G. Chemical Water Quality and Fish Community Characteristics in the Mid- to Downstream Reach of Geum River. *Korean J. Environ. Biol.* **2013**, *31*, 180–188.

95. Choi, J.W.; Han, J.H.; Park, C.S.; Ko, D.G.; Kang, H.I.; Kim, J.Y.; Yun, Y.J.; Kwon, H.H.; An, K.G. Nutrients and sestonic chlorophyll dynamics in Asian lotic ecosystems and ecological stream health in relation to land-use patterns and water chemistry. *Ecol. Eng.* **2015**, *79*, 15–31.

96. An, K.G.; Jung, S.H.; Choi, S.S. An Evaluation on Health Conditions of Pyong-Chang River using the Index of Biological Integrity (IBI) and Qualitative Habitat Evaluation Index (QHEI). *Korean J. Limnol.* **2001**, *34*, 153–165.

97. An, K.G.; Kim, J.H. A Diagnosis of Ecological health Using a Physical Habitat Assessment and Multimetric Fish Model in Daejeon Stream. *Korean J. Limnol.* **2005**, *38*, 361–371.

98. Bae, E.Y.; An, K.G. Stream Ecosystem Assessments, based on a Biological Multimetric Parameter Model and Water Chemistry Analysis. *Korean J. Limnol.* **2006**, *39*, 198–208.

Environmental Factors Structuring Fish Communities in Floodplain Lakes of the Undisturbed System of the Biebrza River

Katarzyna Glińska-Lewczuk, Paweł Burandt, Roman Kujawa, Szymon Kobus, Krystian Obolewski, Julita Dunalska, Magdalena Grabowska, Sylwia Lew and Jarosław Chormański

Abstract: We evaluated the influence of habitat connectivity and local environmental factors on the distribution and abundance of functional fish groups in 10 floodplain lakes in the Biebrza River, northeastern Poland. Fish were sampled by electrofishing, and 15 physico-chemical parameters were recorded at three sampling sites at each lake in the period of 2011–2013. A total of 18,399 specimens, belonging to 23 species and six families, were captured. The relationships between environmental factors and fish communities were explored with the use of canonical correspondence analysis (CCA). Sampling sites were grouped based on fish communities using a hierarchical cluster analysis (HCA). Along a lateral connectivity gradient from lotic to lentic habitats (parapotamic–plesiopotamic–paleopotamic), the proportions of rheophilic species were determined as 10:5:1, whereas the proportion of limnophilic species was determined as 1:2:5. The predominant species were the roach (*Rutilus rutilus*), and pike (*Esox lucius*) in parapotamic lakes, rudd (*Scardinius erythropthalmus*) and pike in plesiopotamic lakes, and sunbleak (*Leucaspius delineates*) and Prussian carp (*Carassius auratus gibelio*) in paleopotamic lakes. The findings indicated that the composition and abundance of fish communities are determined by lake isolation gradient, physico-chemical parameters and water stage. Although intact riverine ecosystems may promote fish biodiversity, our findings suggest that lateral connectivity between the main channel and floodplain lakes is of utmost importance. Thus, the conservation of fish biodiversity requires the preservation of this connectivity.

Reprinted from *Water*. Cite as: Katarzyna Glińska-Lewczuk, Paweł Burandt, Roman Kujawa, Szymon Kobus, Krystian Obolewski, Julita Dunalska, Magdalena Grabowska, Sylwia Lew and Jarosław Chormański Environmental Factors Structuring Fish Communities in Floodplain Lakes of the Undisturbed System of the Biebrza River. *Water* **2016**, *8*, 146.

1. Introduction

Natural river floodplains consist of complex habitats differing in hydrological connectivity, which affects fish community dynamics [1–3]. In the temperate climate zone, large undisturbed floodplains have often been disrupted by channel regulation that exerted direct and indirect effects on habitat heterogeneity, successional trajectories and, ultimately, the ecological integrity of rivers [4,5].

The functional feature of intact alluvial floodplains of meandering rivers is a mosaic of lotic and lentic ecosystems, including the river and its side channels, tributary streams and cut-off channels. Due to the variation in the connectivity, laterally across the floodplain, a distinct zonation of the habitats has been widely reported both for tropical [6–8] and temperate rivers [9,10]. Differences in the connection of floodplain lakes with the river channel determine the availability of nutrients and the degree to which processes such as primary productivity and decomposition are controlled by the river [11–13]. Floodplain ecosystems connected to the river are particularly open to exchange of matter with the river which leads to higher concentrations of macroelements, while ecosystems isolated from the nearby river for most of the year acquire a lentic character that promotes autogenic, mainly organic, matter cycling. An increased flow rate throughout bi-connected water bodies is beneficial to organisms in that it transports food, oxygen, nutrients, and particulate and dissolved organic matter [14,15].

The hydrological integration between river and cut-off channels is a significant habitat parameter for species that require different aquatic microhabitats in the course of their life cycle, e.g., certain species of fish [4,16–18]. Fish communities in European lowland riverine ecosystems are composed of rheophilic (require flowing water to spawn), eurytopic (habitat generalists) and limnophilic (found in stagnant and strongly vegetated floodplain water bodies) fish species guilds [4,19,20] that contribute to the overall high species diversity [4]. For example, the ide *Leuciscus idus* requires flowing water habitats, whereas the crucian carp *Carassius carassius* needs a single stagnant floodplain lake that exists over a long period.

Floodplain lakes are ecosystems with diverse fish species adapted to periods of low and high water stages (flood-pulse), which affect any wetland water quality parameters [11,21–23]. Higher water stages promote greater nutrient availability, aquatic primary production, allochthonous inputs, and secondary production, which are especially beneficial for early life stages of fish in floodplain habitats. In contrast, low-water conditions lead to the contraction of marginal aquatic habitats, decay of aquatic macrophytes, and higher densities of aquatic organisms, including phytoplankton and zooplankton in floodplain water bodies [24]. During low water periods, non-flowing ecosystems have been recognized as having limited conditions for light penetration and thereby limited photosynthesis. In turn, summer oxygen deficits are attributed to shading by emergent and floating vegetation, high biological oxygen demand and limited aeration [25].

The natural hydrological regime is one of the key drivers of ichthyofauna development [2,8]. Floodplain lakes typically serve as nursery for young fish whereas adults live in the main channel or connected side arms [4,18]. Welcomme and Halls [8] observed that extensive flooding increased the area available for spawning sites and provided fish with more food and better shelter opportunities, whereas the duration of the flood influenced the time during which fish could grow and find shelter from predators. In several studies, fish migrated to floodplain water bodies, in particular to lentic habitats, in search of refuge during floods [26–28]. When water levels drop, fish either migrate back to the river and become a source of food for resident piscivores or remain in isolated floodplain water bodies [2,6,29].

An increase in the reductions in landscape connectivity, ecological functioning and ecosystem biodiversity has driven initiatives to improve the ecological status of rivers, e.g., the European Union Water Framework Directive (2000/60/EEC) [30], and to protect biological diversity, e.g., the Habitats Directive (92/43/EEC) [31] and Agenda 21 of the Rio Convention and the Convention on Biological Diversity. According to Welcomme *et al.* [20], fish environmental guilds could be used as a tool for assessing the ecological status of rivers. Further, the knowledge of differences in the responses of functional fish groups to environmental factors is useful for predicting the effects of future environmental manipulations (e.g., changes in hydrology and connectivity) on fish communities in various aquatic systems. Achieving good ecological status by promoting fish abundance and diversity involves the creation of habitats that are functionally similar to natural lowland river-floodplain ecosystems. Nevertheless, this approach requires a relevant reference area to test whether environmental parameters are affecting the qualitative and quantitative structure of hydrobionts. The Biebrza river provides an opportunity to realize this test as it is an undisturbed system presenting variable levels of lateral connectivity with sequential shift in fish community composition from rheophilic to eurytopic to limnophilic fish species guilds, as it was reported *ca.* 30 years ago by Witkowski [14,32]. Moreover, unlike many rivers in Europe, the floodplain lakes in the middle and lower section of the Biebrza River have not been disturbed by hydraulic structures, excessive nutrients or sediments introduced by runoffs from the surrounding farmland.

The aim of this study was to determine whether lateral connectivity and environmental parameters are influencing the qualitative and quantitative structure of fish communities in floodplain lakes. For this purpose, we sampled 10 natural floodplain lakes and the river channel in the middle section of the Biebrza River, (NE Poland) depending on their connectivity and habitat diversity.

2. Materials and Methods

2.1. Study Area and Sampling Sites

Biebrza is a medium-sized low-gradient river in NE Poland. Its catchment occupies a total area of 7057 km^2, and the floodplain covers an area of 1950 km^2. The Biebrza River Valley features the Upper, Middle and Lower basins, which have been classified based on differences in their geomorphologic structure [33,34]. The Middle Basin is approximately 33 km long and spreads along the floodplain between the villages of Sztabin and Osowiec. The river intersects boggy meadows and marshes. Throughout its course, the river forms a large number of old riverbeds and floodplain water bodies in different stages of succession. The catchment was extensively drained in the mid-1970s, but Biebrza's floodplain escaped alteration. In the middle section its natural landscape and flood-pulse pattern have been nearly entirely preserved [24,33]. Excluding a 10-km-long section, the river is part of the Biebrza National Park, and it is protected under the Ramsar Convention.

The hydrological regime in Biebrza's middle course has a natural pattern. Mean annual amplitude of water levels is within the range of 264 cm [34]. The river is characterized by a snowy, distinct flood-pulse regime, with long-term spring floods. During spring floods, the narrow river swells to form a vast shallow impoundment, locally up to 1 km wide, that lasts for several months. The average multiannual flow (1984–2013) measured at a gauge located in Osowiec was 22.78 m$^3 \cdot$s^{-1} (Q_{min} = 3.08 m$^3 \cdot$s^{-1}, Q_{max} = 360.00 m$^3 \cdot$s^{-1}). During our study (2011–2013), frequency of high water stages (HWL) was higher in comparison to the multiannual period of 1984–2010 [24]. Inundation periods in 2011, 2012, and 2013, calculated as the percentage of days in a year when water table exceeded bankfull level (BL = 107.70 m above sea level), were 42%, 35% and 48%, respectively (Figure 1B). Stages below mean low water level (MLW) in 2011 lasted 8% of the year, in 2012 as much as 23% while in 2013 as much as 19%.

2.2. Environmental Description of Study Sites

Sampling sites were located in 10 floodplain lakes and the main river channel of the Biebrza River (Figure 1). The lakes were chosen based on a wide range of lake morphometric characteristics (e.g., area and connectivity to the river channel) as well as hydrological and water quality parameters. Lakes with passable inlets and outlets may have a different hydrologic cycle than lakes without inlets or outlets, which facilitates fish movement. The analysed water bodies were classified into four types with different hydrological connectivity and water retention patterns according to the typology proposed by Amoros and Roux [1] (Figure 1):

Eupotamic—the main river channel (Biebrza 1 to 3);

Parapotamic—lotic side-channels (by-passes) with flowing water: Stara Rzeka (STR), Mostek (MOS) and Czerwony Domek (CZD);

Plesiopotamic—semi-lotic abandoned meanders, permanently connected with the river by a downstream arm: Bocianie Gniazdo (BOC), Klewianka (KLE), Tur (TUR) and Glinki (GLI);

Paleopotamic—lentic side channels and depressions filled with stagnant water and isolated from the river unless flooded: Budne (BUD), Bednarka (BED) and Fosa (FOS).

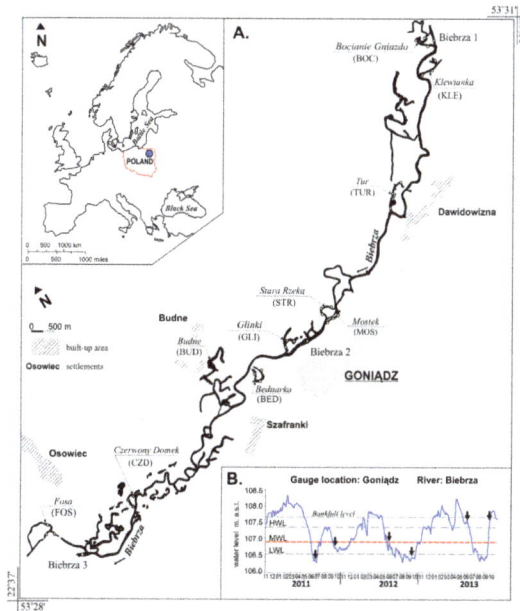

Figure 1. (**A**) Location of the study area in the Biebrza River; (**B**) Sampling events indicated by arrows on the background of hydrological situation during the period of study. Open circles indicate sampling sites located on the surveyed floodplain lakes.

Parapotamic lakes belong to the youngest water bodies that are permanently bi-connected to the river channel. Those lakes form natural by-pass channels, with an area of 0.62–3.31 ha, maximum depths of 1.4–3.4 m and a water table with a relatively low macrophyte cover (15%–25%) with a predominance of the Myriophyllo-Nupharetum plant association (Table 1). Lake banks were occupied by natural marshes with a narrow belt of plants characteristic of Scirpo-Phragmitetum communities and *Phragmites australis* (Cav.)Trin. ex Steud. or *Acorus calamus* L. The group of lakes was characterized by medium to coarse-grain mineral substrates (*i.e.*, sand or gravel), depending on the scouring flow velocity.

357

Table 1. Selected morphological parameters of the studied floodplain lakes in the Middle Basin of the Biebrza River. Denotations: L, length; W, width; h_{av}, mean depth; h_{max}, maximum depth; A, area; Eme, emerged vegetation; Sub, submerged vegetation.

Floodplain Lake (local Name)	Geographical Coordinates	Type of Connection	L (m)	W (m)	A (ha)	Distance from the River Channel			Depth		Sub-Strate	Macrophyte Cover	
						Upstream Arm (m)	Downstream Arm (m)	Max. Distance (m)	h_{av} (m)	h_{max} (m)		Emerged (%)	Submerged (%)
Starra Rzeka	N:53°30'0.26" E:22°44'37.2"	Parapotamic (lotic)	1380	24	3.31	0	0	523	2.2	3.4	mineral (gravel, pebbels)	12	7
Czerwony domek	N:53°29'21.53" E:22°39'54.54"	Parapotamic (lotic)	548	17	1.38	0	0	136	1.8	3.3	mineral (gravel)	14	11
Mostek	N:53°29'35.9" E:22°44'15.95"	Parapotamic (lotic)	306	20	0.62	0	0	30	1.0	1.4	mineral (sand, gravel)	10	5
Bocianie Gniazdo	N:53°31'15.47" E:22°47'55.62"	Plesio-potamic (semi-lotic)	569	25	1.43	104	0	304	2.8	5.5	mineral (sand, gravel), organic	18	29
Klewianka	N:53°30'57.34" E:22°47'55.87"	Plesio-potamic (semi-lotic)	520	22	1.41	80	0	260	2.6	4.1	mineral (sand, gravel), organic	21	36
Tur	N:53°30'15.68" E:22°46'17.54"	Plesio-potamic (semi-lotic)	678	21	2.03	51	0	150	2.2	3.9	organic, mineral (sand)	18	31
Glinki	N:53°29'40.29" E:22°43'28.94"	Plesio-potamic (semi-lotic)	459	32	1.48	150	0	226	2.2	5.5	organic, mineral (sand)	24	59
Bednarka	N:53°29'31.51" E:22°42'43.46"	Paleo-potamic (lentic)	740	29	2.16	50	42	313	1.6	2.6	organic	38	59
Budne	N:53°29'56.68" E:22°42'14.06"	Paleo-potamic (lentic)	1652	26	4.41	407	380	680	1.2	1.8	organic	21	76
Fosa	N:53°29'32.49" E:22°38'36.27"	Paleo-potamic (lentic)	1360	31	4.30	547	792	817	2.9	5.6	organic, mineral (sand)	10	28

2.3. Environmental Data

Water for chemical analyses was sampled twice a year, in June and September of each year of the study (2011–2013), from 10 floodplain lakes (30 sampling points) and the Biebrza River (3 sampling points), simultaneously with fish catches. *In situ* measurements of dissolved oxygen (DO), pH, electrolytic conductivity (EC), total dissolved solids (TDS) and chlorophyll-a levels were performed using the YSI 6600R2™ calibrated multiprobe (Yellow Springs, OH, USA). Water transparency was measured with Secchi disc (20 cm in diameter). The concentrations of phosphates, nitrates, nitrites and ammonium ions were determined in a laboratory with the use of standard analytical methods [35]. Total organic carbon (TOC) levels were determined in unfiltered samples. Dissolved organic carbon (DOC) was quantified after the samples had been passed through nitrocellulose membrane filters with a pore size of 0.45 μm (Millipore). TOC and DOC analyses were conducted by high-temperature combustion (HTC) (Shimadzu TOC 5000 analyzer, Tokyo, Japan), and performed according to the protocol described by Dunalska *et al.* [36].

Lake morphometric variables were estimated once and referred to mean water level. Mean depth measurements of each lake were determined by traveling zigzag patterns across each lake with a portable depth finder (echosonde Humminbird) attached to a GPS receiver (TOPCON) and recorded every 15 s. Maximum depth of each lake was determined from these measurements. In this study, we employed three water stage categories of high (HW), mean (MW) and low (LW) water based on the data provided by the Institute of Meteorology and Water Management in Poland (IMGW) for the Biebrza River at Osowiec. Furthermore, in order to develop stream flow hydrographs for use in flood extend estimation, continuously recording water depth logger (MiniDiver, Van Essen Instruments B.V., Delft, The Netherlands) was installed in the Biebrza River channel at the town of Goniądz. The logger was set to take measurements at approximately 15 min intervals. The measurement protocol was described by Grabowska *et al.* [24].

2.4. Fish Sampling

Fish were sampled by the electrofishing method [37]. Fish were caught from a boat along three 80–200 m transects (upstream arm, downstream arm and the middle section) along each of the 10 floodplain lakes (30 sampling sites) and from the river channel (3 sampling sites) in accordance with standard PL-EN 14011 [38] (Figure 1). Single-pass electrofishing was performed along one bank, in a distance of 2–3 m from the bank, with repeated immersion of anode for 20–30 s. Total time of a catch at one site lasted ~20 min.

Stunned fish were collected with nets and placed in tanks with aerators. Tank water had to be aerated to keep the catch alive and to minimize damage associated with handling and holding. Fish were identified to species level, counted and measured. Total fish length was recorded in mm, and fish weight in g. Large and medium-sized specimens were weighed individually. The weights of small individuals were calculated based on the length-weight ratios determined from the first sampling period. Immediately after the measurements, all fish were released into a calm area near the capture site. For each species in every catch, abundance was determined as the number of fish per ha based on individual fish counts, and biomass was calculated based on weight extrapolations for the estimated area of the electrofishing transect [14]. However, the last two parameters should be considered as rough estimations and used for comparative purposes.

2.5. Data Analysis

Fish biodiversity, taxonomic guilds and functional guilds were examined. Quantitative data were analyzed and converted to biocenotic indices: constancy of occurrence and dominance according to the formulas used by Obolewski *et al.* [39]. The constancy of occurrence (C, %) defines the occurrence of a given species within a single biocensosis. It has been calculated according to the formula $C = 100 \, n_a/N_n$, where: n_a-number of sites where the species was noted; and N_n-total number of sites. The values of constancy of occurrence allowed for the following classification of consecutive species: euconstants > 75.00%; constants 50.01%–75.00%; subconstants 30.01%–50.00%; accesoric taxa 15.00%–30.00% and accidental taxa < 15.00%. Dominance index (D, %) was calculated according to the following formula: $D = 100 \, n_i/N$, where: n_i—number of specimens of species "*I*" in the sample, N-number of all specimens in the sample. The following levels of dominance have been applied: eudominants >10.00%; dominants 5.10%–10.00%; subdominants 2.10%–5.00%; and recedents $\leqslant$ 2.00%. Additionally, the Shannon–Wiener's diversity (H') and the Pielou's evenness (J') indices were calculated. Biodiversity metrics were calculated using the diversity modules available in the PAST ver. 3.02^{TM} software [40]. Significant differences in density and biomass between functional groups of floodplain lakes as well as between seasons and water levels were determined by one-way and two-way ANOVA followed by Duncan's multiple range test at $p \leqslant 0.05$. All of these calculations were performed in Statistica 10.0 PL (StatSoft, Tulsa, OK, USA, 2011).

Canonical correspondence analysis (CCA) was used to determine the extent to which environmental variables (including water quality, lateral hydrological connectivity and water levels) were responsible for variations in the taxonomic composition of fish between lakes. Unimodal ordination was applied because the gradient length along axis 1 in detrended canonical analysis (DCA) exceeded 3.0 SD turnover units. Linear ordination was used in the remaining cases [41]. We used the forward selection procedure to determine the extent to which environmental and community variables explained fish community variations. Conditional effects, which indicate the order of inclusion, and amounts of variance explained in addition to previously added variables, of each environmental variable in the model were calculated and tested for significance using Monte Carlo permutation tests (999 random permutations). Only environmental variables explaining significant amounts of variance ($p \leqslant 0.05$) were retained in the model and tested for significance. We also determined and report the variance attributed to each variable independent of other environmental variables (marginal effects). Fish and environmental data were $\log(x + 1)$ transformed before CCA to reduce the influence of outliers on the results. The relationships between species and the selected environmental variables were examined in CCA ordination plots based on species scores. Since rare species may have strong influence in ordinations, the analyses were also performed on reduced data sets, excluding taxa with dominace index (D) $\leqslant 2\%$ (recedent species). All ordinations were performed in CANOCO version 4.5 [41].

In hierarchical cluster analysis (HCA), objects are classified based on their similarity to other objects in the cluster based on a predetermined selection criterion. HCA was applied to investigate the grouping of the sampling sites. The Euclidean distances were used as a measure of similarity between sampling sites while Ward's error sum of squares hierarchical clustering method were applied to minimize the increase in within-group variance. The spatial variability of fish ecological guilds was determined from HCA using the linkage distance, reported as D_{link}/D_{max}. The quotient is then multiplied by 100 as a way to standardize the linkage distance represented on the Y-axis.

3. Results

3.1. Water Quality

The surveyed floodplain lakes differed significantly in size, depth, range of water level fluctuations and the gradient of lateral connectivity with the adjacent river channel, which produced a dataset with a broad range of physicochemical parameters (Table 2). Water mineralization was average with mean electrolytic conductivity of 542 ± 110 $\mu S \cdot cm^{-1}$. Lower values in the range represented lotic habitats, and higher values were associated with lentic lakes. The pH of water was slightly basic (pH 7.77), and it decreased significantly when water levels were high, in particular in parapotamic water bodies ($r = -0.62$; $p = 0.001$). Floodplain lakes were generally abundant in organic matter with an average TOC content estimated at 12.59 ± 4.64 $mg \cdot L^{-1}$ and a significant share of DOC (10.50 ± 3.18 $mg \cdot L^{-1}$). In addition to autogenic organic matter, the accumulation of humic compounds and decomposed organic matter from adjacent peatlands significantly contributed to an increase in DOC in isolated lakes (48.92 ± 21.50 $mg \cdot L^{-1}$). High concentrations of suspended solids in paleopotamic habitats reduced water transparency to 1.1 ± 0.3 m.

In para- and plesiopotamic lakes, Secchi disc visibility was significantly higher at 1.5 $\pm$ 0.6 and 2.0 $\pm$ 0.9 m, respectively. Overall productivity of the ecosystems was related to the concentrations of chlorophyll-a, which, in paleopotamic and plesipotamic lakes, was twice as much (~17.50 $\mu g \cdot L^{-1}$) as in parapotamic habitats (8.04 $\pm$ 4.03 $\mu g \cdot L^{-1}$) or in the river (7.31 $\pm$ 0.80 $\mu g \cdot L^{-1}$).

The mean concentrations of total phosphorus in the floodplain lakes were determined at 0.35 $mg \cdot L^{-1}$ and no significant differences were observed between the evaluated water bodies. Phosphate phosphorus levels were higher in parapotamic lakes (0.12 $\pm$ 0.05 $mg \cdot L^{-1}$; *post-hoc* Duncan's test, ANOVA; $p \leqslant 0.05$) than in paleopotamic lakes (0.07 $\pm$ 0.04 $mg \, L^{-1}$). Total nitrogen concentrations increased along the isolation gradient from 1.07 $mg \cdot L^{-1}$ in parapotamic lakes to 1.39 $mg \cdot L^{-1}$ in paleopotamic lakes. The content of total inorganic nitrogen (TIN= NO_2–N + NO_3–N + NH_4–N) was similar among the studied floodplain lakes (0.34–1.55 $mg \cdot L^{-1}$), but the contribution of each N-form differed significantly relative to water retention time: ammonium nitrogen was the predominant form in stagnant water of paleopotamic habitats (69%), whereas nitrate nitrogen in flowing water in parapotamic water bodies (62%).

The correlations between TP and TN concentrations *vs.* water levels varied across the examined lakes (correlations are not shown in the tables). TP values were positively correlated with water levels in plesiopotamic ($r = 0.41, p = 0.02$) and parapotamic lakes ($r = 0.45, p = 0.02$), but no such correlations were observed in paleopotamic lakes. Unlike TP, the rise in water levels significantly reduced TN concentrations in all lake types ($r = -0.58; p = 0.001$). DO was significantly higher and more stable in parapotamic lakes (7.45–1.54 $mg \cdot L^{-1}$; *post-hoc* Duncan's test, ANOVA; $p \leqslant 0.05$) than in plesiopotamic (6.51–2.12 $mg \cdot L^{-1}$) and paleopotamic habitats (6.13–2.31 $mg \cdot L^{-1}$).

Table 2. Water quality parameters of the studied floodplain lakes (mean ± SD). Values followed by different superscripts ([a], [b], [c]) in rows are significantly different among types of floodplain lakes in Duncan's multiple range test (*post-hoc*, ANOVA; $p \leqslant 0.05$). pH: * median; # Q_{25}–Q_{75} quartiles.

Type of Hydrological Connectivity	Unit	Parapotamic Lotic $\bar{x}$	±SD	Plesiopotamic Semi-Lotic $\bar{x}$	±SD	Paleopotamic Lentic $\bar{x}$	±SD	All floodplain Lakes $\bar{x}$	±SD	Biebrza River $\bar{x}$	±SD
Parameter											
Temperature	(°C)	17.13	±3.38	16.61	±3.06	17.40	±3.55	17.01	±3.31	17.4	3.42
pH *	pH units	7.83 [b]	7.67 #	7.77 [a,b]	7.56 #	7.69 [a]	7.47 #	7.77	7.58 #	7.87	7.79 #
		7.94		7.95		7.90		7.92		7.92	
DO	(mg·L^{-1})	7.45 [b]	±1.54	6.51 [a]	±2.12	6.13 [a]	±2.31	6.71	±2.08	7.83	±1.47
SEC	(uS·sm^{-1})	536	±90	553	±115	535	±122	542	±110	525	±84
ChL_a	(µg·L^{-1})	8.04 [a]	±1.03	17.61	±12.85	17.47	±8.31	14.70	±10.25	7.31	±0.80
COD$_{Cr}$	(mg·L^{-1})	37.34 [a]	±9.63	41.84 [a,b]	±9.65	48.92 [b]	±21.48	42.63	±14.96	36.51	±10.01
TN	(mg·L^{-1})	1.07	±0.38	1.27	±0.38	1.39	±0.57	1.24	±0.46	0.98	±0.32
NO$_2$-N	(mg·L^{-1})	0.013 [b]	±0.013	0.006 [a]	±0.006	0.005 [a]	±0.004	0.008	±0.009	0.010	±0.71
NO$_3$-N	(mg·L^{-1})	0.17 [c]	±0.08	0.12 [b]	±0.06	0.10 [a]	±0.03	0.13	±0.07	0.17	±0.08
NH$_4$-N	(mg·L^{-1})	0.14 [a]	±0.07	0.19 [b]	±0.08	0.27 [c]	±0.22	0.20	±0.14	0.16	±0.09
TP	(mg·L^{-1})	0.33	±0.17	0.34	±0.16	0.38	±0.21	0.35	±0.18	0.33	±0.14
PO$_4$-P	(mg·L^{-1})	0.12 [b]	±0.05	0.08 [a]	±0.05	0.07 [a]	±0.04	0.09	±0.05	0.12	±0.05
TOC	(mg·L^{-1})	10.70 [a]	±2.94	12.71 [a,b]	±4.58	14.34 [a]	±5.42	12.59	±4.64	10.30	±1.81
DOC	(mg·L^{-1})	9.26	±2.29	10.75	±3.38	11.43	±3.37	10.50	±3.18	9.10	±1.31
Water transparency	(m)	2.0 [b]	±0.9	1.5 [b]	±0.6	1.1 [a]	±0.3	1.6	±0.8	2.5	±0.7

3.2. Fish Assemblages

A total of 18,399 fish from 23 species belonging to 6 families were sampled but the number of specimens varied significantly across species. Cyprinidae was the most abundant and diverse family represented by 15 species (Table 3). The family Cobitidae was represented by three species, Percidae by two species and the families Esocidae, Siluridae and Gadidae by one species each. The roach *R. rutilus* define and the define rudd *S. erythrophthalmus* were eudominant species that accounted for 36% and 12% of all captured specimens, respectively. The pike *E. lucius*, bitterling *Rhodeus sericeus amarus*, silver bream *Blicca bjoerkna* and sunbleak *L. delineates* were dominants. Other species, including the perch *Perca fluviatilis*, Prussian carp *C. auratus gibelio*, tench *Tinca tinca*, bream *Abramis brama* and bleak *Alburnus alburnus*, were less abundant (3%–5%), but they had a stable share of the assemblage.

Among eurytopic fish, the roach *R. rutilus* and the pike *E. lucius* were encountered most frequently (euconstants found in >75% of the sites). Other eurytopic constant species (50% < occurrence < 75%) were the perch and silver bream, as well as the rudd, bitterling and tench in the group of limnophilic fish. Rheophilic species were far less abundant and amounted to 2.3% in plesiopotamic, 3.1% in paleopotamic and 8.2% in parapotamic water bodies.

Amount of rheophils in lotic habitats was 50% lower than that in the Biebrza river channel. Along the lateral connectivity gradient of parapotamic–plesiopotamic–paleopotamic lakes, the proportions of rheophilic species were determined at 10:5:1, and limnophilic species at 1:2:5.

Five out of the 23 identified species have been placed on the IUCN Red List of Threatened Species, and four are listed in Annex II to the EU Habitat Directive (92/43/EEC). They include three rheophilic species: the asp *Aspius aspius*, spined loach *Cobitis taenia*, and stone loach *Barbatula barbatula* and two limnophylic taxa: weatherfish *Misgurnus fossilis* and the bitterling *R. sericeus amarus*. The above species had less than 2.5% share of the fish assemblage, excluding the bitterling, which was a dominant and frequently observed species. The Prussian carp was the only non-native species identified in the study. However, the study of van Damme *et al.* [42] showed evidence of the bitterling as non-native for the Polish watercourses.

As many as six species colonizing the Biebrza floodplain lakes were piscivores (Tables 3 and 4) that accounted for nearly 12.4% of all specimens. Pike represented 60% and perch 35.7% of the identified piscivores. Other predatory species, such as the wels catfish *Silurus glanis*, burbot *Lota lota*, and the chub *Squalius cephalus*, were less populous.

Table 3. Ecological guilds of fish in the studied floodplain lakes of the Biebrza River. Denotations: I, invertivore; O, omnivore; P, piscivore; H, herbivore; E, exotic species; N, number of specimens caught; HD-AII, annex II of EU Habitat Directive; IUCN, IUCN Red List of Threatened Species; n.t., not threatened species.

Family Species	Abbreviation	Common Name	Feeding Guild	Reproductive Guild	Stream Velocity Preference	Constancy of Occurrence (C, %)	Conservation Status
Esocidae							
Esox lucius	Esox	pike	P	phytophils	eurytopic	88	n.t.
Cyprinidae							
Abramis brama	Abr_br	bream	O	phytophils	eurytopic	36	n.t.
Alburnus alburnus	Alb_al	bleak	I	phytophils	eurytopic	26	n.t.
Aspius aspius	Asp_as	asp	P	lithophils	rheophilic	2	HD-AII
Blicca bjoerkna	Bli_bj	silver bream	I	phyto-lithophils	eurytopic	62	n.t.
Carassius auratus gibelio	Car_gi	Prussian carp	O, E	phytophils	limnophilic	14	n.t.
Carassius carassius	Car_ca	crucian carp	O	phytophils	limnophilic	31	n.t.
Gobio gobio	Gobio	gudgeon	I	psammophils	rheophilic	10	n.t.
Leucaspius delineatus	Leu_del	sunbleak	I	phytophils	limnophilic	14	n.t.
Squalius cephalus	Squ_cep	chub	O/P	lithophils	rheophilic	3	n.t.
Leuciscus idus	Leu_idu	ide	O	phyto-lithophils	rheophilic	21	n.t.
Leuciscus leuciscus	Leu_le	dace	I	phyto-lithophils	rheophilic	2	n.t.
Rhodeus sericeus amarus	Rho_se	bitterling	H	ostracophils	limnophilic	55	HD-AII, IUCN
Rutilus rutilus	Rut_ru	roach	O	phyto-lithophils	eurytopic	88	n.t.
Scardinius erythrophthalmus	Sc_ery	rudd	O	phytophils	limnophilic	73	n.t.
Tinca tinca	Ti_tin	tench	I	phytophils	limnophilic	60	n.t.
Cobitidae							
Cobitis taenia	Cob_tae	spined loach	I	phytophils	rheophilic	14	HD-AII, IUCN
Misgurnus fossilis	Misg_fo	weatherfish	I	phytophils	limnophilic	33	HD-AII, IUCN
Barbatula barbatula	Bar_ba	stone loach	I	psammophils	rheophilic	1	IUCN
Siluridae							
Silurus glanis	Sil_gl	wels catfish	P	phytophils	eurytopic	2	n.t.
Gadidae							
Lota lota	Lota_lo	burbot	I/P	litho-pelagophils	rheophilic	19	n.t.
Percidae							
Gymnocephalus cernua	Gym_ce	ruffe	I	phyto-lithophils	eurytopic	2	n.t.
Perca fluviatilis	Perca	perch	I/P	phyto-lithophils	eurytopic	69	n.t.

Table 4. Abundance of fish guilds in floodplain lake sites (specimens· ha⁻¹) as well as biomass (kg· ha⁻¹) and biodiversity metrics of ichthyofauna in the present study in the Biebrza River in the period of 2011–2013. Denotations: * total; ** biodiversity metrics calculated as totals for the river, sites, and types of lakes; $\bar{x}$, mean.

Site Fish guilds or Metrics	Biebrza River	Floodplain Lakes												All Floodplain Lakes $\bar{x}$
		Parapotamic				Plesiopotamic				Paleopotamic				
		Upstream	Middle	Downstream	$\bar{x}$	Upstream	Middle	Downstream	$\bar{x}$	Upstream	Middle	Downstream	$\bar{x}$	
Rheophils, (%)	16.3	9.3	4.0	12.2	8.2	1.9	2.0	2.9	2.3	1.5	0	0.1	0.5	3.1
Eurytops, (%)	65.7	68.3	76.3	68.2	71.1	73.2	67.7	75.8	72.3	34.6	31.2	44.9	36.4	59.4
Limnophils, (%)	18.0	22.4	19.7	19.6	20.7	24.9	30.3	21.3	25.5	63.9	68.8	55.0	63.1	37.5
Total (specimens ha⁻¹)	2164	2275	2083	1628	1995	2782	2623	2655	2687	2603	3503	2703	2936	2552
Spec. contribution (%)*	–	9.1	8.3	6.5	23.9	14.2	13.4	13.5	41.0	10.4	14.0	10.8	35.1	100
Biomass (kg· ha⁻¹)	94.3	70.9	79.2	85.1	78.4	82.0	74.8	98.3	85.0	62.3	92.7	53.1	69.4	77.6
Biomass contribution (%) *	–	10.2	11.3	12.2	33.7	11.7	10.7	14.1	36.5	8.9	13.3	7.6	29.8	100
Species richness, S**	18	18	18	18	21	18	19	19	23	13	14	16	17	23
Shannon's biodiversity index, H'**	2.11	1.49	1.44	1.41	2.45	1.47	1.59	1.51	2.51	1.20	1.29	1.18	2.23	2.41
Pielou's evenness index, J'**	0.91	0.74	0.71	0.73	0.79	0.71	0.78	0.72	0.80	0.73	0.77	0.73	0.82	0.80

3.3. Fish Response to Habitats along the Connectivity Gradient

The composition of fish species in parapotamic, plesiopotamic, and paleopotamic floodplain lakes relative to the location of the sampling site is presented in Table 4.

Parapotamic lakes (lotic side-channels) were inhabited by 21 of the 23 identified fish species, and the only missing taxa were the Prussian carp and the stone loach (Appendix A). Mean fish density in the lakes amounted to 1995 specimens per ha, which was lower in comparison with other types of lakes, and accounted for 23.9% of captured fish (Table 4). Fish species contribution decreased between upstream arms (9.1%) and downstream arms (6.5%). The biodiversity index of parapotamic habitats was determined at H' = 2.45, and species evenness at J = 0.79. Fish biomass in lakes of that type amounted to 78.4 kg· ha^{-1}, which constitutes ~34% of total fish biomass, but it was the highest near the downstream connections with the river (85.1 kg ha^{-1}).

Active exchange of water in habitats provides similar conditions for ichthyofauna to those noted in the Biebrza River. Eurytops accounted for 71%, limnophils for 21%, and rheophils for only 8% of the species in parapotamic lakes. In the group of parapotamic lakes, the roach and pike were eudominants, whereas the rudd, perch and silver bream were dominants (Appendix A). No rheophilic species were eudominants or dominants in the studied group of water bodies. Among rheophilic species, only L. idus was a subdominant.

The ichthyofauna in plesiopotamic lakes was represented by all of the 23 identified fish species. As many as 19 species were found in both upstream and middle arms, whereas 18 taxa were observed near the connection with the river channel. The diversified habitats along plesiopotamic lakes showed decreased water exchange, aeration and depth gradients, and they were colonized by 41% of captured fish and characterized by the highest biodiversity index (H' = 2.51). One-side connected lakes showed mean fish density determined at 2687 specimens ha^{-1}. Fish biomass in plesiopotamic lakes amounted to 85 kg· ha^{-1} and was the highest (36.5% of total biomass) among the studied lakes types. Although eurytops prevailed in the species composition (72.3%), an increase in limnophils by 5%, when compared to parapotamic lakes, has been noted. The number of limnophilic taxa (rudd, bitterling and tench) increased with a distance from the river channel. Eudominants in the plesiopotamic lakes were represented by roach, rudd, and pike, while dominants by silver bream bitterling, and perch. Due to stagnant water, no psammophils were identified, but the presence of ostracophils (bitterling) and litho-pelagophils (burbot) was noted in the downstream section in the vicinity of the river connection.

Fish assemblages in paleopotamic lakes were represented by the lowest number of species (17) and, consequently, the lowest biodiversity index (H' = 2.23) in comparison with the lakes connected to the river (Table 4). Simultaneously, isolated water bodies were characterized by the highest evenness index (J') of 0.82. A clear increase in the share of limnophilic species (up to 63.1%) could be attributed to lakes' isolation from the river channel. The roach was the eudominant species with the highest D index. Paleopotamic lakes provided particularly supportive habitats for sunbleak and Prussian carp (Appendix A). Among dominant species were pike, rudd, bitterling and tench. The number of specimens captured was almost 50% higher than that caught in bi-connected side-channels. Fish density reached 2936 specimens· ha^{-1}. Despite high fish abundance, fish biomass in isolated habitats was lower (70 kg· ha^{-1}, ~29% of total) than in lakes connected to the river channel.

3.4. Influence of Environmental Factors on Fish Abundance and Diversity

Three types of floodplain lakes (parapotamic, plesiopotamic, and paleopotamic), classified based on hydrological connectivity, differed significantly in physicochemical parameters of water and fish responses to varied habitat conditions. The canonical correspondence analysis (CCA) biplot (Figure 2A) demonstrates the correlations between environmental factors as lateral connectivity, water level, water quality parameters and the abundance of fish fauna in the studied floodplain lakes. The results of the CCA (Figure 2A) showed that eigenvalues of the first (λ_{CC1} = 26.2%) and second (λ_{CC2} = 11.1%) CCA axes accounted for 70.2% of the cumulative variation in the environmental data. The final model accounted for 22.6% of the total variance in fish composition and all canonical axes were significant (Monte Carlo test, p = 0.002).

Ten out of 17 input variables were retained as significant contributors to the CCA model (Table 5). The non-retained seven were redundant or did not increase the significance. The hydrological set of variables (including isolation gradient and water-level variability) accounted for $\lambda 1$ = 18% of the species variability, while retained physical and chemical variables, including SEC, DO, COD$_{Cr}$, temperature and TOC explained in total 36%. Trophic variables (NO$_3$–N and PO$_4$–P) explained 11% of the variability.

The ordination space of factors CCA1 and CCA2 were clearly distinguished by environmental variables and fish species composition, confirming previous analyses. The first factor (CCA1) showed a gradient from highly eutrophicated lentic sites to less eutrophicated parapotamic sites (Figure 2A). It correlated negatively with NO$_3$–N, SEC and PO$_4$–P while positively with water level. Organic matter content, expressed by COD$_{Cr}$ and TOC, was positively correlated, whereas DO and water transparency were negatively correlated with CCA2 (20.9% of the variance). Significant variance explained by the lake isolation gradient ($\lambda 1$ = 15%) was confirmed by three clusters of samples comprising parapotamic, plesiopotamic, and paleopotamic habitats (Figure 2B). Lakes situated near the river with transparent and well-aerated waters were characterized by a predominance of rheophils, including the burbot, ide, wels catfish, asp, gudgeon, bleak and dace. Stagnant water habitats attracted mostly the rudd, Prussian carp, crucian carp, tench and weatherfish.

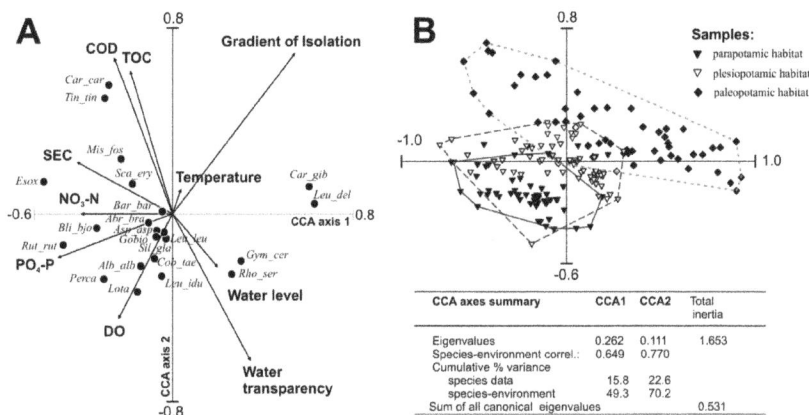

Figure 2. (A) Biplot of CCA relating score, fish, taxa abundance and environmental variables correlated with axes. Environmental variables are represented by arrows that approximately point towards the factor direction of maximum variation. The length of an arrow is proportional to the importance of that variable in assemblage ordination; (B) CCA ordination plot of samples grouped according to hydrological connectivity. Inserted table shows summary of the results of the CCA including eigenvalues, correlations and percentage of variation explained by the two canonical axes (CCA1 and CCA2). For fish species abbreviations, see Table 2.

Forward selection revealed that in the group of input environmental variables, hydrochemical parameters were significant in explaining fish occurrence and abundance patterns in the studied floodplain lakes. White bream and pike were related to habitats with higher NO_3–N concentrations and lower water levels. Limnophilic species (crucian carp and tench) preferred habitats with higher COD and TOC and lower DO content, which are characteristic of lakes isolated from the river. In contrast, rheophilic (burbot) and eurytopic (perch and bleak) species preferred transparent and well aerated water. The availability of dissolved oxygen (DO) is the key contributor to fish yield in lakes. The highest share of fish specimens in water containing > 7 mg DO L^{-1} confirmed distinct preferences of fish for inhabiting water bodies supplied with river water; this is presented in the form of species pies charts in Figure 3A. The diagrams of pies charts based on water levels, as presented in Figure 3B, showed that water stage is a significant factor that changes the quantitative structure of fish species in floodplain lakes.

As typical inhabitants of isolated lakes, limnophils differed significantly (ANOVA, *post-hoc* Duncan's test, $p \leqslant 0.05$) from other guilds during low water stages (Figure 4). In the group of rheophilic species, the chub predominated at low water (50% of specimens), mainly in lakes of high water exchange. High water (*potamophase*) did not contribute to an increase in species abundance. Only two species, the ide and asp, adapted to higher water stages, and their share during that period reached 65% and 60%, respectively (Figure 3A). The above taxa contributed the group of rheophils that were characterized by significant differences (ANOVA, *post-hoc* Duncan's test, $p \leqslant 0.05$) during high water. Eurytops were more opportunistic and had no water stage preferences, but their share in lentic lakes decreased during floods (Figure 4).

Table 5. Marginal effect (absolute) and conditional effect (additional) explained by each environmental variable in the constrained ordination listed after the automatic forward selection. The *P* values and F-statistics were obtained by Monte Carlo test (999 permutations).

Variable	Marginal Effects		Conditional Effects		
	λ1 *	λA **	P	F-Value
Isolation from the river	0.15	0.13	0.002	14.92
Water transparency	0.06	0.07	0.002	8.86
SEC	0.05	0.05	0.002	6.06
DO	0.03	0.05	0.002	5.88
COD$_{Cr}$	0.05	0.03	0.006	3.91
Temperature	0.02	0.03	0.004	3.77
Water level	0.03	0.02	0.004	3.53
PO$_4$–P	0.07	0.03	0.004	3.12
NO$_3$–N	0.04	0.01	0.020	2.33
TOC	0.04	0.01	0.020	2.32

*λ1 indicates the percentage of the variability explained by a single variable. ** λA indicates the percentage of the variability explained by a variable after the forward selection starting from the best variable (marginal effects). Each subsequent variable is ranked on the basis of the fit that the variable gives in conjunction with the variables already selected (conditional effects).

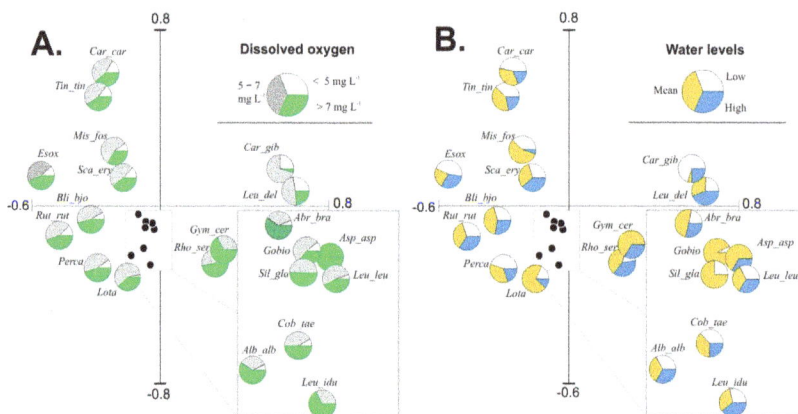

Figure 3. Biplots of CCA with species pies classes distinguished for: (**A**) water aeration (dissolved oxygen, DO in mg·L^{-1}); and (**B**) zones of water levels: low; mean and high water. Each pie chart in the figure represents the percentage of individuals of a given species in selected classes of dissolved oxygen and water levels. Rare species were down weighted. For fish abbreviations, see Table 2.

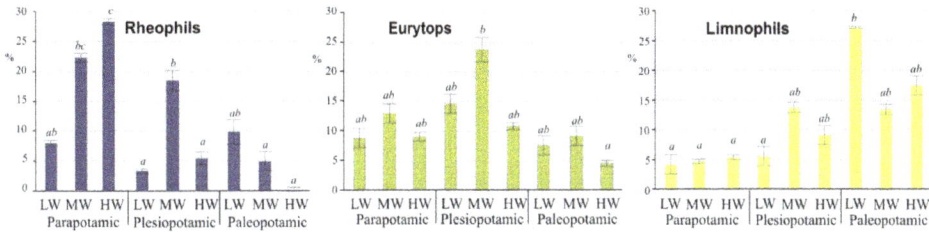

Figure 4. Effect of water level on the ecological fish guilds in floodplain lakes differed by hydrological connectivity: LW, low water; MW, mean water; HW, high water. Different letters denote groups of means, statistically different in the Duncan's test, *post-hoc*, two-way ANOVA, at $p \leqslant 0.05$.

Hierarchical cluster analysis (HCA) performed on the percentage of species abundance in nine lake-sites produced four clusters of objects (Figure 5). Cluster I covered lentic sites with a high percentage of limnophils (56%) and complementary species, mainly eurytops. The cluster comprised the predominant species in lentic habitats: the Prussian carp, crucian carp, sunbleak, bitterling and weatherfish. Cluster II was characterized by a significant share (58%) of eurytopic species. Clusters III and IV had a similar share of rheophils (40% in total), including the asp, gudgeon, chub, dace and spine loach.

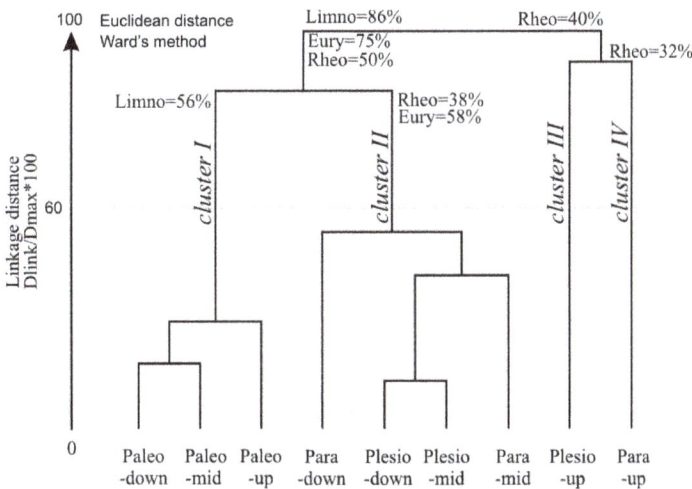

Figure 5. Dendrogram of Hierarchical Cluster Analysis (HCA) for upstream (-up), middle parts (-mid) and downstream arms (-down) of floodplain lakes types based on % of fish abundance data, obtained by using the Ward's method as linkage rule and Euclidean distance as a metric for distance calculation. Statistically significant clusters when $(D_{link}/D_{max}) \cdot 100 < 60$.

371

4. Discussion

The present study showed that variations in fish abundance and community structure were related to environmental variables in floodplain lakes of the Middle Basin of the Biebrza River. Furthermore, the diversity of environmental conditions in the lakes produced distinctive fish guilds. Major environmental gradients related to the structuring of fish communities involved degree of isolation from the river, water transparency, SEC, DO, COD, temperature and water levels, whereas nutrients such as PO_4–P, NO_3–N and TOC played a minor role.

Lateral connectivity was found to be a key driver in shaping fish assemblage attributes in the studied floodplain lakes. It promotes fish migration, and thus species exchange, when the surface pathway is maintained between river channel and water ecosystems. Kwak [43] reported that the increase in lateral movement of fishes between the river channel and floodplain habitats is enhanced by the increasing river discharge. In the case of the Biebrza River, the fish colonization of isolated water bodies probably occurs annually in spring when frequency and duration of connection is greatest. Flood-pulse of water is an essential factor that determines nutrient cycling, which is consistent with the hypothesis postulated by, e.g., Amoros and Bornette [44] or Junk and Watzen [23]. The connection of lakes with the main river channel during periods of high water provides the input of well-aerated river water, which protects the lakes against severe oxygen deficits. It is a period of somewhat "deep breath", which enables many organisms to survive in isolation, and may be crucial for maintaining their populations [15]. During low water periods, particularly in summer, the decreasing gradient of oxygen resources is the most distinct: from optimal aeration (*ca.* 7 mg DO·L^{-1}) in parapotamic lakes towards paleopotamic lakes with significant DO deficits (<2 mg·L^{-1}). For example, DO measurements performed in the summer of 2011 in vertical profiles of isolated lakes during a low water stage revealed a shortage of oxygen already 0.4 m below the water table. According to estimates, more than two-thirds of the water volume in isolated lakes could experience anaerobic conditions in summer (not shown in the figure drawing), which could have significant implications for fish populations in those ecosystems.

Lateral gradient of lake connectivity is directly related to fish densities, which increased from *ca.* 2000 in lotic to 3000 specimens·ha^{-1} in lentic habitats. Parapotamic lakes were inhabited by 24%, plesiopotamic by 41% and paleopotamic lakes by 35% of the evaluated fish population. Similar to the study of Winemiller *et al.* [15], we observed higher values of the biodiversity index (H') in floodplain lakes than in the Biebrza River. In our study, H' was higher in plesiopotamic lakes than in parapotamic and paleopotamic lakes. The evaluated lake types were arranged in the following sequence based on the values of H': plesiopotamic > parapotamic > paleopotamic. The results of this study confirm the findings of Tockner *et al.* [21]

or Guti [9] for the Danube, where the structure of ecological fish guilds changed from rheophilic to limnophilic and the diversity of fish species decreased in the lateral transect along with the increased distance from the main river channel and a decrease in hydrological connectivity. Along the lateral connectivity gradient of parapotamic–plesiopotamic–paleopotamic lakes, the proportions of rheophilic species were determined at 10:5:1, and limnophilic species-at 1:2:5. High species diversity in the analyzed floodplain lakes resulted from the co-occurrence of eurytopic, limnophilic and rheophilic guilds.

The ichthyofauna in the Biebrza floodplain lakes was most abundant in eurytops (59% of all specimens), that live under both lotic and lentic conditions. A wide range of microhabitats in semi-lotic lakes creates greater opportunities for eurytopic species, which are capable of living in varied habitats, than lotic or lentic environments. The predominant eurytopic species were the roach, pike, rudd, bream and bleak. Long-term isolation allows for a succession of the fish assemblage towards floodplain specialist limnophilic species [10]. Limnophilic species accounted for 37% of the evaluated fish guilds, where the Prussian carp was the eudominant, and the crucian carp, bitterling and tench were the dominants. The above species are tolerant of habitats with low oxygen resources and high organic matter content (e.g., Prussian carp) that are found in isolated water bodies. Limnophils abundance was highly correlated with water quality descriptors (COD, TOC, DOC, NH_4–N and TN) characteristic of fertile and productive ecosystems. Other variables, including pH, DO, turbidity and macrophyte cover, also influenced fish assemblages inhabiting isolated water bodies as reported in temperate floodplains [15,17,45]. Adaptation to anoxic conditions of the floodplain specialists is often accompanied by poor competitive abilities and predation avoidance, resulting in low population densities and high mortalities in multispecies fish assemblages, as in crucian carp [46]. Thus, long-term isolated floodplain lakes in advanced successional stages with a tendency to temporary anoxia provide important refuge habitats for these species [45]. Advanced successional stages of such lakes provide most suitable habitats for still-water species and specialist species that have evolved physiological adaptations and strategies to survive hypoxic and anoxic conditions [10]. Accordingly, limnophilic fish are essential elements of the typical fish community of floodplain aquatic ecosystems and thus can serve as an indicator of their ecological integrity. We share the opinion of Welcomme *et al* [20] that isolated lakes are at least as important as the lakes with high rates of water exchange as fish habitats. Isolated lakes offer a rich array of habitat types at different stages of succession, and they are in direct need of statutory protection.

Our findings are consistent with the results of previous studies on fish assemblage metrics among floodplain lakes. In a study of fish assemblage structure among oxbow lakes of the Brazos River (Texas) species richness, diversity, and evenness were greatest in the connected oxbow lake [15]. Similarly, Galat *et al* [47], Miranda [12,45], and Petry *et al* [48] noted more species in oxbow lakes that are connected to the Mississippi, Missouri, and the Parana River (Brazil), respectively, than in lakes permanently separated from the river. They observed greater species richness in connected floodplain water bodies than in isolated water bodies. Also Dembkowski and Miranda [13], based on the monitoring of fish assemblages in two segments of an oxbow lake, one connected to and the other isolated from the Yazoo River, stated that greater species richness, diversity, and evenness were observed in the connected segment than isolated one.

Although the environmental variables play an initial role in structuring the fish assemblage, biological interactions are then superimposed on that structure. Thus, the influence of biotic factors, such as predator–prey interactions [18,49], which are significant determinants of assemblage composition, cannot be ruled out. The present study has suggested associations between environmental variables and fish abundance, but both interspecific and intra-specific biotic interactions remain to be studied in this system. Although the analyzed floodplain area is characterized by nearly pristine conditions as part of the Biebrza National Park, our results could be underestimated because legal angling and the significant increase in the number of poached fish exert pressure on the local fish assemblage. Unfortunately, fish loss is difficult to estimate. It should be noted, however, that fish abundance and biomass exceed the values reported in other temperate floodplain lakes, such as the lower section of the Vistula River [50], but any comparisons should be performed with great caution due to differences in the applied equipment and calculations methods.

To protect and improve the biotic integrity of these relict ecosystems, knowledge about the influence of historic and contemporary connectivity with adjacent rivers on fish species richness, diversity, and assemblage composition is essential [15]. Nevertheless, long-term data about connectivity are seldom available as these lakes are often located in remote areas and lack continued monitoring. An interesting background to a trend assessment can be drawn up from previous studies preformed within the Biebrza River. A comparison of our results with the findings of the most recent study that investigated fish communities in the Biebrza River in 1997 and 1998 [51] indicates that the composition of various fish guilds in the quantitative structure of fish communities was not affected by anthropogenic changes. An analysis of historical [14,30] and present data revealed that the fish assemblage in the Middle Biebrza floodplain is still characterized by high species diversity (23 species at present, 26 species in 1983) and even higher fish abundance (4% increase in connected backwaters and 23% increase in disconnected backwaters) and fish biomass (10% increase in connected backwaters and 54% increase in disconnected backwaters) than in the past. The results point to the effectiveness of the nature conservation program covering the floodplain area, which has been included in the Biebrza National Park (established in 1993), the largest nature reserve in Poland. Results of that comparison support our hypothesis that fish populations are important bioindicators of ecological integrity in the river network under natural flood-pulsing conditions. We suggest that maintenance or restoration of connection should be an integral part of the fluvial ecosystem management plans.

5. Conclusions

The heterogeneity of floodplain habitats resulting from variations in hydrological connectivity driven by the flood pulse in the Middle Basin of the Biebrza River provides optimal habitat condition for fish lifecycle in the temperate climate zone. Habitat diversity within that natural river floodplain shows that a mosaic of habitats within a single floodplain can provide fish assemblages with shelter and supportive conditions for spawning, breeding and feeding. These findings should be taken into consideration to maximize the success of future restoration projects in regulated river floodplains. The Biebrza River is one of the few surviving natural watercourses that present us with a rare opportunity to explore ecological interactions under natural river conditions.

Our results indicate that the Biebrza River could represent reference conditions for promoting fish species diversity. Pristine riverine ecosystems in the Middle Basin of the Biebrza River contribute to the diversity of fish species, but effective conservation of fish resources requires the preservation of variously aged lakes that provide a wide range of habitats for diverse aquatic biota.

Acknowledgments: This study was financially supported by the Polish Ministry of Education and Science, grant no. NN 304 317440. We thank the anonymous reviewers for their constructive comments that helped us improve an earlier version of the manuscript.

Author Contributions: Katarzyna Glinska-Lewczuk developed the concept of the study, field study design, and wrote the manuscript and also provided funding for editorial support. Paweł Burandt prepared the manuscript draft with important intellectual input. Paweł Burandt and Szymon Kobus performed field research including data collection and data analysis. Roman Kujawa was responsible for fish catches and fish survey. Krystian Obolewski performed statistical analyses and contributed to interpretation of data. Julita Dunalska performed hydrochemical analyses and interpretation of data and verified laboratory database. Magdalena Grabowska and Sylwia Lew assisted in manuscript preparation and acquisition of hydrobiological data. Jarosław Chormański contributed to acquisition of hydrological data. The first and second authors contributed equally to this work. All authors approved the final version before submission.

Conflicts of Interest: The authors declare no conflict of interest.

Appendix A

Table A1. Abundance of fish species in floodplain lake sites (specimens.ha^{-1}) in the floodplain lakes the Biebrza River in the period of 2011–2013. Dominance of species is based on sum of specimens: ●—eudominant; ▼—dominant; □—subdominant; $\bar{x}$—mean.

Family and Species/Site	Bie-Brza River	Floodplain Lakes												Mean $\bar{x}$
		Parapotamal				Plesiopotamal				Paleopotamal				
		Up-Stream	Mid-Dle	Down-Stream	$\bar{x}$	Up-Stream	Mid-Dle	Down-Stream	$\bar{x}$	Up-Stream	Mid-Dle	Down-Stream	$\bar{x}$	
Esocidae														
Esox lucius	225●	156	178	208	181●	234	199	233	222●	160	163	107	143□	182▼
Cyprinidae														
Abramis brama	36□	46	74	48	56□	60	103	104	89□	14	79	29	41	62□
Alburnus alburnus	192▼	97	140	26	88□	111	101	82	98□	0	36	0	12	66□
Aspius aspius	3	3	3	0	2	1	0	0	0	0	0	0	0	1
Blicca bjoerkna	58□	120	139	86	115▼	302	268	298	289▼	115	65	158	113	173▼
Carassius auratus gibelio	0	0	0	0	0	0	1	0	0	560	1076	363	666●	222▼
Carassius carassius	0	0	14	2	6	23	15	18	19	101	60	47	69	31
Gobio gobio	33	40	3	8	17	40	26	36	34	38	0	0	13	21
Leucaspius delineatus	0	7	0	0	2	61	36	2	33□	568	649	568	595●	210▼
Squalius cephalus	11	6	1	4	4	1	0	1	1	0	0	0	0	1
Leuciscus idus	183▼	80	32	137	83□	7	16	25	16	0	0	0	0	33
Leuciscus euciscus leuciscus	13	33	0	0	11	0	1	1	1	0	0	0	0	4
Rhodeus sericeus amarus	126□	194	66	75	112□	212	222	187	207▼	200	243	243	229▼	182▼

Table 1. Cont.

Family and Species/Site	Bie-Brza River	Floodplain Lakes												Mean x̄
		Parapotamal				Plesiopotamal				Paleopotamal				
		Up-Stream	Mid-Dle	Down-Stream	x̄	Up-Stream	Mid-Dle	Down-Stream	x̄	Up-Stream	Mid-Dle	Down-Stream	x̄	
Rutilus rutilus	676●	995	936	599	843●	1186	942	1164	1097●	546	692	889	709●	883●
Scardinius erythrophthalmus	264▼	276	189	140	202▼	329	379	301	337●	118	274	126	173▼	237●
Tinca tinca	19	17	24	16	19	49	77	41	56□	76	93	76	82	52
Cobitidae														
Cobitis taenia	17	6	17	8	10	0	3	7	3	3	0	3	2	5
Misgurnus fossilis	57□	16	117	86	73□	18	65	15	33	39	14	64	39	48
Barbatula barbatula	0	0	0	0	0	1	0	0	0	0	0	0	0	0
Siluridae														
Silurus glanis	4	0	1	3	1	0	1	0	0	0	0	0	0	1
Gadidae														
Lota lota	94▼	43	28	41	37	3	5	8	5	0	1	0	0	14
Percidae														
Gymnocephalus cernua	0	0	0	1	0	1	0	0	0	1	0	0	0	0
Perca fluviatilis	133▼	140	121	138	133▼	142	160	133	145▼	64	58	29	50	110□

References

1. Amoros, C.; Roux, A.L. Interactions between water bodies within floodplains of large rivers: Function and development of connectivity. *Munst. Geogr. Arb.* **1988**, *29*, 125–130.

2. Penczak, T.; Galicka, W.; Glowacki, Ł.; Koszalinski, H.; Kruk, A.; Zięba, G.; Kostrzewa, J.; Marszal, L. Fish assemblage changes relative to environmental factors and time in the Warta River, Poland, and its oxbow lakes. *J. Fish Biol.* **2004**, *64*, 483–501.

3. Bolland, J.D.; Nunn, A.D.; Lucas, M.C.; Cowx, I.G. The importance of variable lateral connectivity between artificial floodplain waterbodies and river channels. *River Res. Appl.* **2012**, *28*, 1189–1199.

4. Copp, G.H. The habitat diversity and fish reproductive function of floodplain ecosystems. *Environ. Biol. Fish* **1989**, *26*, 1–26.

5. Hohausová, E.; Jurajda, P. Restoration of a river backwater and its influence on fish assemblage. *Czech J. Anim. Sci.* **2005**, *50*, 473–482.

6. Fernandes, C.C. Lateral migration of fishes in Amazon floodplains. *Ecol. Freshw. Fish* **1997**, *6*, 36–44.

7. Agostinho, A.A.; Gomes, L.C.; Zalewski, M. The importance of floodplains for the dynamics of fish communities of the upper River Paraná. *Ecohydrol. Hydrobiol.* **2001**, *1*, 209–217.

8. Welcomme, R.L.; Halls, A. Dependence of tropical river fisheries on flow. In Proceedings of the Second International Symposium on the Management of Large Rivers for Fisheries, Phnom Penh, Kingdom of Cambodia, 11–14 February 2003; Welcomme, R., Petr, T., Eds.; FAO Regional Office for Asia and the Pacific, RAP Publ: Bangkok, Thailand, 2004; pp. 267–283.

9. Guti, G. Significance of side-tributaries and floodplains for the Danubian fish populations. Large Rivers. *Arch. Hydrobiol. Suppl.* **2002**, *13*, 151–163.

10. Schomaker, C.; Wolter, C. The contribution of long-term isolated water bodies to floodplain fish diversity. *Freshw. Biol.* **2011**, *56*, 1469–1480.

11. Tockner, K.; Pennetzdorfer, D.; Reiner, N.; Schiemer, F.; Ward, J.V. Hydrological connectivity and the exchange of organic matter and nutrients in a dynamic river-floodplain system (Danube, Austria). *Freshw. Biol.* **1999**, *41*, 521–535.

12. Miranda, L.E. Fish assemblages in oxbow lakes relative to connectivity with the Mississippi River. *Trans. Am. Fish Soc.* **2005**, *134*, 1480–1489.

13. Dembkowski, D.J.; Miranda, L.E. Hierarchy in factors affecting fish biodiversity in floodplain lakes of the Mississippi Alluvial Valley. *Environ. Biol. Fish* **2012**, *93*, 357–368.

14. Witkowski, A. An analysis of the ichthyofauna of the Biebrza River system. Part II. Materials to the knowledge of the ichthyofauna and review of species. *Fragm Faun* **1984**, *28*, 137–184. (In Polish)

15. Winemiller, K.O.; Terim, S.; Shormann, D.; Cotner, J.B. Fish assemblage structure in relation to environmental variation among Brazos River oxbow lakes. *Trans. Am. Fish Soc.* **2000**, *129*, 451–468.

16. Schiemer, F.; Spindler, T. Endangered fish species of the Danube River in Austria. *Regul. River Res. Manag.* **1989**, *4*, 397–407.

17. Lubinski, B.J.; Jackson, J.R.; Eggleton, M.A. Relationships between floodplain lake fish communities and environmental variables in a large river-floodplain ecosystem. *Trans. Am. Fish Soc.* **2008**, *137*, 895–908.

18. Rodriguez, M.A.; Lewis, W.M., Jr. Structure of fish assemblages along environmental gradients in floodplain lakes of the Orinoco River. *Ecol. Monogr.* **1997**, *67*, 109–128.

19. Schiemer, F.; Waidbacher, H. Strategies for Conservation of Danubian Fish Fauna. In *River Conservation and Management*; Boon, P.J., Calow, P., Petts, G.J., Eds.; Wiley-Blackwell: London, UK, 1992; pp. 363–382.

20. Welcomme, R.L.; Winemiller, K.O.; Cowx, I.G. Fish environmental guilds as a tool for assessment of ecological condition of rivers. *River Res. Appl.* **2006**, *22*, 377–396.

21. Tockner, K.; Baumgartner, C.; Schiemer, F.; Ward, J.V. Biodiversity of a Danubian Floodplain: Structural, Functional and Compositional Aspects. In *Biodiversity in Wetlands: Assessment, Function and Conservation*; Gopal, B., Junk, W.J., Davis, J.A., Eds.; Backhuys Publishers: Leiden, The Netherlands, 2000; pp. 141–159.

22. Ward, J.V.; Tockner, K.; Schiemer, F. Biodiversity of floodplain river ecosystems: Ecotones and connectivity. *Regul. Rivers Res. Manag.* **1999**, *15*, 125–139.

23. Junk, W.J.; Wantzen, K.M. The flood pulse concept: New aspects, approaches and applications—An update. In Proceedings of the Second International Symposium on the Management of Large Rivers for Fisheries, Phnom Penh, Kingdom of Cambodia, 11–14 February 2003; Welcomme, R., Petr, T., Eds.; FAO Regional Office for Asia and the Pacific, RAP Publication: Bangkok, Thailand, 2004; pp. 117–140.

24. Grabowska, M.; Glińska-Lewczuk, K.; Obolewski, K.; Burandt, P.; Kobus, S.; Dunalska, J.; Kujawa, R.; Goździejewska, A.; Skrzypczak, A. Effects of Hydrological and Physicochemical Factors on Phytoplankton Communities in Floodplain Lakes. *Pol. J. Environ. Stud.* **2014**, *23*, 713–725.

25. Hein, T.; Baranyi, C.; Reckendorfer, W.; Schiemer, F. The impact of surface water exchange on the nutrient and particle dynamics in side-arms along the River Danube, Austria. *Sci. Total Environ.* **2004**, *328*, 207–218. PubMed]

26. Molls, F.; Neumann, D. Fish abundance and fish migration in gravel-pit lakes connected with the River Rhine. *Water Sci. Technol.* **1994**, *29*, 307–309.

27. Allouche, S.; Thevenet, A.; Gaudin, P. Habitat use by chub (*Leuciscus cephalus*) in a large river, the French Upper Rhone, as determined by radiotelemetry. *Arch. Hydrobiol.* **1999**, *145*, 219–236.

28. Schwartz, J.S.; Herricks, E.E. Fish use of stage-specific fluvial habitats as refuge patches during a flood in a low-gradient Illinois stream. *Can. J. Fish. Aquat. Sci.* **2005**, *62*, 1540–1552.

29. Schmutz, S.; Kaufmann, M.; Vogel, B.; Jungwirth, M.; Muhar, S. *A Multi-Level Concept for Fish-Based, River-Type-Specific Assessment of Ecological Integrity*; Springer: Berlin, Germany, 2000; pp. 279–289.

30. WFD, EU. *Directive 2000/60/EC of the European Parliament and of the Council Establishing a Framework for the Community Action in the Field of Water Policy*; The European Parliament and the Council of the European Union: Brussels, Belgium, 2000.

31. House, A. *Council Directive 92/43/EEC on the Conservation of Natural Habitats and of Wild Fauna and Flora*; European Commission: Brussels, Belgium, 2014.

32. Witkowski, A. Structure of communities and biomass of ichthyofauna in the Biebrza River, its old river beds and affluents. *Pol. Ecol. Stud.* **1984**, *10*, 447–474.

33. Wassen, M.J.; Peeters, W.H.M.; Veneterink, H.O. Patterns in vegetation, hydrology and nutrient availability in an undisturbed river floodplain in Poland. *Plant Ecol.* **2002**, *165*, 27–43.

34. Chormański, J.; Okruszko, T.; Ignar, S.; Batelaan, O.; Rebel, K.T.; Wassen, M.J. Flood mapping with remote sensing and hydrochemistry: A new method to distinguish the origin of flood water during floods. *Ecol. Eng.* **2011**, *37*, 1334–1349.

35. *Standard Methods for the Examination of Water and Wastewater*, 20th ed.; American Public Health Association: Washington, DC, USA, 1998.

36. Dunalska, J.A.; Górniak, D.; Jaworska, B.; Evelyn, E.; Gaiser, E.E. Effect of temperature on organic matter transformation in a different ambient nutrient availability. *Ecol. Eng.* **2012**, *49*, 27–34.

37. Persat, H.; Copp, G.H. Electrofishing and Point Abundance Sampling for the Ichthyology of Large Rivers. In *Developments in Electrofishing. Fishing New Books*; Cowx, I.G., Ed.; Blackwell Scientific Publications Ltd: Oxford, UK, 1989; pp. 205–215.

38. *European Standard CSN-EN 14011-Water Quality-Sampling of Fish with Electricity*; European Committee for Standardization: Brussels, Belgium, 2006.

39. Obolewski, K. Epiphytic macrofauna on water soldiers (*Stratiotes aloides* L.) in Slupia River oxbows. *Oceanol. Hydrobiol. Stud.* **2005**, *34*, 37–54.

40. Hammer, Ø.; Harper, D.A.T.; Ryan, P.D. PAST: Paleontological statistics software package for education and data analysis. *Palaeontol. Electron.* **2001**, *4*, 1–9.

41. Ter Braak, C.J.F.; Šmilauer, P. *Reference Manual and CanoDraw for Windows User's Guide*; Version 4.5; Software for Canonical Community Ordination—Microcomputer Power; Biometrics: Ithaca, NY, USA, 2002.

42. Damme, D.V.; Bogutskaya, N.; Hoffmann, R.C.; Smith, C. The introduction of the European bitterling (*Rhodeus amarus*) to west and central Europe. *Fish Fish.* **2007**, *8*, 79–106.

43. Kwak, T.J. Lateral movement and use of floodplain habitat by fishes of the Kankakee River Illinois. *Am. Midl. Nat.* **1988**, *120*, 241–249.

44. Amoros, C.; Bornette, G. Connectivity and biocomplexity in waterbodies of riverine floodplains. *Freshw. Biol.* **2002**, *47*, 761–776.

45. Miranda, L.E.; Andrews, C.S.; Kröger, R. Connectedness of land use, nutrients, primary production, and fish assemblages in oxbow lakes. *Aquat. Sci.* **2014**, *76*, 41–50.

46. Holopainen, I.J.; Tonn, W.M.; Paszkowski, C.A. Tales of two fish: The dichotomous biology of crucian carp (*Carassius carassius* (L.)) in northern Europe. *Ann. Zool. Fenn.* **1997**, *34*, 1–22.

47. Galat, D.L.; Fredrickson, L.H.; Humburg, D.D.; Bataille, K.J.; Bodie, J.R.; Dohrenwend, J.; Gelwicks, G.T.; Havel, J.E.; Helmers, D.L.; Hooker, J.B.; *et al.* Flooding to restore connectivity of regulated, large-river wetlands. *BioScience* **1998**, *48*, 721–733.

48. Petry, A.C.; Agostinho, A.A.; Gomes, L.C. Fish assemblages of tropical floodplain lagoons: Exploring the role of connectivity in a dry year. *Neotropical Ichthyol.* **2003**, *1*, 111–119.

49. White, J.L.; Harvey, B.C. Effects of an introduced piscivorous fish on native benthic fishes in a Coastal River. *Freshw. Biol.* **2001**, *46*, 987–995.

50. Wiśniewolski, W.; Ligęza, J.; Prus, P.; Buras, P.; Szlakowski, J.; Borzęcka, I. Znaczenie łączności rzeki ze starorzeczami dla składu ichtiofauny na przykładzie środkowej i dolnej Wisły. *Nauka Przyr. Technol.* **2009**, *3*, 2–10. (In Polish)

51. Wiśniewolski, W.; Szlakowski, J.; Buras, P.; Klein, M. Ichtiofauna (Ichthyofauna). In *Kotlina Biebrzańska i Biebrzański Park Narodowy*; Banaszuk, H., Ed.; WE&S: Białystok, Poland, 2004; pp. 455–489. (In Polish)

An Efficiency Analysis of a Nature-Like Fishway for Freshwater Fish Ascending a Large Korean River

Jeong-Hui Kim, Ju-Duk Yoon, Seung-Ho Baek, Sang-Hyeon Park,
Jin-Woong Lee, Jae-An Lee and Min-Ho Jang

Abstract: Using traps and passive integrated transponder (PIT) telemetry, we investigated the effectiveness of the nature-like fishway installed at Sangju Weir on the Nakdong River, Korea. In 11 regular checks over the study period, 1474 individuals classified into 19 species belonging to 5 families were collected by the traps, representing 66% of the species inhabiting the main channel of the Nakdong River. PIT tags were applied to 1615 individuals belonging to 22 species, revealing fishway attraction and passing rates of 20.7% and 14.5%, respectively. Interspecific differences were also shown. For 63.2% of fishes, it took more than a day to pass through the fishway. Some individuals spent a longer time (>28 days) inside the fishway, suggesting the fishway was also being used for purposes other than passage. In this study, we verified species diversity of fish using a nature-like fishway installed in a large river in Korea. The results of this study provide a useful contribution to the development of fishways suitable for fish species endemic to Korea and for non-salmonid fish species worldwide.

Reprinted from *Water*. Cite as: Kim, J.-H.; Yoon, J.-D.; Baek, S.-H.; Park, S.-H.; Lee, J.-W.; Lee, J.-A.; Jang, M.-H. An Efficiency Analysis of a Nature-Like Fishway for Freshwater Fish Ascending a Large Korean River. *Water* **2016**, *8*, 3.

1. Introduction

A large number of structures such as dams and weirs have been constructed across rivers and streams for management of water resources. These structures alter the physical properties of rivers and riverine environments, and affects water quality by slowing down its flow. They also affect the movement of migratory fish by reducing the longitudinal connectivity of the river [1,2]. Adverse effects of such artificial structures across rivers have been reported for the local aquatic ecosystems, especially in relation to decreases in the number of anadromous fish communities, such as salmonid and lamprey fish, due to the obstruction of movement to their respective spawning habitats. For freshwater fish species, genetic discontinuity between the upstream and downstream populations can also occur [3,4].

Fishways (fish ladders, fish passes, or fish steps) are the most effective solution to the problems related to the blocked downstream and upstream movements of aquatic fauna caused by manmade structures in rivers [5]. Early fishway construction targeted only a few fish species with strong swimming abilities, such as adult salmonids [6], but recent trends are directed at making fishways available for the passage of all species in all life stages [7]. Among the various types of fishways, nature-like fishways are constructed with boulders, large wooden debris, and riparian vegetation to imitate natural environments, instead of concrete or steel, thus producing hydrodynamic and morphological properties similar to those of natural rivers [8,9]. Owing to these characteristics, unlike other fishways, nature-like fishways can be used by fish species with a wide range of sizes and swimming abilities [10]. Best-suited for the original purpose of fishways, nature-like fishways are being constructed across the world in increasing numbers [11]. In Korea too, nature-like fishways are attracting increasing amounts of attention, and 8 nature-like fishways have been constructed in large rivers since 2010.

By regularly monitoring the use of a fishway after its installation, not only can its attraction and passage efficiencies be checked, but useful data for its efficient management can also be obtained to better address fish movement issues. The most common conventional monitoring method, which is still frequently used [12], is to count the number of individuals passing through the exit of the fishway. In particular, customized traps are often used owing to their advantages in identifying the species, numbers, and sizes of individuals using fishways, and can also be used for collection purposes. On the other hand, fish telemetry using passive integrated transponder (PIT) tags, radio tags, or acoustic tags is mostly used for assessing the attraction and passage efficiencies of a fishway [7]. Such tagging methods deliver detailed information on fish behavior and movement, enabling an accurate quantitative evaluation of its attraction and passage efficiencies [13]. However, telemetry methods can only selectively monitor the tagged species, and thus cannot deliver the data for identifying all fish species using the fishway. These disadvantages can be overcome by the combined use of these two methods to obtain accurate high-quality information for fishway management.

This study aimed to obtain accurate data necessary for evaluating the efficiency of a nature-like fishway constructed on a large river by applying a refined method combining customized traps and a PIT telemetry system to both identify the species and sizes of fishes using the fishway and evaluate the attraction and passage efficiencies. Additionally, we presented measures to improve the efficiency of the fishway by analyzing the correlation between the upstream water level and fishway use data.

2. Materials and Methods

2.1. Study Site

The Nakdong River (525.15 km in length) is one of the longest rivers in Korea. Its main channel has a midstream width of at least 100 m and up to 1 km in some downstream areas. Eight large weirs have been constructed in the main channel to manage water resources and control water flow (Figure 1). Of the 8 weirs in the Nakdong River, Sangju Weir lies the furthest upstream. It is 335 m long and 11 m high, with a nature-like fishway installed on the right side and a small hydropower plant on the left side. The upstream water level is constantly maintained at its management water level of 47.00 m AMSL, except during the flooding season.

2.2. Fishway

The study site was the nature-like fishway installed on the right side of Sangju Weir (Figure 1). The fishway is 700 m long in total, at a slope of 1/100. It has a zigzag path designed to make optimal use of the narrow space. Its width ranges from 6 m to 18 m (mean = 7.4 m), with slight differences per section. Although the water depth of the fishway is maintained at an average of 50 cm, it fluctuates depending on the inflow rate into the fishway and the location within the fishway. The inflow rate into the fishway varies in accordance with the upstream water level. The fishway was thus designed to induce an optimal inflow rate, provided that Sangju Weir is maintained at its intended water level.

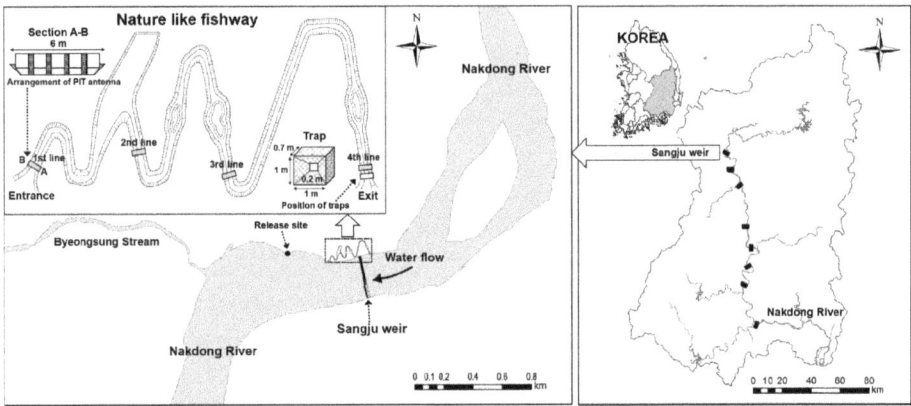

Figure 1. The location of Sangju Weir on the Nakdong River in Korea, the positions of the 4 PIT antenna lines along the fishway, and 6 traps at the exit of the fishway used during the sampling period.

2.3. Ichthyofauna of the Nakdong River

To determine the fish species that might use the fishway, the ichthyofaunal assemblages of communities found in the main channel of the Nakdong River were analyzed. Fish were collected at one upstream site and one downstream site, both at a 1 km distance from Sangju Weir. A cast net (mesh: 7 mm; area [πr^2]: 16.6 m^2) and a fyke net (mesh: 5 mm) were used as active and passive fishing gear, respectively. The cast net was thrown 15 times for each survey, and the fyke net was set for 48 h. After measuring the total length (TL, mm) of the collected fish on site, they were released back to the stream at the same point as collection. Surveys were carried out 4 times (July and October 2012, March and May 2013) at 2–3 month intervals, excluding the winter months.

2.4. Fishway Monitoring

2.4.1. Trap Monitoring

To identify the species and number of individuals using the fishway, 6 traps (dimensions: $1 \times 1 \times 0.7$ m; mesh: 4 mm) were installed at the exit of the fishway (Figure 1), which blocked the entire fishway exit. All ascending fish were thus collected. The traps were installed for monitoring once per month for 11 months between June 2012 and July 2013. Due to the limited mobility of fish at low water temperatures, the winter months (December 2012 through February 2013) were excluded from monitoring. The traps were installed at 16:00 and collected after 24 h. According to the study by Lee *et al.* [14], which evaluated the same type of trap we used in this study, the installment of the trap did not significantly affect the water velocity within the fishway. Therefore, we assume the installment of the trap did not affect the fish using the fishway. However, when a trap is installed for a long period, it may come loose, or debris may accumulate. Therefore, we checked the status of the trap every 6 h following the installment. Before being released upstream of the weir, the number of collected individuals were counted per species and TLs were measured.

2.4.2. PIT Telemetry

We applied PIT telemetry to monitor the efficiency of the fishway from July 2012 to July 2013, on a total of 1615 individuals of 22 species from 8 families (Table 1). Samples were collected using a cast net (mesh, 7 mm) and fyke net (mesh, 5 mm) within 1 km downstream of the weir. Collected fish were immediately moved to an aerated plastic tank ($1 \times 1 \times 0.7$ m) and stabilized for 30 min prior to tag implantation. Each fish without visible injuries was subsequently anaesthetized using 0.1 g·L^{-1} ethyl 3-aminobenzoate methanesulfonate salt (Sigma-Aldrich, Munich, Germany), after which its TL and body weight (BW) were measured (Table 1). A 3-mm-long incision was made on the ventral body surface and a PIT tag (12 mm in length and 2.12 mm in diameter, FDX-B, Finfotech, Korea), was inserted into the abdominal cavity. A biological bond (Vetbond; 3M, Minnesota, USA) was used to prevent water from entering the cavity. After tag insertion, the fish was then moved to an aerated plastic tank (1 × 1 × 0.7 m) for recovery. Tagged fish were released at the release site 300 m downstream from the fishway entrance, and their fishway usage patterns were tracked.

The total length of the nature-like fishway installed at Sangju Weir is 700 m. With the first 50 m from the entrance (lower end) constantly immersed, the actual length used by the fish is 650 m. Four antenna lines (1st line at the entrance, 2nd and 3rd lines in the middle, 4th line at the exit) were installed at intervals of 200–230 m. We used rectangular PIT antennas (1 (w) × 0.4 (h) m, F-12030, Finfotech, Korea), and each antenna line comprised 5 antennas. Since a signal could be detected whenever a fish passes inside an antenna and up to 15 cm outside of it, the total detection range of each antenna was 130 cm. By installing them at 30 cm intervals to ensure detections free of inter-antenna interference, each antenna line consisting of 5 antennas had a signal detection range of 6.5 m. In other words, the antennas were arranged so that any signal emitted within the fishway at a given moment could be detected by one antenna. The 4 antenna lines were numbered from the entrance to the exit for easy identification. Tests were performed every 2 weeks to check the operation and detection range of each antenna in the following manner. Each antenna was connected to a reader (Finfotech, Korea), and readers were connected to a data logger (Finfotech, Korea). Detection data that was stored in a data logger was subsequently transferred to a database [15] by the code division multiple access (CDMA) method every 10 min. More detailed information about the PIT antenna are available in the study of Yoon *et al.* [16] using the same antenna system.

To determine the factors that affect the efficiency of the fishway, the water velocity in the fishway and the upstream water level were measured. Water velocity was measured with a handheld velocity meter (FP111, Global Water, TX, USA) at the exit of the fishway. The flow rate in the fishway was calculated on the basis of the water depth and velocity at the exit. The water velocity and flow rate were monitored 16 times during the study period in accordance with upstream water level variations. Upstream water levels were extracted from the data collected by the Nakdong River Flood Control Office at 10-min intervals.

Table 1. Information on the tagged and detected fish in the nature-like fishway at Sangju Weir. The "movement" of a detected fish indicates number of fishes that reached each antenna line. "Return" refers to the number of individuals that turned back from the exit after being detected by the 4th antenna line. "RA" means relative abundance.

Family	Species	Tagged Fish						Detected fish							
				TL (mm)		BW (g)		Attraction		Movement				Passing	
		N	RA (%)	Mean (SD)	Range	Mean (SD)	Range	N	Rate (%)	1st	2nd	3rd	4th (return)	N	Rate (%)
Cyprinidae	Cyprinus carpio	8	<1	189 (25.8)	149–228	96.2 (42.5)	42.8–166.9	1		1					
	Carassius auratus	16	<1	179 (54.2)	76–312	118.9 (98.8)	5.8–441.4	1	6.3						50
	Carassius cuvieri [2]	20	1.2	159 (70.5)	86–346	104.4 (151.6)	10.4–554.9	2	10.0		1	1	1	1	
	Acheilognathus lanceolatus	7	<1	71 (24.9)	60–131	6 (7.7)	2.2–24.8	2	28.6						
	Pungtungia herzi	2	<1	74 (8.0)	66–82	3.5 (1.2)	2.3–4.7								
	Squalidus chankaensis tsuchigae [1]	109	6.7	89 (9.4)	72–114	6 (2.2)	2.3–13.3	11	10.1	7	4				
	Hemibarbus labeo	28	1.7	212 (144.3)	92–488	191.1 (289.6)	6.6–866.4	11	39.3	4	1	4	2 (1)	1	9.1
	Hemibarbus longirostris	66	4.1	142 (40.1)	70–208	28.1 (19.0)	2.6–74.1	26	39.4	6	1	9	10	10	38.5
	Pseudogobio esocinus	212	13.1	153 (26.6)	80–214	29.5 (16.1)	3.7–117.1	33	15.6	7	10	6	10 (1)	9	27.3
	Zacco platypus	507	31.4	122 (23.1)	70–184	17.9 (12.1)	2.3–138	147	29.0	64	22	42	19 (2)	17	11.6
	Opsariichthys uncirostris amurensis	427	26.4	145 (67.7)	77–359	41.6 (62.2)	2.8–451.6	75	17.6	44	8	16	7 (2)	5	6.7
	Erythroculter erythropterus	11	<1	360 (83.9)	234–493	281.8 (200.8)	59.2–615.8	4	36.4	2	1		1	1	25
	Hemiculter eigenmanni [1]	84	5.2	114 (11.8)	92–150	8.8 (3.4)	4.1–24.8	2	2.4	2					
Bagridae	Pseudobagrus fulvidraco	41	2.5	195 (28.3)	132–241	75.9 (26.7)	25–121.3	9	22.0		3	4	2	2	22.2
	Leiocassis ussuriensis	2	<1	132 (6.5)	125–138	14.9 (1.6)	13.3–16.5	1	50.0				1	1	100

387

Table 1. *Cont.*

Family	Species	N	RA (%)	Tagged Fish						Detected fish					
				TL (mm)		BW (g)		Attraction		Movement				Passing	
				Mean (SD)	Range	Mean (SD)	Range	N	Rate (%)	1st	2nd	3rd	4th (return)	N	Rate (%)
Siluridae	*Silurus asotus*	23	1.4	291 (52.1)	203–425	145.9 (78.1)	46.1–346.9	3	13.0	2			1	1	33.3
Centropomidae	*Siniperca scherzeri*	20	1.2	173 (60.0)	112–310	86.4 (108.1)	16–394.3	1	5.0	1					
	Coreoperca herzi[1]	21	1.3	156 (15.9)	108–180	57.2 (15.0)	16.4–87.7	4	19.0	2		1	1	1	25
Centrarchidae	*Micropterus salmoides*[2]	6	<1	199 (26.5)	161–240	112.2 (32.9)	59.5–154.3	1	16.7	1					
Odontobutidae	*Odontobutis platycephala*[1]	3	<1	166 (6.4)	161–175	61.7 (7.9)	55.3–72.8	1	33.3		1				
Gobiidae	*Rhinogobius giurinus*	1	<1	70		4.1									
Channidae	*Channa argus*	1	<1	451		663.1									
	Total	1615		140 (59.0)	60–493	37.5 (72.1)	2.2–866.4	334	20.7	144	52	83	55 (6)	49	14.5

1: Endemic species; 2: Exotic species.

2.5. Data Analysis

To evaluate the efficiency of the fishway, the attraction rate, passing rate, passing time, and movement time were calculated for each species. Attraction rate was the percentage of all tagged individuals that were detected by antennas installed in the fishway. Passing rate was the percentage of the total number of individuals that were detected at the 4th antenna line and that did not move back to the range of the other (1st–3rd) antenna lines [13,17]. Additionally, we tracked fish movements inside the fishway and presented the distances covered by the individuals after entering the fishway. Passing time, which is the time taken for an individual to move from the entrance to the exit (650 m), reflects its swimming speed. The movement time between the antennas (3 sections) was calculated and compared by section (low section: 1st→2nd; middle section: 2nd→3rd; upper section: 3rd→4th). A Spearman's rank correlation analysis was carried out on *Zacco platypus*, which passed the fishway most frequently, to determine the correlation between passing time and TL. To investigate the interspecific differences in fishway usage, a one-way ANOVA test was performed to analyze the upstream water level at the time of fishway usage by the 4 most frequently passing species (*Z. platypus*, *Opsariichthys uncirostris amurensis*, *Pseudogobio esocinus*, and *Hemibarbus longirostris*). Additionally, the daily movement pattern of each species was compared and analyzed. All analyses were performed using SPSS ver. 20.0 (SPSS Inc., Chicago, IL, USA).

3. Results

3.1. Species Assemblages of the Fish Collected in The Nakdong River and the Fishway Traps

Twenty-eight species from nine families were collected upstream and downstream in the Nakdong River's main channel. Of these, 19 species belonged to the Cyprinidae family, and their relative abundance was very high (97.8%) compared to other families. Nineteen species from five families were collected from the traps installed at the exit of the fishway. Hence, 66.0% of the fish species inhabiting the main channel of the Nakdong River used the fishway (Table 2). The species that used the fishway most frequently were *O. u. amurensis* (RA, 39.9%), followed by *Hemibarbus labeo* (39.2%), *Z. platypus* (8.1%), *P. esocinus* (4.2%), and *Squalidus chankaensis tsuchigae* (3.4%), which were also the dominant species among those collected in the Nakdong River's main channel. Four of the seven endemic species (57.0%) collected from the main channel were also collected in the fishway. Up to 11 species were found to be using the fishway on the same day (in July and August 2012).

Table 2. Species assemblage of the fish collected in the main channel of the Nakdong River and the number of individuals, relative abundance (RA) and total length (TL) of fishes that were collected by the traps. TLs are expressed as mean values (±standard deviations), and the ranges are also presented.

Family	Species	River N	River RA (%)	2012 June	July	August	September	October	November	2013 March	April	May	June	July	Total	RA (%)	TL (mm)
Cyprinidae	*Cyprinus carpio*	24	1.5			1	1				3			1	6	<1	343 ± 127 (216–500)
	Carassius auratus	54	3.5	1	3	1	2						2		9	<1	167 ± 92 (46–340)
	Carassius cuvieri[3]	1	<1		5	1	1								7	<1	169 ± 60 (45–230)
	Acheilognathus lanceolatus	4	<1								1				1	<1	115
	Acanthorhodeus macropterus	18	1.1														
	Pseudorasbora parva	4	<1														
	Pungtungia herzi	11	<1		1										1	<1	60
	Sarcocheilichthys variegatus[2]	2	<1														
	Squalidus japonicus coreanus[2]	4	<1									1			1	<1	114
	Squalidus chankaensis tsuchigae[2]	61	3.9	9	21	3						9	8		50	3.4	89 ± 13 (70–123)
	Hemibarbus labeo	41	2.6	1	4	1				4	3	1	5	6	25	1.7	325 ± 151 (50–515)
	Hemibarbus longirostris	7	<1	1			1					3	1	1	7	<1	149 ± 14 (133–168)
	Pseudogobio esocinus	74	4.7	8		1		1			5	3	33	11	62	4.2	169 ± 16 (96–218)
	Microphysogobio yaluensis[2]	2	<1														
	Zacco platypus	631	40.1	21	5	13	27	36			3	8	4	2	119	8.1	121 ± 34 (48–175)

Table 2. *Cont.*

Family	Species	River N	River RA (%)	2012 June	2012 July	2012 August	2012 September	2012 October	2012 November	March	April	2013 May	2013 June	2013 July	Total	Total RA (%)	TL (mm)
	Opsariichthys uncirostris amurensis	504	32.0	84	68	108	26	7		8	36	56	81	113	587	39.9	175 ± 67 (48–330)
	Erythroculter erythropterus	75	4.8	2	15	15	24	4			6	77	96	339	578	39.2	356 ± 53 (136–550)
	Culter brevicauda [1]	9	<1														
	Hemiculter eigenmanni [2]	2	<1		1	1									2	<1	84 ± 53 (46–121)
Cobitidae	*Misgurnus anguillicaudatus*	1	<1		1										1	<1	53
Bagridae	*Pseudobagrus fulvidraco*	11	<1														
Siluridae	*Silurus asotus*	2	<1														
Centropomidae	*Siniperca scherzeri*	4	<1									8	1	1	10	<1	218 ± 61 (158–300)
	Coreoperca herzi [2]	4	<1										1		1	<1	155
Centrarchidae	*Micropterus salmoides* [3]	2	<1			1									1	<1	168
Odontobutidae	*Odontobutis platycephala* [2]	1	<1														
Gobiidae	*Rhinogobius brunneus*	5	<1					3							5	<1	46 ± 6 (39–50)
	Tridentiger brevispinis	4	<1														
Channidae	*Channa argus*	1	<1														
No. of individuals		1564		127	127	146	82	51	0	12	57	166	232	474	1474		
No. of species		29		8	11	11	7	5	0	2	7	9	10	8	19		

1: Endangered species; 2: Endemic species; 3: Exotic species.

The frequency of fishway usage decreased during the low water temperature season (none in November; two species in March). The TL distribution of fish species inhabiting the main channel of the Nakdong River demonstrated that individuals ranging in sizes from 50 to 100 mm accounted for approximately 50% of the total, and the ratio gradually decreased as the size of the individuals increased. The range of the TL distribution of fish that used the fishway was 39–550 mm. Moreover, the size distribution of individuals using the fishway was not limited to a certain size. Fish of various sizes utilized the fishway evenly. This differs from the main channel of the Nakdong River (Figure 2).

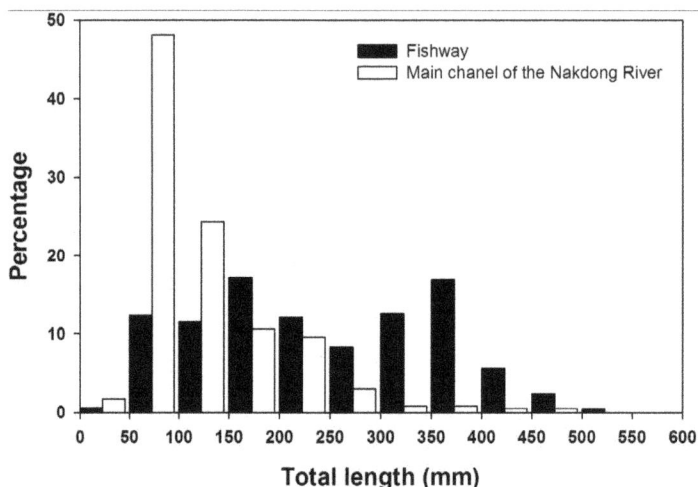

Figure 2. The size distribution of the fish captured by the traps in the nature-like fishway and collected from the main channel of the Nakdong River. Size class intervals were 50 mm.

3.2. Efficiency of the Fishway

Of the 1615 PIT-tagged individuals, 334 individuals were detected in the fishway, showing an attraction rate of 20.7% (Table 1). The species-dependent attraction rate varied widely (0%–50%), with the attraction rates of *Z. platypus, O. u. amurensis,* and *P. esocinus,* three species predominantly tagged, being 17.6%, 29.0%, and 15.6%, respectively. Only 49 of the 334 individuals attracted to the fishway succeeded in passing it and moving upstream of Sangju Weir, showing a passing rate of 14.5%. *Z. platypus* showed the highest number of individuals attracted to the fishway, but a below-average passing rate (11.6%, 17/147), whereas *Leiocassis ussuriensis* showed a passing rate of 100% because the only individual of this species that was attracted to the fishway passed it. Some individuals of *H. labeo, P. esocinus, Z. platypus,* and *O. u. amurensis* attracted to the fishway were detected by the 4th antenna line at the exit, but turned back into the fishway instead of moving upstream into the main channel of the Nakdong River.

The analysis of passing time through the fishway revealed that 36.8% of the individuals passed the fishway within 1 day, while the rest took longer (Table 3), 26.3% of which stayed in the fishway for more than 10 days. The average passing time ranged from 1.2 h to 1559.4 h, showing species-dependent differences. Considerable intraspecific variations in passing time were also verified. For Z. *platypus*, for example, the shortest and longest stays in the fishway were 31.7 h and 2135.0 h. To analyze the effect of body length on passing time, we performed a Spearman's rank correlation analysis between the TL and the passing time of Z. *platypus*, which passed the fishway most frequently, but no statistically significant correlation was confirmed ($r_s = 0.071$, $p > 0.05$). The average attraction time it took each tagged fish to move 300 m from the release site to the fishway entrance was 627.6 h, which amounts to more than 26 days. Additionally, the attraction time varied among individuals, ranging from a minimum of 0.1 h to 7532.2 h. No specific pattern was observed when comparing the data of different species.

The movement time in the Sangju Weir nature-like fishway varied according to the sections (Table 4). The average length of time from the detection at the entrance (1st antenna line) to the detection at the exit (4th antenna line) was 316.7 h, or approximately 13 days. On the other hand, analysis of movement time by section revealed that the longest time (mean = 284.9 h) was spent passing the first section (from the 1st to the 2nd antenna line), which was then increasingly shortened as the individuals approached the exit.

The results of the analysis of upstream water levels at the time of fish attraction and passing revealed that both attraction and passing occurred more frequently when the water level was higher than the intended management water level (47.00 m AMSL), at 47.08 m and 47.19 m AMSL, respectively (Figure 3). By substituting the mean upstream water levels at the attraction and passing times for the changes in water velocity and flow rate in the fishway into the regression analysis formula (Figure 4), it was found that attraction occurred most frequently at a water velocity and flow rate of 0.90 m·s^{-1} and 1.21 m^3·s^{-1}, respectively, and passing at 1.05 m·s^{-1} and 1.57 m^3·s^{-1}, respectively. The results of the analysis of water temperature-dependent attraction and passing rates showed that attraction and passing occurred only at temperatures higher than 4 °C and 9 °C, respectively, and no tagged fish were detected in the months of January and February, when the water temperature dropped below 4 °C (Figure 3).

Table 3. Passing time of the species that passed the fishway. Among the species passing through the fishway, the 1st antenna line detection time of *Carassius cuvieri*, *Pseudobagrus fulvidraco*, *Silurus asotus*, and *Coreoperca herzi* were not uploaded to the database (internet server), due to a miscommunication between the data logger and the database of the PIT telemetry system; therefore, we could not present the passing time of these individuals.

Family	Species	N	Time (h) Mean (SD)	Proportion (%) <1 day	1–5 days	5–10 days	>10 days
Cyprinidae	*Hemibarbus labeo*	1	168.9			100	
	Hemibarbus longirostris	2	8.7 (7.9)	100			
	Pseudogobio esocinus	2	127.4 (172.8)	50		50	
	Zacco platypus	8	470.8 (646.0)		37.5	12.5	50
	Opsariichthys uncirostris amurensis	4	62.4 (107.4)	75		25	
	Erythroculter erythropterus	1	1.2	100			
Bagridae	*Leiocassis ussuriensis*	1	1559.4				100
	Total	19		36.8	15.8	21.1	26.3

The detection times of the 4 major species using the fishway were checked against water levels upstream of Sangju Weir. Results showed that all 4 species were using the fishway when the upstream water level ranged from 46.80–47.30 m AMSL (Figure 5), without statistically significant interspecific differences (one-way ANOVA test, $p > 0.05$). The analysis of detection times to compare fishway usage time by species revealed species-specific patterns (Figure 6). *Z. platypus* and *O. u. amurensis* were using the fishway during both daytime (06:00–18:00) and nighttime (18:00–06:00) hours, most frequently during sunset (17:00–19:00) (Figure 6a,b). In contrast, *P. esocinus* and *H. longirostris* used the fishway mostly after sunset, avoiding the daytime hours (Figure 6c,d).

Table 4. Attraction time and movement time by fishway section. Attraction time indicates the time spent from the release site to the first antenna. We also presented the movement speed (m/h) in each section.

		Attraction		Movement		Passing
		Release Site→1st	1st→2nd	2nd→3rd	3rd→4th	1st→4th
Distance (m)		300	220	230	200	650
Area (m²)		–	2695	1562	1235	5492
Time (h)	Mean (SD)	627.6 (1687.9)	284.9 (443.8)	114.2 (185.6)	33.3 (84.0)	316.7 (570.5)
	Range	0.1–7532.2	0.3–2012.4	0.8–740.2	0.7–434.9	1.2–2135.0
Speed (m/h)		0.5	0.8	2.0	6.0	2.1

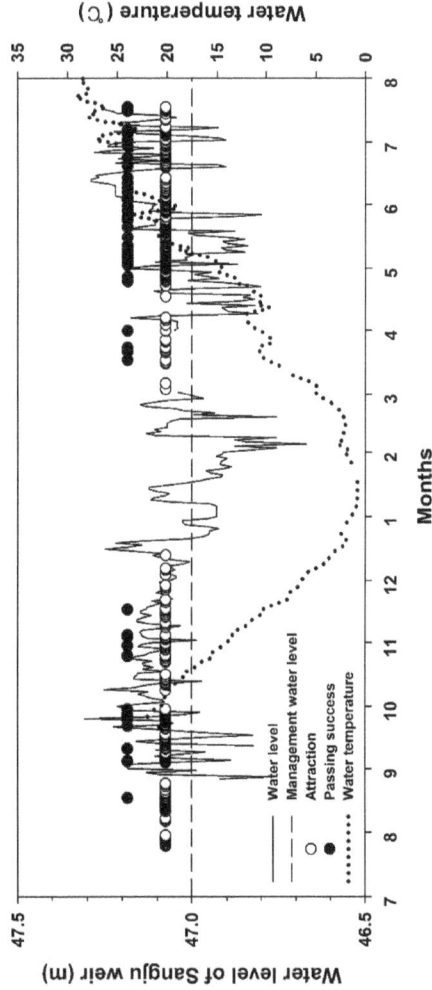

Figure 3. Correlations among upstream water level, water temperature change, and tag detection. Attraction and passing success occurred when the upstream water level was higher than the management water level (47.00 m AMSL), at 47.08 m AMSL and 47.19 m AMSL, respectively. No detection in the fishway was made between mid-December 2012 and March 2013, when water temperature was low.

395

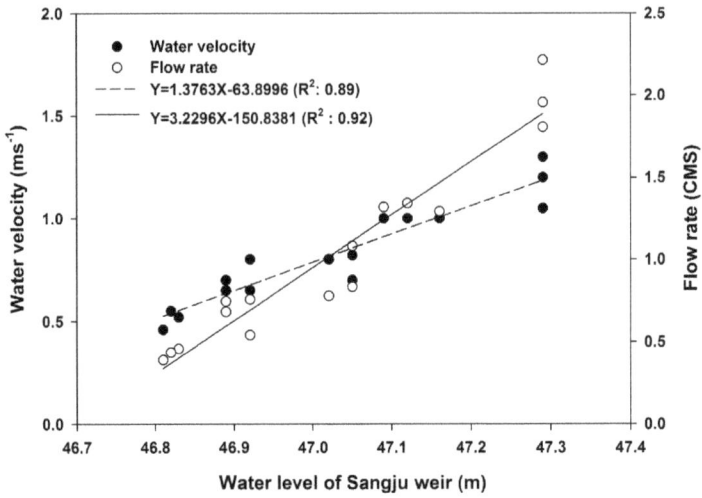

Figure 4. Changes in water velocity and flow rate in the fishway according to the water level upstream of the Sangju Weir.

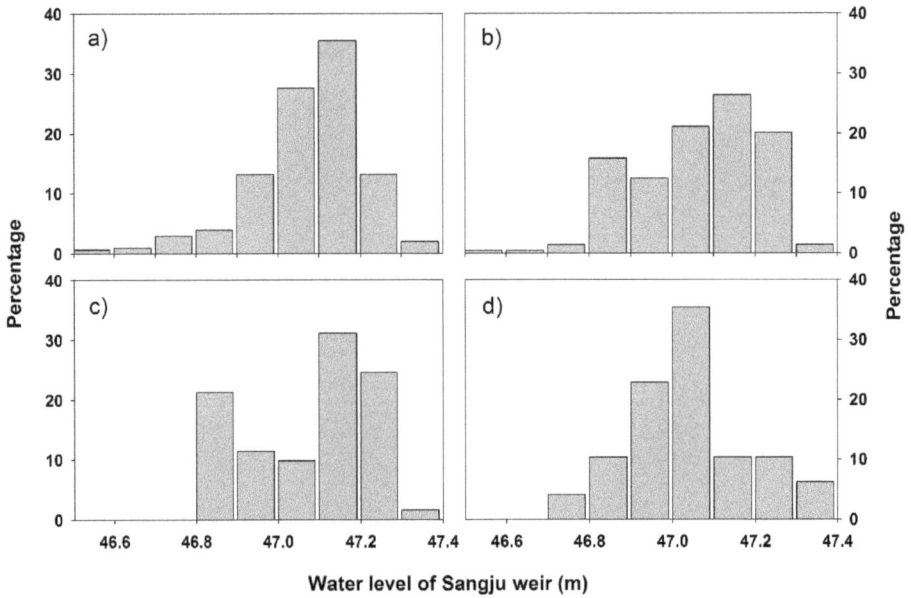

Figure 5. Species-specific patterns of fishway usage depending on upstream water level: (**a**) *Zacco platypus*; (**b**) *Opsariichthys uncirostris amurensis*; (**c**) *Pseudogobio esocinus*; (**d**) *Hemibarbus longirostris*.

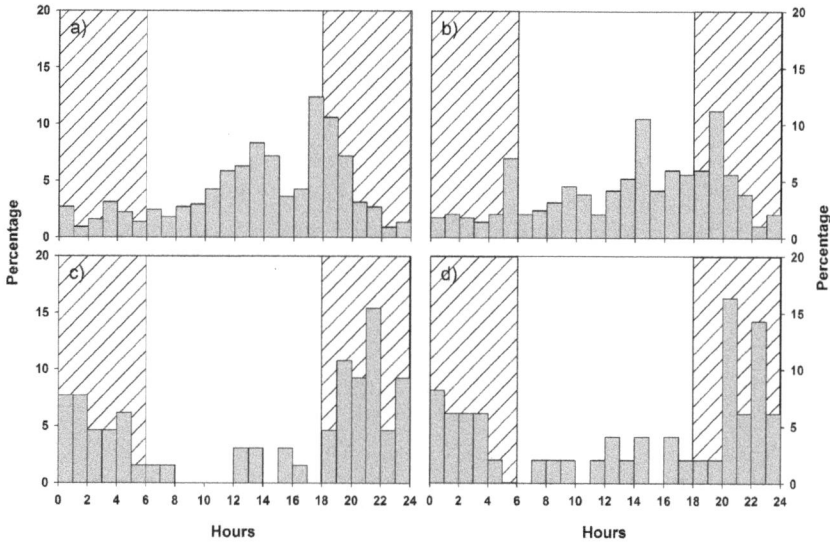

Figure 6. Species-specific patterns of fishway usage depending on the time of the day: (**a**) *Zacco platypus*; (**b**) *Opsariichthys uncirostris amurensis*; (**c**) *Pseudogobio esocinus*; and (**d**) *Hemibarbus longirostris*. For the analysis, the first attraction data of each individual were utilized. Time of day was divided into daytime (06:00–18:00) and nighttime (18:00–16:00, oblique line) hours.

4. Discussion

The recent trend of fishway design reflects the goal of attracting a wide variety of fish species rather than targeted species [18,19]. In Korea too, fishway design features focus on the species diversity of individuals using fishways. A comparison of the fish species using the fishway with those inhabiting the main channel of the Nakdong River showed that a high diversity (66%) of species inhabiting the main channel used the nature-like fishway at Sangju Weir. This included four endemic species and one species protected under the Endangered Species Act. Thirty-four percent of the fish species that inhabit the main channel of the Nakdong Rivier were not observed in the fishway. Unlike the species that utilized the fishway, these species lack the motivation to use the fishway; their usage therefore was not confirmed. The main biological aim of installing a fishway is to enable movement of anadromous and potamodromous fish to their spawning grounds and to help freshwater fish downstream of the weir move upstream (and vice versa), preventing isolation between the two areas [20]. Genetic differences between upstream and downstream species, due to long-term isolation resulting from a structure constructed across the river, have been reported [21]. When such situations persist and genetic diversity decreases, certain species may become endangered or extinct. The Sangju

Weir fishway can be used by a variety of fish species of the ichthyofaunal communities in the main channel of the Nakdong River.

The fishway was being used by individuals of all sizes without any size-dependent tendency, with a TL ranging from 39 to 550 mm. The nature-like fishway has a very low slope of 1%, providing lower water velocity and turbulence compared to other fishways [22]. Water velocity is an important parameter for the attraction and passage efficiencies of a fishway because a high water velocity prevents small fish with weak swimming ability from using the fishway [18]. Moreover, Stuart and Mallen-Copper [23] reported that turbulence can also be an adverse factor for the movement of small individuals. In this regard, the Sangju Weir fishway was verified as having an appropriate water velocity and turbulence for a nature-like fishway, facilitating its usage by small individuals. Furthermore, while technical fishways have a constant water velocity and water depth due to their homogenous structure, nature-like fishways have lower slopes and nature-mimicking irregular substrates, producing various water depths and velocities [22]. Such nature-like fishway environments allow fish of different sizes and swimming abilities to use the fishway [9].

Bunt et al. [13] performed a meta-analysis using efficiency evaluation results for various types of fishways, and noted that attraction efficiency was lowest for nature-like fishways (mean 48%), while passage efficiency was highest (mean 70%). At 20.7% and 14.5%, respectively, the actual attraction and passage efficiencies of the nature-like fishway at Sangju Weir were lower than the values presented by Bunt et al. [13]. From a structural point of view, the extremely low water velocity at the fishway entrance is probably responsible for the low attraction efficiency [24,25]. Moreover, the fishway length is inversely proportional to passage efficiency [5]. Compared to the slopes (1.7%–7.1%) and lengths (12–370 m) of the 7 nature-like fishways analyzed by [13], the Sangju Weir nature-like fishway has a lower slope (1%) and is much longer (650 m), which are both considered structural characteristics unfavorable for attraction and passage efficiencies.

According to the results of the trap survey, despite the high species diversity of the individuals using the Sangju Weir nature-like fishway, all species were freshwater fish, without any anadromous fish species. Since anadromous species have to ascend the stream to return to their spawning grounds, attraction and passage efficiencies for such species are high in most fishways [26]. On the other hand, as freshwater fish species migrate without such a natural drive, their attraction and passage efficiencies may vary among fishways [27]. Consequently, the final results of attraction and passage efficiencies of the nature-like fishway at Sangju Weir can be considered to have been affected by the ecological characteristics of the endemic fish species and the structural characteristics of the fishway.

The benefit of the telemetry method lies in the quantitative evaluation of the efficiency of the attraction and passage of the fishway, which enables the provision of plans to increase the efficiency of the corresponding fishway. In this regard, environmental factors such as water depth, velocity, temperature, and discharge rate were reported to affect the upstream migration of fish [28–31]. In the case of the nature-like fishway at Sangju Weir, the flow rate and water velocity varied as the upstream water level changed, as did its attraction and passage efficiencies. If the optimal conditions for attraction and passage efficiencies are different, a question of priority arises. Attraction efficiency can be increased by changing the position and structure of the entrance without necessarily changing the flow rate [32,33], but this has little influence on the passage efficiency. Thus, when setting standards, passage efficiency should be prioritized and measures to increase attraction efficiency should also be considered. The management water level of Sangju Weir is maintained at 47.00 m AMSL, and it is difficult to maintain a higher upstream water level. Therefore, to increase the passage efficiency of the nature-like fishway at Sangju Weir, the structure of the fishway exit should be altered to induce an optimal flow rate of $1.57 \text{ m}^3 \cdot \text{s}^{-1}$, which was found to ensure the highest passage efficiency, provided that the management water level upstream of Sangju Weir is maintained. This structural alteration is expected to improve the fishway passage efficiency.

Many studies in which telemetry-based fishway evaluation was performed have measured the attraction and passage efficiencies of fishways, but very few have made actual passing time measurements. Compared to ice-harbor-type fishways constructed at larger rivers, where 83% of passing individuals passed the fishway within 6 h [16], for 63.2% of the individuals that passed the nature-like fishway at Sangju Weir, it took more than one day. Even assuming that this difference may be attributed to the difference in TL and the length of the fishway, some individuals spent up to 28 days (2135.0 h) in the fishway. As the nature-like fishway has a river-like shape, it can function as a habitat as well as a passage [34]. In other words, it may be suspected that those individuals that stayed within the nature-like fishway at Sangju Weir for a prolonged period of time were using the fishway as their temporary habitat. Moreover, although the distance between each section is similar, individuals stayed longest in the 1st→2nd section, which offers the largest area (habitat space), thus supporting the assumption that the nature-like fishway at Sangju Weir also acts as habitat. On the other hand, Agostinho *et al.* [35] reported the problematic usage of the fishway by certain predator fish as their foraging area in Brazil. In this study, we were able to observe the usage of the fishway by predator fish such as *Siniperca scherzeri*, *Coreoperca herzi*, and *Micropterus salmoides* by trapping. If these species use the fishway as their habitat and forage fish within the fishway, this could affect the efficiency of the fishway. Therefore, further studies on the usage of the fishway as a habitat by predator fish should be performed in the near future.

Z. platypus, O. u. amurensis, P. esocinus, and *H. longirostris* were the major fish species using the nature-like fishway at Sangju Weir, and the frequency of usage increased without any significant interspecific differences when the upstream water level was 47.00–47.20 m AMSL. The water velocity and inflow rate into the fishway suitable for the passage of these fish species ranged from 0.79–1.06 $m \cdot s^{-1}$ and from 0.95–1.60 $m^3 \cdot s^{-1}$, respectively. These values should be considered as design factors for future fishways likely to be used by these species.

Interspecific differences in fishway usage patterns depending on the time of the day were observed among these 4 species. *Z. platypus* and *O. u. amurensis* were observed to use the fishway at both daytime and nighttime hours, more frequently during the daytime until sunset. *P. esocinus* and *H. longirostris* mostly used the fishway during nighttime hours, with a very low daytime usage frequency. Such time-dependent usage patterns are related to different ecological characteristics (diurnal or nocturnal) of the species [36] and the survival strategy of avoiding visual predators [37]. However, given the similar sizes of the species and the absence of strong predators among cohabiting species in the fishway, it is difficult to understand the nocturnal pattern in terms of visual predator avoidance. Thus, this difference is assumed to be ascribable to the differences in ecological characteristics. Given the lack of studies on these species, however, further research is required to determine the exact reason for the time-dependent interspecific differences.

Acknowledgments: This study was performed under the project of "Nationwide Aquatic Ecological Monitoring Program" and "Establishing an Aquatic Ecosystem Health Network" in Korea, and was supported by the Ministry of Environment and the National Institute of Environmental Research, Korea. The authors are grateful to survey members involved in the project. The authors also thank the anonymous reviewers for their help in improving the scientific content of the manuscript.

Author Contributions: Jeong-Hui Kim, Ju-Duk Yoon, and Min-Ho Jang conceived and designed the study. All authors collected the data and Jeong-Hui Kim wrote the manuscript. All authors read and approved the final version of the manuscript.

Conflicts of Interest: The authors declare no conflict of interest.

References

1. Reyes-Gavilán, F.G.; Garrido, R.; Nicieza, A.G.; Toledo, M.M.; Brana, F. Fish community variation along physical gradients in short streams of northern Spain and the disruptive effect of dams. *Hydrobiologia* **1996**, *321*, 155–163.
2. Rosenberg, D.M.; Berkes, F.; Bodaly, R.A.; Hecky, R.E.; Kelly, C.A.; Rudd, J.W. Large-scale impacts of hydroelectric development. *Environ. Rev.* **1997**, *5*, 27–54.
3. Northcote, T.G. Migratory behaviour of fish and its significance to movement through riverine fish passage facilities. In *Fish Migration and Fish Bypasses*; Jungwirth, M., Schmutz, S., Weiss, S., Eds.; Fishing News Book: Cambridge, MA, USA, 1998; pp. 3–18.

4. Yamamoto, S.; Morita, K.; Koizumi, I.; Maekawa, K. Genetic differentiation of white-spotted charr (*Salvelinus leucomaenis*) populations after habitat fragmentation: Spatial-temporal changes in gene frequencies. *Conserv. Genet.* **2004**, *5*, 529–538.

5. Clay, C.H. *Design of Fishways and Other Fish Facilities*, 2nd ed.; CRC Press: Boca Raton, FL, USA, 1994.

6. Laine, A.; Jokivirta, T.; Katopodis, C. Atlantic salmon, *Salmo salar* L., and sea trout, *Salmo trutta* L., passage in a regulated northern river—fishway efficiency, fish entrance and environmental factors. *Fish. Manag. Ecol.* **2002**, *9*, 65–77.

7. Lucas, M.C.; Baras, E. *Migration of Freshwater Fishes*; Blackwell Science: Oxford, UK, 2001.

8. Jungwirth, M. Bypass channels at weirs as appropriate aids for fish migration in rhithral rivers. *Regul. Rivers* **1996**, *12*, 483–492.

9. Eberstaller, J.; Hinterhofer, M.; Parasiewicz, P. The effectiveness of two nature-like bypass channels in an upland Austrian river. In *Migration and Fish Bypasses*; Jungwirth, M., Schmutz, S., Weiss, S., Eds.; Fishing News Books: Cambridge, MA, USA, 1998; pp. 363–383.

10. Mallen-Cooper, M.; Stuart, I.G. Optimising Denil fishways for passage of small and large fishes. *Fisheries Manag. Ecol.* **2007**, *14*, 61–71.

11. Parasiewicz, P.; Eberstaller, J.; Weiss, S.; Schmutz, S. Conceptual guidelines for nature-like bypass channels. In *Migration and Fish Bypasses*; Jungwirth, M., Schmutz, S., Weiss, S., Eds.; Fishing News Books: Cambridge, MA, USA, 1998; pp. 348–362.

12. Cada, F.G. Fish passage migration at hydroelectric power projects in the United States. In *Migration and Fish Bypasses*; Jungwirth, M., Schmutz, S., Weiss, S., Eds.; Fishing News Books: Cambridge, MA, USA, 1998; pp. 208–219.

13. Bunt, C.M.; Castro-Santos, T.; Haro, A. Performance of fish passage structures at upstream barriers to migration. *River Res. Appl.* **2012**, *28*, 457–478.

14. Lee, J.W.; Yoon, J.D.; Kim, J.H.; Park, S.H.; Baek, S.H.; Yoon, J.H.; Jang, M.H. Efficiency analysis of the ice harbor type fishway installed at the Gongju Weir on the Geum River using traps. *Korean J. Environ. Biol.* **2015**, *33*, 75–82.

15. Finfortech Internet Database. Available online: http://finfotech.kr/admin/ (assessed on 2 August 2013).

16. Yoon, J.D.; Kim, J.H.; Yoon, J.H.; Baek, S.H.; Jang, M.H. Efficiency of a modified Ice Harbor-type fishway for Korean freshwater fishes passing a weir in South Korea. *Aquat. Ecol.* **2015**, *49*, 1–13.

17. Aarestrup, K.; Lucas, M.C.; Hansen, J.A. Efficiency of a nature-like bypass channel for sea trout (*Salmo trutta*) ascending a small Danish stream studied by PIT telemetry. *Ecol. Freshw. Fish* **2003**, *12*, 160–168.

18. Mallen-Cooper, M. Swimming ability of adult golden perch, *Macquaria ambigua* (Percichthyidae), and adult silver perch, *Bidyanus bidyanus* (Teraponidae), in an experimental vertical-slot fishway. *Mar. Freshw. Res.* **1994**, *45*, 191–198.

19. Barrett, J.; Mallen-Cooper, M. The Murray River's 'Sea to Hume Dam' fish passage program: Progress to date and lessons learned. *Ecol. Manag. Rest.* **2006**, *7*, 173–183.

20. Porcher, J.P.; Travade, F. Fishways: Biological basis, limits and legal considerations. *B. Fr. Pêche. Piscic.* **2002**, *364*, 9–20.

21. Neraas, L.P.; Spruell, P. Fragmentation of riverine systems: The genetic effects of dams on bull trout (*Salvelinus confluentus*) in the Clark Fork River system. *Mol. Ecol.* **2001**, *10*, 1153–1164.

22. Bretón, F.; Baki, A.B.M.; Link, O.; Zhu, D.Z.; Rajaratnam, N. Flow in nature-like fishway and its relation to fish behaviour. *Can. J. Civ. Eng.* **2013**, *40*, 567–573.

23. Stuart, I.G.; Mallen-Cooper, M. An assessment of the effectiveness of a vertical-slot fishway for non-salmonid fish at a tidal barrier on a large tropical/subtropical river. *Regul. Rivers* **1999**, *15*, 575–590.

24. Larinier, M.; Chanseau, M.; Bau, F.; Croze, O. The use of radio telemetry for optimizing fish pass design. In Aquatic Telemetry: Advances and Applications, Proceedings of the Fifth Conference on Fish Telemetry, Europe, Ustica, Italy, 9–13 June 2003; Spedicato, M.T., Lembo, G., Marmulla, G., Eds.; FAO/COISPA: Rome, Italy, 2003; pp. 53–60.

25. Sprankle, K. Interdam movements and passage attraction of American shad in the lower Merrimack River main steam. *N. Am. J. Fish. Manag.* **2005**, *25*, 1456–1466.

26. Calles, E.O.; Greenberg, L.A. Evaluation of nature-like fishways for re-establishing connectivity in fragmented salmonid populations in the river Emån. *River Res. Appl.* **2005**, *21*, 951–960.

27. Calles, E.O.; Greenberg, L.A. The use of two nature-like fishways by some fish species in the Swedish River Emån. *Ecol. Freshw. Fish* **2007**, *16*, 183–190.

28. Jensen, A.J.; Aass, P. Migration of a fast-growing population of brown trout (*Salmo trutta L.*) through a fish ladder in relation to water flow and water temperature. *Regul. Rivers* **1995**, *10*, 217–228.

29. Fernandez, D.R.; Agostinho, A.A.; Bini, L.M.; Gomes, L.C. Environmental factors related to entry into and ascent of fish in the experimental ladder located close to Itaipu Dam. *Neotropical Ichthyol.* **2007**, *5*, 153–160.

30. Kemp, P.S.; Russon, I.J.; Vowles, A.S.; Lucas, M.C. The influence of discharge and temperature on the ability of upstream migrant adult river lamprey (*Lampetra fluviatilis*) to pass experimental overshot and undershot weirs. *River Res. Appl.* **2011**, *27*, 488–498.

31. Santos, J.M.; Branco, P.J.; Silva, A.T.; Katopodis, C.; Pinheiro, A.N.; Viseu, T.; Ferreira, M.T. Effect of two flow regimes on the upstream movements of the Iberian barbel (*Luciobarbus bocagei*) in an experimental pool-type fishway. *J. Appl. Ichthyol.* **2013**, *29*, 425–430.

32. Bunt, C.M. Fishway entrance modifications enhance fish attraction. *Fisheries Manag. Ecol.* **2001**, *8*, 95–105.

33. Lindberg, D.E.; Leonardsson, K.; Andersson, A.G.; Lundström, T.S.; Lundqvist, H. Methods for locating the proper position of a planned fishway entrance near a hydropower tailrace. *Limnologica* **2013**, *43*, 339–347.

34. Pander, J.; Mueller, M.; Geist, J. Ecological functions of fish bypass channels in streams: Migration corridor and habitat for rheophilic species. *River Res. Appl.* **2013**, *29*, 441–450.

35. Agostinho, A.A.; Agostinho, C.S.; Pelicice, F.M.; Marques, E.E. Fish ladders: Safe fish passage or hotspot for predation? *Neotrop. Ichthyol.* **2012**, *10*, 687–696.

36. Santos, J.M.; Ferreira, M.T.; Godinho, F.N.; Bochechas, J. Efficacy of a nature-like bypass channel in a Portuguese lowland river. *J. Appl. Ichthyol.* **2005**, *21*, 381–388.
37. Jonsson, N. Influence of water flow, water temperature, and light on fish migration in rivers. *Nord. J. Freshwat. Res.* **1991**, *66*, 20–35.

Field Evaluation of a Stormwater Treatment Train with Pit Baskets and Filter Media Cartridges in Southeast Queensland

Darren Drapper and Andy Hornbuckle

Abstract: Field monitoring of a stormwater treatment train has been underway between November 2013 and May 2015 at a townhouse development located at Ormiston, southeast Queensland. The research was undertaken to evaluate the effectiveness of a 200 micron mesh pit basket in a 900 square format and an 850 mm high media filtration cartridge system for removing total suspended solids and nutrients from stormwater runoff. The monitoring protocol was developed with Queensland University of Technology (QUT), reflecting the Auckland Regional Council Proprietary Device Evaluation Protocol (PDEP) and United States Urban Stormwater BMP Performance Monitoring Manual with some minor improvements reflecting local conditions. During the 18 month period, more than 30 rain events have occurred, of which nine comply with the protocol. The Efficiency Ratio (ER) observed for the treatment devices are 32% total suspended solids (TSS), 37% for total phosphorus (TP) and 38% total nitrogen (TN) for the pit basket, and an Efficiency Ratio of 87% TSS, 55% TP and 42% TN for the cartridge filter. The performance results on nine events have been observed to be significantly different statistically ($p < 0.05$) for the filters but not the pit baskets. The research has also identified the significant influence of analytical variability on performance results, specifically when influent concentrations are near the limits of detection.

Reprinted from *Water*. Cite as: Drapper, D.; Hornbuckle, A. Field Evaluation of a Stormwater Treatment Train with Pit Baskets and Filter Media Cartridges in Southeast Queensland. *Water* **2015**, *7*, 4496–4510.

1. Introduction

The release of the Queensland State Planning Policy (SPP) requires local planning schemes to integrate the state's interest in water quality by applying stormwater management objectives relevant to the climatic region, or demonstrating current best practice environmental management for urban developments. The SPP seeks to facilitate innovative and locally appropriate solutions to achieve the stormwater management design objectives typically 80% total suspended solids (TSS), 60% total phosphorus (TP), and 45% total nitrogen (TN) [1].

Several documents have been released in Australia over the past decade providing guidance on the design, modelling, construction, implementation and maintenance of stormwater quality management measures to achieve these objectives [2–4]. These guidelines have typically focussed on the constructed "natural" treatment measures including swales, biofiltration and wetlands.

Few of the guidelines include sections for demonstrating the performance of other types of stormwater treatment solutions. When compared with international evaluation protocols, it is apparent that the necessary detail to demonstrate performance to local conditions is omitted from these local guidelines. This paper presents a protocol developed for local Australian conditions by local universities in conjunction with SPEL Environmental (SPEL), a stormwater technology supplier, and applies it to testing an innovative stormwater treatment train in southeast Queensland, with discussion of the performance results observed.

2. Local Field Testing Site Details

Testing has been under way for more than 18 months at a townhouse complex at Ormiston, Queensland. The site is about 28 km east of the Brisbane Central Business District. Runoff from the site enters the local drainage network via grated inlets and is transported to an underground chamber for further treatment and detention prior to its discharge into the Council network. The site has a total area of 2028 m^2 with approximately 1140 m^2 of roof area (56%), 500 m^2 of impervious driveway (25%) and the balance, 388 m^2 (19%) of pervious area. The stormwater treatment train includes rainwater tanks for roof water, 900 mm square pit baskets (also known as catch basin inserts) with a 200 micron mesh bag in each of the gully pits (catch basins), and an underground vault with two 850 mm high media cartridge filters. The surface runoff from the site drains through the pit baskets into the pipe network, whereas the roofwater overflow from the rainwater tanks enters the pipe network beneath the pit baskets. This configuration is a typical stormwater treatment train for a medium- to high-density residential development in southeast Queensland. The site is also representative of typical applications for the pit basket and media filter treatment train.

The pit baskets are designed to capture the gross pollutants and coarse sediment leaving the pervious and impervious ground surfaces and are installed in the gully pits (catch basins). Figure 1 shows a plan view of a typical installation. The cartridge filters utilise a perlite, zeolite and activated alumina media to provide physical filtration and adsorption of stormwater pollutants, including nutrients. Overflow from the small rainwater tanks (3 kL per dwelling) enters the pipe drainage network beneath the pit baskets, and hence will provide significant dilution to the stormwater water quality exiting the pit baskets. Figure 2 is a photograph of the monitoring site. Figure 3 is a schematic of the catchment, and Figure 4 is a schematic cross-section of the filter vault and monitoring installation.

(a)	**(b)**

Figure 1. Typical pit basket (catch basin insert) installation plan view (**a**); side view (**b**).

Figure 2. Townhouse development at Ormiston, QLD, showing driveway area, landscaping and filter cartridge for installation.

Runoff samples are collected by four ISCO GLS auto-samplers at the locations shown in Figure 4. Runoff is sampled as it leaves the driveway surface and enters the pit basket (1). A second sampler to determine the water quality of runoff in the conveyance pipe is installed because the roofwater enters the pipe drainage network beneath the pit baskets and provides dilution to the surface runoff. A third sampler collects water from the outlet pipe of the cartridge media filters, which are located upstream of the 850 mm baffle wall in the detention chamber (3). A fourth sampler collects filtered water from a tray beneath the pit basket (4). A photograph of the StormSack collection point is presented in Figure 5.

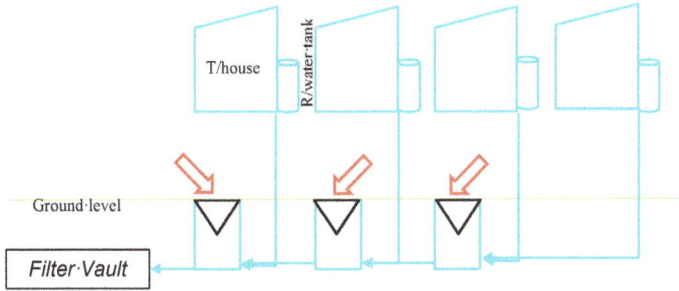

Figure 3. Schematic of the runoff pathways for the townhouse development at Ormiston. The red arrows indicate stormwater runoff entering the pit baskets. Excess rainfall (blue lines) enters the central drainage pipe, which then passes into the filter vault.

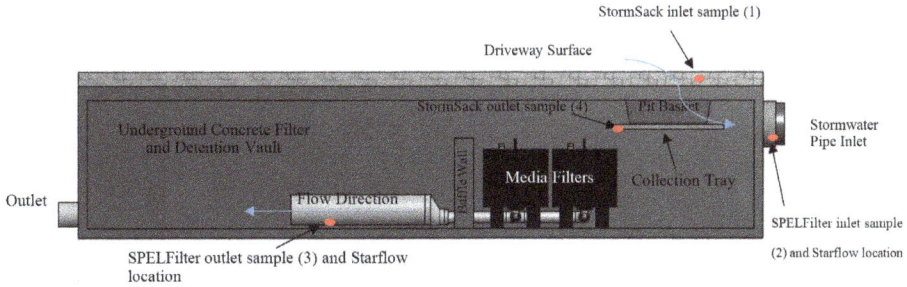

Figure 4. Schematic cross section of the filter vault with the stormwater treatment system.

Figure 5. StormSack sample collection photograph (inlet circled).

407

3. Local Field Testing Methodology

Due to the lack of formalised testing protocols in Australia, the University of the Sunshine Coast (USC), Queensland University of Technology (QUT), Griffith University (GU) and SPEL have formulated testing protocols based on the Auckland Council Proprietary Device Evaluation Protocol (PDEP), Washington Department of Ecology (WDOE) and Stormwater Equipment Manufacturers Association (SWEMA) protocols [5–7]. The protocols have been formalised to deliver a robust, scientifically defensible outcome. Even so, the protocols developed at the initiation stage have needed refinement once actual site data was observed, influenced by local hydrological conditions and equipment constraints. The protocol applied at the Ormiston test location, and monitored by QUT is detailed in Table 1.

Much of these protocol criteria appear in the Stormwater Quality Improvement Device Evaluation Protocol (SQIDEP) released as a consultation draft by Stormwater Australia [8].

Table 1. QUT-SPEL Field Testing Protocol Requirements for Ormiston.

Parameter	Ormiston
Minimum Storm Duration	5 min
Catchment type	Medium density townhouse property
Stormwater Treatment Device Type	Full scale—200 micron mesh pit basket and radially-wound media filter combination
Target Number of Storm events	15
Minimum rainfall depth per event	5 mm
Minimum inter-event period	48–72 h, depending on influent concentrations > Limit of Detection (LOD)
Minimum hydrograph sampling	First 60% of hydrograph
Flow rates tested	At least 3 events >75% of the treatable flow rate (TFR) with 1 exceeding the TFR.
Minimum number of water sub-samples collected per event	Minimum 8 influent and 8 effluent subsamples for each event. (Based on advice from the laboratory regarding minimum sample amount)
Sampling method	Auto-sampler, flow-weighted in 1000 L intervals (pipe network) and 0.5 mm rainfall for pit basket samples
Data Management	Campbell Scientific CR800 Data logger with Ethernet Modem
Particle Size Distribution (PSD) analysis via Laser Diffraction	Continuously stirred, without chemical dispersion or sonication
Total Suspended Solids (TSS)	American Public Health Association (APHA) (2005) 2540 D [9]
Total Nitrogen and species (water samples only)	APHA (2005) 4500 N, APHA (2005) 4500 NH3, APHA (2005) 4500 NO3
Total Phosphorus and Orthophosphate (water samples only)	APHA (2005) 4500 P
pH and Electrical Conductivity (EC)	Handheld probe, calibrated to manufacturer's specifications

The sampling program listed in Table 1 is triggered by two criteria. Firstly, >2 mm of rainfall over a rolling 30-min window must occur, based on field experience. This was programmed into the datalogger, to ensure sufficient water depth was available in the pipe to collect samples. Rainfall is measured onsite by a 0.2 mm waterlog tipping bucket rain gauge. The second criteria is flow volume, where a sample is initiated after 1000 L of stormwater discharge past each of the two pipe sampling points shown in Figure 4. Flow rate/volume was measured by two Starflow ultrasonic probes installed at the inlet and outlet pipes of the concrete chamber shown in Figure 4. For the pit basket where flow measurement was impracticable, sample intervals were triggered at 0.5 mm rainfall intervals. As the basket effectively has zero residence time, the inlet and outlet samples were triggered simultaneously.

Ultrasonic probes were selected for flow measurement due to a reported accuracy of ±2% for flow and ±0.25% for depth [10]. This accuracy is comparable to flumes and weirs but without the associated interference with water quality, especially TSS, observed with the latter.

A 1000 L volume of water was chosen as the sampling interval as this is 50% of the cartridge vault volume. This volume also corresponded to 0.5 mm of runoff over the site, assuming zero losses. Analysis of a smaller flow volume trigger indicated that it could challenge the physical limitations of the samplers' purge/collection cycle (about a 90 s cycle). All the subsamples collected during a runoff event were composited within the sampler in a 9 L bottle. Each subsample was 200 mL to ensure sufficient volume was available for the suite of subsequent chemical analyses (listed in Table 1). This flow-weighted sampling protocol provides an Event Mean Concentration (EMC).

As has been noted previously [11], the physical limitations of the equipment and analysis process can subsequently affect the protocol. Therefore, any nominated protocol needs flexibility to respond to these potential constraints. For example, to collect eight subsamples practically restricts the minimum time for a "qualifying" storm to greater than 12 min, even though flow may occur quicker. Hence, for this site, a storm event less than ten minutes duration is unlikely to provide sufficient time to collect eight aliquots even if sufficient volume were present. As the project progressed, the laboratory advised that analyses could be performed on much smaller volumes, thereby permitting as few as three aliquots to be sufficient from short duration events. The intent of the monitoring program, however, is to collect a spread of subsamples across the hydrograph of every event regardless of the duration.

On the other hand, a maximum number of subsamples can be collected before the container is full, and therefore an analysis of the likelihood of rainfall events exceeding the maximum capacity of the containers was undertaken to identify the likely upper event size. As the ISCO sampler can collect a maximum of 9 L of sample, 45 sub samples, each of 200 mL, are possible. For Ormiston, this equates

409

to approximately 22.5 mm total rainfall. Statistical analysis of rainfall events for Brisbane between July 2000 and July 2010 (assuming no runoff losses) indicates that this 9 L capacity would allow capture of >90% of the daily runoff events.

The inter-event period (antecedent duration) was set in the protocol to 48–72 h between rainfall events, as previous QUT research into pollutant build-up and wash-off on urban surfaces indicated that this was the optimal point at which pollutants reach a detectable level in runoff [12,13]. This research has shown that low intensity, low-volume events do not produce detectable concentrations for antecedent periods (ADP) less than 72 h. However, the Ormiston project has shown that high intensity events with less than 72 h ADP can produce detectable concentrations. Therefore, the protocol has been adjusted to include events where ADP might be <72 h, if pollutant concentrations are measurable. The minimum rainfall depth for a qualifying storm will vary between monitoring sites, depending on the catchment characteristics. For the medium density Ormiston townhouse site with a high fraction of impervious area, the minimum daily rainfall for monitoring has been set to 5 mm, as this is the level at which observable runoff can be measured. Other monitored sites with larger, more pervious catchments will require more rainfall to produce sufficient runoff for sampling. Therefore, we caution against setting a rigid minimum rainfall volume for qualifying storms in monitoring protocols, as this is inherently site-specific.

The draft SQIDEP also requires a minimum of three flow events >75% of the maximum treatable flow rate (TFR), with at least 1 event greater than the TFR [8]. It should be noted that a requirement for *all* events be at the TFR, or >75% of the TFR, may be statistically rare. For example, an evaluation of the hydrology for the Ormiston site across 10 years of historical data, indicates that this may be achieved less than three times annually. Hence to achieve 15 qualifying events at the TFR, would require a minimum of five years of sampling.

The monitoring equipment and sample collection were independently undertaken by staff from QUT, and analysed in NATA registered laboratories. Reports on the findings were prepared by QUT [14]. This maintains independence and integrity of the sampling, collection and analysis process. As there is a range of possible metrics used to assess performance data, this paper presents several of them.

Average Concentration Removal Efficiency (CRE) is calculated from the function:

$$Avg.\ CRE = \frac{\Sigma \left[\frac{\{EMCin - EMCout\}}{EMCin} \right]}{no.\ of\ events} \tag{1}$$

Efficiency Ratio (ER) is calculated from the function:

$$ER = 1 - \frac{Mean\ EMCout}{Mean\ EMCin} \tag{2}$$

To briefly paraphrase the above, the CRE is the average of the removal ratios (percentages) for every event, whereas the ER is the removal ratio of the average inflow and outflow concentrations for all events. The ER weights EMCs (flow-weighted concentrations) from all storms equally, regardless of the pollutant concentration or runoff volume, and minimises the potential impacts of smaller, cleaner events on performance calculations. The ER can, however, be influenced by a small number of high influent concentration events that skew average concentration results. Therefore, other metrics, including Average CRE, should also be considered [15]. The Average CRE quantifies the percent removal for each event, and calculates an average value of the percentages, allowing the smaller, cleaner events to have greater influence on the average CRE, and hence minimise the influence of the few, large influent concentrations.

4. Results and Discussion

A report on 18 months of monitoring has been released by QUT [14]. Of sixteen (16) captured rainfall events > 5 mm, nine events are qualifying. Where the results have been less than the limits of detection (LOD), they have been shown as 50% of the LOD. All events reported in this paper had flows >75% of the TFR, with one event exceeding the TFR.

Table 2 presents the water quality data observed at the pit basket (catch basin insert) and shows influent concentrations for TSS similar to those reported as typical by guidelines for urban residential catchments, whereas the TN and TP concentrations are mostly below guideline figures. The preliminary results indicate that the relatively simple 200 micron filter bag removes about 32% of the suspended solids and 37% and 38% of the TP and TN concentrations respectively based on the ER metric. The performance indicated by the CRE metric, is strongly influenced by very low inflow concentrations and slightly higher outflow concentrations, generating a negative ratio.

Water quality data from the media filter samples is presented in Table 3. It can be seen that the pollutant concentrations observed in the pipe inflow (inlet to the filters) is significantly lower than the pit basket outflow concentrations shown in Table 2. This is a direct result of stormwater dilution by overflow from the rainwater tanks entering the network at the base of the gully pits (catch basins). Even so, the data indicates that the filters are removing TSS, TN and TP, at very low concentrations. Mean ERs of 87%, 55% and 42% for TSS, TP and TN respectively are observed. Of particular note, the outflow TSS concentrations from the media filter are consistently below detection limits (<5 mg/L), for most events. Similarly the outflow TP concentrations are very close to the limits of detection.

Table 2. Pit Basket Water Quality Results.

Parameter	TSS		TN		TP	
LOD (mg/L) [1]	5		0.1		0.01	
Event	In (mg/L)	Out (mg/L)	In (mg/L)	Out (mg/L)	In (mg/L)	Out (mg/L)
23 June 2014	10	NC [2]	0.60	NC	0.04	NC
16 August 2014	122	40	0.90	0.40	0.13	0.10
18 August 2014	12	2.5	0.20	0.20	0.05	0.05
23 August 2014	9	2.5	0.20	0.10	0.05	0.15
26 September 2014	346	253	2.40	1.90	0.58	0.36
9 December 2014	202	186	3.85	2.20	0.40	0.10
18 December 2014	2.5	10	0.30	0.30	0.03	0.07
20 Febuary 2015	34	6	0.70	0.30	0.09	0.02
30 April 2015	90	58	0.90	0.50	0.13	0.07
Average Conc.	102.19	69.75	1.18	0.74	0.18	0.11
Median Conc.	34.00	25.00	0.70	0.35	0.09	0.08
Efficiency Ratio (Avg)	-	32%	-	38%	-	37%
Average CRE	-	9%	-	34%	-	−9%
Efficiency Ratio (Median)	-	26%	-	50%	-	7%
Median CRE	-	51%	-	44%	-	31%

Notes: [1] LOD = Limits of Detection of the analytical method; [2] As Outflow samples were not collected (NC) from this event, it has been excluded from the calculations.

Table 3. Media filter cartridge Water Quality Results.

Parameter	TSS		TN		TP	
LOD (mg/L) [1]	5		0.1		0.01	
Event	In (mg/L)	Out (mg/L)	In (mg/L)	Out (mg/L)	In (mg/L)	Out (mg/L)
22 June 2014	2.72	2.50	0.70	0.60	0.02	0.01
16 August 2014	22.69	2.50	0.40	0.30	0.04	0.02
18 August 2014	24.95	2.50	0.39	0.20	0.06	0.04
23 August 2014	2.50	2.50	0.20	0.10	0.03	0.03
25 September 2014	74.39	2.50	0.75	0.30	0.08	0.02
8 December 2014	31.59	15	0.47	0.20	0.03	0.02
18 December 2014	2.50	2.50	0.44	0.20	0.02	0.01
20 Febuary 2015	20.65	2.50	0.41	0.40	0.01	0.01
30 April 2015	65.65	0.50 *	0.31	0.05	0.03	0.01
Average Conc.	24.7	3.67	0.45	0.26	0.04	0.02
Median Conc.	12.6	2.5	0.41	0.2	0.03	0.02
Efficiency Ratio (Avg)	-	87%	-	42%	-	55%
Average CRE	-	58%	-	44%	-	56%
Efficiency Ratio (Median)	-	89%	-	52%	-	33%
Median CRE	-	88%	-	49%	-	64%

Notes: [1] LOD = Limits of Detection of the analytical method; * LOD = 1 mg/L for this event.

As can be seen in Tables 2 and 3, the ER and CRE metrics vary, though both use the same concentration data. This is the result of the two methods using different mathematical logic. For example, the pit basket result for TP on 23 August 14 indicates a CRE of −200% that subsequently causes the average CRE to be negative, even though all the other events show positive CRE values. Results near the limits of detection, such as that for 23 August 14, can skew the average CRE metric. A recorded inflow concentration of 0.01 mg/L, for example, and an outflow concentration of 0.02 mg/L will provide an individual CRE of −100% and influence the average CRE, yet be as a result of analytical error. Results on duplicate samples from Ormiston have been observed to differ by 0.3 mg/L for TN and 0.02 mg/L for TP, and result in a "theoretical export" of pollutants of ~200% for these very low influent concentrations. This large negative percent removal then has a knock-on effect on the average CRE value, and so we suggest CRE is not an appropriate metric when influent concentrations are close to the LOD.

We suggest that ER is the better metric for evaluation of this dataset. However, in the instances that high concentration influent outliers are recorded (for example, above the Water by Design MUSIC modelling guidelines [4]) as the dataset grows, we suggest that Average CRE, Comparison of Medians, and statistical analyses should all be used to validate performance. In the dataset observed by this research, there are no outliers based on the Water by Design MUSIC Modelling Guidelines for storm concentrations from an urban residential catchment [4]. In fact, the observed concentrations at Ormiston are low in comparison with the guideline values, as shown in Table 4. We therefore maintain that ER is the more suitable metric to be used at this point in time, for this site.

Table 4. Comparison of Ormiston Surface Water Quality Results with Brisbane MUSIC Guidelines for urban residential areas.

Parameter	MUSIC Guideline Values (Lumped Urban Residential Catchment) [1]			Ormiston Surface Influent (Pit Basket Inflow)		
	−1SD	Mean	+1SD	−1SD	Mean	+1SD
TSS (mg/L)	61.7	151	372	0	102.2	222.52
TP (mg/L)	0.162	0.339	0.708	0	0.182	0.380
TN (mg/L)	1.07	1.82	3.09	0	1.181	2.474

Note: [1] Reference: [4].

Significant debate continues as to the "best" method to calculate device performance. Statistical validation (Paired *t*-test) of the dataset is also recommended to confirm significant differences between the influent and effluent sample sets [5]. The Auckland PDEP also indicates that where the median and the mean of the performance metric (e.g., ER) vary by more than 10%, additional sampling events are recommended. The median concentrations presented in Tables 2 and 3, when used to calculate an Efficiency Ratio, result in a difference of more than 10% for TN and TP when compared with a Mean Efficiency Ratio. In comparison, however, the Median CRE values appear to be generally consistent with the Average ER.

5. Normality and Log-Normality Tests

Environmental monitoring data is typically log-normally distributed, therefore, the data sets recorded for this site were evaluated against a normal distribution, in a log-transformed basis. The Anderson-Darling Normality test identifies normal distributions where the p-value is >0.05 (alpha). As can be seen from Table 5, all of the datasets are log-normally-distributed, except for TSS outflow and TP outflow on the media filters.

Table 5. Anderson-Darling Normality tests on log-transformed dataset, statistically significant results shown in bold.

Treatment Device	p-value					
	TSS in	TSS out	TN in	TN out	TP in	TP out
StormSack	0.748	0.497	0.506	0.328	0.518	0.615
SPELFilter	0.054	0.005	0.418	0.413	0.906	0.015

This is likely a result of the outflow concentrations converging on similar results (*i.e.*, Below or at Detection limits). This information is relevant for confirming that a Student's paired *t* test is a valid method to compare the influent and effluent datasets. Figure 6 presents the Q-Q plots for the StormSack pollutant data in log-transformed format. Figure 7 presents the Q-Q plots for the SPELFilter pollutant data in log-transformed format.

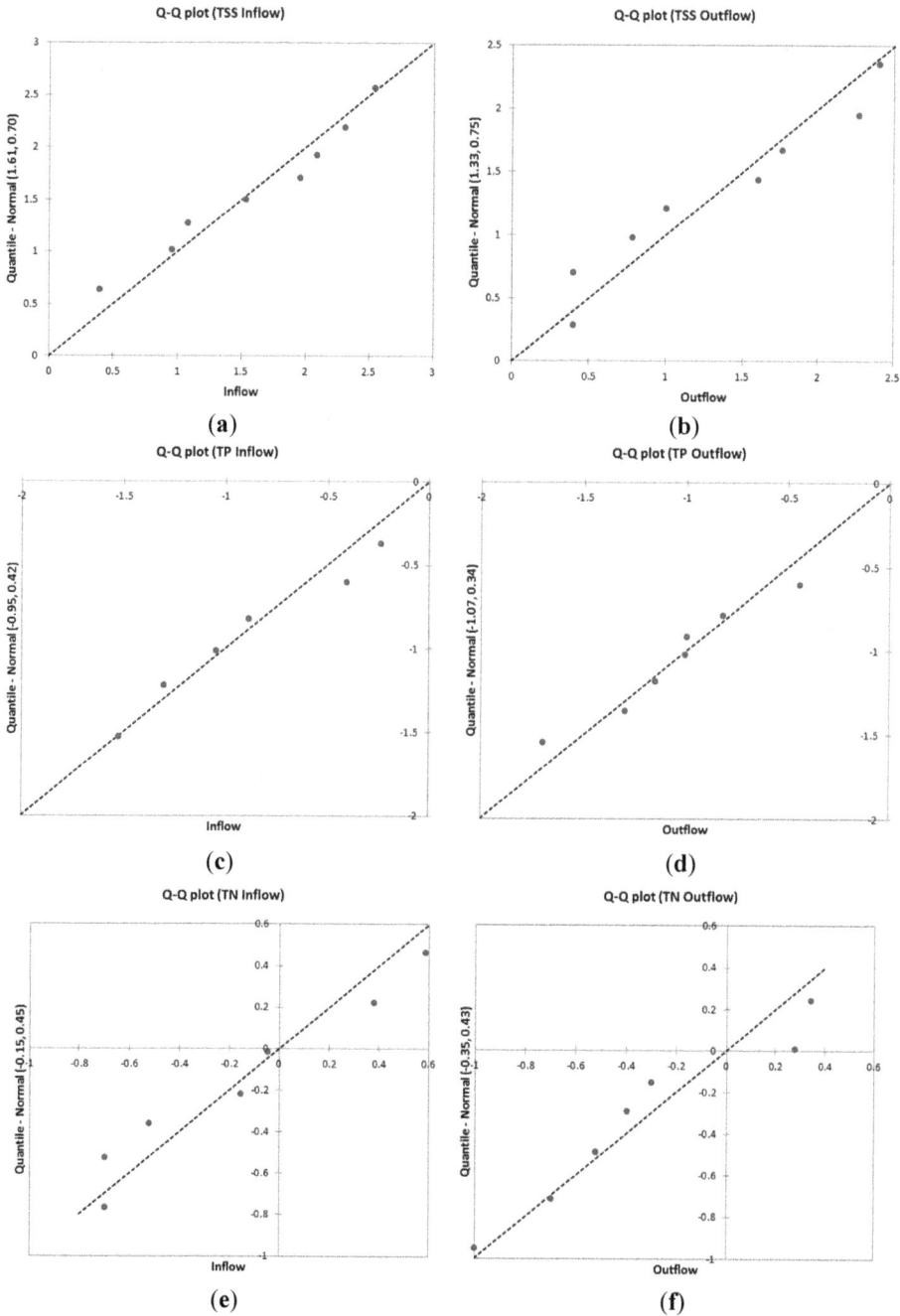

Figure 6. Q-Q Plots of log-normal distributions—Pit Basket (**a**) TSS Inflow; (**b**) TSS Outflow; (**c**) TP inflow; (**d**) TP outflow; (**e**) TN inflow; (**f**) TN outflow.

415

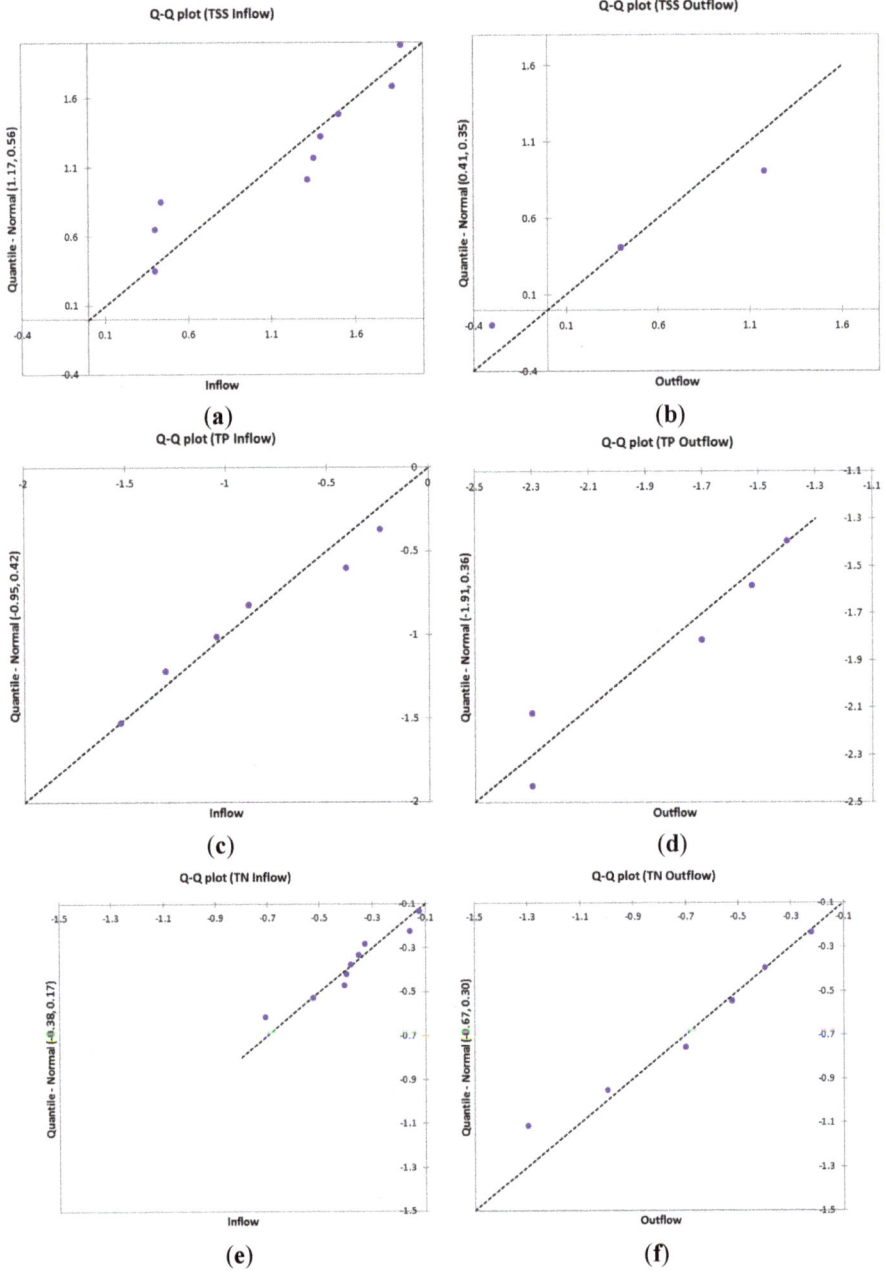

Figure 7. Q-Q Plots of log-normal distributions—SPELFilter Inflow and Outflow, TSS, TN and TP respectively. (**a**) TSS Inflow; (**b**) TSS Outflow; (**c**) TP Inflow; (**d**) TP Outflow; (**e**) TN Inflow; (**f**) TN Outflow.

6. Statistical Significance Tests

To evaluate whether the data demonstrate that the treated flow is statistically different from the inflow, statistical tests were performed on the log-transformed datasets. The paired Student's t test evaluates whether the two datasets have the same mean. Therefore, if the datasets are considered statistically to be significantly different, they are shown in bold below. As can be seen from t tests on the log-transformed data, all pollutants in and out of the media filter are statistically significantly different. The pit basket concentrations data is not statistically significantly different, according to this test. The t test results are presented in Table 6. Given the inherent variability of environmental data, a further statistical test was performed on the raw dataset. The Mann-Whitney (Wilcoxon) Rank-Sum test was performed on the raw datasets for the pit basket and media filter. As can be seen in Table 7, the media filter results are confirmed to be statistically significantly different, however, the pit basket datasets are not.

Table 6. Student's t tests on log-transformed dataset, statistically significant results shown in bold.

Treatment Device	p-value (Two-Tailed)		
	TSS	TN	TP
StormSack	0.117	0.006	0.412
SPELFilter	0.015	0.005	0.002

Table 7. Wilcoxon-Mann-Whitney Rank-Sum test on raw data, statistically significant results shown in bold.

Treatment Device	TSS		TN		TP	
	p-value	Sig.	p-value	Sig.	p-value	Sig.
StormSack	0.496	no	0.429	no	0.711	no
SPELFilter	0.008	yes	0.021	yes	0.031	yes

These results confirm international observations that environmental data may require large datasets that are economically unviable to demonstrate statistical significance [15]. Further, the concentrations observed on the catchment are beyond the control of the researcher. For example, an estimation of the number of samples required for a paired comparison on the pit basket dataset, as indicated by the equation described by Burton and Pitt [16] shown below, suggests that 160 samples are required for TSS, 103 samples are required for TP and 220 samples are necessary for TN.

$$n = 2 \left[\frac{Z_{1-\alpha} + Z_{1-\beta}}{\mu_1 - \mu_2} \right]^2 \sigma^2 \qquad (3)$$

where n = number of sample pairs needed; α = false positive rate ($1-\alpha$ is the degree of confidence. A value of α of 0.05 is usually considered statistically significant, corresponding to a $1-\alpha$ degree of confidence or 95%); β = false negative rate ($1-\beta$ is the power. If used, a value of β of 0.2 is common but it is frequently ignored, corresponding to a β of 0.5); $Z_{1-\alpha}$ = Z score (associated with area under normal curve) corresponding to $1-\alpha$; $Z_{1-\beta}$ = Z score corresponding to $1-\beta$ value; μ_1 = mean of dataset one; μ_2 = mean of dataset two; σ = standard deviation (same for both datasets, assuming normally-distributed).

SPEL and Drapper Environmental Consultants (DEC) are monitoring seven research sites across southeast Queensland, and have observed that for each qualifying event there are three others discarded for non-conformance with the protocol. Continuing a monitoring program to achieve 220 qualifying events required for statistical certainty (>600 events overall) would be financially prohibitive for any research program and delay outcomes for many years.

7. Conclusions

Evaluation of alternate stormwater treatment devices has been under way for decades internationally and, appears to be gaining momentum in Australia. While a number of existing guidelines stipulate that performance of alternate stormwater treatment devices must be demonstrated for local and regional conditions, the guidelines generally do not define how this should be accomplished. USC, QUT, GU, DEC and SPEL have worked together to adapt international protocols to suit local and regional conditions on a variety of sites and treatment measures in southeast Queensland. This paper details the protocol being implemented on one of the monitoring sites at Ormiston, Southeast Queensland. A report published by QUT on the nine complying events at the site indicate Efficiency Ratios of 32% TSS, 37% TP and 38% TN for the 900 square StormSack pit basket, and 87% TSS, 55% TP and 42% TN for the 85mm high, radially-wound, multi-media SPELFilter cartridge. Given the dataset analyses on the field testing of this treatment train indicates that the performance of the SPELFilter is statistically proven, and, when combined in a treatment train, it will comply with the QLD SPP water quality objectives of 80% TSS, 60% TP and 45% TN removal.

Acknowledgments: The authors acknowledge the contribution to this research by QUT, including Ashantha Goonetilleke, Prasanna Egodawatta, and Buddhi Srinath Wijesiri Mahappu Kankanamalage.

Author Contributions: Darren Drapper has been engaged to project manage the research and supervise the ongoing operation of the monitoring system; Andy Hornbuckle is the National Manager for SPEL Environmental and pioneered the site selection, approvals, financing and regulator liaison for the project. Both authors have contributed to the preparation of this journal article.

Conflicts of Interest: The research behind this paper was independently undertaken by QUT under a funding arrangement with SPEL Environmental.

References

1. State of Queensland, Department of State Development, Infrastructure and Planning. *State Planning Policy*; State of Queensland: Brisbane, Australia, 2013.
2. Mackay Regional Council. *Mackay Regional Council MUSIC Guidelines*; Version 1.1; Mackay Regional Council: Mackay, Australia, 2008.
3. Melbourne Water. *MUSIC Guidelines; Recommended Input Parameters and Modelling Approaches for MUSIC Users*; State Government of Victoria: Melbourne, Australia, 2010.
4. Water by Design. *MUSIC Modelling Guidelines*; SEQ Healthy Waterways Partnership: Brisbane, Australia, 2010.
5. Auckland Regional Council. *Proprietary Devices Evaluation Protocol (PDEP) for Stormwater Quality Treatment Devices*; Version 3; Auckland Regional Council: Auckland, New Zealand, 2012.
6. Washington Department of Ecology. *Technical Guidance Manual for Evaluating Emerging Stormwater Treatment Technologies; Technology Assessment Protocol—Ecology (TAPE)*; Washington Department of Ecology: Washington, WA, USA, 2011.
7. Stormwater Equipment Manufacturer's Association (SWEMA). *Evaluation of Hydrodynamic Separators*; Stormwater Equipment Manufacturer's Association: St. Paul, MN, USA, 2010.
8. Stormwater Australia. Stormwater Quality Improvement Device Evaluation Protocol (SQIDEP). Available online: http://stormwater.asn.au/images/SQID/SQIDEP_Release_version_December_2014.pdf (accessed on 10 August 2015).
9. American Public Health Association (APHA). *Standard Methods for the Examination of Water and Wastewater*, 21st ed.; American Public Health Association Press: Washington, DC, USA, 2005.
10. Unidata. *Manual Starflow Ultrasonic Doppler Instrument with Micrologger, Model 6526, Issue 4.0*; Unidata Pty Ltd: O'Connor, Australia, 2013.
11. Kelly, C. *Equipment, Set-up and Water Quality Analyses for Propriety Stormwater Quality Improvement Device (SQID) Field Testing*; Stormwater Australia Bulletin; Stormwater Australia: Melbourne, Australia, 2014.
12. Liu, A. Influence of Rainfall and Catchment Characteristics on Urban Stormwater Quality. Ph.D. Thesis, Queensland University of Technology, Brisbane, Queensland, Australia, 30 October 2011.
13. Parker, N. Assessing the Effectiveness of Water Sensitive Urban Design in Southeast Queensland. Master of Engineering Thesis, Queensland University of Technology, Brisbane, Queensland, Australia, 13 August 2010.
14. Goonetilleke, A.; Egodawatta, P. *Evaluation of Treatment Performance of the StormSack and SPELFilter Installations at Ormiston*; Progress Report; QUT: Brisbane, Australia, 2015.

15. Geosyntec Consultants and Wright Water Engineers, Inc. *Urban Stormwater BMP Performance Monitoring*; Geosyntec Consultants and Wright Water Engineers, Inc.: Portland, OR, USA, 2009.

16. Burton, A.; Pitt, R. Stormwater Effects Handbook: A Toolbox for Watershed Managers, Scientists and Engineers. Available online: http://unix.eng.ua.edu/~rpitt/Publications/ BooksandReports/Stormwater%20Effects%20Handbook%20by%20%20Burton% 20and%20Pitt%20book/hirezhandbook.pdf (accessed on 10 August 2015).

MDPI AG

St. Alban-Anlage 66

4052 Basel, Switzerland

Tel. +41 61 683 77 34

Fax +41 61 302 89 18

http://www.mdpi.com

Water Editorial Office

E-mail: water@mdpi.com

http://www.mdpi.com/journal/water

www.ingramcontent.com/pod-product-compliance
Lightning Source LLC
Chambersburg PA
CBHW051925190326
41458CB00026B/6413